doodle notes

learning benefits of visual note-taking

stronger **focus**

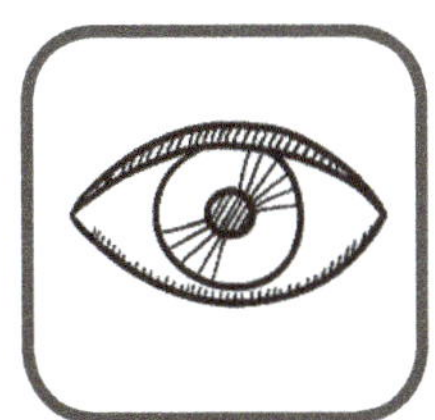

retention through dual coding

mental connections

memory **boost**

communication between **brain** hemispheres

building long-term memories

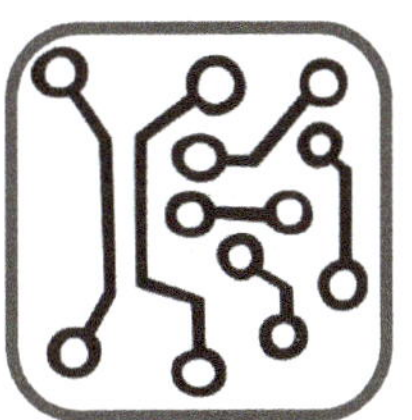

activated **neural** pathways

increased **creativity** & alertness

associative recognition

picture superiority **effect**

relaxation **benefits**

problem solving skills boost

Thanks so much for purchasing this book!

Thanks and credit to the following artists:

To learn more about the doodle note® method and download your free Doodle Note Handbook, visit

doodlenotes.org

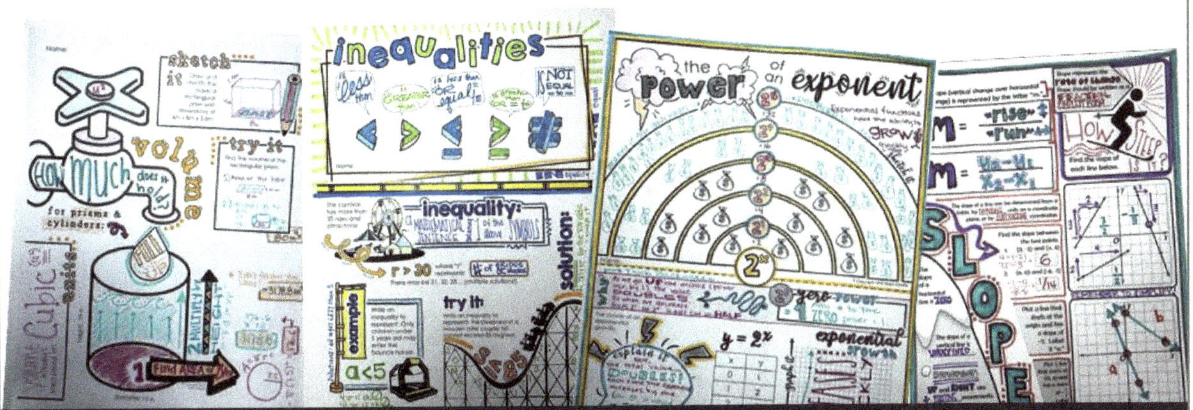

Teacher Guide: How to Use Your doodle notebooks

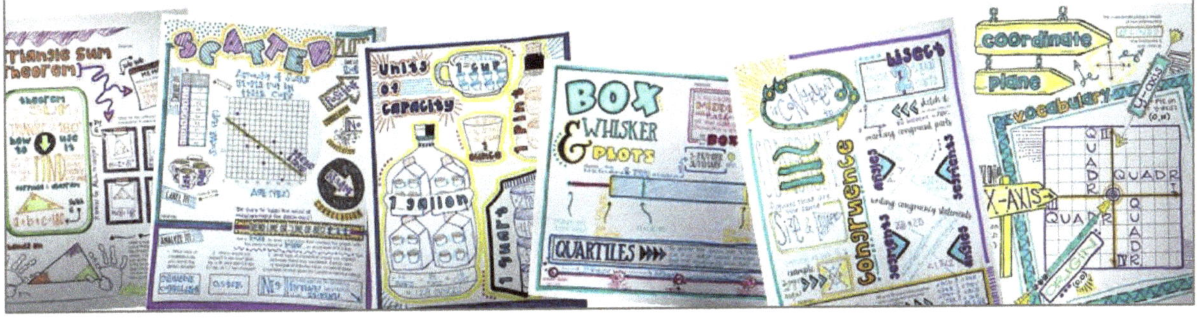

doodle notes®

learning benefits of visual note-taking

The brain processes linguistic input (teacher voice, written text, words) in a completely different area from all graphic input. But when visual or graphic input is BLENDED with the words, the two regions of the brain connect.

The referential connections between the two zones allow the information to actually be stored and become long-term memory!

This is called Dual Coding Theory.

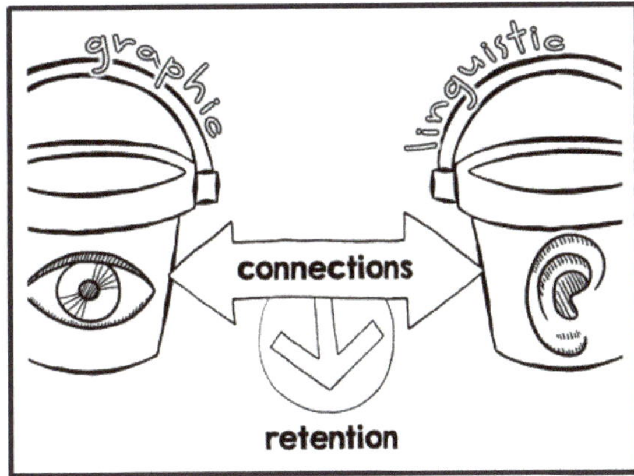
graphic
linguistic
connections
retention

It has been proven that students can retain more information when they connect the linguistic and visual centers of the brain.

doodling
takes just enough
attention to keep the brain from daydreaming without allowing it to become distracted.
doodlenotes.org

Doodling while listening has also been proven to improve focus! Visual note methods have numerous benefits for student learning.

doodle notebook

What's Included?

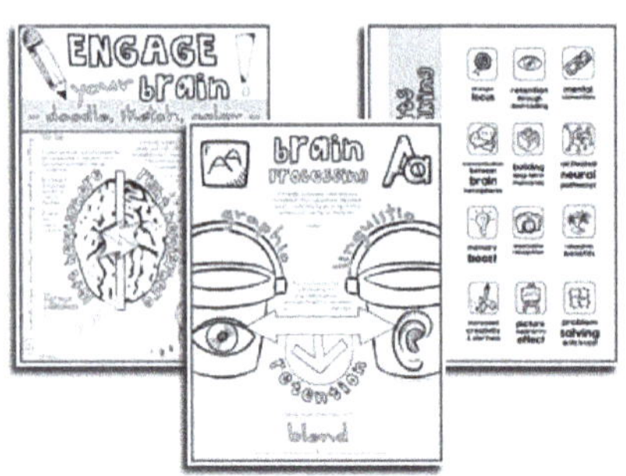

Introductory Pages:
to model the method while showing students WHY visual note-taking engages the brain

Chapter Cover Pages
Just for fun & organization!
(Students can color if they want to)

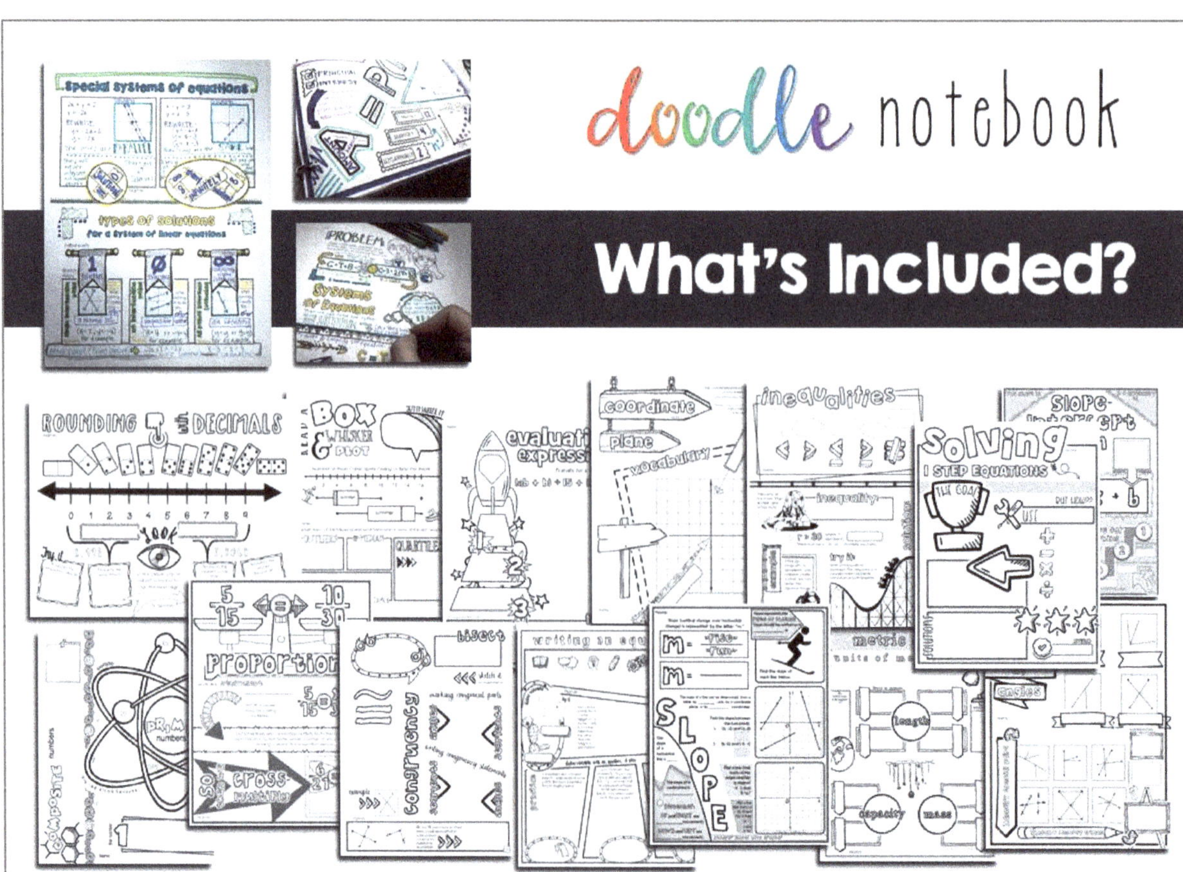

doodle notebook

What's Included?

Visual, Interactive "Doodle Note" Lesson Pages
for all the key ideas in middle school math / Pre-Algebra coursework

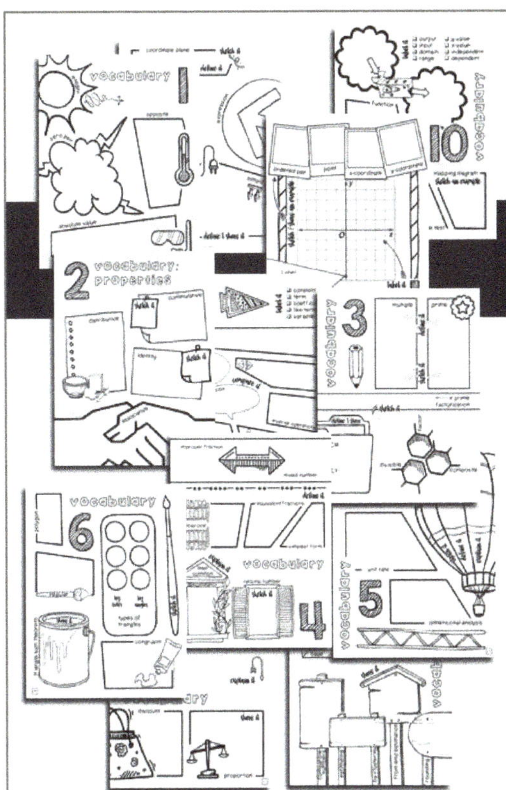

doodle notebook

What's Included?

Vocabulary Reviews
doodle-friendly sheets for students to review key terms after each chapter (26 total pages of vocab review included)

BONUS coloring / fill-in pages
a few extras here and there for students to review key ideas

doodle notebook

What's Included?

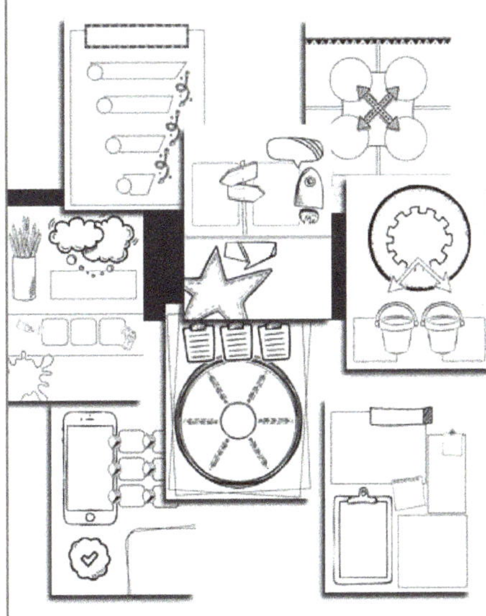

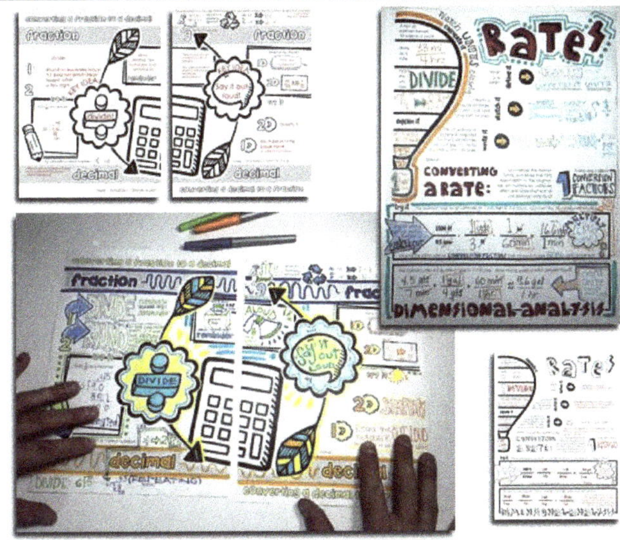

10 Blank Templates
for adding any extra lesson topics you'd like students to have included

This way everything will be in one place instead of additional notes being on loose paper (and still with the brain-based benefits of the visual note format!)

Answer Keys & Photo Samples
for guided note-taking

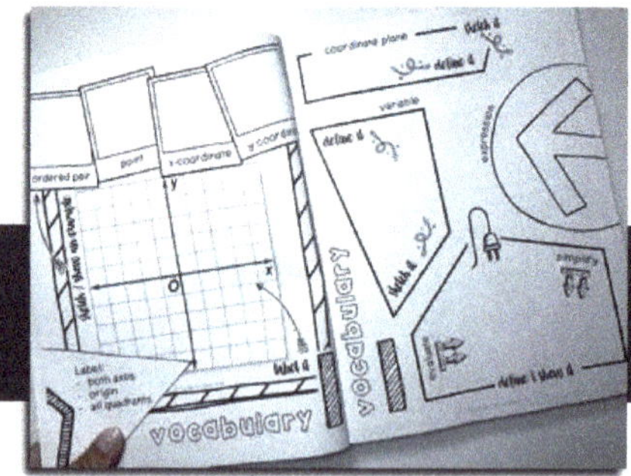

doodle notebook

How to Use

Use the Bonus Pages

Assign the vocabulary review sheets at the end of each chapter as independent work. Students can develop their own sketches, definitions, and examples to review each key term. These pages have plenty of flexibility built in so that students can review the key ideas in their own creative ways. The chapter cover pages and bonus coloring pages can be used by students who want to go above and beyond to create a complete doodle notebook, packed with color and embellishment. Coloring the extra "key concept" pages will help students remember the big ideas.

Use the Templates

For any additional topics you'd like to cover, use the templates in the back of the book. These are included so that students can have a complete book with any other lessons you want included in your curriculum. They'll still get all the brain-based benefits of visual notes when they write the lesson topic themselves and complete the page with their own sketches and written notes.

Add Plenty of Practice

The doodle notebook is intended to accompany lectures and direct instruction. It does NOT include a full curriculum of practice / homework. You will still need to supplement and follow up each lesson as you normally do. You can do this with problems from the textbook, worksheets, online practice, etc. Students need to be exposed to each type of problem. The doodle note pages are intended to just introduce the key ideas and be used as a reference.

doodle notebook

What makes the "doodle note" method effective?

A unique blend of...

- visual memory triggers / graphic analogies

visual memory trigger

a blend of teacher input and plenty of space for student input

graphic analogy

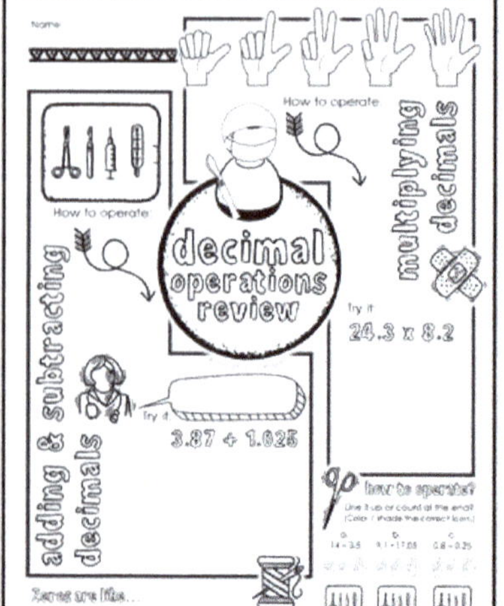

visual memory trigger

interactive task

opportunities to color, doodle, embellish, and do hand-lettering

doodle notebook

contents

contents

contents

contents

ENGAGE your brain!

- doodle, sketch, color -

TO DO:

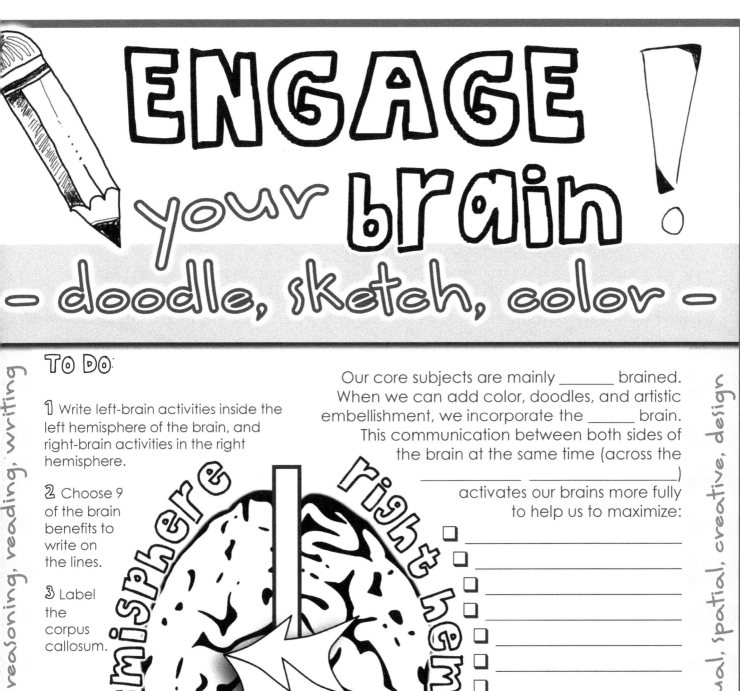

1 Write left-brain activities inside the left hemisphere of the brain, and right-brain activities in the right hemisphere.

2 Choose 9 of the brain benefits to write on the lines.

3 Label the corpus callosum.

Our core subjects are mainly _______ brained. When we can add color, doodles, and artistic embellishment, we incorporate the ______ brain. This communication between both sides of the brain at the same time (across the ____________ ________________) activates our brains more fully to help us to maximize:

- ❑ ______________________
- ❑ ______________________
- ❑ ______________________
- ❑ ______________________
- ❑ ______________________
- ❑ ______________________
- ❑ ______________________
- ❑ ______________________
- ❑ ______________________

left hemisphere

right hemisphere

b-a-l-a-n-c-e

The **corpus callosum** is...

visual memory triggers...

LEFT: words, logic, numbers, reasoning, reading, writing

RIGHT: color, music, art, visual, spatial, creative, design

Name:

brain processing

The brain processes new learned material in two completely separate areas! Label the types of input that enter each center of the brain.

Name:

graphic

linguistic

But, to convert the information to long-term memory (actually LEARN it), we need to ________________ the two!

retention

Taking visual notes helps us to

blend

_________________________ & _________________________

learning benefits of visual note-taking

stronger **focus**

retention through dual coding

mental connections

communication between **brain** hemispheres

building long-term memories

activated **neural** pathways

memory **boost**

associative recognition

relaxation **benefits**

increased **creativity** & alertness

picture superiority **effect**

problem solving skills boost

1
Expressions
& Integers

Expressions & Variables

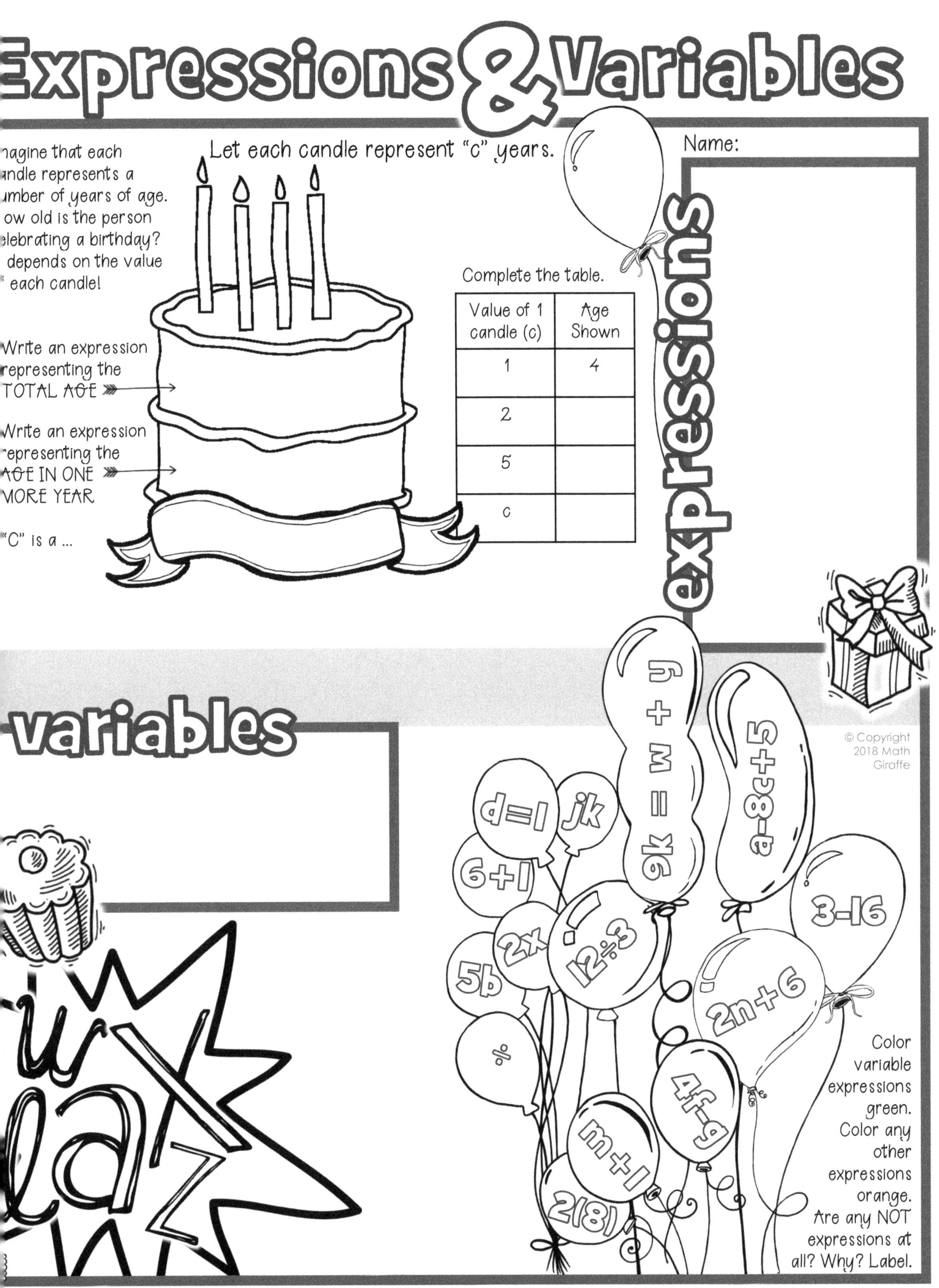

Value of 1 candle (c)	Age Shown
1	4
2	
5	
c	

writing

Name:

Writing expressions is like TRANSLATING betwee
English Language (words) and Math Language (symbols

Three more than the product of a number and the "quantity" two less than

expressions

Simplify:

$12 + 7b - 4$

$8(6) - 3 + mn$

simplifying

Think like a car crusher!

same material (equivalent), but in its smallest, most compact form!

evaluating

Evaluate for $h = 3$, $j = 1$, and $k = 5$

$2h + k$

$8hjk + 17$

Order of Operations

When we **simplify** an ______________, we must complete the steps in the correct order, based on the operations.

This **sequence** of steps is called the ______________ of operations.

If the order is not followed correctly, our simplified version will not be ____________________!

To remember the order, try memorizing "**PEMDAS**".

Sample: Simplify.

For each step, write the operation that allows the simpler, equivalent expression to be written.

$$27 \div (4 - 1)^2 - 1 + 6$$

$$27 \div 3^2 - 1 + 6$$

$$27 \div 9 - 1 + 6$$

$$3 - 1 + 6$$

$$2 + 6$$

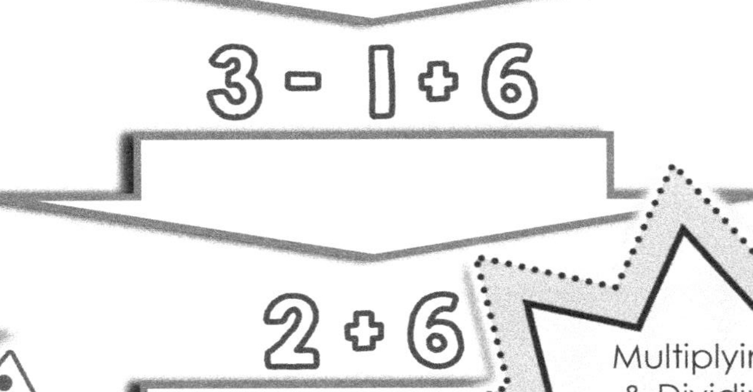

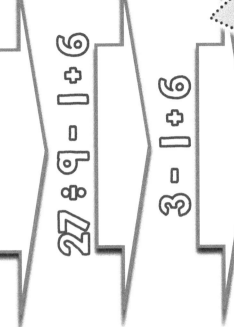

Multiplying & Dividing happen TOGETHER from ____ to ____ (same with _______ & ________).

Name:

Order of Operations Practice & Reminders

Name:

Types of Grouping Symbols

Multiplying and Dividing & Adding and Subtracting

How to Show Your Work

Simplify each expression. Show only one step per line.

1. $24 - 2^3 \cdot 2 + 5$
2. $5 - 2 + 9 - 8$
3. $35 - [12 + 3 \cdot (1 + 2)]$
4. $(4 + 1)^2 \div 1^4$
5. $2 + \dfrac{5 - 2}{6 + 9 \div 3}$

Evaluate for $a = 4$, $b = 6$

$$(ab + b) \div 15 + (2b - a)$$

1

2

3

try it

Evaluate for $x = 1$, $y = 2$, and $z = 5$

$$\frac{y^2}{x+z}$$

$$\frac{x + 2yz}{2(y+4) - z}$$

are greater
than zero
0
are
less
than
zero

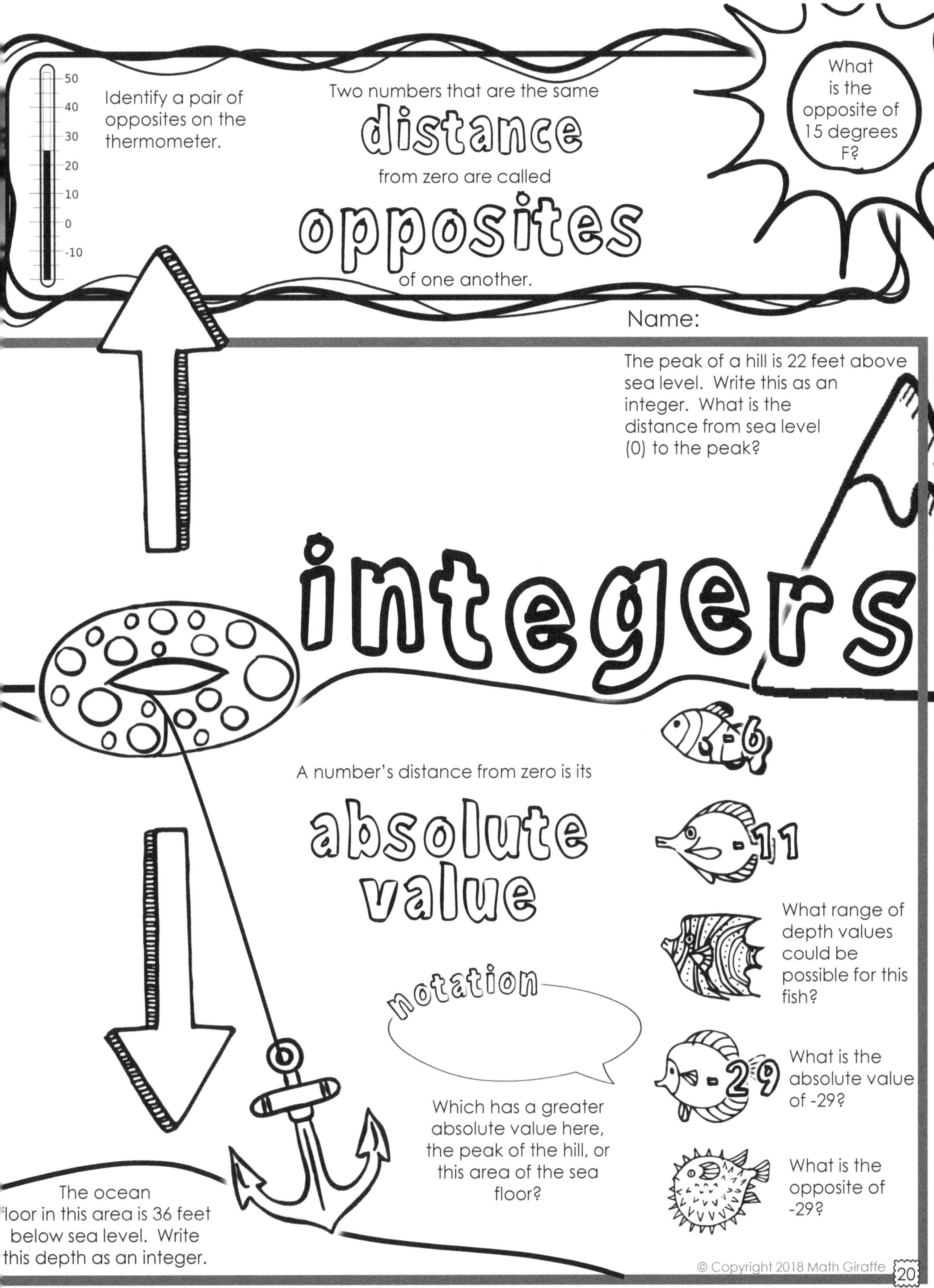
50
40
30
20
10
0
-10
Identify a pair of opposites on the thermometer.
Two numbers that are the same
distance
from zero are called
opposites
of one another.
What is the opposite of 15 degrees F?
Name:
The peak of a hill is 22 feet above sea level. Write this as an integer. What is the distance from sea level (0) to the peak?
integers
-6
A number's distance from zero is its
absolute value
-11
What range of depth values could be possible for this fish?
notation
-29
What is the absolute value of -29?
Which has a greater absolute value here, the peak of the hill, or this area of the sea floor?
What is the opposite of -29?
The ocean floor in this area is 36 feet below sea level. Write this depth as an integer.

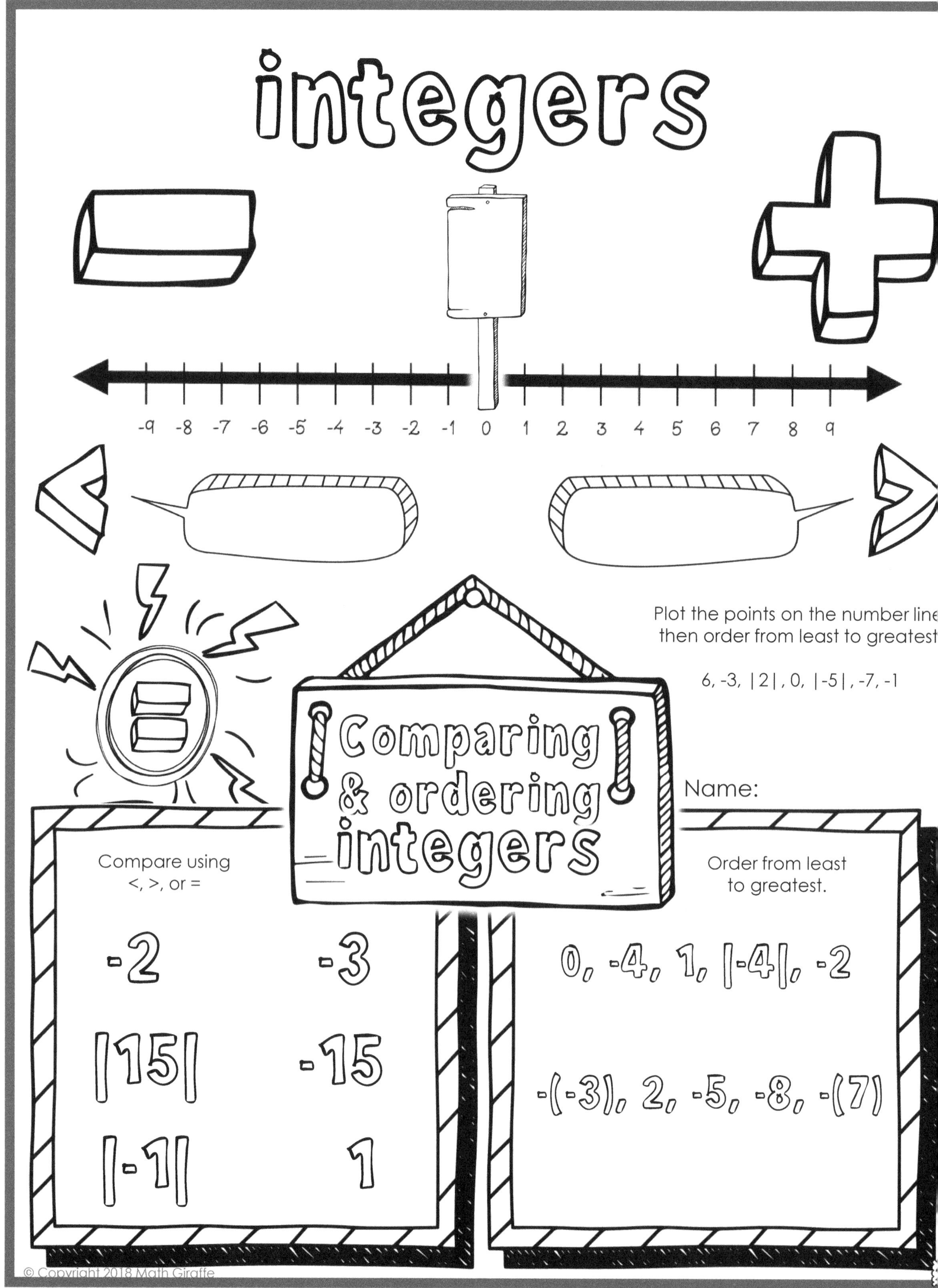
integers
-9 -8 -7 -6 -5 -4 -3 -2 -1 0 1 2 3 4 5 6 7 8 9
Plot the points on the number line then order from least to greatest
6, -3, |2|, 0, |-5|, -7, -1
Comparing & ordering integers
Name:
Compare using <, >, or =
-2 -3
|15| -15
|-1| 1
Order from least to greatest.
0, -4, 1, |-4|, -2
-(-3), 2, -5, -8, -(7)

ABSOLUTE
VALUE BARS
operate as
GROUPING
SYMBOLS

A scuba diver swam...

then

then

then

then

then

Wher did h end u

INTEGER

(positive and negative whole numbers & ____)

operations

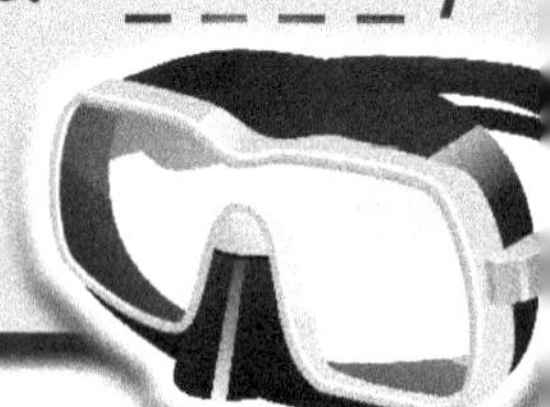

adding

Think: Adding two DESCENTS together goes EVEN DEEPER. Adding two ascents together is rising EVEN HIGHER.

pos + pos
neg + neg

same sign

Think: An ascent can partially cancel out or overcome a descent, or...

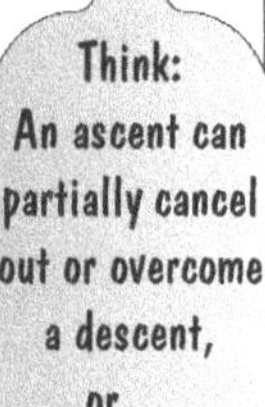

different sign

subtracting

Remember:

Another great analogy for understanding integers is money.

Think about adding and removing (subtracting) credits and debts. Example: Removing a debt is a positive movement!

Think: Removing a recent descent is actually rising again, removing a recent ascent is the same as descending again.

Name:

PRACTICE WITH INTEGER OPERATIONS

Simplify each expression.

multiplying & dividing

Example 1
$(-4) + (-16) =$

Example 2
$9 + (-12) =$

Example 3
$(-24) \div (-3) =$

Example 4
$15 \cdot (-3) =$

Think:
Multiple descents of the same depth work like repeatedly adding those negative movements.
Ex: 3 descents of 6 meters each can be written:
$3 \cdot (-6)$ or
$(-6) + (-6) + (-6)$

Example 5
$(-17) - 21$

Example 6
$4 - (-30)$

Example 7
$(-35) - (-40)$

CHALLENGE

$3 + (-5) + 8 =$

$(-2) + (-5) + (-4) =$

$(-4)(5)(-1)(3) =$

$4 \cdot (-2)(-3)(-5) \cdot 2 =$

Word Problem

Alicia withdrew 45 dollars from her bank account. Later that day, she deposited \$80, then wrote a check for \$250. Last, she spent another \$55 from the account while shopping with her direct debit card. What was her total gain or loss in funds from the account that day? Write an integer expression, then simplify.

Note:
When multiplying or dividing 3 or more integers,
if the number of negatives is
EVEN,
but if it is
ODD,

Name:

WHEN MODELING INTEGERS, THINK IN TERMS OF
zero pairs

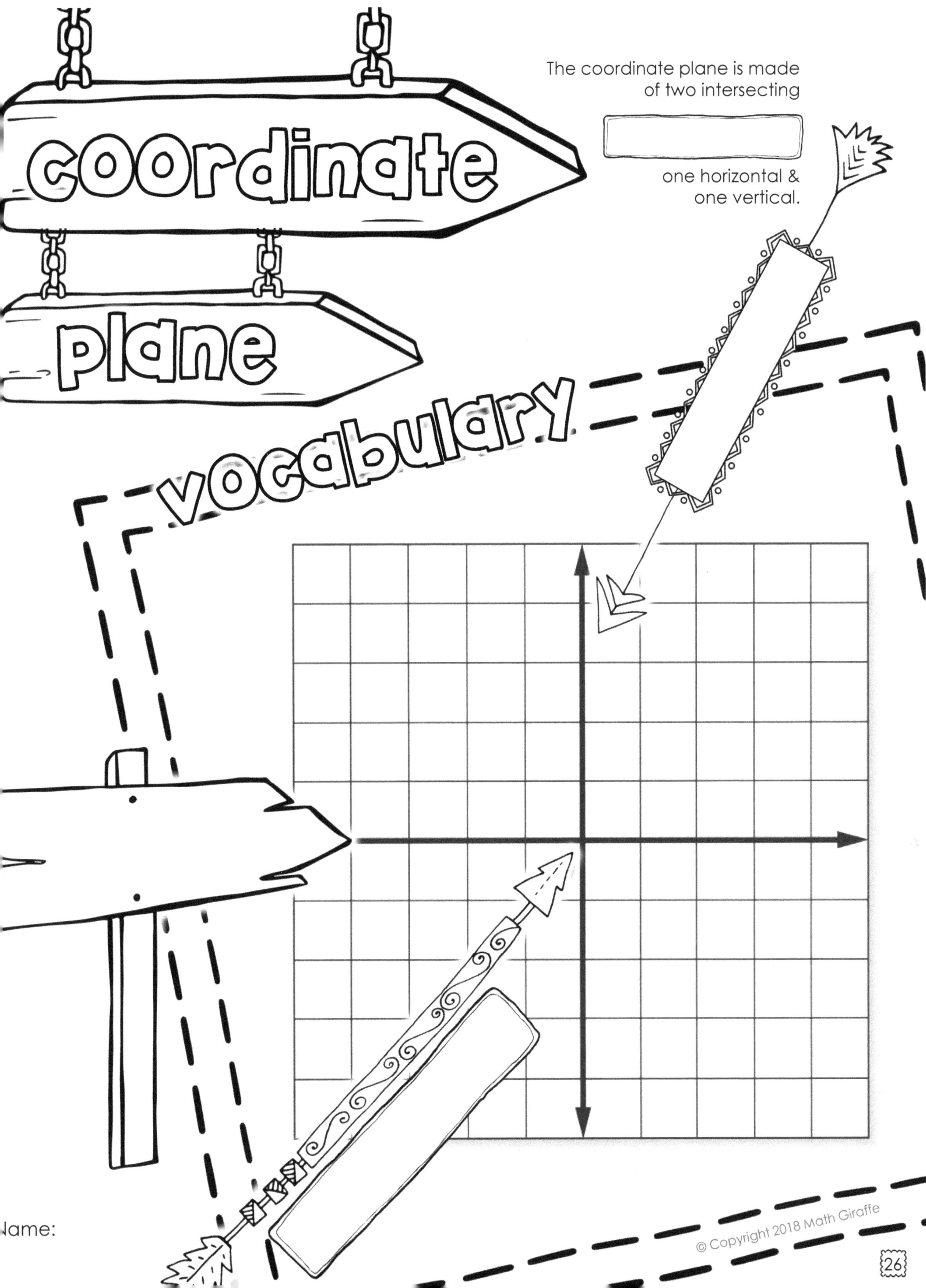
coordinate
plane
vocabulary
The coordinate plane is made of two intersecting
one horizontal & one vertical.
Name:

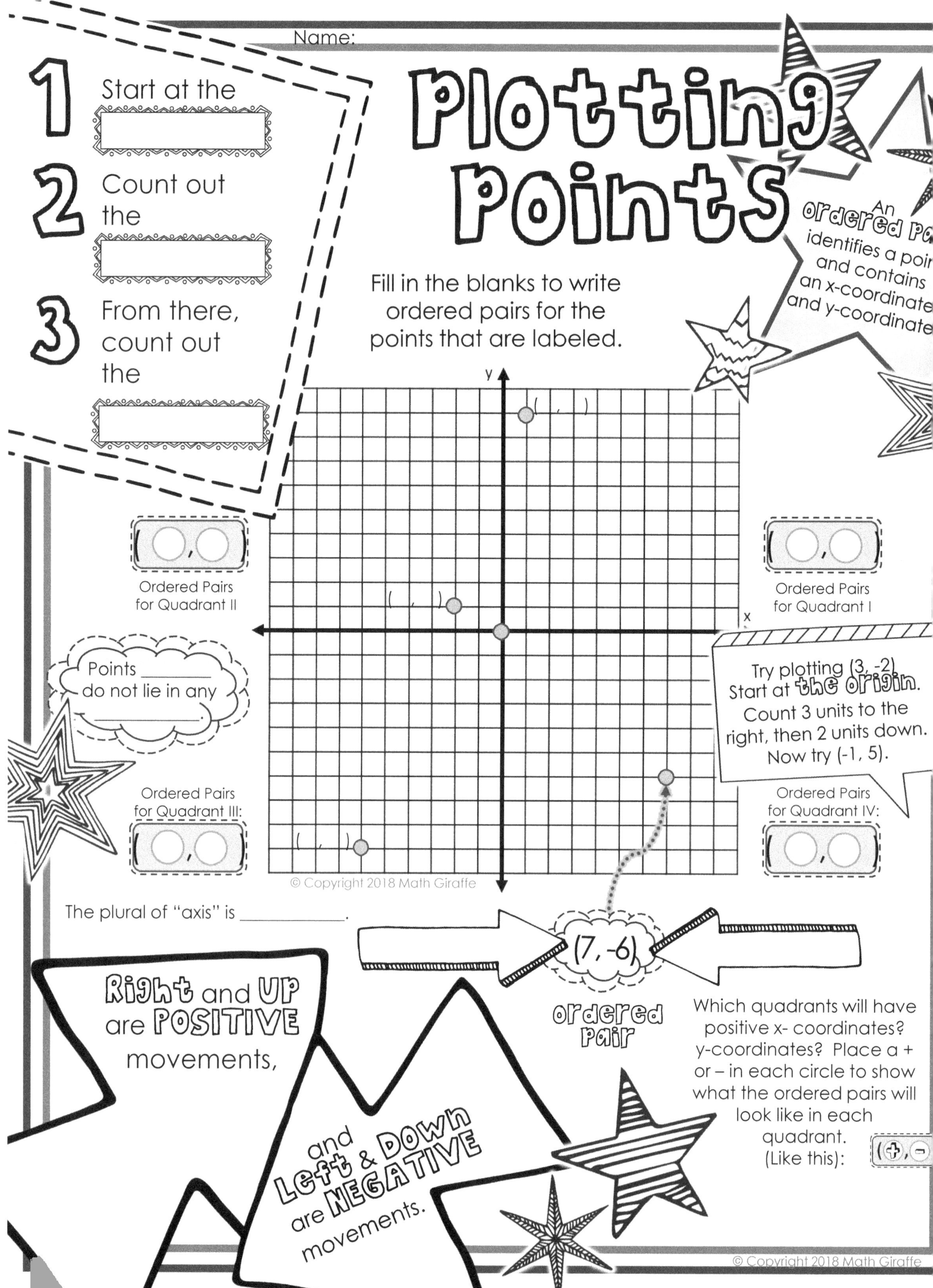
Name:
Plotting Points
1 Start at the
2 Count out the
3 From there, count out the
An ordered pa[ir] identifies a poi[nt] and contains an x-coordinate and y-coordinate
Fill in the blanks to write ordered pairs for the points that are labeled.
y
x
(,)
Ordered Pairs for Quadrant II
Ordered Pairs for Quadrant I
Points ________ do not lie in any ____________.
Try plotting (3, -2). Start at the origin. Count 3 units to the right, then 2 units down. Now try (-1, 5).
Ordered Pairs for Quadrant III:
Ordered Pairs for Quadrant IV:
© Copyright 2018 Math Giraffe
The plural of "axis" is ____________.
(7, -6)
ordered pair
Right and Up are POSITIVE movements,
and Left & Down are NEGATIVE movements.
Which quadrants will have positive x- coordinates? y-coordinates? Place a + or – in each circle to show what the ordered pairs will look like in each quadrant. (Like this):
(+, –)
© Copyright 2018 Math Giraffe

ORDERED
PAIRS
represent
POINTS
AND CONTAIN
&
AN X-COORDINATE
A Y-COORDINATE

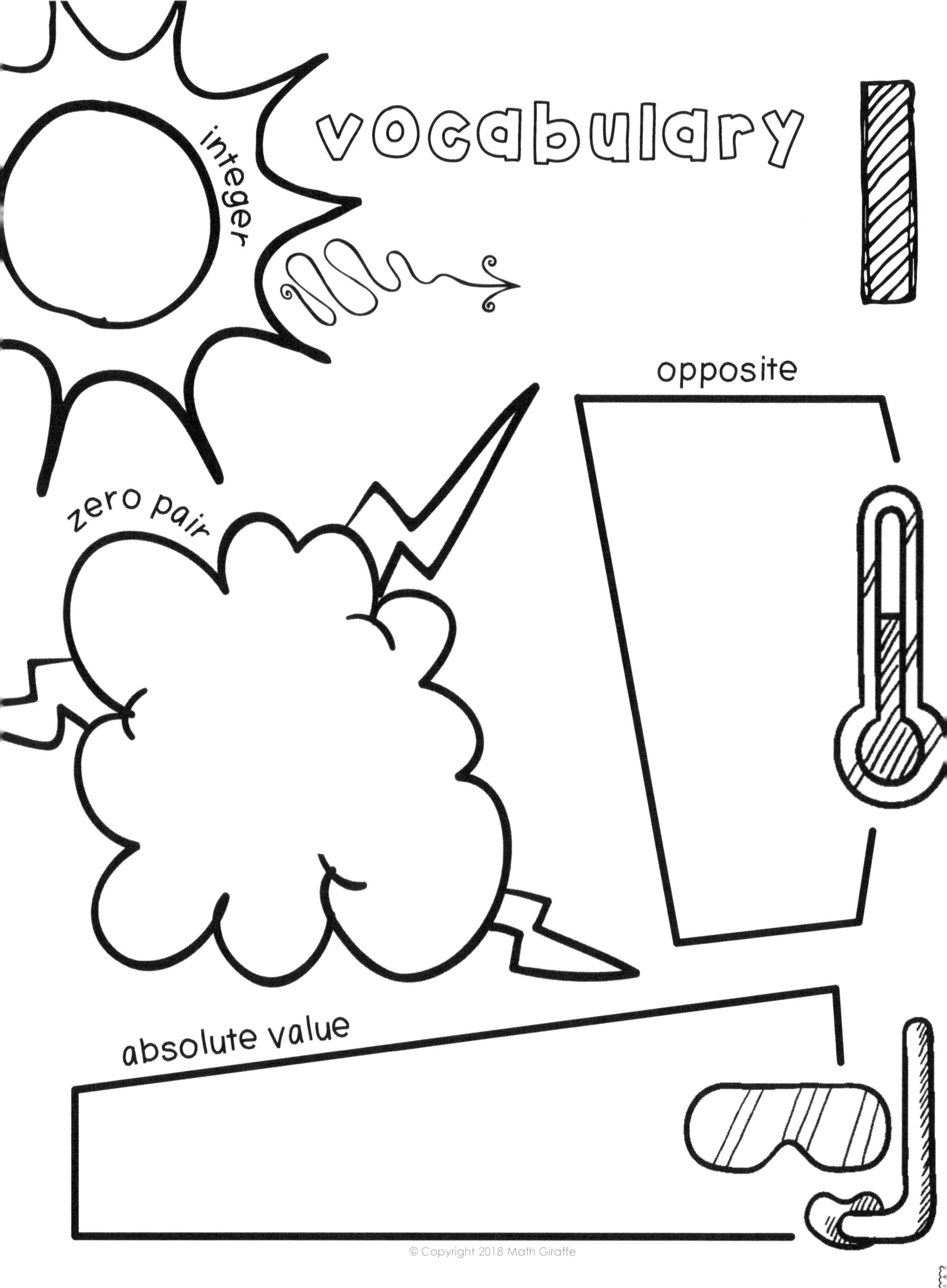
vocabulary
integer
opposite
zero pair
absolute value

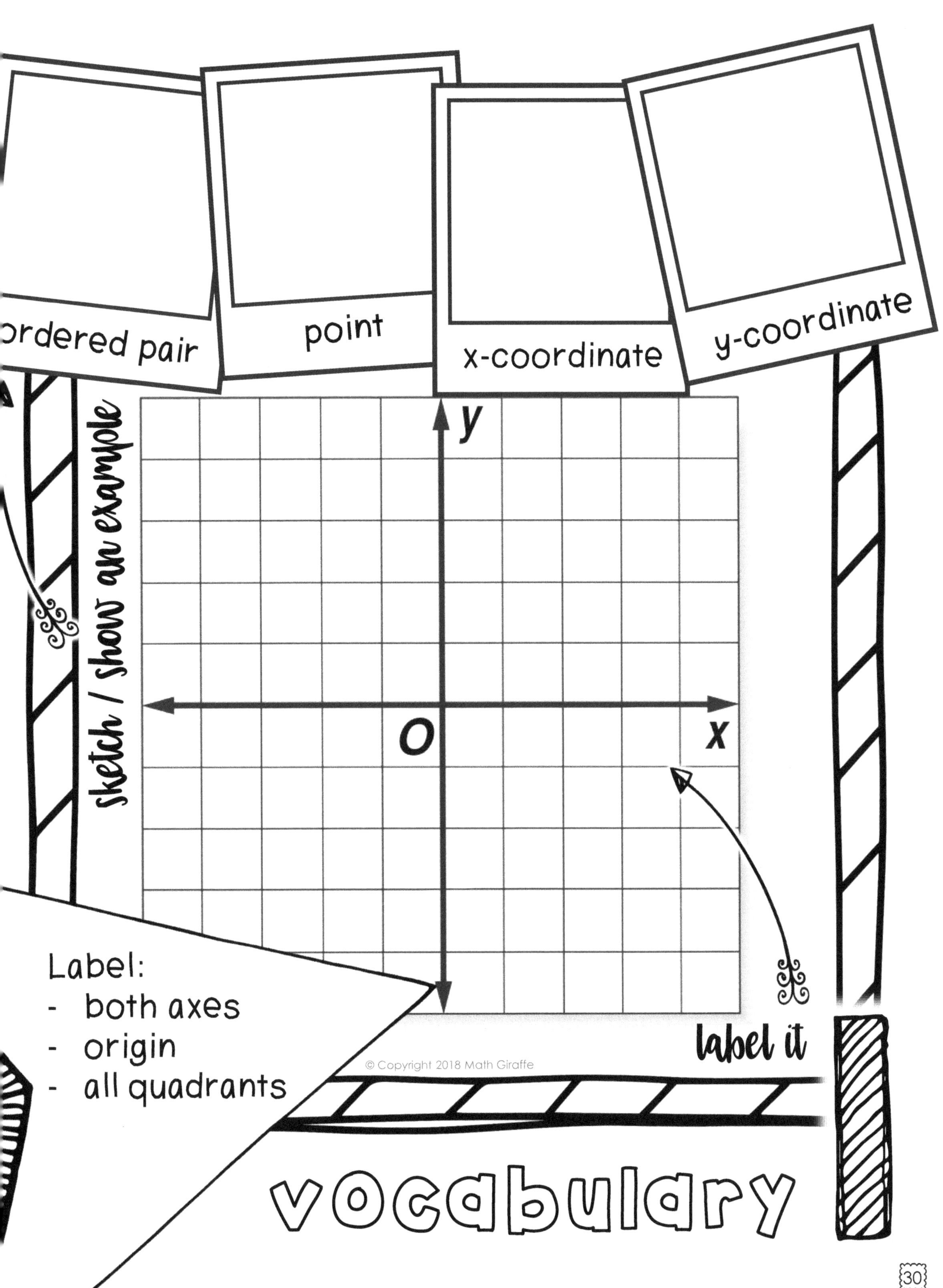
ordered pair
point
x-coordinate
y-coordinate
sketch / show an example
y
O
x
Label:
- both axes
- origin
- all quadrants
label it
vocabulary

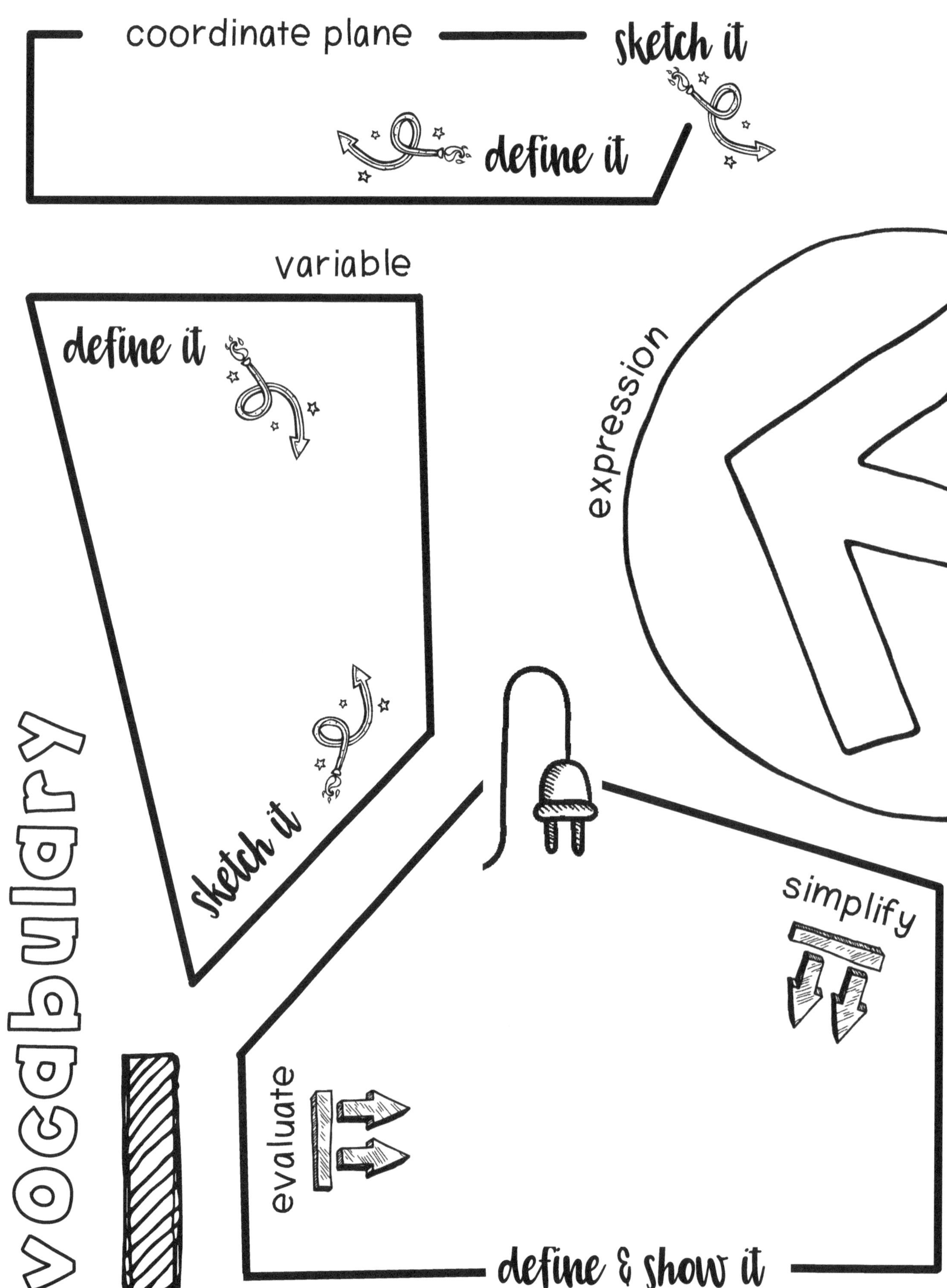
coordinate plane
sketch it
define it
variable
define it
sketch it
expression
simplify
evaluate
define & show it
vocabulary

Equations & Inequalities

These number properties, or rules, allow us to:

1

EXAMPLE

$7 + 6 = 6 + \_\_$

$f \cdot 144 = \_\_\_\_ f$

Name:

2

"Associate" means...

Associative Property of Addition

Associative Property of Multiplication

EXAMPLES

$8g(4) = 8(\_\_\_ \cdot \_\_\_)$

$+ (12 + 3) = (\_\_\_ + \_\_\_) + 3$

Name:

3

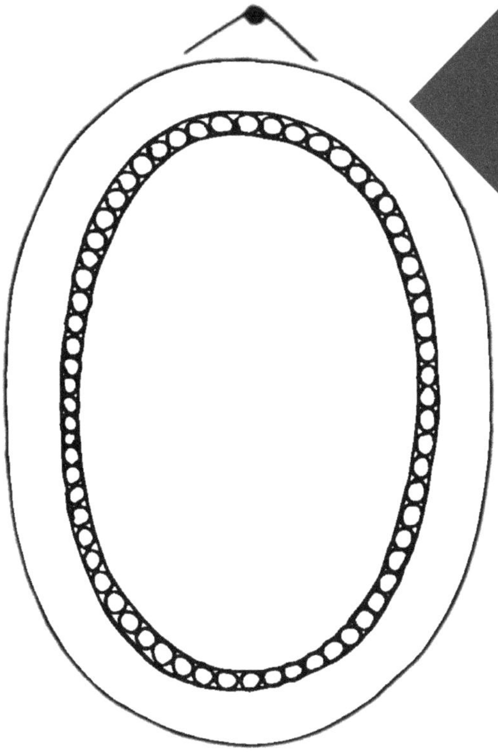

Identity Property of Addition

Identity Property of Multiplication

"Identity" means...

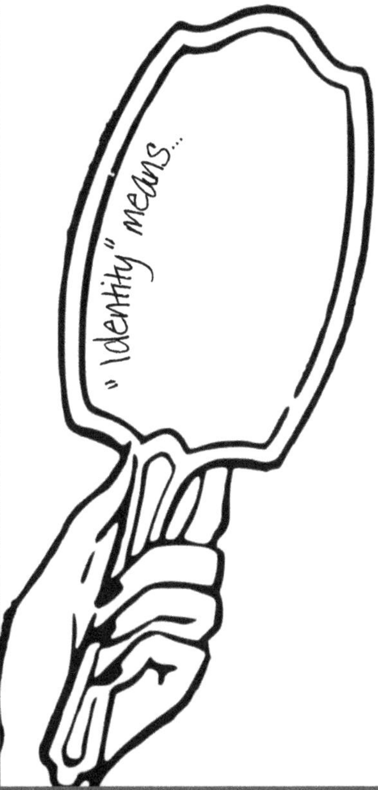

EXAMPLES

$p + 0 = \_\_\_$

$154 \cdot 1 = \_\_\_$

Practicing Properties

Identify each property.

1. $3v + w = w + 3v$

2. $0 + 45 = 45$

3. $5(15 \cdot 8) = (5 \cdot 15) \cdot 8$

4. $1(93) = 93$

5. $(5 + 6) + (8 + 1) = (8 + 1) + (5 + 6)$

Write an equivalent expression.

1. $25 + 44 =$
(Use the commutative property.)

2. $f(g \cdot h) =$
(Use the associative Property.)

3. $319 + (x + 7) =$
(Use the associative Property.)

Simplify each expression.

1. $156 \cdot 1 + 0 =$

2. $10(2 \cdot 21) =$

3. $20 + (110 + 58) =$

4. $10 + 39 + 10 + 20 + 1 =$

Name:

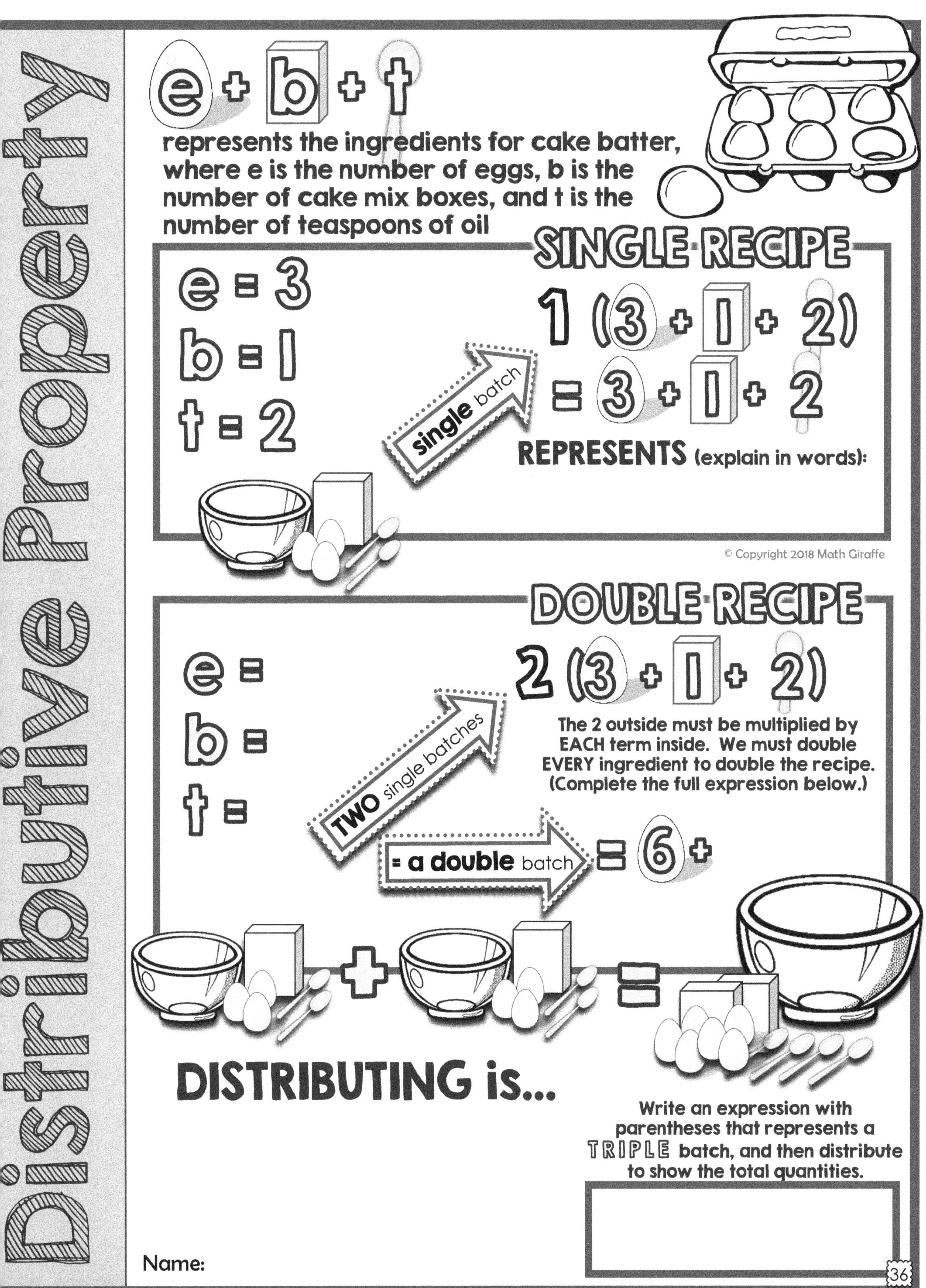
Distributive Property
e + b + t
represents the ingredients for cake batter, where e is the number of eggs, b is the number of cake mix boxes, and t is the number of teaspoons of oil
SINGLE RECIPE
e = 3
b = 1
t = 2
single batch
1 (3 + 1 + 2)
= 3 + 1 + 2
REPRESENTS (explain in words):
© Copyright 2018 Math Giraffe
DOUBLE RECIPE
e =
b =
t =
TWO single batches
2 (3 + 1 + 2)
The 2 outside must be multiplied by EACH term inside. We must double EVERY ingredient to double the recipe. (Complete the full expression below.)
= a double batch
= 6 +
DISTRIBUTING is...
Write an expression with parentheses that represents a TRIPLE batch, and then distribute to show the total quantities.
Name:
36

Name:

THE DISTRIBUTIVE PROPERTY

$a(b + c) =$

Next, we are preparing a recipe that requires 5 frozen bananas per batch, plus an unknown amount of chocolate (x).

Write an expression with parentheses that represents a quadruple recipe.

NOTE

When distributing a **NEGATIVE...**

QUADRUPLE RECIPE

This should work for ANY value for ___ .

Simplify your expression, and write the expression for a quadruple batch in the bowl.

DISTRIBUTING A NEGATIVE

Samples:

$-2(x - 5)$

$-(4e + 3)$

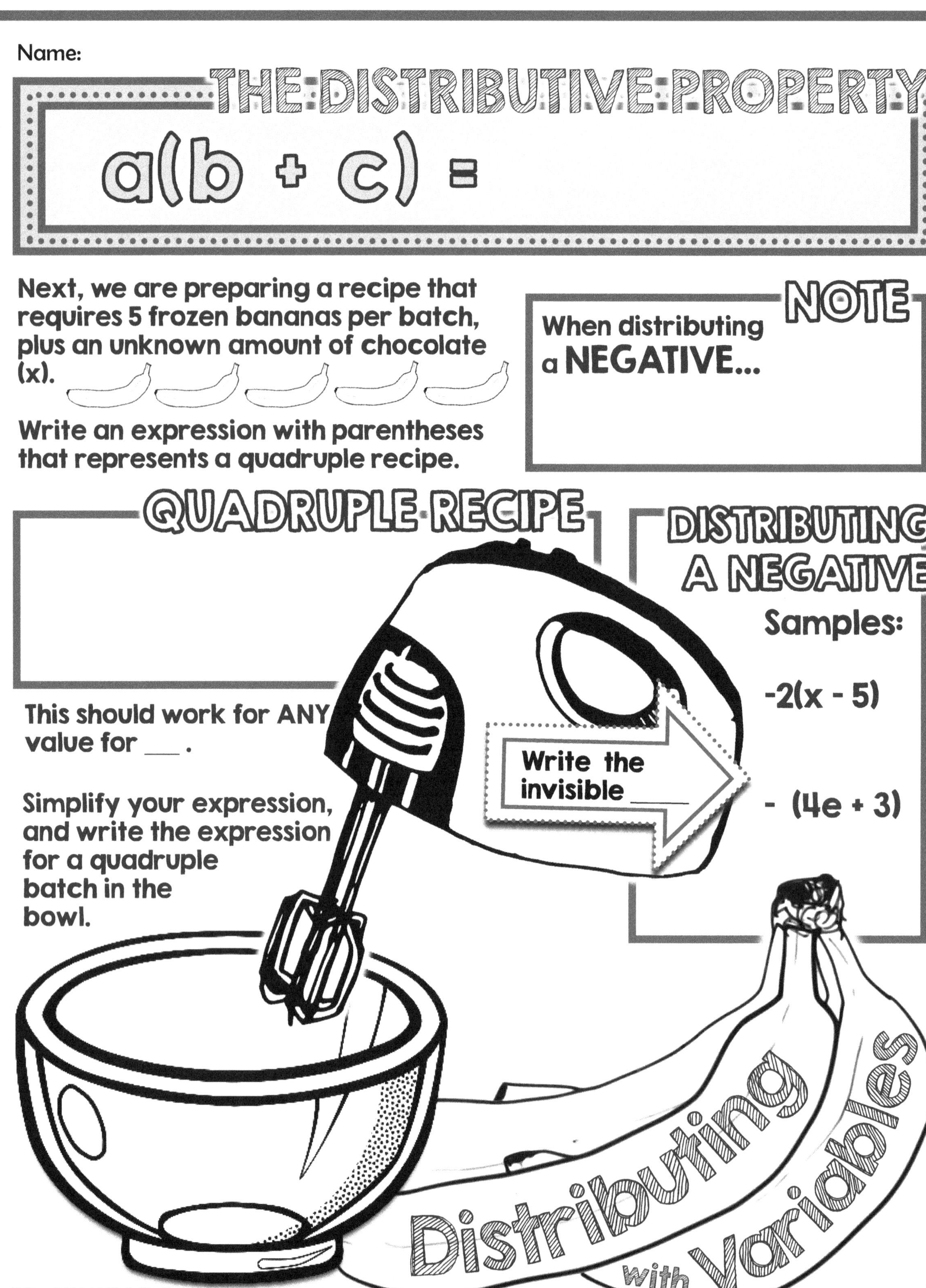

DISTRIBUTIVE PROPERTY - PRACTICE

Name:

Simplify each expression mentally using the distributive property.

Example 1

3(100 + 4)

Example 2

8(4 + 50)

Example 3

5(8 + 8 + 8)

Example 4

15(10 - 2)

TIPS & REMINDERS

In the left column, simplify using order of operations. In the right column, distribute, then simplify. Verify that the two are equivalent.

Example 5

4(12 - 5)

4(12 - 5)

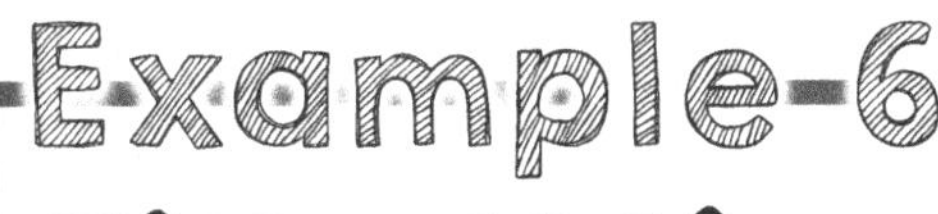

Example 6

5(10 + 100)

5(10 + 100)

DISTRIBUTIVE PROPERTY - PRACTICE

Name:

Simplify each expression using the distributive property.

Example 1

$-6(2x + 3)$

Example 2

$5(m - n)$

Example 3

$3c(a + b + 8)$

Example 4

$-(3f - 4g + 5)$

Example 5

$-8u(3u - w + 7)$

Example 6

$3h(j + 9k - 2)$

Example 7

$tv(t - 4v + 1)$

Example 8

$-3yz(2yz - 3y - 1)$

Example 9

$-x(14x - 9) + 3(x - 1)$

Example 10

$-(2m + 2) - 5(m - 1)$

Name: 

vocabulary

$$3x^2 - x + 2 - 4x$$

Combining
-LIKE TERMS-

Like terms have the same ______________ and the same ________________.

"To DO" List

- ❑ Add invisible "+" signs.
- ❑ Add invisible "1" coefficients.
- ❑ Color like terms with matching patterns/colors.
- ❑ Color, doodle, & embellish key ideas!

add like terms (while sticking with the +/- signs in front of each term)

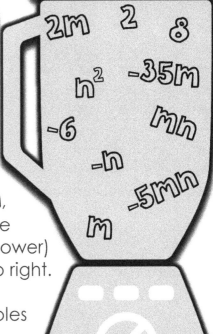

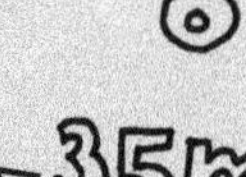

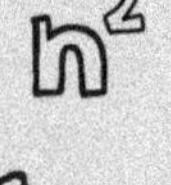

To write an expression in **standard form**, order terms so that the ________________ (power) decreases from left to right. Constants will be ____________. Variables should be in ____________________ order.

Try It

Simplify.

1. $3y - x - 6y - 1 + 12xy + 4 + xy$
2. $4ab - a^2 + 3a - b + a^2 - 2b - ab$

Combining like terms - Practice

Simplify each expression. Write each answer in standard form.

$$3 - 5w + 12 + 2w - 3$$

Example 2

$$7 - 4j - 2j^2 - 5j^3 + j^2 - j$$

Example 3

$$4(x - 3) + x - (3x + 7)$$

Example 4

$$8a - a(b + 6) + 2b(a - b) + a$$

An equation is like a complete sentence, while an expression is like a phrase.
6a = 12
"equals"
Six times a "IS" 12
3x + 2
© Copyright 2018 Math Giraffe

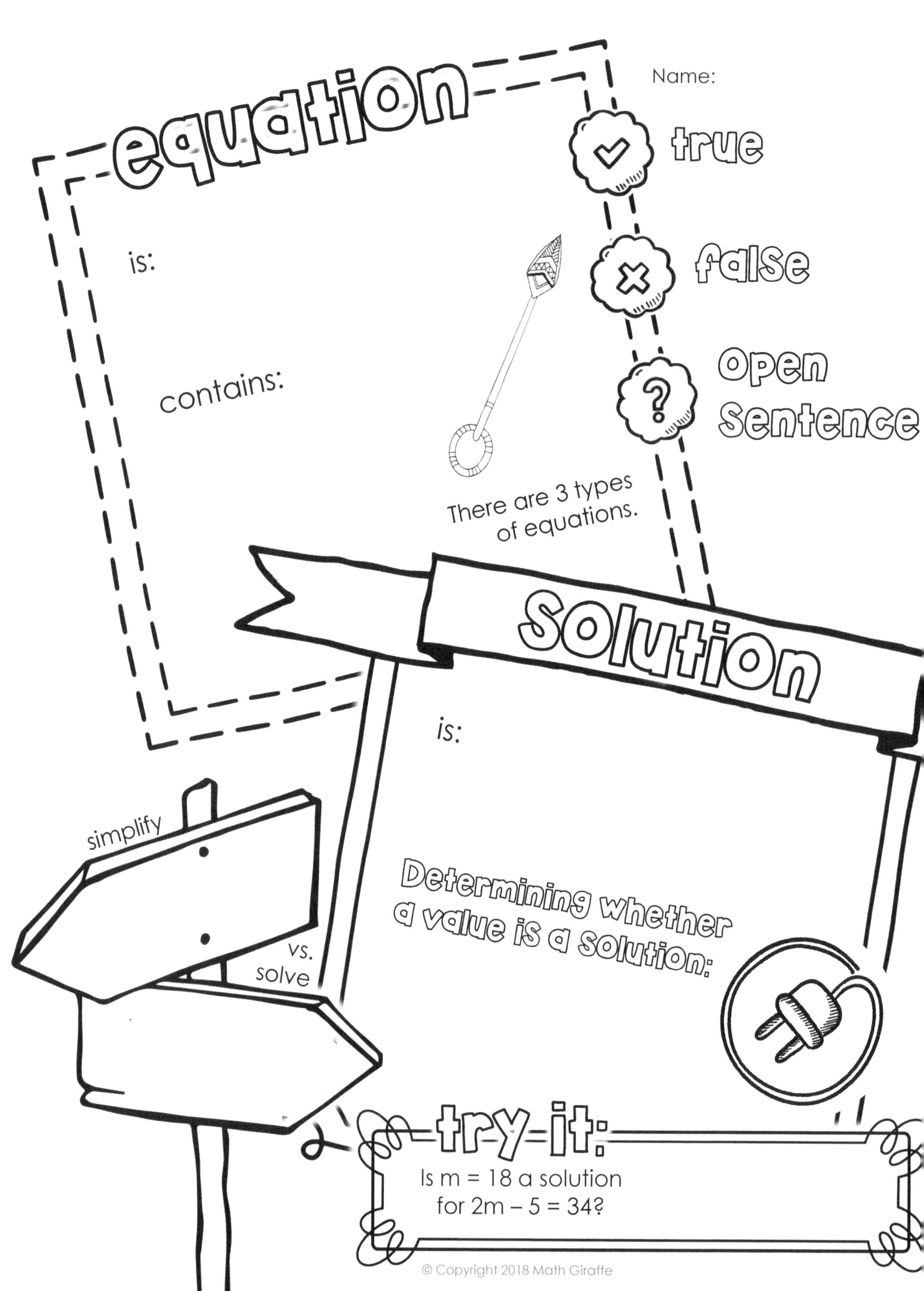
Name:
equation
is:
contains:
true
false
Open Sentence
There are 3 types of equations.
Solution
is:
simplify
vs.
solve
Determining whether a value is a solution:
try it:
Is m = 18 a solution
for 2m – 5 = 34?

Name:

Substitution:
7k = k + 24
Is 4 a solution?
Classify each as true, false, or an open sentence.
3m – 2 = 14
15 = 7(2) + 3
72 ÷ 3 = 8(3)
Solution
Solutions for equations
How many solutions?
w + l = 6
3b = 3b
A team has 9 players on the field plus a group on the bench. If their roster shows 23 players total, how many are on the bench? Write an equation. Can there be 15 on the bench?

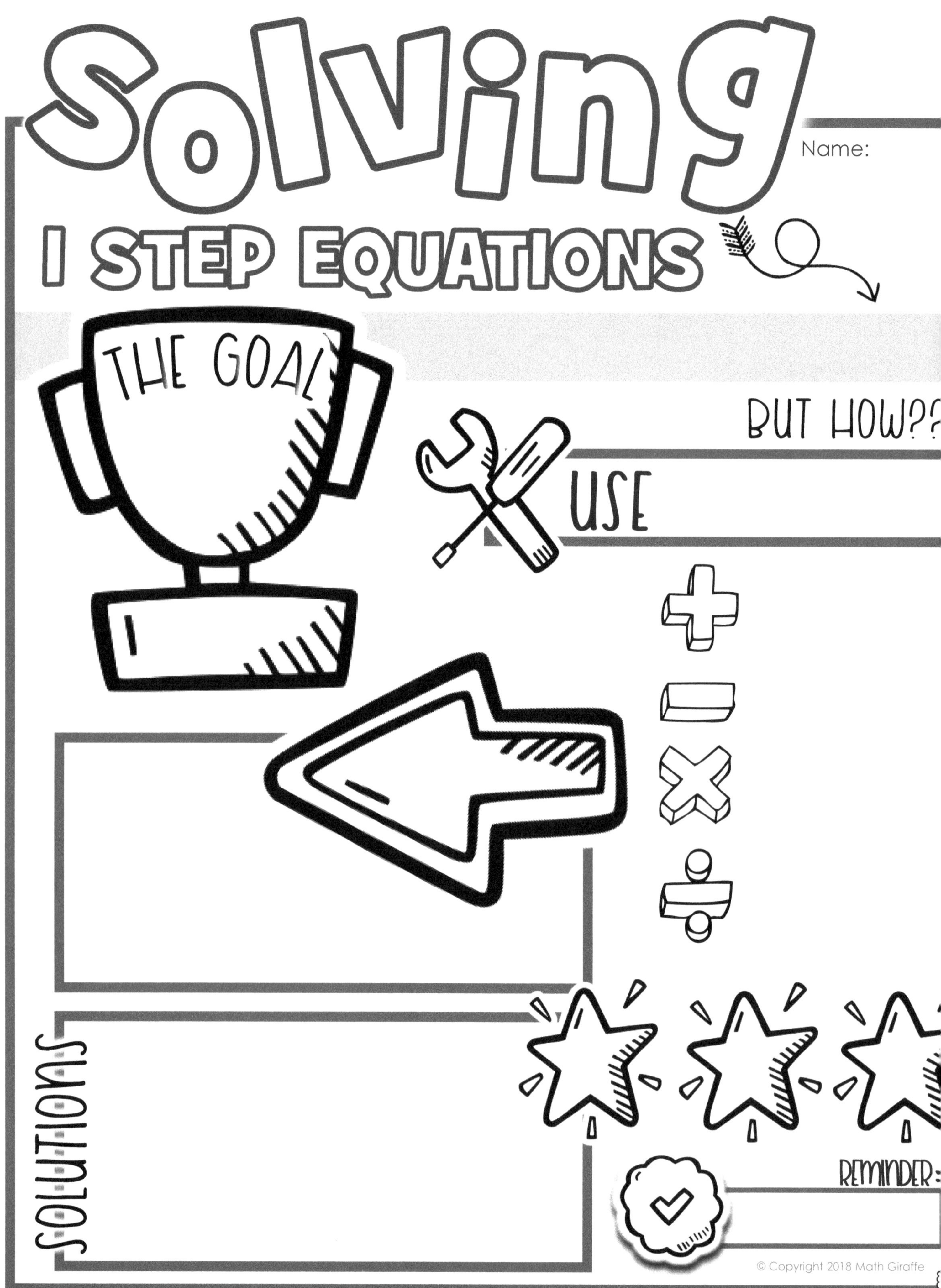
Solving
Name:
1 STEP EQUATIONS
THE GOAL
BUT HOW??
USE
SOLUTIONS
REMINDER:
© Copyright 2018 Math Giraffe

Name:

HOW TO... SHOW YOUR WORK

SOLVING 1 STEP EQUATIONS

examples

SOLVE FOR J:
J − 12 = 68

FIRST:

IDENTIFY THE OPERATION AND ITS ________.

THEN:

SHOW WORK TO _____ IT, KEEPING BOTH SIDES _____ / ________.

LAST:

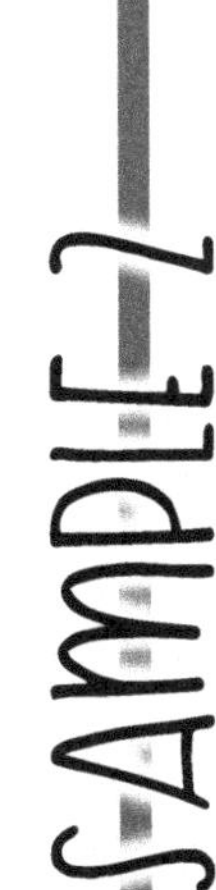

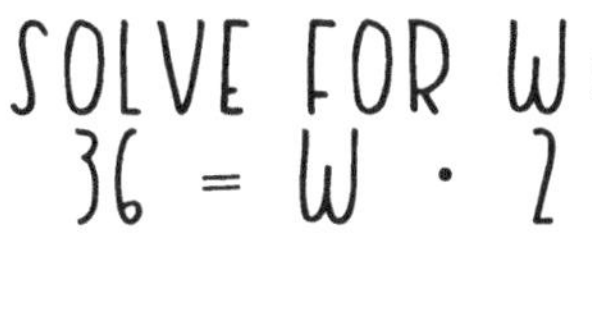

solving 1 step equations

SOLVE FOR X:

$18 = x \div 4$

Name:

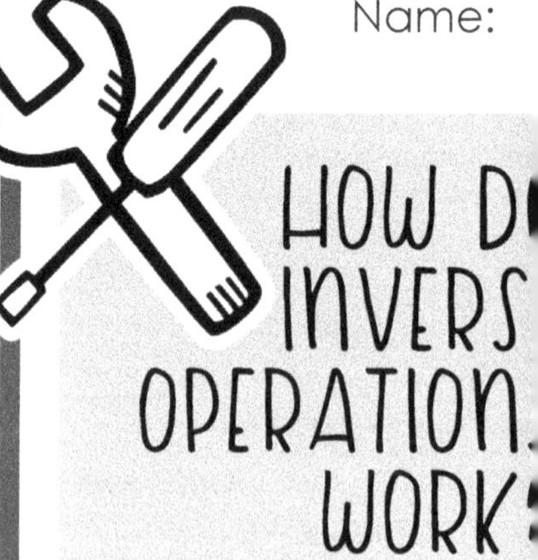

HOW D
INVERS
OPERATION
WORK

A SOLUTION IS...

SOLVE FOR M

$m + 9 = 38$

WHEN SOLVING, WHAT IS THE

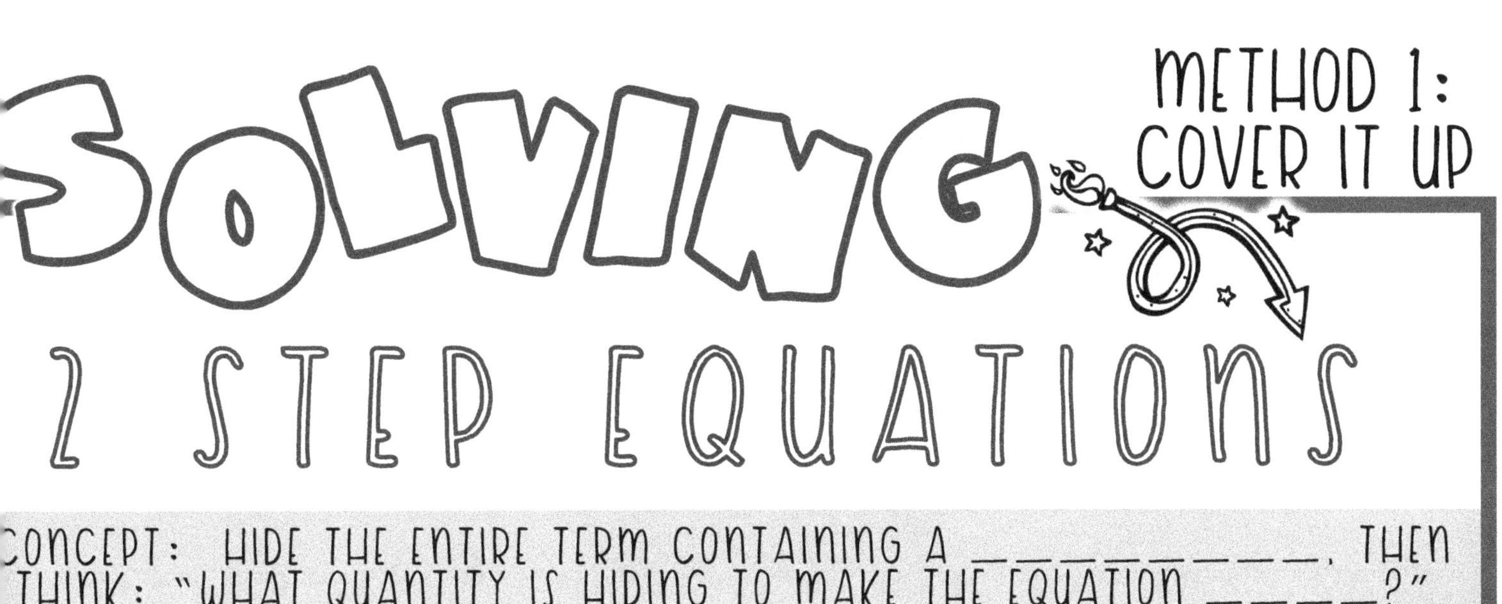

CONCEPT: HIDE THE ENTIRE TERM CONTAINING A __________, THEN THINK: "WHAT QUANTITY IS HIDING TO MAKE THE EQUATION ____?"

HOW IT WORKS

5 + 4X = 33

THIS TIME, WE WILL DO IT TOGETHER, AND COVER IT UP WITH A MAILBOX! WHEN YOU TRY ON YOUR OWN, YOU CAN CHOOSE ANY CONTAINER TO "HIDE" YOUR VARIABLE TERM.

5 + [mailbox] = 33

So 4x = a total of ◯ hiding inside.

LET EACH ENVELOPE REPRESENT X. HOW MUCH IS EACH WORTH?

=

=

REMINDER:

Name:

A solution is...

TRY IT

SOLVE FOR X:
29 − 2X = 13

THE  REPRESENTS:

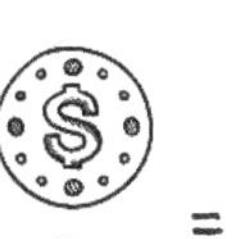 =

=

TRY IT

SOLVE FOR B:
8B + 12 = 84

SOLVING

2 STEP EQUATIONS

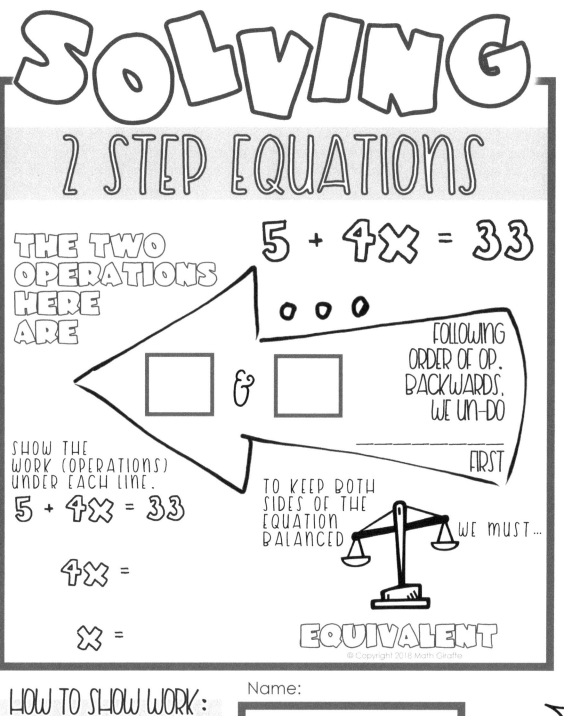

THE TWO OPERATIONS HERE ARE

$5 + 4x = 33$

☐ & ☐

FOLLOWING ORDER OF OP. BACKWARDS, WE UN-DO ______________ FIRST

SHOW THE WORK (OPERATIONS) UNDER EACH LINE.

$5 + 4x = 33$

$4x =$

$x =$

TO KEEP BOTH SIDES OF THE EQUATION BALANCED WE MUST...

EQUIVALENT

METHOD 2: INVERSE OPERATIONS

CONCEPT:

REMEMBER, WHEN WE SIMPLIFY OR EVALUATE, WE DO A STANDARD ORDER OF OPERATIONS.

BUT

WHEN SOLVING, WE ARE UN -DOING THE OPERATIONS, SO WE PERFORM THE ORDER OF OPERATIONS

BACK WARDS

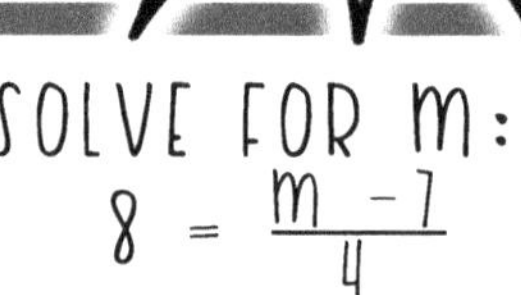

HOW TO SHOW WORK:

REMINDER:

Name:

TRY IT

SOLVE FOR H:

$\frac{H}{5} + 6 = 15$

ON YOUR OWN

SOLVE FOR m:

$8 = \frac{m - 7}{4}$

PRACTICE

SOLVING 2 STEP EQUATIONS

A SOLUTION IS...

USE INVERSE OPERATIONS

SOLVE FOR Y:
$5Y - 2 = 18$

ORDER OF OPERATIONS

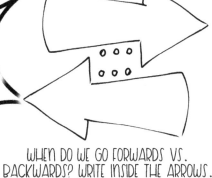

WHEN DO WE GO FORWARDS VS. BACKWARDS? WRITE INSIDE THE ARROWS.

COVER IT UP:

SOLVE FOR K:
$24 = 5K - 1$

YOUR CHOICE

SOLVE FOR P:
$\frac{P}{3} + 9 = 11$

SOLVE EACH FOR X

1. $\frac{X - 4}{4} = 2$

2. $6 = 4 + \frac{X}{2}$

3. $13 = 5 + 2X$

Name:

IN COMPLETE SENTENCES!

EXPLAIN

HOW & WHY THE "COVER IT UP" METHOD WORKS:

HOW & WHY TO KEEP AN EQUATION "BALANCED."

inequalities

less than	More than	less than or equal to	More than or equal to	Not Equal
$<$	$>$	$\leq$	$\geq$	$\neq$
		no more	at least	

Name:

The carnival has more than 30 rides and attractions!

inequality:

$r > 30$ where "r" represents:

There may be 31, 32, 33… (multiple solutions!)

solution:

example

Write an inequality to represent: Only children under 5 years old may enter the bounce house.

try it:

Write an inequality to represent: The steepness of a wooden roller coaster hill cannot exceed 85 degrees.

graphing solutions of an inequality

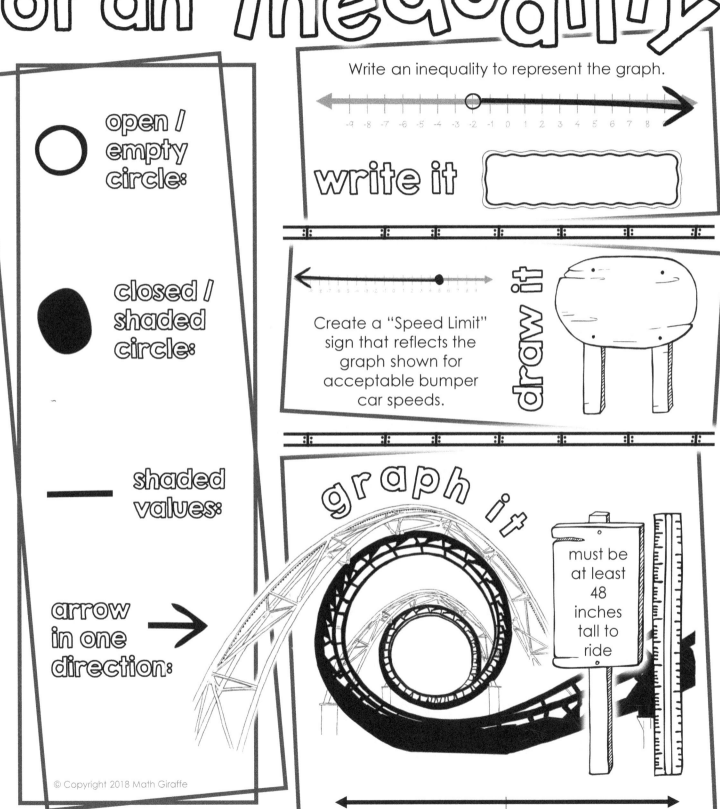

open / empty circle:

closed / shaded circle:

shaded values:

arrow in one direction:

Write an inequality to represent the graph.

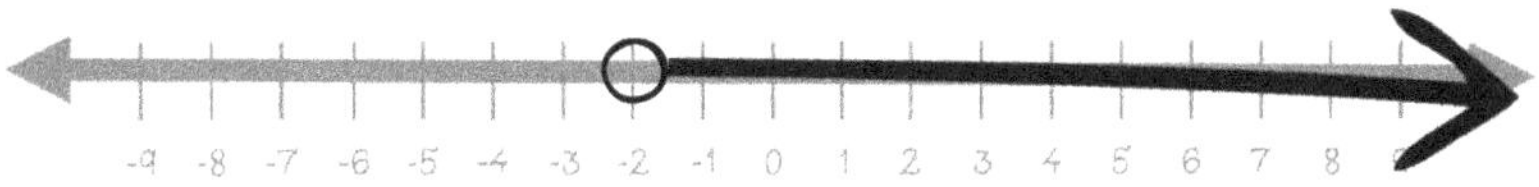

-9 -8 -7 -6 -5 -4 -3 -2 -1 0 1 2 3 4 5 6 7 8

write it

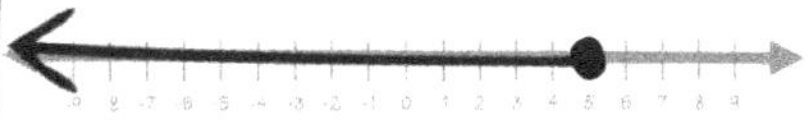

Create a "Speed Limit" sign that reflects the graph shown for acceptable bumper car speeds.

draw it

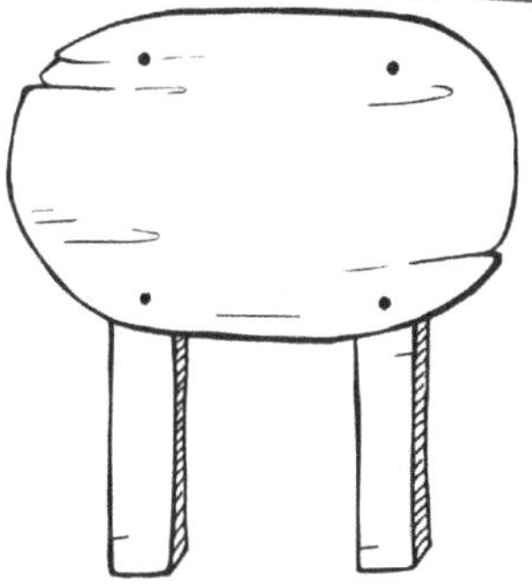

graph it

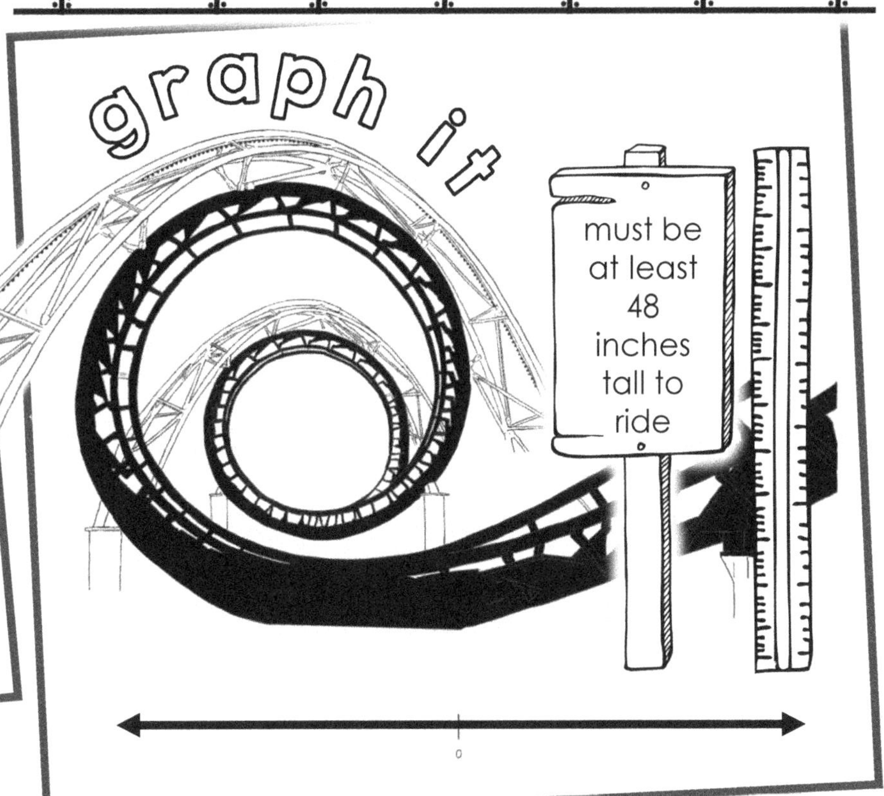

must be at least 48 inches tall to ride

0

Name:

Remember that you can
imagine
the inequality sign
"eating"
the greater number to determine which direction it should point.
862 >

Solving one step inequalities

by adding or subtracting

inverse operations

goal

FINISH

Name:

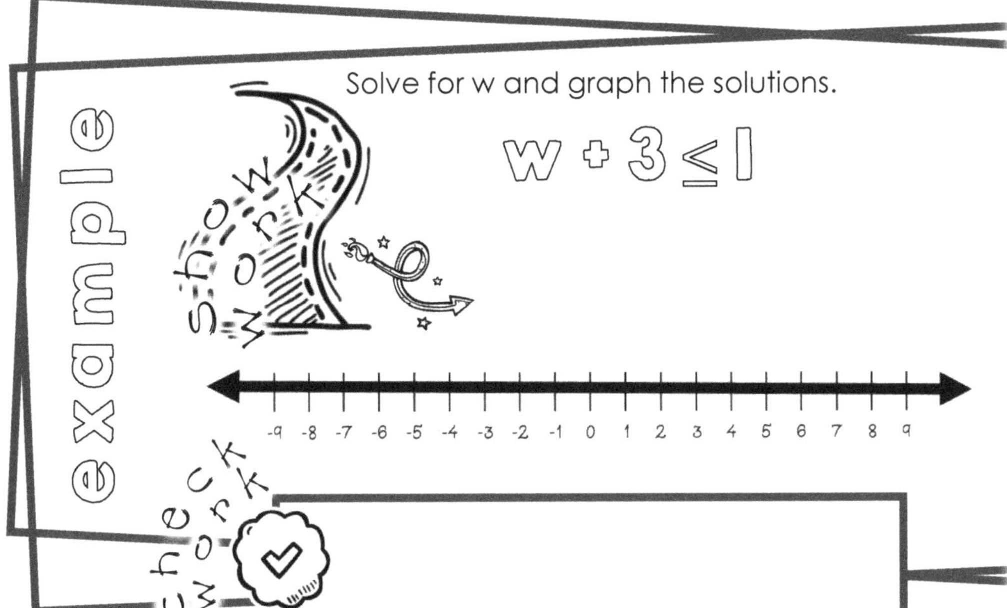

example
Solve for w and graph the solutions.
w + 3 ≤ 1
show work
-9 -8 -7 -6 -5 -4 -3 -2 -1 0 1 2 3 4 5 6 7 8 9
check work

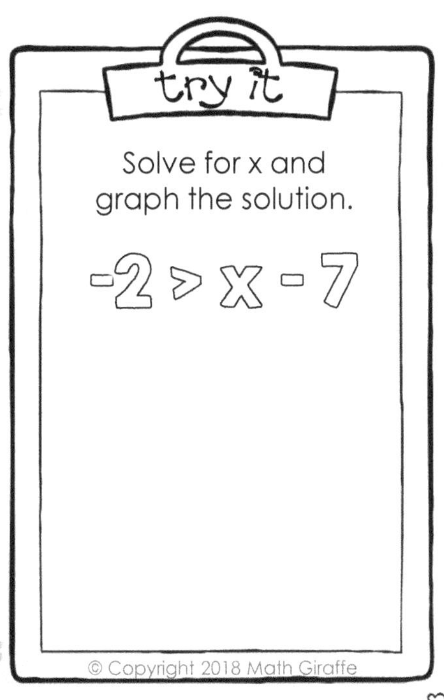

try it
Solve for x and graph the solution.
-2 > x - 7

Solving
inequalities
by multiplying
or dividing
Name:
watch out
when to flip
-6> -x
Divide both sides by -1
Remember: ONLY flip the inequality symbol when you...
otherwise, just solve normally.

Try one that requires a **flip:**

$-7x < 42$

Why do we need to flip this symbol?

to flip or not to flip?

$15 < 2b$

$-4 + m > 2$

$\frac{g}{-8} \geq 2$

$p(3) \leq -2$

$-12 < \frac{v}{4}$

Name:

solve $-64 > 16p$

flip?

graph

-9 -8 -7 -6 -5 -4 -3 -2 -1 0 1 2 3 4 5 6 7 8 9

solve $\frac{m}{-3} \leq 1$

flip?

graph

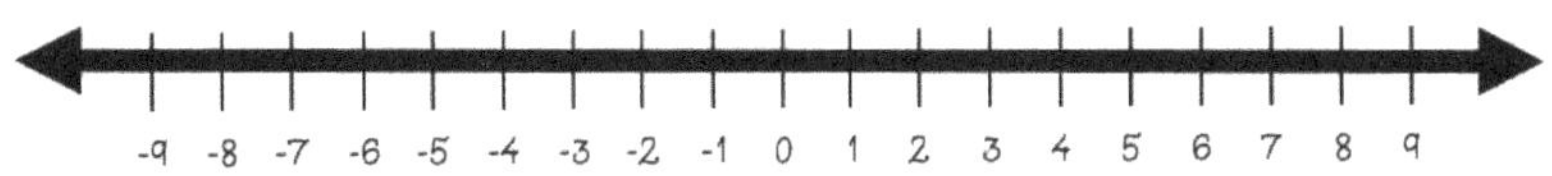

-9 -8 -7 -6 -5 -4 -3 -2 -1 0 1 2 3 4 5 6 7 8 9

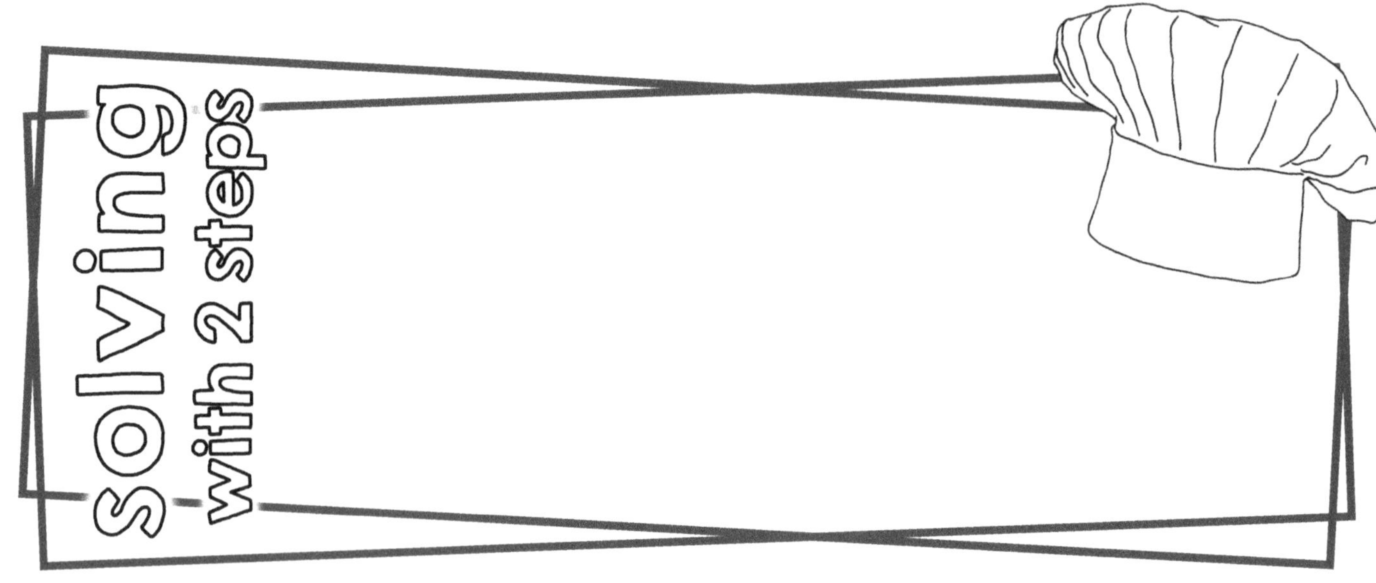

Name:

Solve, check, and graph solutions.

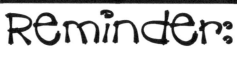

watch out

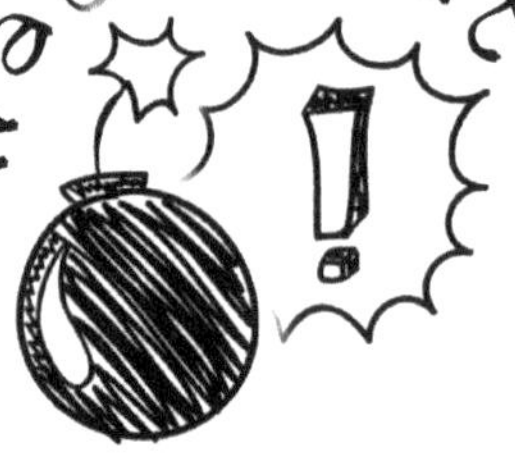

If during any step of solving ...

$2x - 5 > 1$

$2 - 6r \geq 3$

$3 < \frac{k - 8}{-5}$

practice

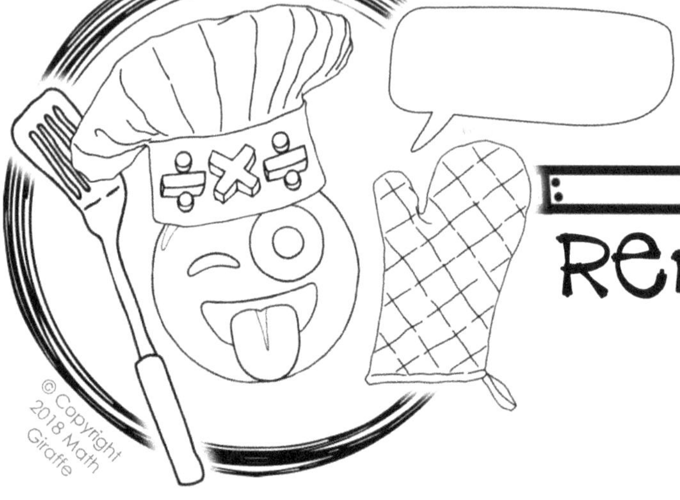

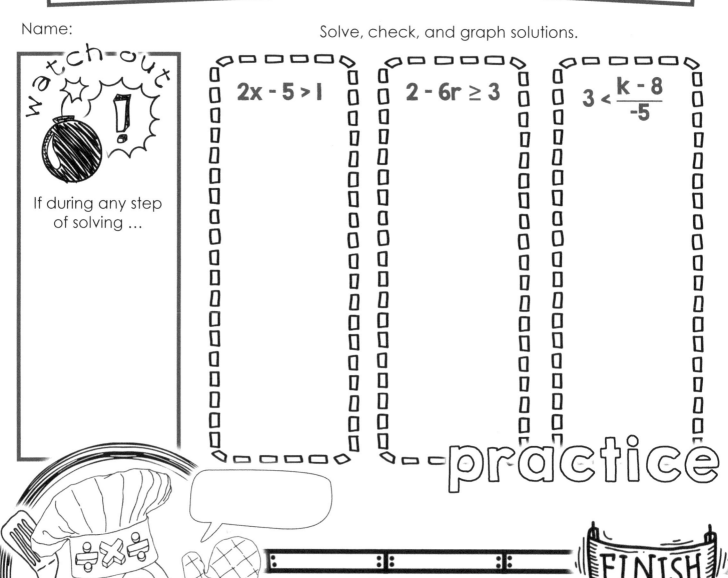

Reminder:

2 vocabulary: properties

write an *expression* in the box below that includes each:

label it

- ❑ constant
- ❑ term
- ❑ coefficient
- ❑ like terms
- ❑ variable

compare it

equation

vs.

expression

< >

inequality

inverse operations

open

solution

open sentence

vocabulary 2

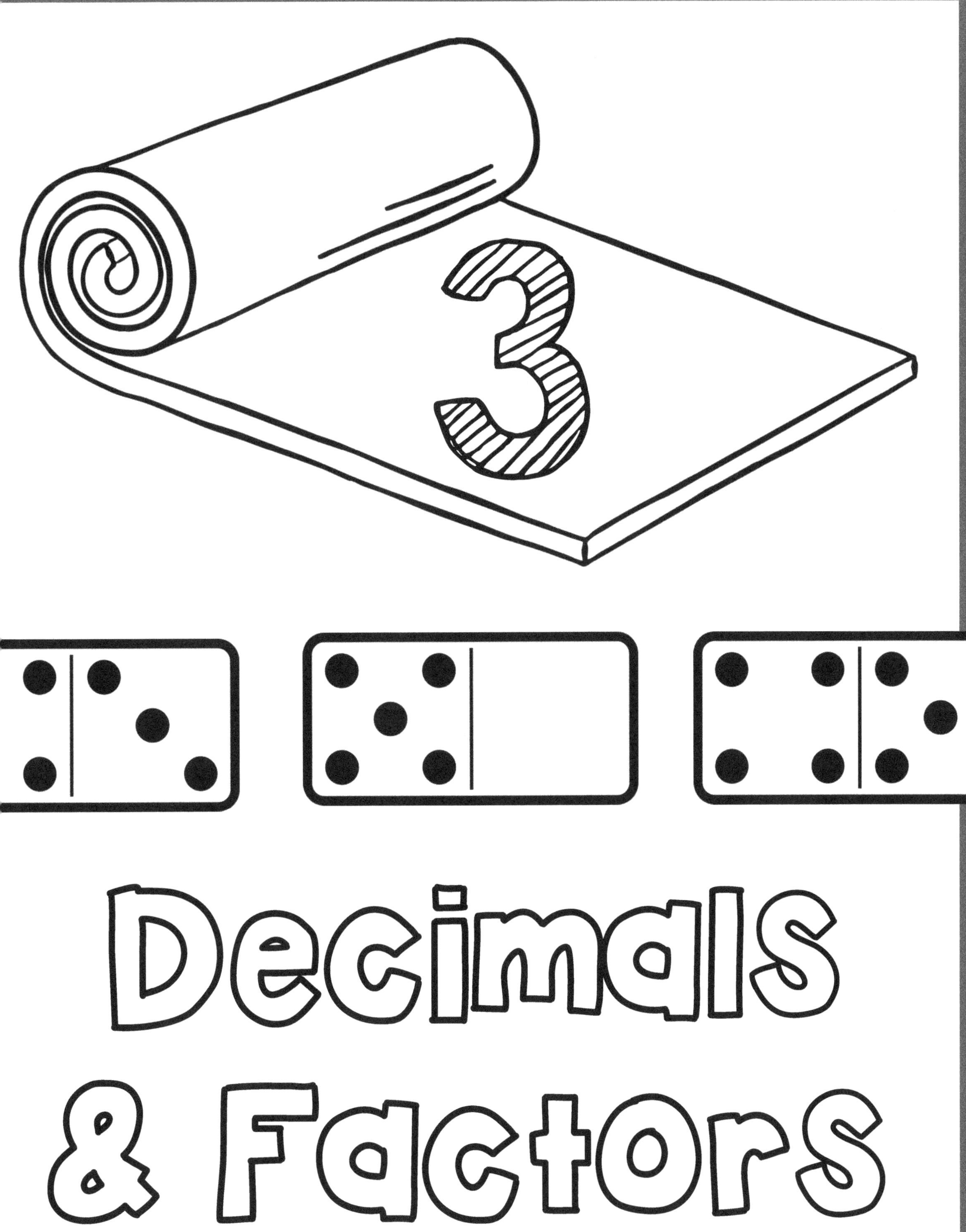

Decimals & Factors

Name:

PLACE VALUE REVIEW

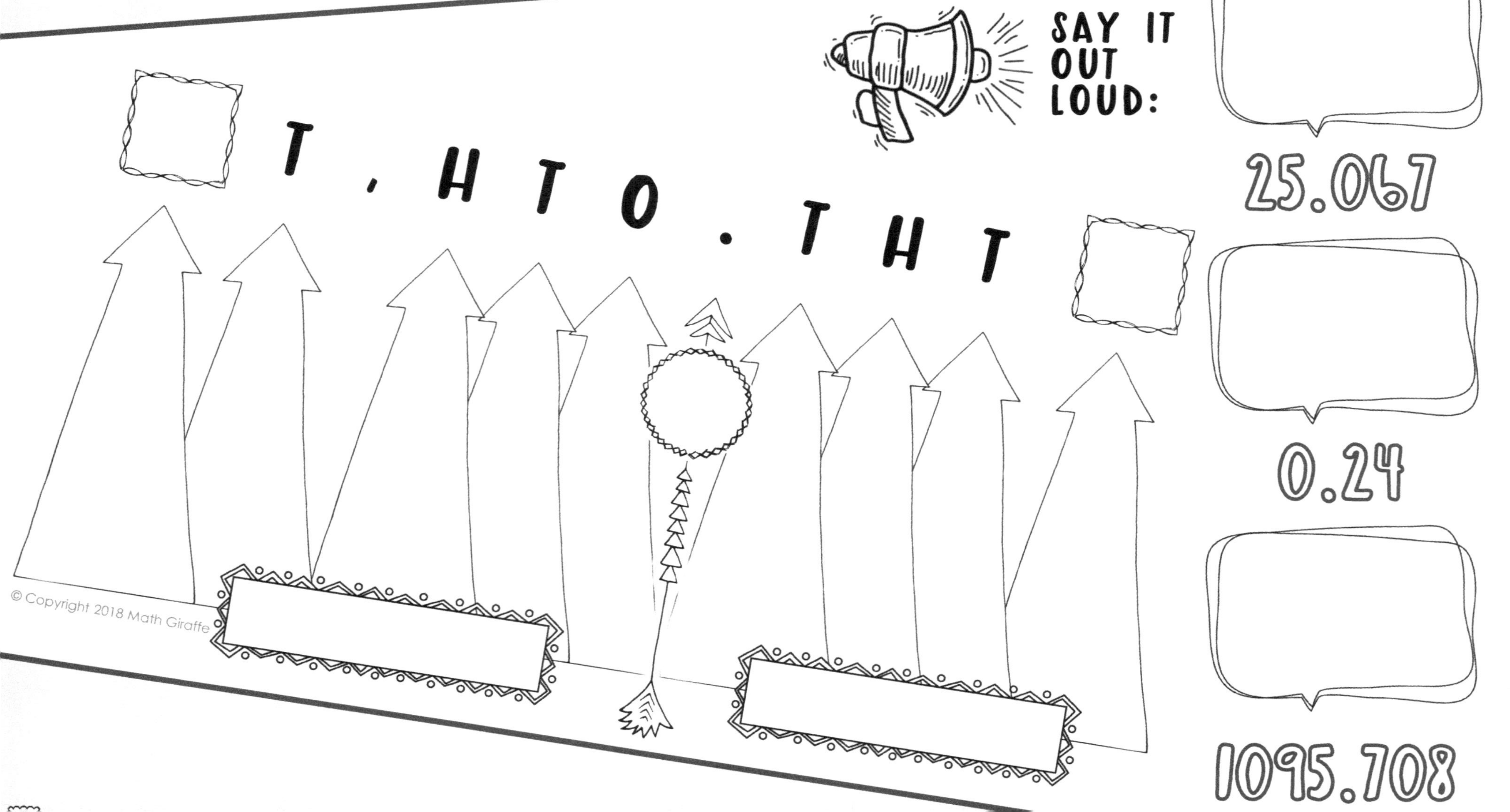

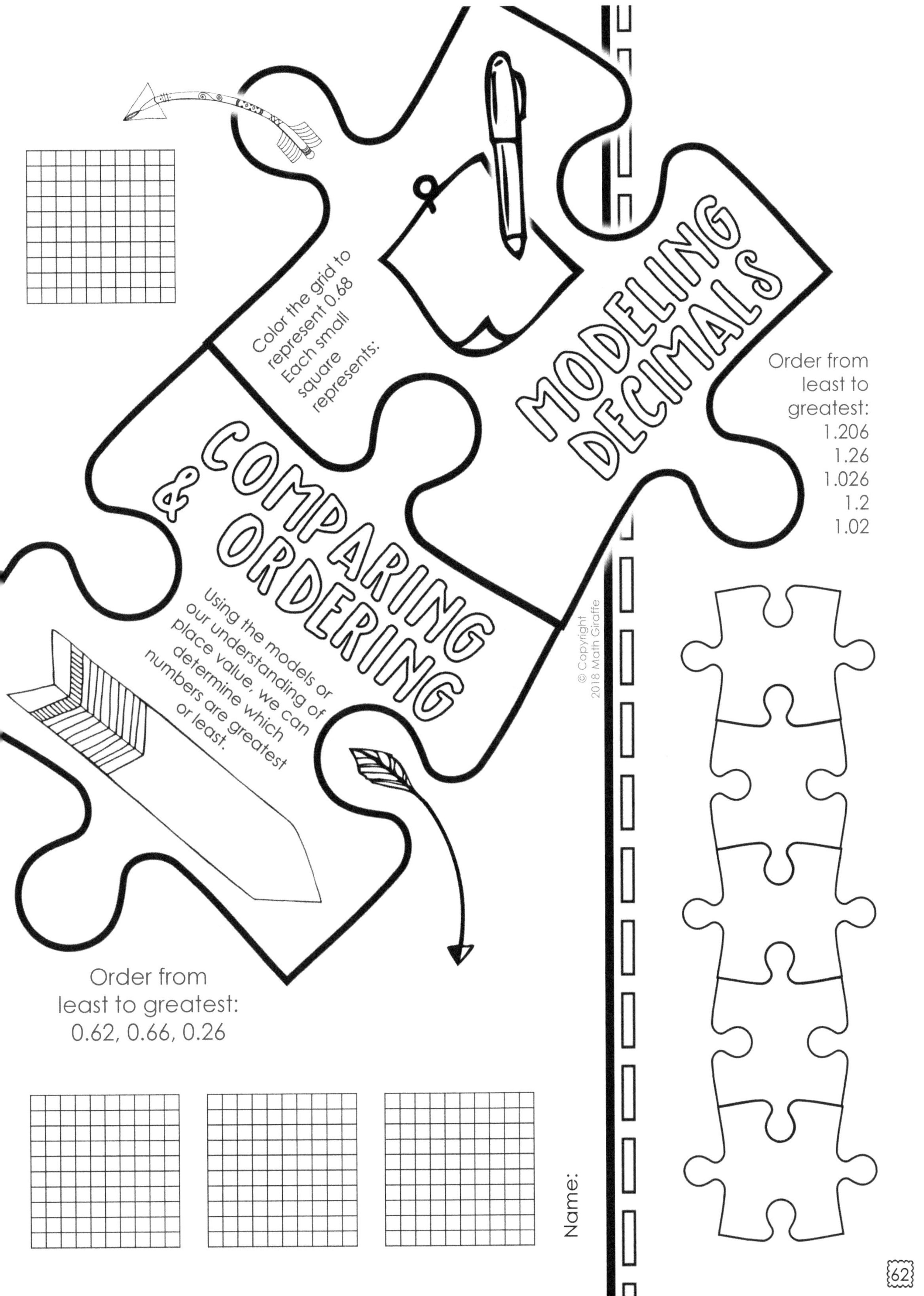

MODELING DECIMALS
& COMPARING ORDERING
Color the grid to represent 0.68 Each small square represents:
Using the models or our understanding of place value, we can determine which numbers are greatest or least.
Order from least to greatest: 1.206 1.26 1.026 1.2 1.02
Order from least to greatest: 0.62, 0.66, 0.26
© Copyright 2018 Math Giraffe
Name:
62

Rounding with Decimals

Name:

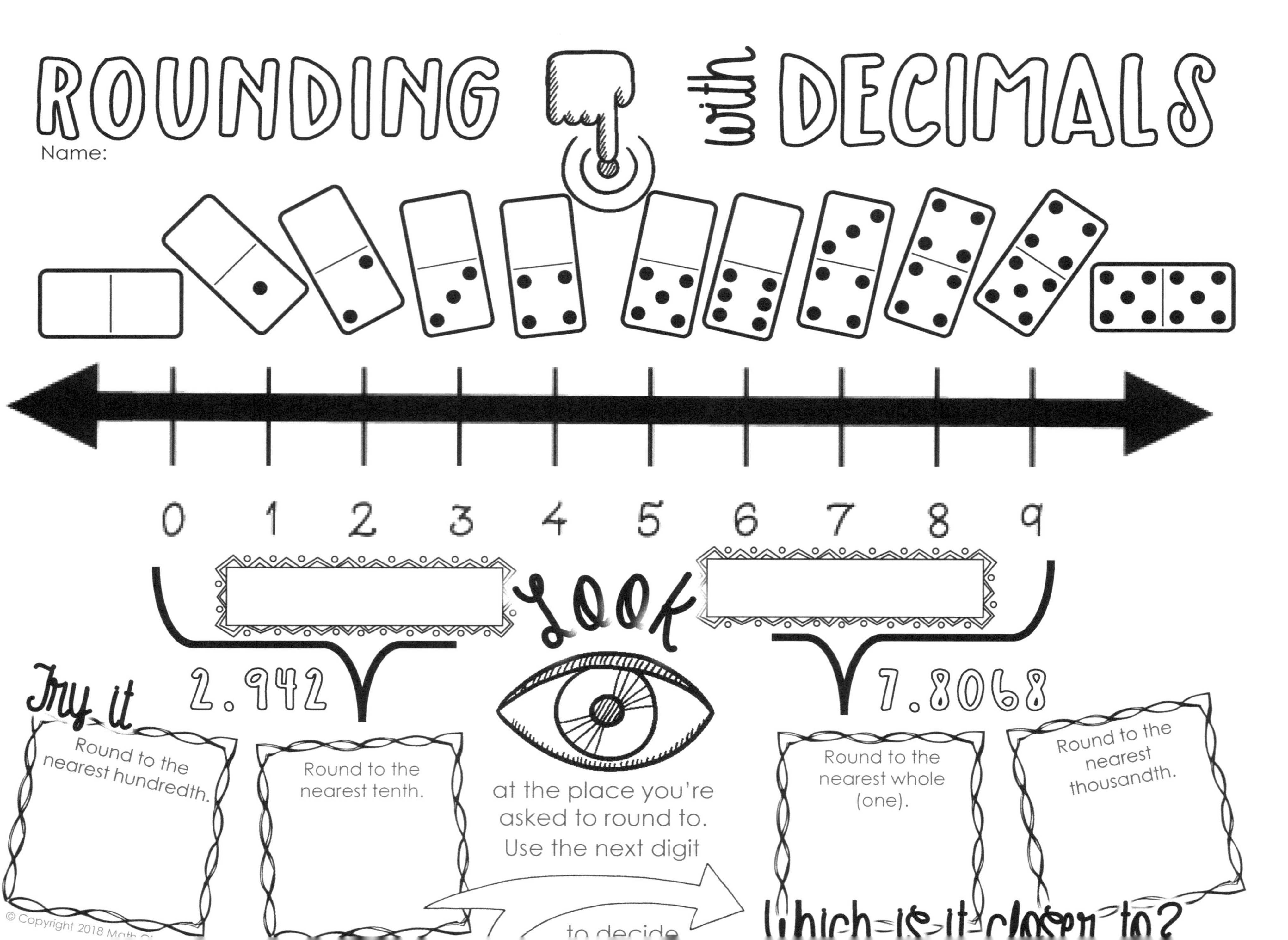

Name:

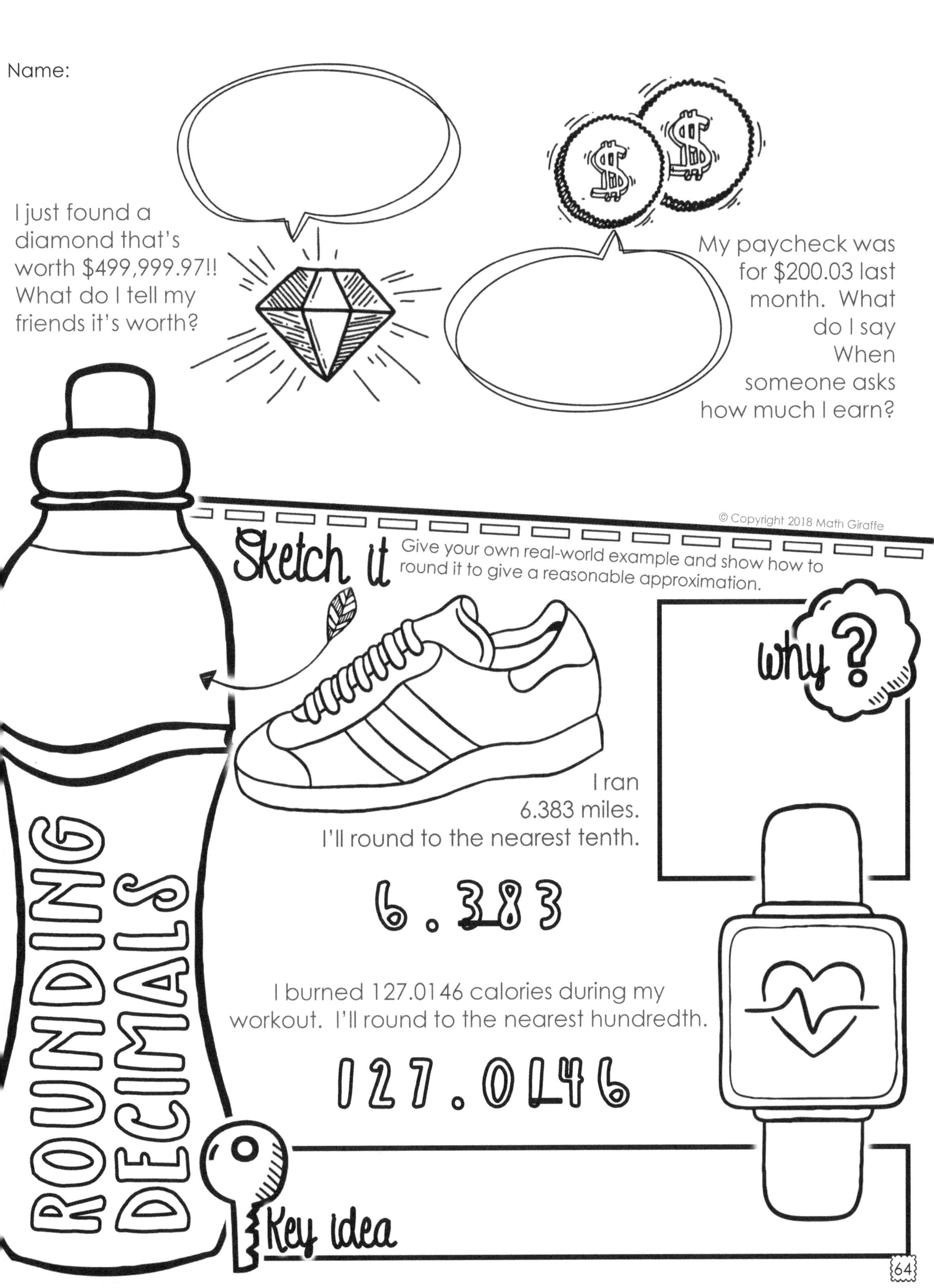
I just found a diamond that's worth $499,999.97!! What do I tell my friends it's worth?
My paycheck was for $200.03 last month. What do I say When someone asks how much I earn?
Sketch it
Give your own real-world example and show how to round it to give a reasonable approximation.
why?
I ran 6.383 miles. I'll round to the nearest tenth.
6.383
I burned 127.0146 calories during my workout. I'll round to the nearest hundredth.
127.0146
ROUNDING DECIMALS
Key idea

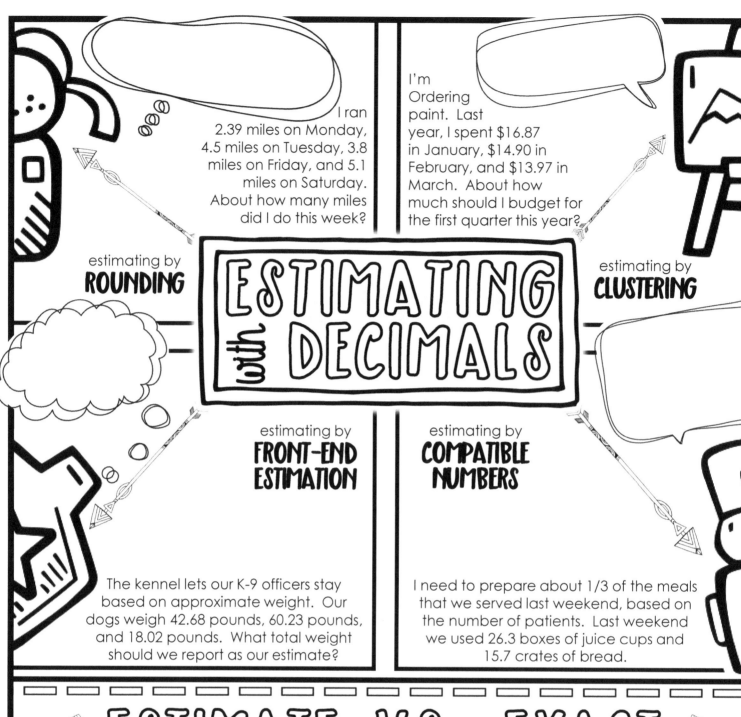

I ran 2.39 miles on Monday, 4.5 miles on Tuesday, 3.8 miles on Friday, and 5.1 miles on Saturday. About how many miles did I do this week?

I'm Ordering paint. Last year, I spent $16.87 in January, $14.90 in February, and $13.97 in March. About how much should I budget for the first quarter this year?

ESTIMATING with DECIMALS

estimating by **ROUNDING**

estimating by **CLUSTERING**

estimating by **FRONT-END ESTIMATION**

estimating by **COMPATIBLE NUMBERS**

The kennel lets our K-9 officers stay based on approximate weight. Our dogs weigh 42.68 pounds, 60.23 pounds, and 18.02 pounds. What total weight should we report as our estimate?

I need to prepare about 1/3 of the meals that we served last weekend, based on the number of patients. Last weekend we used 26.3 boxes of juice cups and 15.7 crates of bread.

ESTIMATE VS. EXACT

which is better?

Color the correct icon for each!

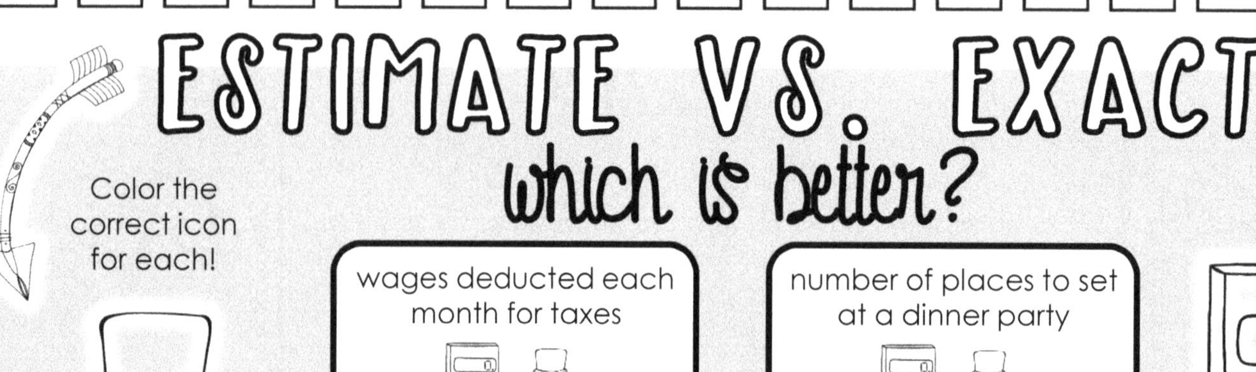

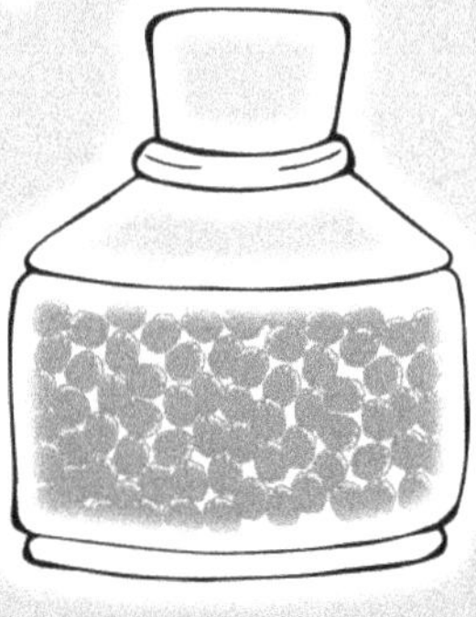

wages deducted each month for taxes

number of places to set at a dinner party

the distance between Kenya and Colorado

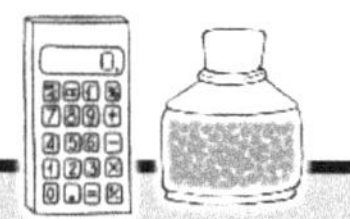

the population of the world

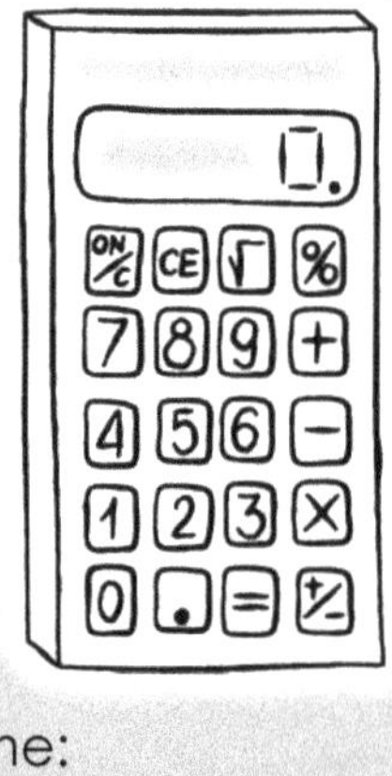

Name:

Name:

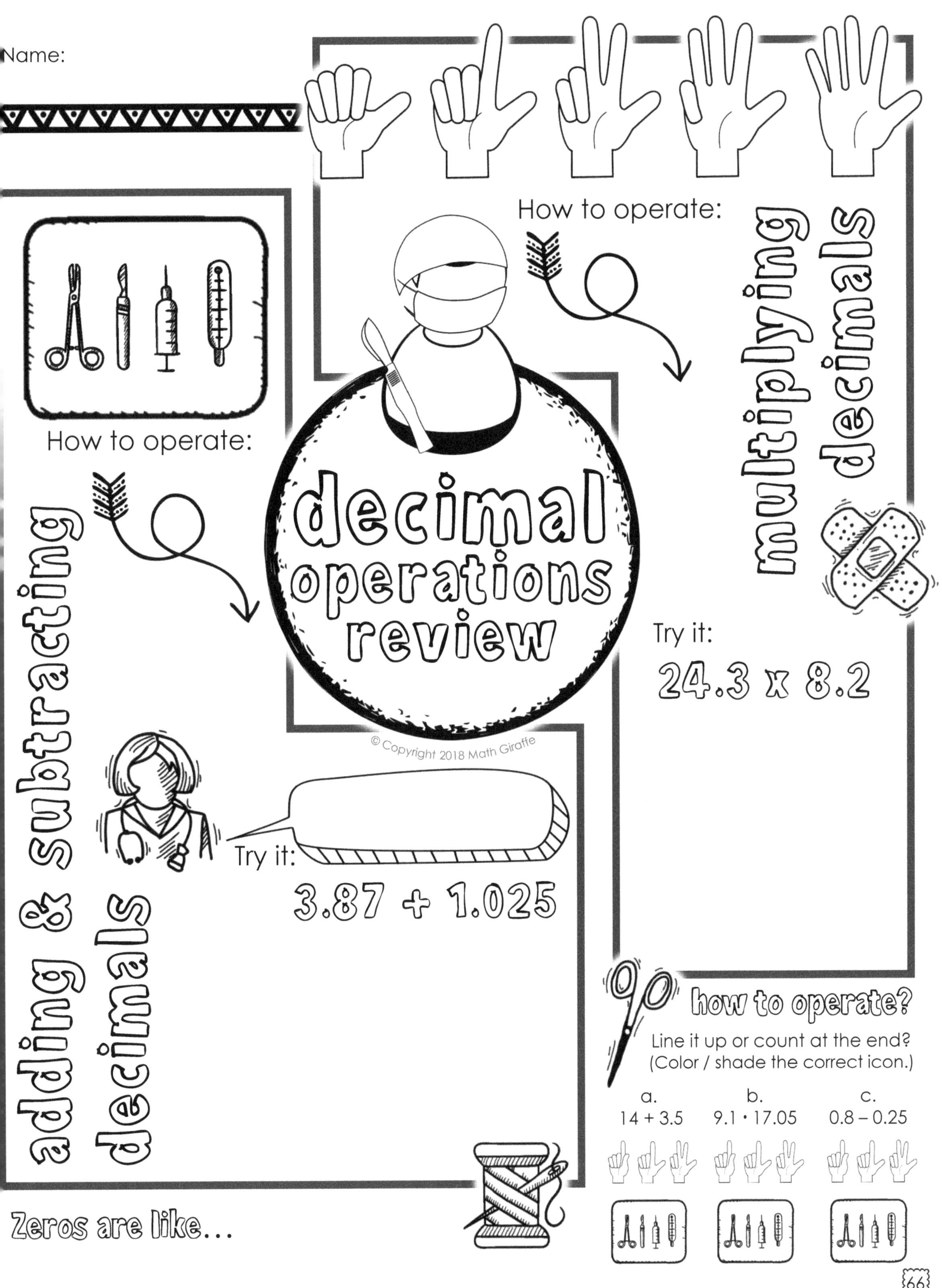

How to operate:
multiplying decimals
decimal operations review
How to operate:
adding & subtracting decimals
Try it:
24.3 x 8.2
Try it:
3.87 + 1.025
how to operate?
Line it up or count at the end?
(Color / shade the correct icon.)
a.
14 + 3.5
b.
9.1 • 17.05
c.
0.8 – 0.25
Zeros are like...

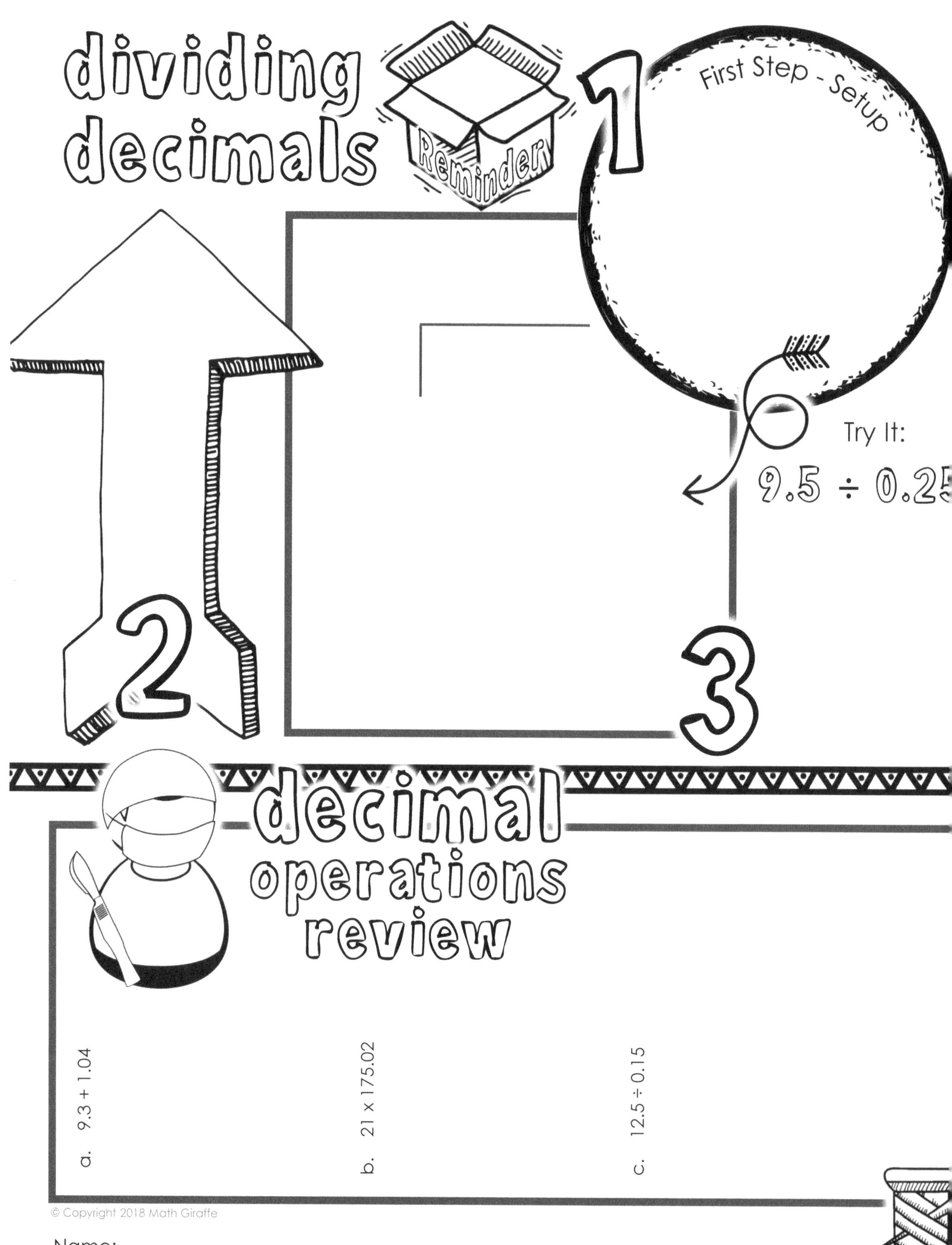
dividing decimals
Reminder
1
First Step - Setup
Try It:
9.5 ÷ 0.25
2
3
decimal operations review
a. 9.3 + 1.04
b. 21 x 175.02
c. 12.5 ÷ 0.15

Name:

So now, we can blend our understanding of

with our knowledge of

DECIMAL OPERATIONS

Try it: 3.2 (x) = 48

EXPLAIN:

SOLVE:

DIVISIBILITY RULES:

2 3 4 5 6 7 8 9

DIVISIBLE

FACTOR:

definition

DIVISIBILITY

MULTIPLES

There are a ____________________ of multiples (infinitely many).

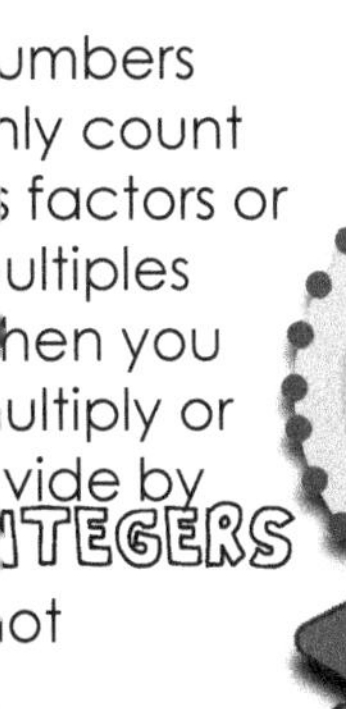

umbers
nly count
s factors or
ultiples
hen you
ultiply or
vide by
ITEGERS
ot

To find multiples, we

and many more...

16

To find factors, we
© Copyright 2018 Math Giraffe

FACTORS

Factors come in

There are a ____________________ of factors (limited #).

Name:

Try It

List all factors and the first four multiples of 72.	List the factors of 45.	List five multiples of 8.

LCM

The LEAST COMMON multiple of two or more numbers is the...

Try It

Find the LCM of:

9 & 12

1 LIST the

____________________,

2 LOOK for

____________________,

3 FIND the

____

of the matches as requested.

The GREATEST COMMON factor of two or more numbers is the...

Try It

Find the GCF of:

40 & 64

GCF

Try it

1. Find the GCF of 9, 15, and 18.

2. Find the LCM of 7, 9, and 21.

3. Find the GCF of 14 and 28.

Name:

TRY IT
Color all prime numbers blue & composite numbers red.
examples
examples:
62
31
19
11
9
16
PRIME
numbers
167
83
are like atoms because...
8
105
COMPOSITE
numbers
the number
1
Name:
© Copyright 2018 Math Giraffe
72

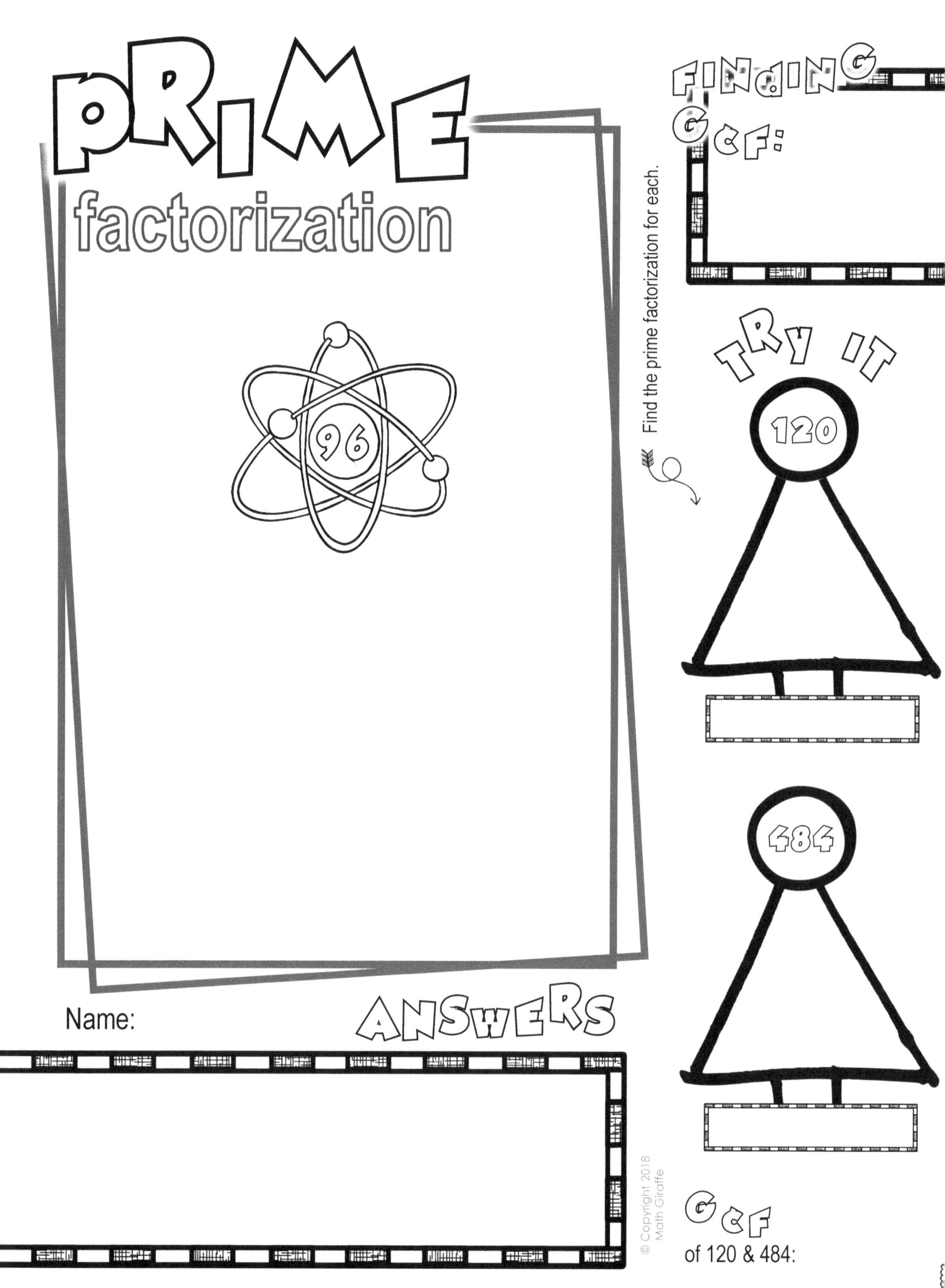
PRIME
factorization
96
Name:
ANSWERS
FINDING GCF:
Find the prime factorization for each.
TRY IT
120
484
GCF
of 120 & 484:

Name:

When you first learned multiplication, you probably represented multiplication using repeated addition in groups. Basic exponent operations can be represented by repeated ______________________

$4^3 = 4 \cdot 4 \cdot 4 = 64$

$1^5 = 1 \cdot 1 \cdot 1 \cdot 1 \cdot 1 =$

$3^2 =$

This means "b" multiplied by ______ "n" times.

b^n

When we say an exponent out loud, we call a number with a power of two "______________" and a power of three "______________."

Simplify each expression.

1. 2^5

2. 5^3

3. 10^3

4. 12^1

5. 10^5

6. 10^{13}

Any number to the ____ power is 1.

Any number to the ____ power is itself.

Powers of ____ will only contain 1s and 0s.

BEWARE: The base is only what is **directly** in front of the power. In order for a negative symbol to be included in the base, it must be grouped with parentheses.

The negative symbol is NOT part of the base.

$-2^4 = -[2 \cdot 2 \cdot 2 \cdot 2] = -16$

The negative symbol IS part of the base.

$(-2)^4 = (-2) \cdot (-2) \cdot (-2) \cdot (-2) = 16$

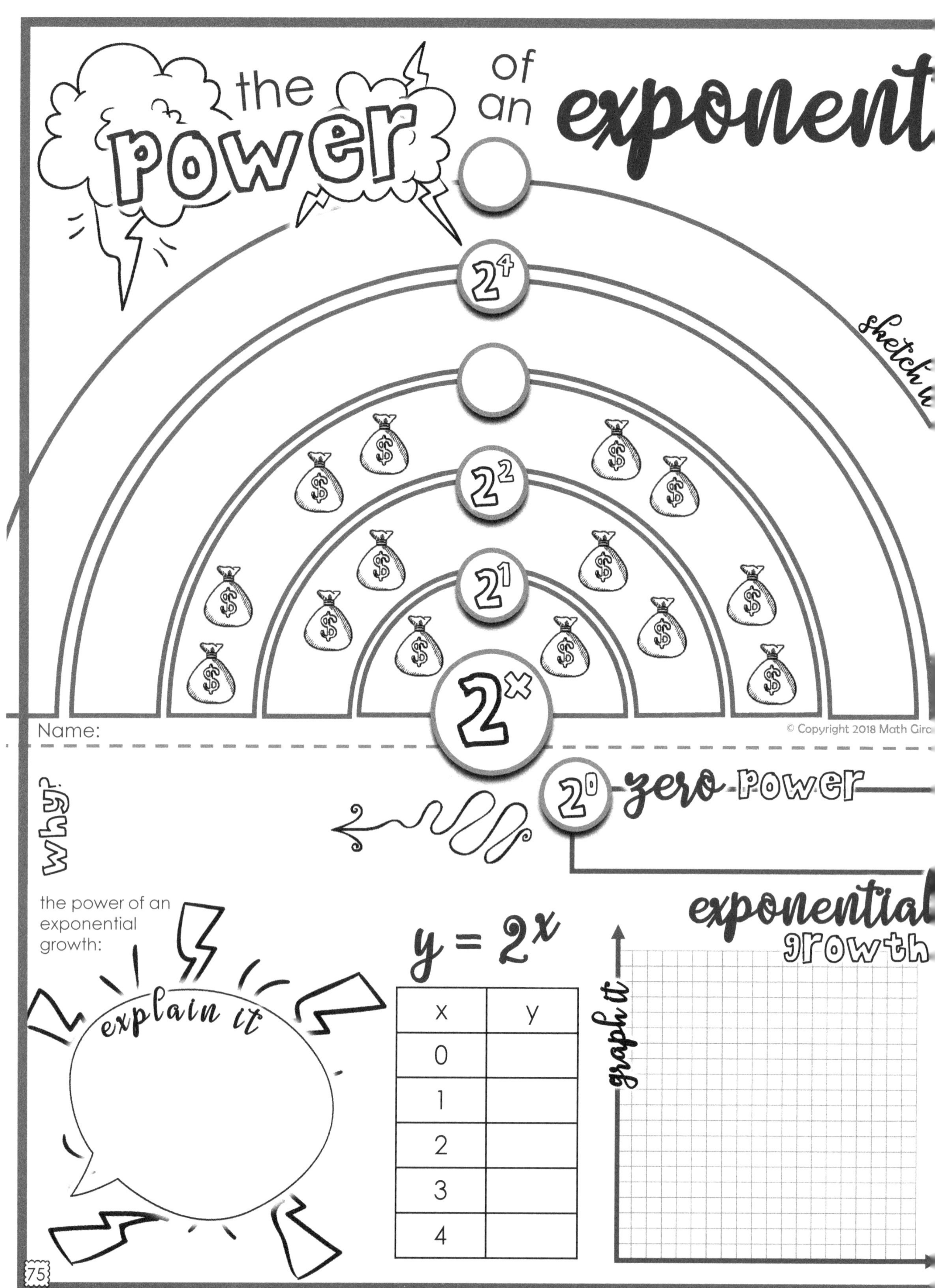
the power of an exponent
2^4
2^2
2^1
2^x
sketch it
Name:
© Copyright 2018 Math Gir
2^0 zero power
why?
exponential growth
the power of an exponential growth:
explain it
$y = 2^x$
graph it
75

x	y
0	
1	
2	
3	
4	

stretch it out
$2^4 \cdot 2^5$
$m^a \cdot m^b =$
write a rule
heads up!
must be the same ____
zero as an exponent
$m^0 =$
$15^0 =$
$-3^0 =$
Name:
e - x - p - a - n - d
$(4^2)^3$
$(m^a)^b =$
write a rule
reminder
what is the base?
explain it:
Identify which rule applies to each expression.
exponent rules
$(xy)^0$
$(b^f)^g$
$7^3(7^{11})$
$(9^8)^{12}$
$w^2 w^{10}$
18^0

expand

it out

$\frac{3^7}{3^3}$ >

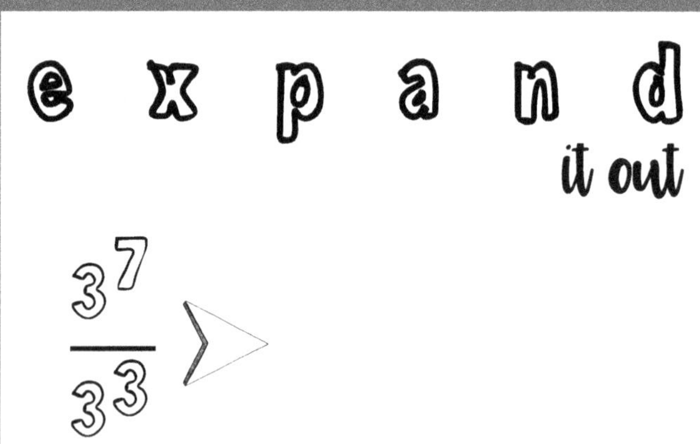

cancel

write a rule

$\frac{m^a}{m^b} =$ ☐

Name:

heads up!

must be the same ____

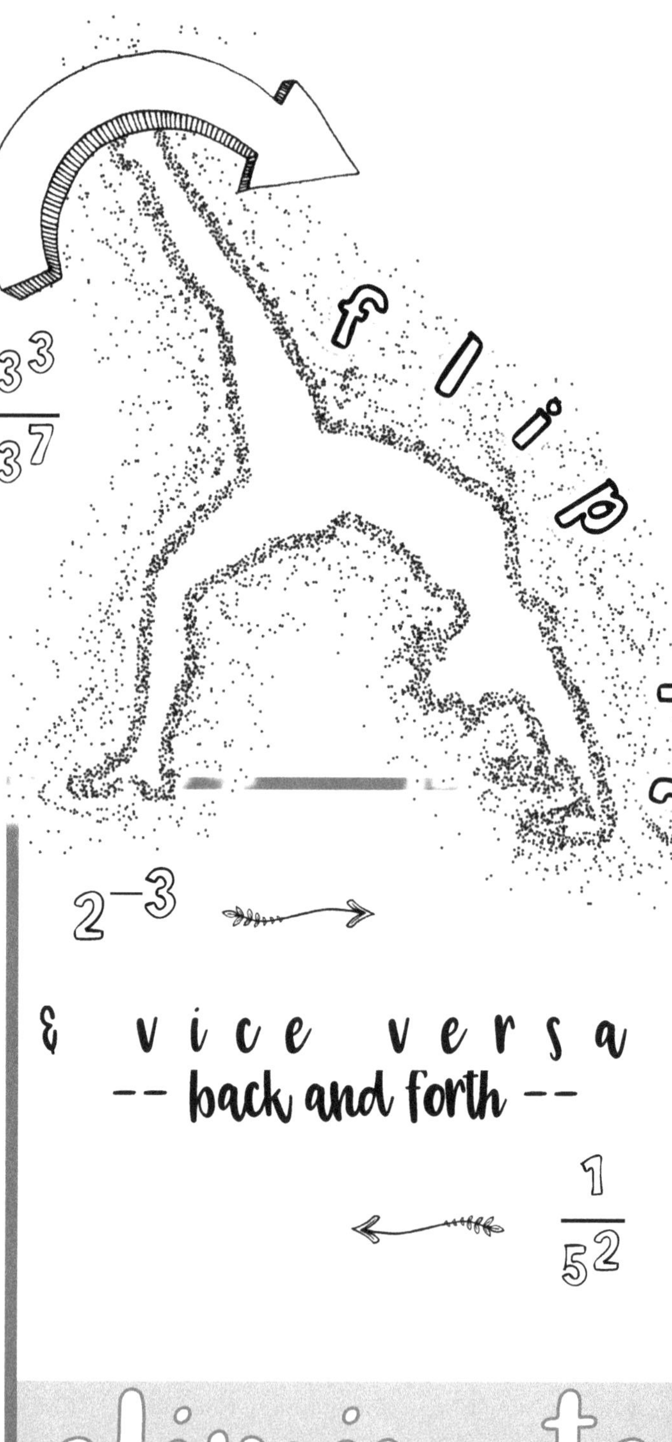

$\frac{3^3}{3^7}$

flip it

2^{-3} →

& vice versa

-- back and forth --

← $\frac{1}{5^2}$

eliminate

$m^{-a} =$ ☐

write a rule

2,540,000

so many digits!

Sometimes with numbers this small or this large, it can be helpful to rewrite without the extra zeroes.

0.008264

So, we ____ the 1st non-zero digit, place a ______ ____ before the next, and use powers of ___ to correctly represent how we moved the decimal point.

The rewritten format is called

SCIENTIFIC NOTATION

2,540,000

becomes

0.008264

becomes

When converting a number into scientific notation,

If you moved a decimal to the

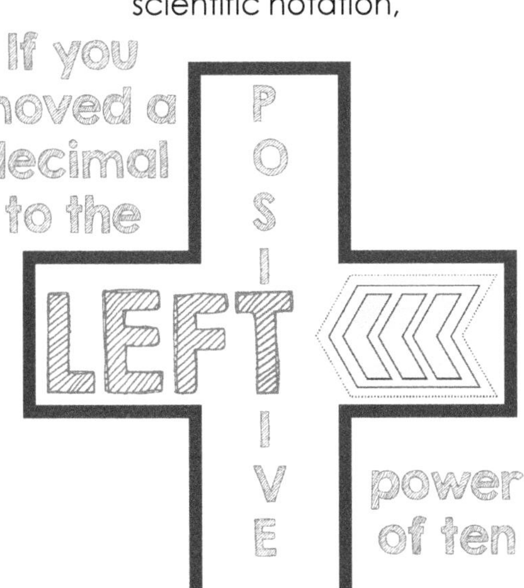

power of ten

Moving the decimal point to the LEFT represents...

Moving the decimal point to the RIGHT represents...

When writing in scientific notation,

If you moved the decimal to the RIGHT NEGATIVE power of ten

Try It

Write 0.0337 in scientific notation.

Name:

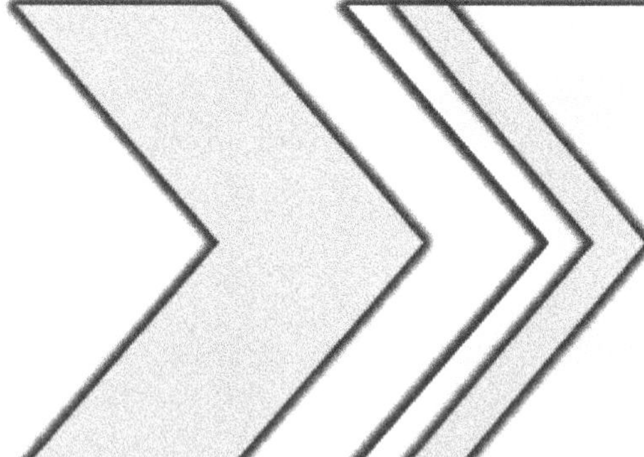

Powers of Ten

$8 \times 100 =$

$5 \div 100 =$

$0.6 \times 1000 =$

$0.9 \div 1000 =$

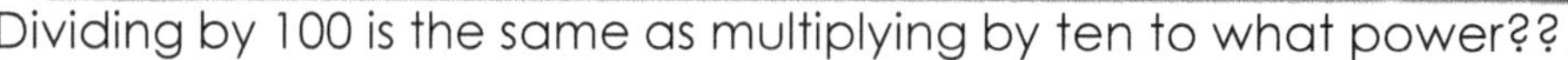

Dividing by 100 is the same as multiplying by ten to what power?? ______

SCIENTIFIC NOTATION -- PRACTICE --

When we UN-DO, we do the OPPOSITE (inverse operations)!

STANDARD FORM	SCIENTIFIC NOTATION
56,790	
	2.4×10^{-2}
7,000	
	1.65×10
0.93806	
	1.469×10^{8}

Working Backwards:

When converting FROM scientific notation INTO standard form, remember that the decimal is moving in the opposite direction. So if the power is **NEGATIVE**, you will move it to the ______, and if the power is **POSITIVE** you will move it to the ______.

Name:

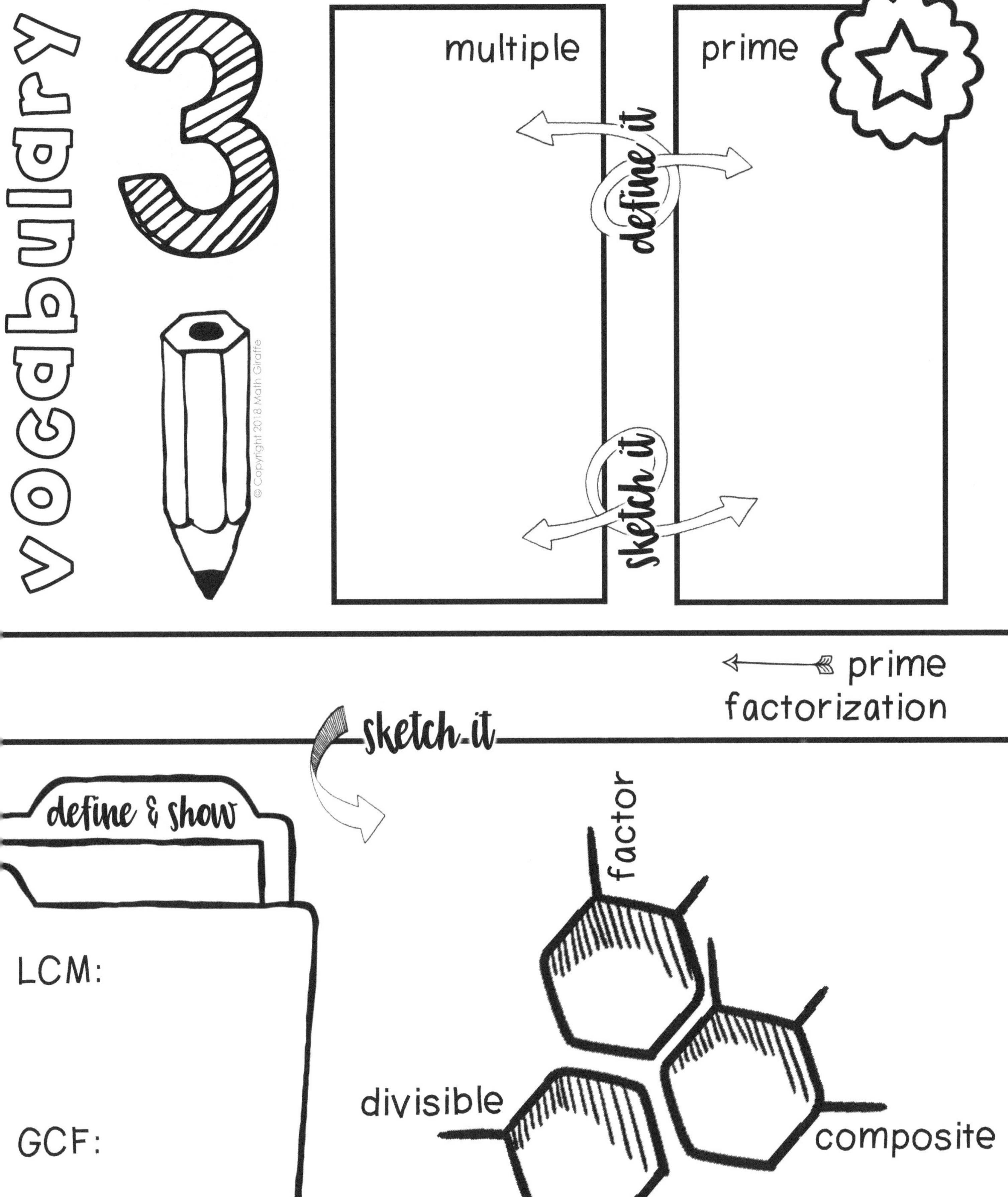
vocabulary
3
multiple
prime
define it
sketch it
© Copyright 2018 Math Giraffe
prime factorization
sketch it
define & show
LCM:
GCF:
factor
divisible
composite

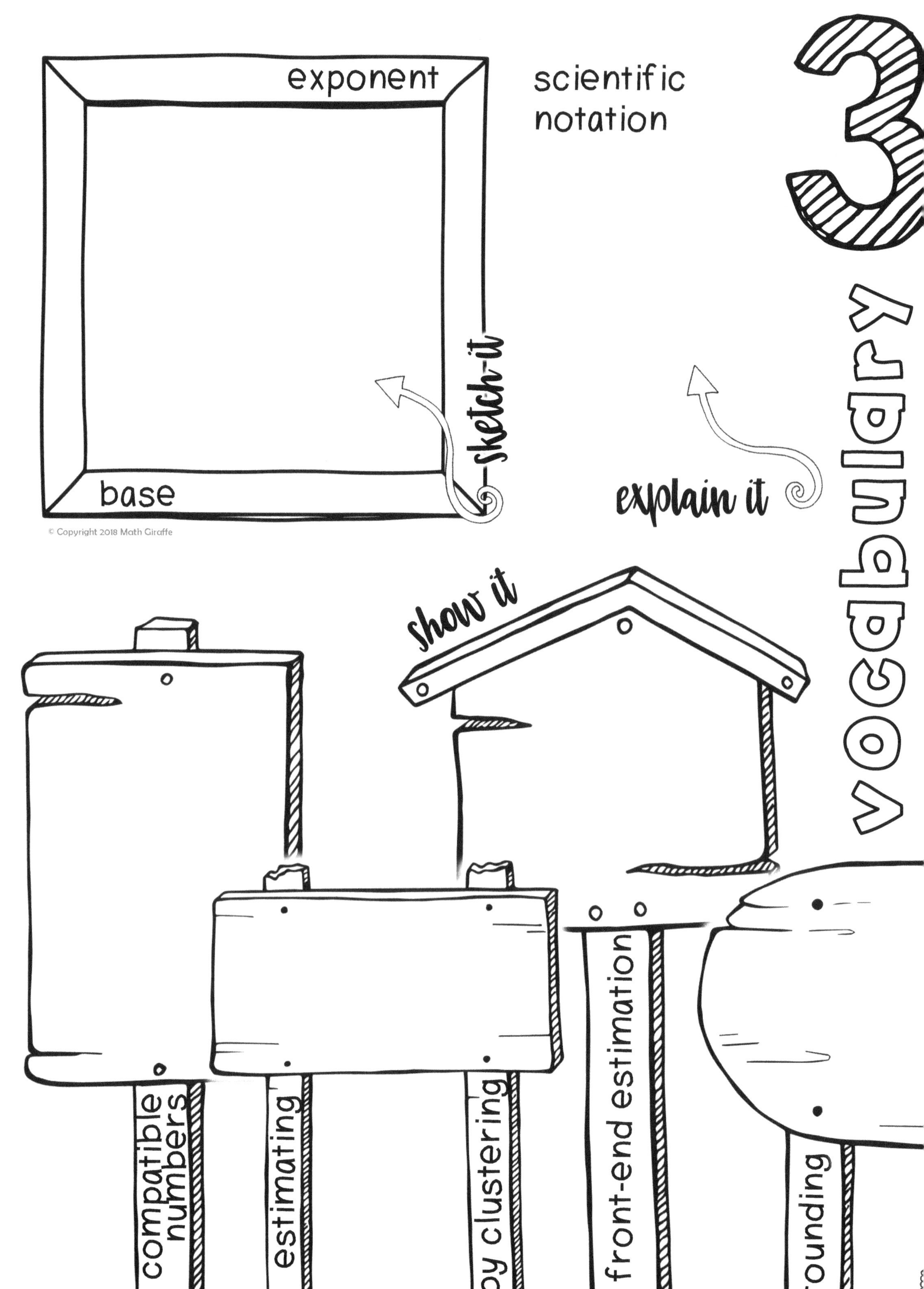

exponent
scientific notation
sketch it
base
explain it
show it
vocabulary
3
compatible numbers
estimating
by clustering
front-end estimation
rounding

Fractions

the real number system

Can you give definitions
and examples for each?

FRACTIONS

\ fraction represents ______________________________.
n a proper fraction, ______________ < ________________.
n an improper fraction, ______________ < ________________.
Shade/color below to show the fraction $\frac{4}{7}$.)

-SIMPLIFYING-

› simplify a fraction, divide the _____________ and
_____________ by ___________________________.
hade/color to show that the simplest form is equivalent to the original.)

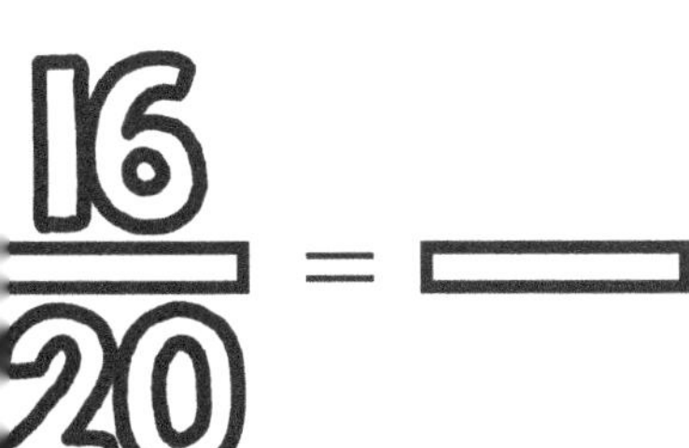

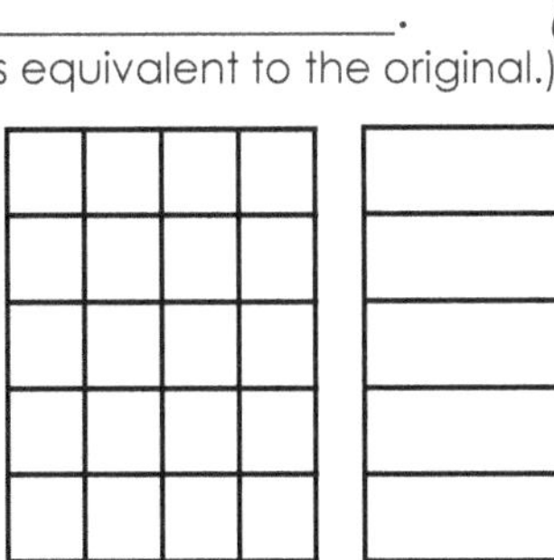

-EQUIVALENT-

To write an equivalent fraction...

(Go back to the top and draw a dividing line on the grid for $\frac{4}{7}$ to show an equivalent fraction. Write the fraction here: ____)

TRY IT

For each fraction, write the simplest form plus one additional equivalent fraction.

$\frac{8}{20}$ = ___ = ___

$\frac{9}{21}$ = ___ = ___

$\frac{6}{18}$ = ___ = ___

$\frac{10}{12}$ = ___ = ___

Name:

-COMPARING & ORDERING-

FRACTIONS

Name:

COMMON DENOMINATORS

TRY IT

Rewrite using the LCD.

$\frac{5}{6}$ and $\frac{7}{15}$

$\frac{1}{3}$ and $\frac{4}{27}$

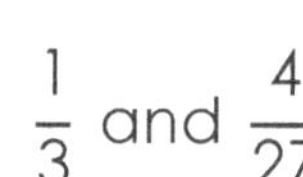

Use LCD

MODELING

compare

$\frac{3}{8}$ & $\frac{1}{2}$

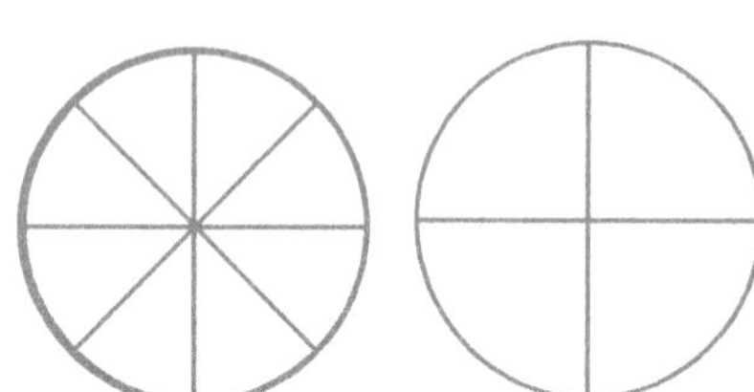

COMPARE

using <, >, or =

$\frac{7}{8}$ and $\frac{29}{32}$

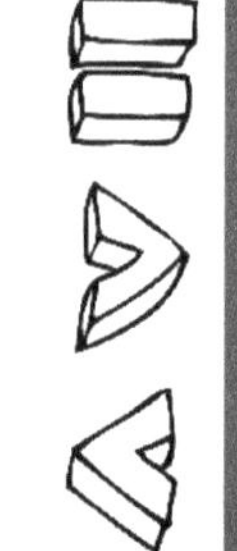

ORDER

from least to greatest

$\frac{5}{9}$, $\frac{7}{12}$, $\frac{5}{6}$, $\frac{2}{3}$

Name:
$2\frac{1}{3}$
MULTIPLY
the striped,
then
ADD
the
dotted.
2
1
3
WHY? MODEL IT
MIXED NUMBERS
What are they good for?
IMPROPER FRACTIONS
7
3
DIVIDE.
remainder
denominator
becomes
the
fractional
part
7
3
What are they good for?

WORKING WITH RATIONAL NUMBERS

Put your new skills together to order each set from least to greatest!

1

$-1\frac{6}{7}, \frac{26}{14}, -\frac{16}{9}, 1\frac{3}{4}$

2

$\frac{1}{8}, \frac{1}{9}, \frac{2}{12}, \frac{3}{25}, \frac{3}{28}, \frac{3}{23}$

3

$-2\frac{3}{5}, -2\frac{4}{11}, -\frac{16}{6}, -\frac{9}{4}, -1, 2\frac{1}{3}$

Now, put this set on a number line.

Can you give definitions and examples for each?

TERMINATING DECIMAL

REPEATING DECIMAL

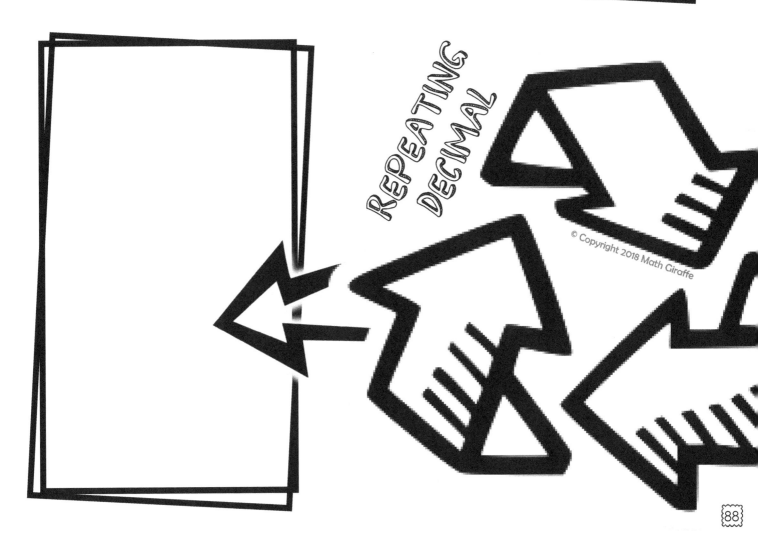

converting a fraction to a decimal

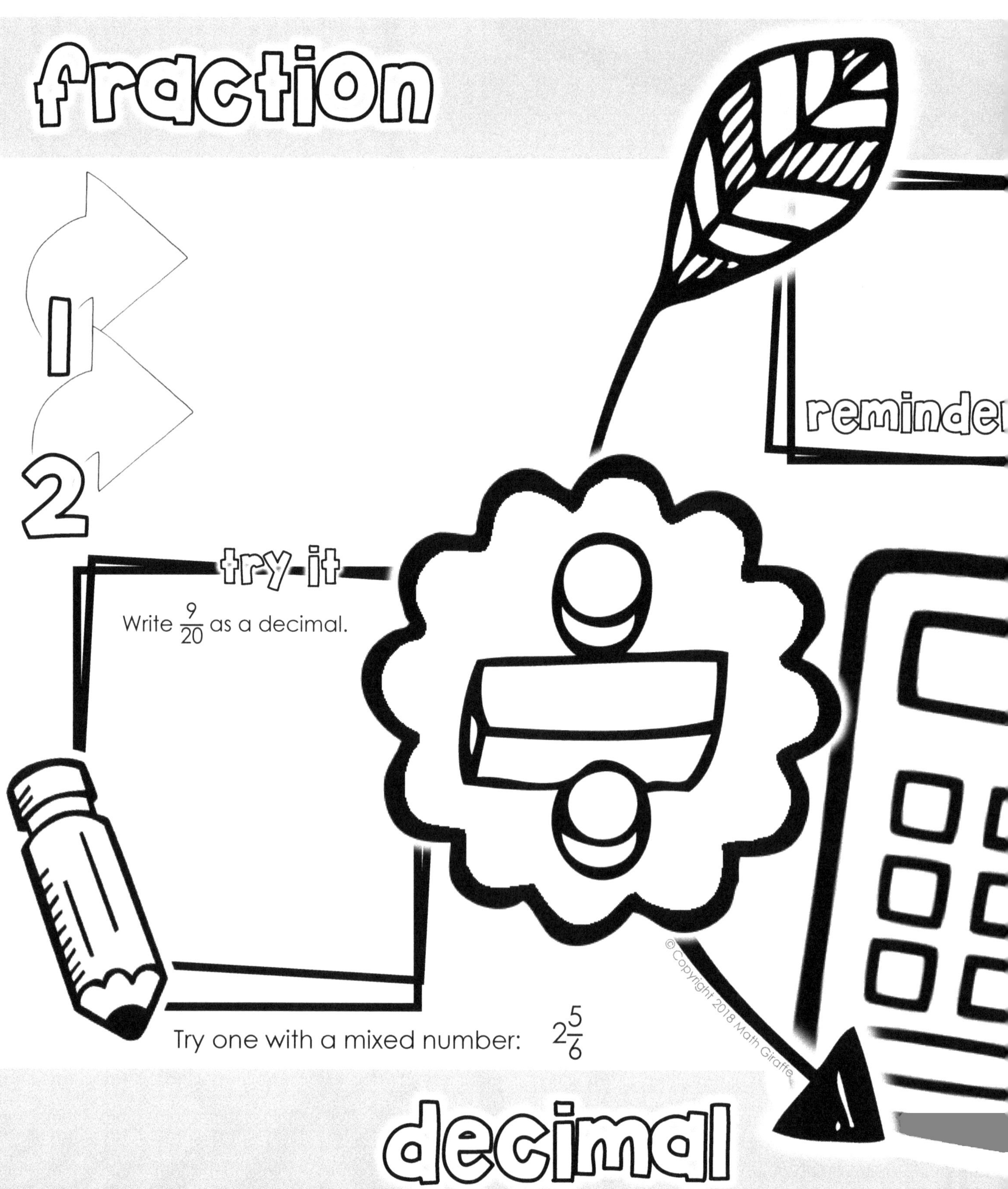
fraction
1
2
reminder
try it
Write $\frac{9}{20}$ as a decimal.
Try one with a mixed number: $2\frac{5}{6}$
decimal

Name:

converting a decimal to a fraction

try it: $1\frac{8}{15} - \frac{5}{6}$

a fraction & a whole number

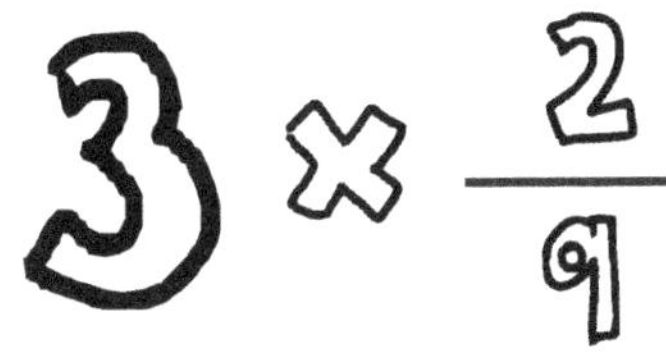

$3 \times \frac{2}{9}$

A whole number can always be written as a fraction - Put it over ___.

$\frac{3}{1} \times \frac{2}{9}$

Simplify!

multiplying fractions

two fractions

$\frac{4}{5} \times \frac{3}{8}$

Show the cancellations that allow the problem to be rewritten as follows:

$\frac{1}{5} \times \frac{3}{2}$

Finish it up. What if you had not cancelled?

answers

All answers must be in ______________________________ and converted from improper fractions to mixed numbers if necessary.

canceling

To "cancel," divide a number within a ________________________ and a number within a ____________________ by their _______ (greatest common factor)

Name:

Fraction Multiplication:

Sometimes, multiplication can be represented by the word "____.
Color the models using two colors to represent each problem.

Concepts

Visual Models

What is ¾ x 2?

Rewrite as a problem with two fractions, then solve.

Two copies OF a model for ¾

What is ½ of ¼?

Write as a multiplication problem, then solve.

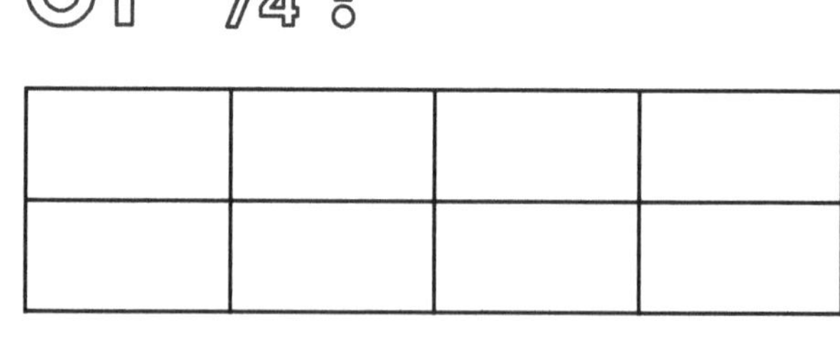

The model represents ½ x ¼

Multiplying Mixed Numbers

Convert into ____________________________ first, then ...

$2\frac{2}{3} \cdot 1\frac{3}{4}$ rewrite →

A _____ can replace the x to symbolize multiplication.

Name:

Examples and Practice

multiplying fractions

fraction & whole number

$\frac{3}{5} \cdot 6$

2 fractions

$\frac{5}{8} \cdot \frac{3}{4}$

cancel first

$\frac{16}{21} \cdot \frac{7}{12}$

3 fractions

$\frac{2}{15} \cdot \frac{9}{14} \cdot \frac{5}{7}$

mixed numbers

$4\frac{6}{7} \cdot 1\frac{1}{2}$

TIPS

- Put whole numbers over ____.
- Simplify and/or cancel ____________ multiplying, because it is easier than at the end.
- Rewrite mixed numbers as ____________ ____________, then convert back when you have an answer if needed.

Try it

Complete each problem, showing all work.

1. $\frac{5}{8} \cdot \frac{2}{3}$
2. $\frac{3}{5} \cdot \frac{15}{24}$
3. $\frac{4}{12} \cdot \frac{1}{8} \cdot \frac{9}{10}$
4. $2\frac{2}{7} \cdot \frac{1}{4}$
5. $6 \cdot \frac{8}{9}$
6. $1\frac{25}{48} \cdot \frac{36}{50}$

Name:

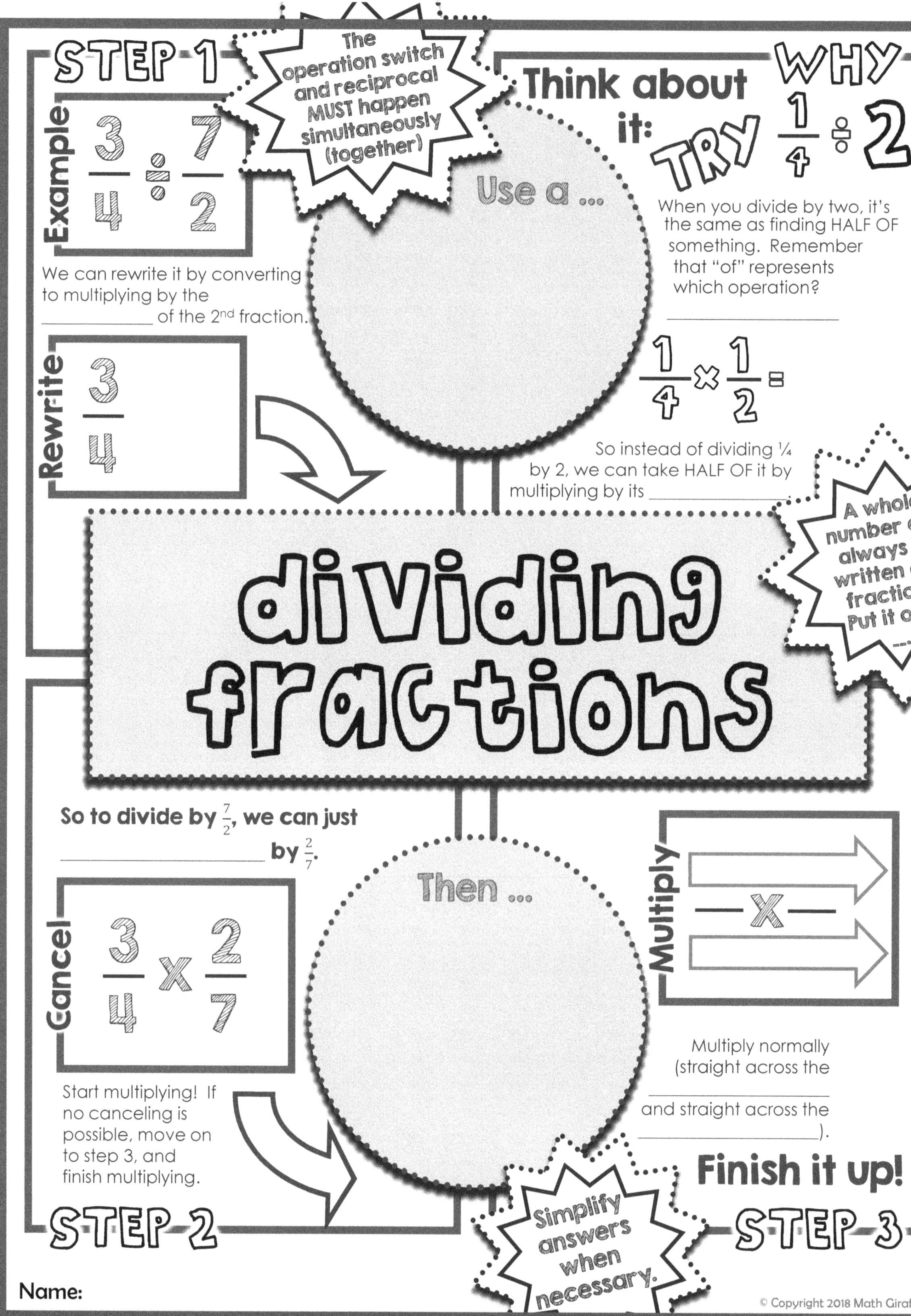
STEP 1
The operation switch and reciprocal MUST happen simultaneously (together)
Example
3/4 ÷ 7/2
We can rewrite it by converting to multiplying by the ____________ of the 2nd fraction.
Use a ...
Rewrite
3/4
Think about it: TRY 1/4 ÷ 2
WHY
When you divide by two, it's the same as finding HALF OF something. Remember that "of" represents which operation?
____
1/4 x 1/2 =
So instead of dividing ¼ by 2, we can take HALF OF it by multiplying by its ____________
A whole number ca always b written as fraction Put it ov ___.
dividing fractions
So to divide by 7/2, we can just ____________________ by 2/7.
Then ...
Cancel
3/4 x 2/7
Start multiplying! If no canceling is possible, move on to step 3, and finish multiplying.
STEP 2
Multiply
___ X ___
Multiply normally (straight across the ____________________ and straight across the ____________________).
Finish it up!
Simplify answers when necessary.
STEP 3
Name:
© Copyright 2018 Math Giraffe

Fraction Division

Remember to convert mixed numbers into improper fractions while working, then convert back when finished.

Remember that division is represented by determining how many _____________ a number can be divided up into. Divide up and color or shade the models to represent each problem.

Concepts

Visual Models

What is $\frac{1}{2} \div \frac{1}{6}$?

Rewrite as a multiplication problem using a reciprocal, then solve.

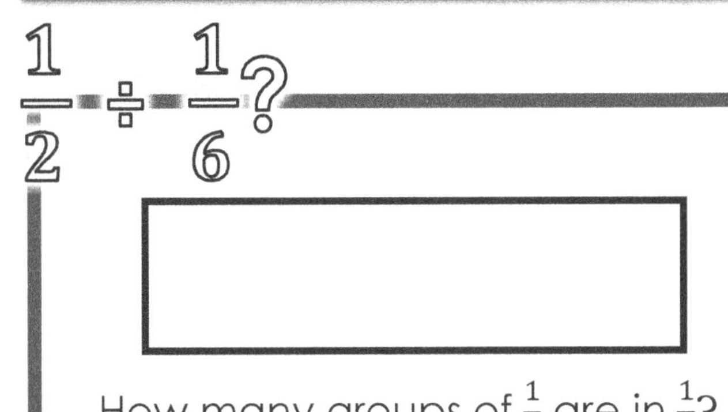

How many groups of $\frac{1}{6}$ are in $\frac{1}{2}$?

What is $1\frac{1}{2} \div \frac{1}{8}$?

Convert the mixed number into an improper fraction, write as a multiplication problem using a reciprocal, then solve.

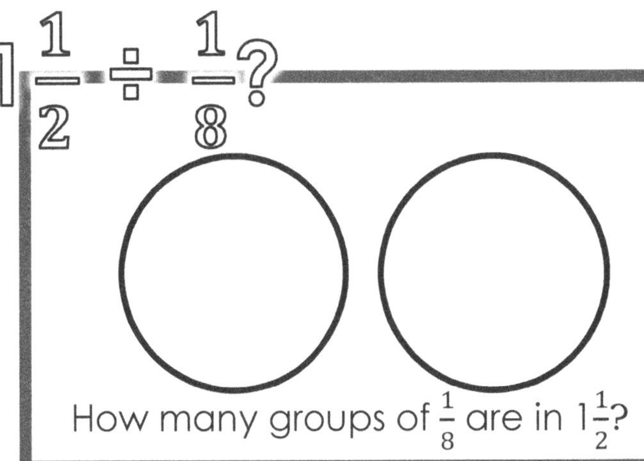

How many groups of $\frac{1}{8}$ are in $1\frac{1}{2}$?

answers

All answers must be in ______________________________ and converted from improper fractions to mixed numbers if necessary.

Name:

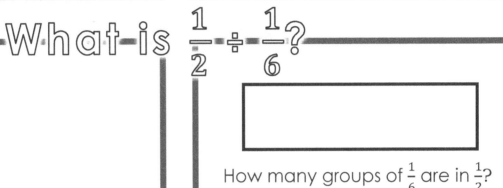

Fraction Division- Practice

Show all work. Write each answer in standard form.

Example 1

$\frac{2}{3} \div 8$

Rewrite

Example 2

$\frac{3}{4} \div \frac{1}{8}$

Rewrite

Model It

Example 3

$1\frac{9}{10} \div \frac{4}{15}$

Rewrite

Example 4

$3 \div \frac{2}{3}$

Rewrite

Model It

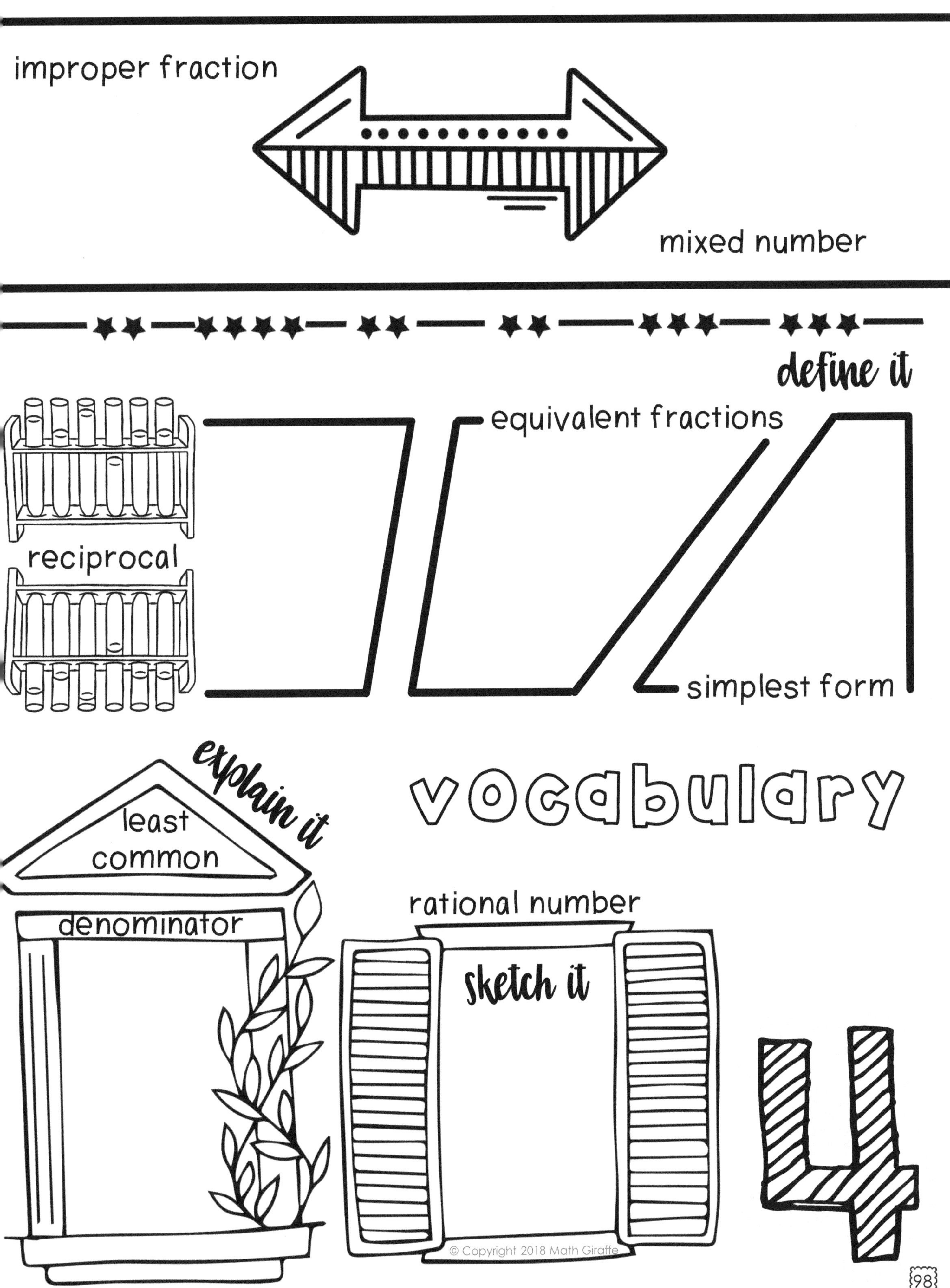
improper fraction
mixed number
define it
equivalent fractions
reciprocal
simplest form
explain it
least common denominator
vocabulary
rational number
sketch it
4

5
Ratios,
Proportions
& Percents

WORKING WITH RATIOS:
Name:
THREE WAYS:
1
2
3
ratio:
simplest form:
In the balloon, show with drawings how to divide up your own sketched 3:9 ratio to simplify it.
model it
write it
a ratio representing hot air balloons with some type of stripe to hot air balloons with dots
sketch it
a 3:9 ratio
simplify it
number of dotted balloons to total number of balloons
© Copyright 2018 Math Giraffe
analyze it
the root word for "ratio"
Write a ratio for balloons to clouds all three ways, then simplify it.
try it
compare it
RATIOS
part-to-part ratio vs. part-to-whole ratio

RATES

A hot air balloon travels 33 miles in 4 hours.

calculate it

Write as a rate.

Find the unit rate.

explain it

define it What is a rate?

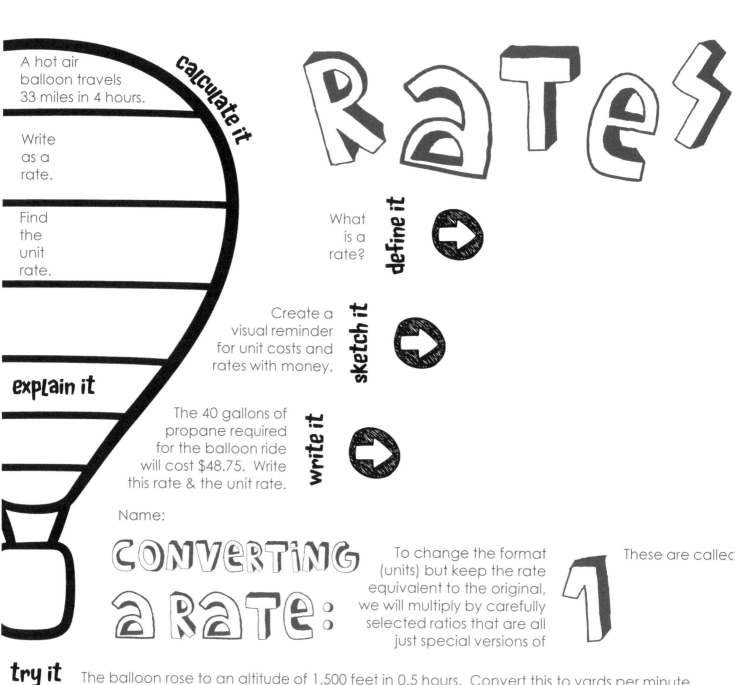

sketch it Create a visual reminder for unit costs and rates with money.

write it The 40 gallons of propane required for the balloon ride will cost $48.75. Write this rate & the unit rate.

Name:

CONVERTING A RATE:

To change the format (units) but keep the rate equivalent to the original, we will multiply by carefully selected ratios that are all just special versions of 1

These are callec

try it The balloon rose to an altitude of 1,500 feet in 0.5 hours. Convert this to yards per minute.

Start by setting units up so that they will …

$$\frac{1500\text{ ft}}{0.5\text{ hour}} \cdot \frac{\_\_\_}{\text{ft}} \cdot \frac{\text{hr}}{\_\_\_} = \_\_\_$$

MULTIPLY

Choose the other units for CONVERSION FACTORS that will result in your __________ units at the end.

The balloon uses 4.5 quarts of gas in 7 minutes. How many gallons will it use per hour?

UNIT RATES

Are ...

Try it

Determine which shampoo bottle on the page offers the best deal per ounce.

$5.19

32 OZ

How-it-works:

Try writing a unit price (cost per ounce) in the bottle.

#1

Karl drove 40 miles in 50 minutes. Teresa drove 35 miles in 42 minutes. Who had a higher average speed?

EMEMBER UNITS IN ANSWERS!

always...

Reminders

Think "blank per blank" and then write the first item in the ________________ and the second in the ________________.

Then,

Show the faster driver's average speed on the speedometer.

$3.95

20 oz

Example

A box of 100 hexagonal floor tiles covers 16 square feet of floor. How much area does a single tile cover?

Rephrase:
We want to know _______ per ________.

Set Up:
The quantity in the first blank belongs in the ______________. The quantity in the second blank belongs in the ____________.

Set Up

Calculate:
Divide! (top ÷ bottom)

Answer

Check:
Is it reasonable?

#2

A crew can pave 850 feet of road in an 8-hour day. Find the unit rate.

$4.69

27.5 oz

ame:

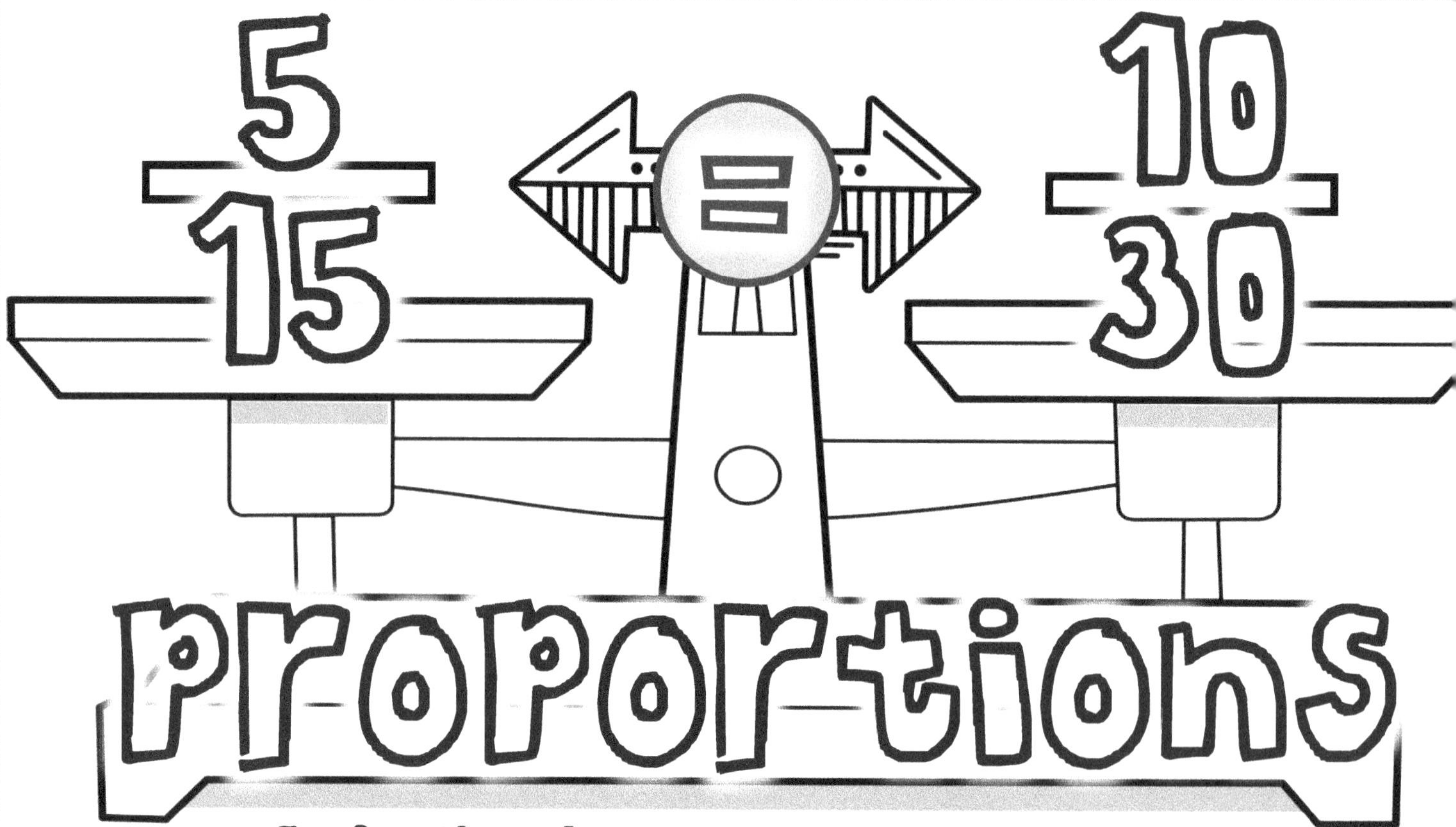

$$\frac{5}{15} = \frac{10}{30}$$

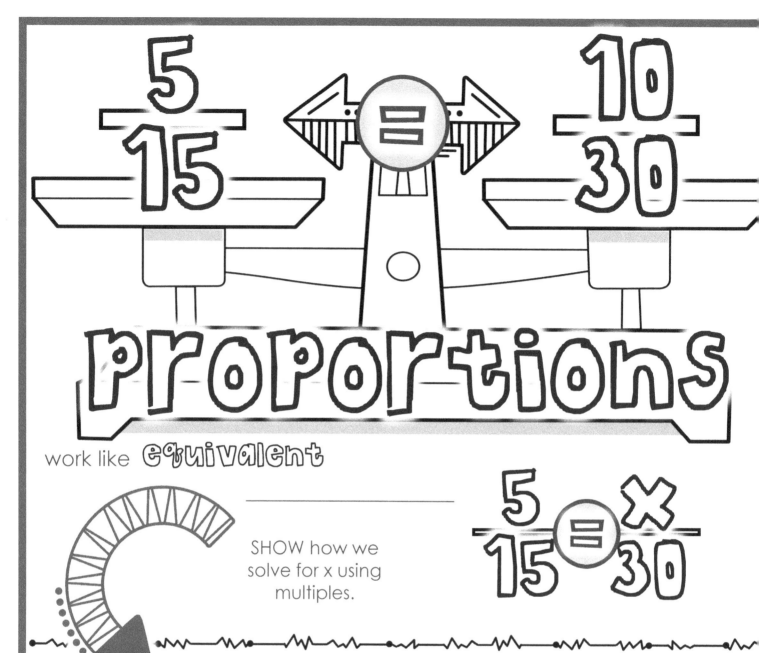

proportions

work like equivalent ____________________

SHOW how we solve for x using multiples.

$$\frac{5}{15} = \frac{x}{30}$$

... but sometimes we cannot just multiply both the numerator and denominator by the same whole number, as we can to find a missing value in the proportion above...

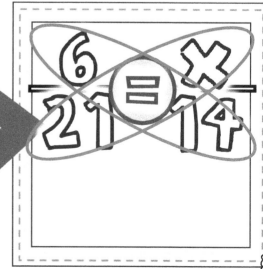

$$\frac{6}{21} = \frac{x}{14}$$

Name:

Solving proportions

Complete the equation to the right to show how cross-multiplication works. Then, solve each proportion. Show your method.

Name:

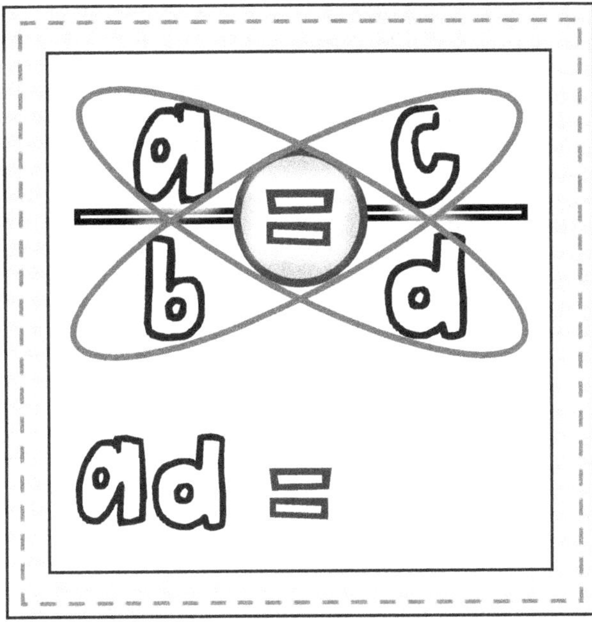

Determine whether each pair is proportional.

1. $\frac{2}{3} = \frac{22}{33}$ y/n
2. $\frac{35}{42} = \frac{63}{84}$ y/n
3. $\frac{8}{45} = \frac{12}{67.5}$ y/n
4. $\frac{3}{7.5} = \frac{6}{15}$ y/n

$$\frac{w}{16} = \frac{8}{10}$$

w =

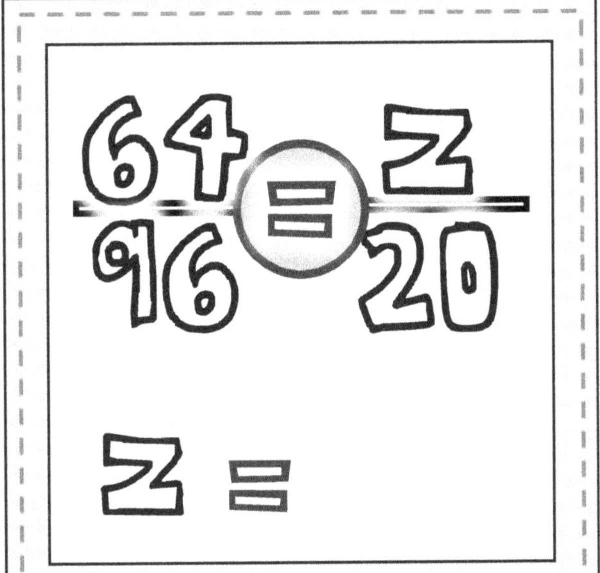

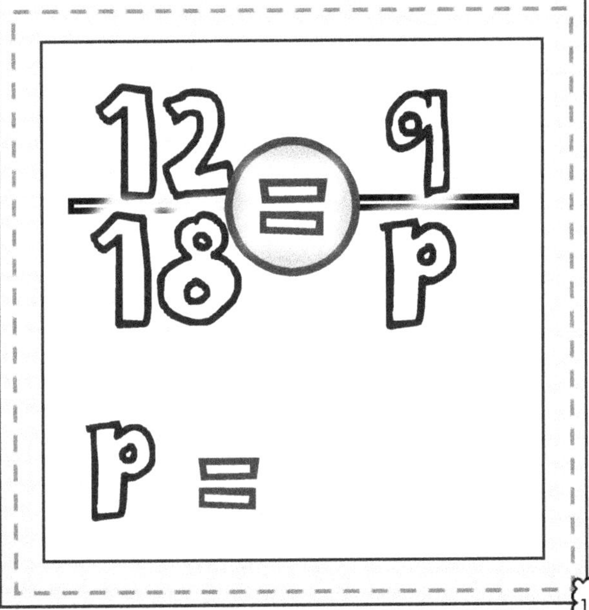

57%

Write $\frac{1}{5}$ as a percent:

Name:

root word

"cent"

GIVE EXAMPLES OF WORDS WITH "CENT" ROOT

SKETCH THE MEANING

Move the decimal point 2 places to the right.

But sometimes it's not so easy!

EXAMPLE:

Write $\frac{5}{6}$ as a percent.

SKETCH A REMINDER

TRY IT

Why?

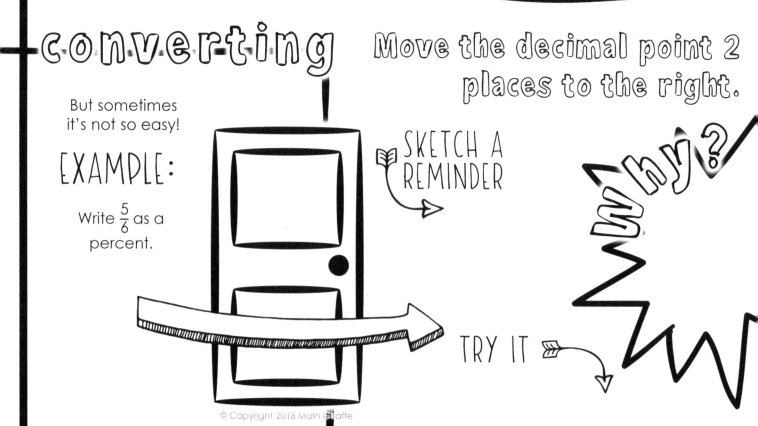

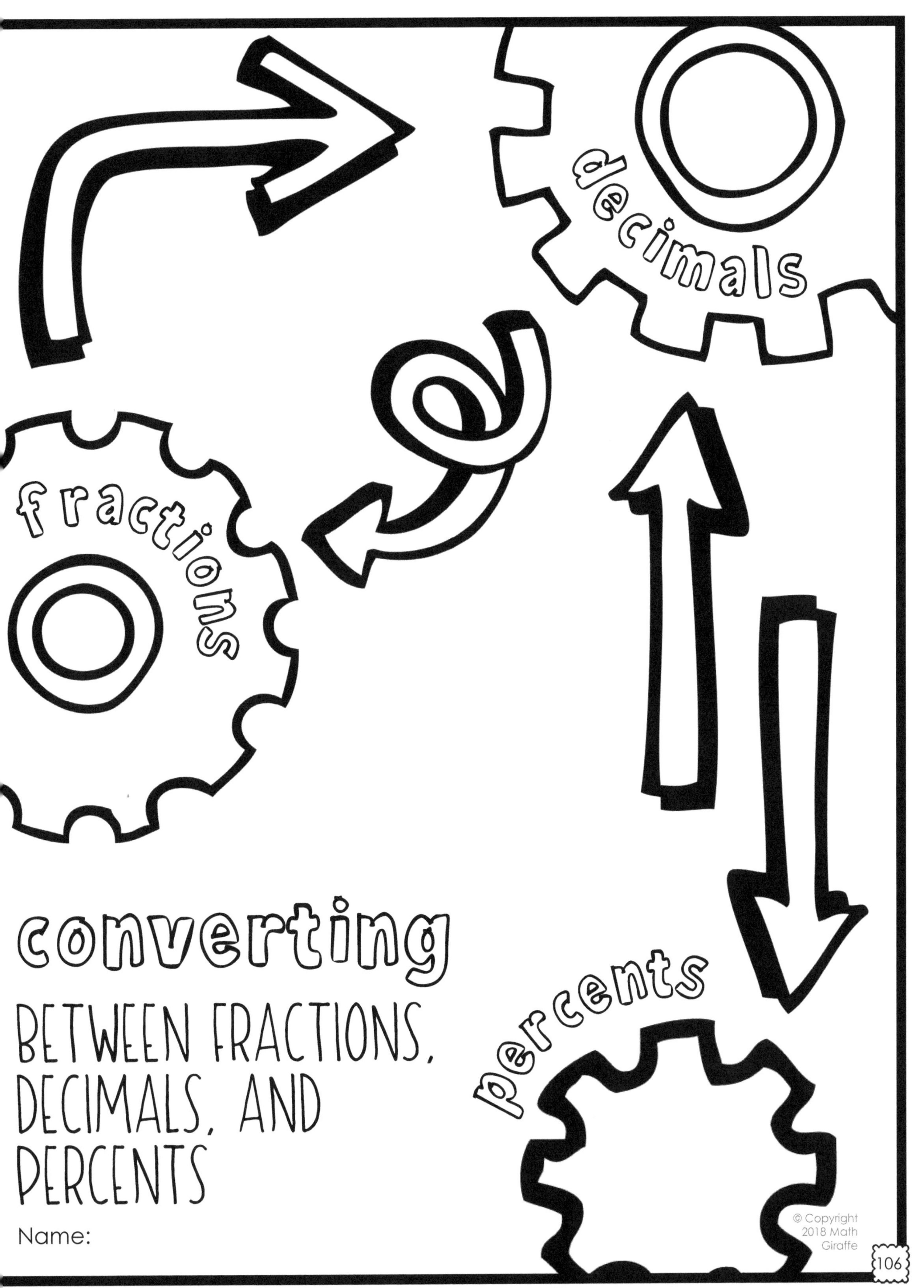

converting

BETWEEN FRACTIONS, DECIMALS, AND PERCENTS

Name:

PERCENT EQUATIONS

Within and around each symbol, show or write how you convert each piece of a question into a symbol or different format.

When you see...
"what"
use a ...

When you see...
"is"
use an ...

When you see...
"of"
use...

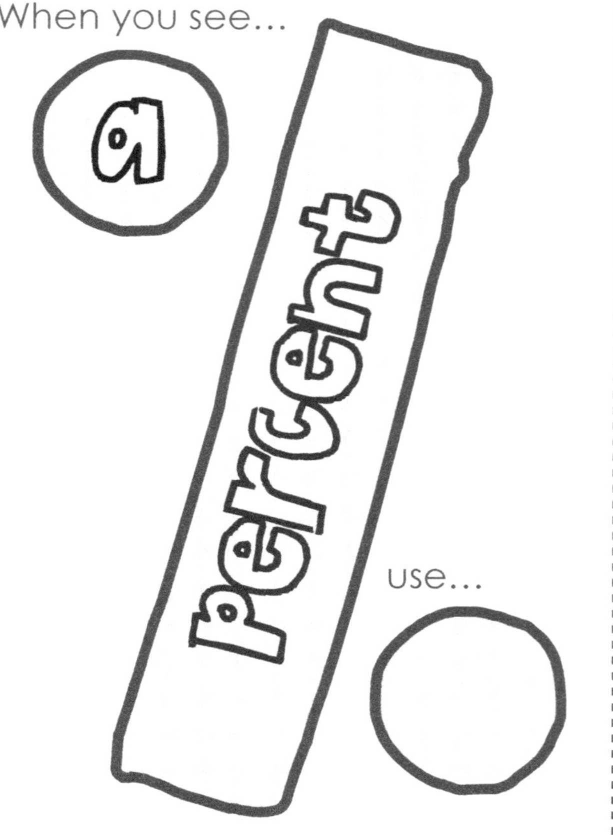

When you see...
a
percent
use...

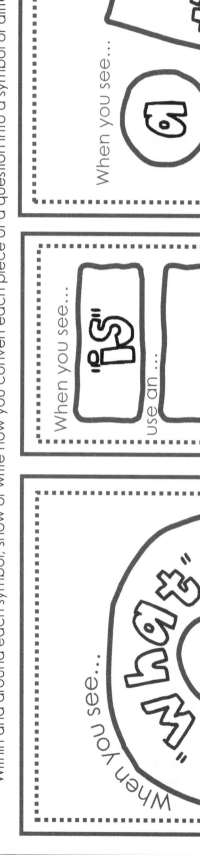

ame:

Practice with Percent Equations

Write an equation, then solve.

What is 15% of 350?

18 is 20% of what number?

What percent of 145 is 35?

Since the question asked what PERCENT, we have to convert the answer back into percent form.

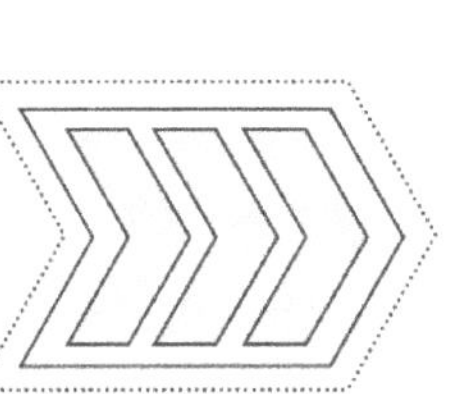

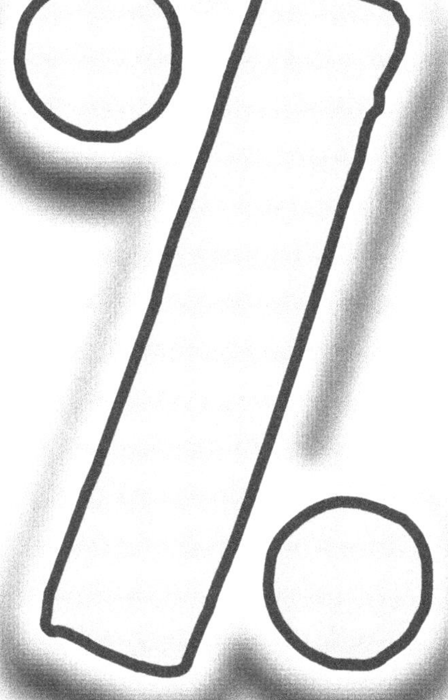

15% of the class of 40 is going home early. How many students are going home early?

Liz has added 20% of the total amount of acid that she needs in the beaker for her experiment. She has put in 10 mL so far. How much more does she need to add?

percent change =

Name:

A credit card bill that was $346 went up to $380 the next month after not being paid. By what percent did it increase?

Find the cost for a consumer if a coffee shop marks up a $0.40 cup of coffee by 260%.

define it

markup

A grocery store purchases crates of oranges for their produce section for $3.00 per crate of a dozen. They sell the oranges individually for $0.49 each. Find the percent of markup.

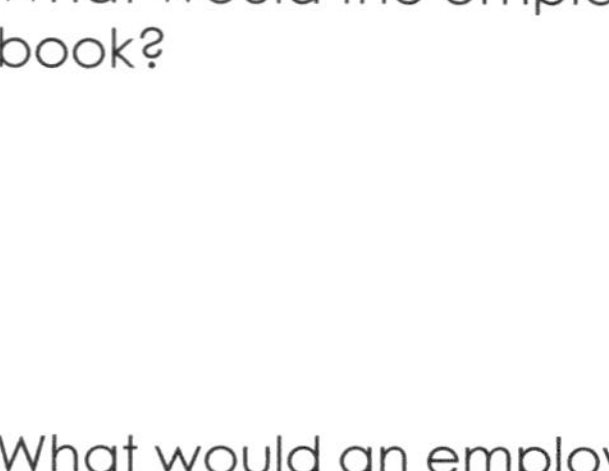

discount

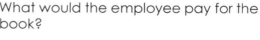

A store offers a 15% discount to its own employees.

How much is the employee discount on a book with a price of $18.50?

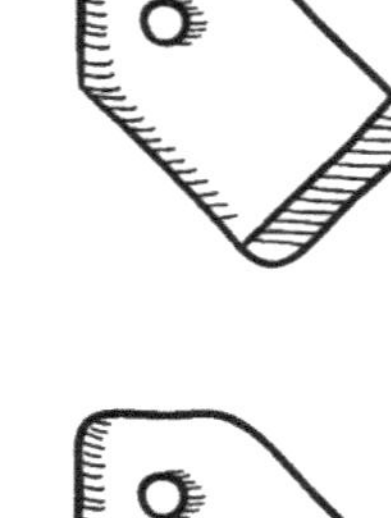

What would the employee pay for the book?

What would an employee pay for a computer with a price of $489.95?

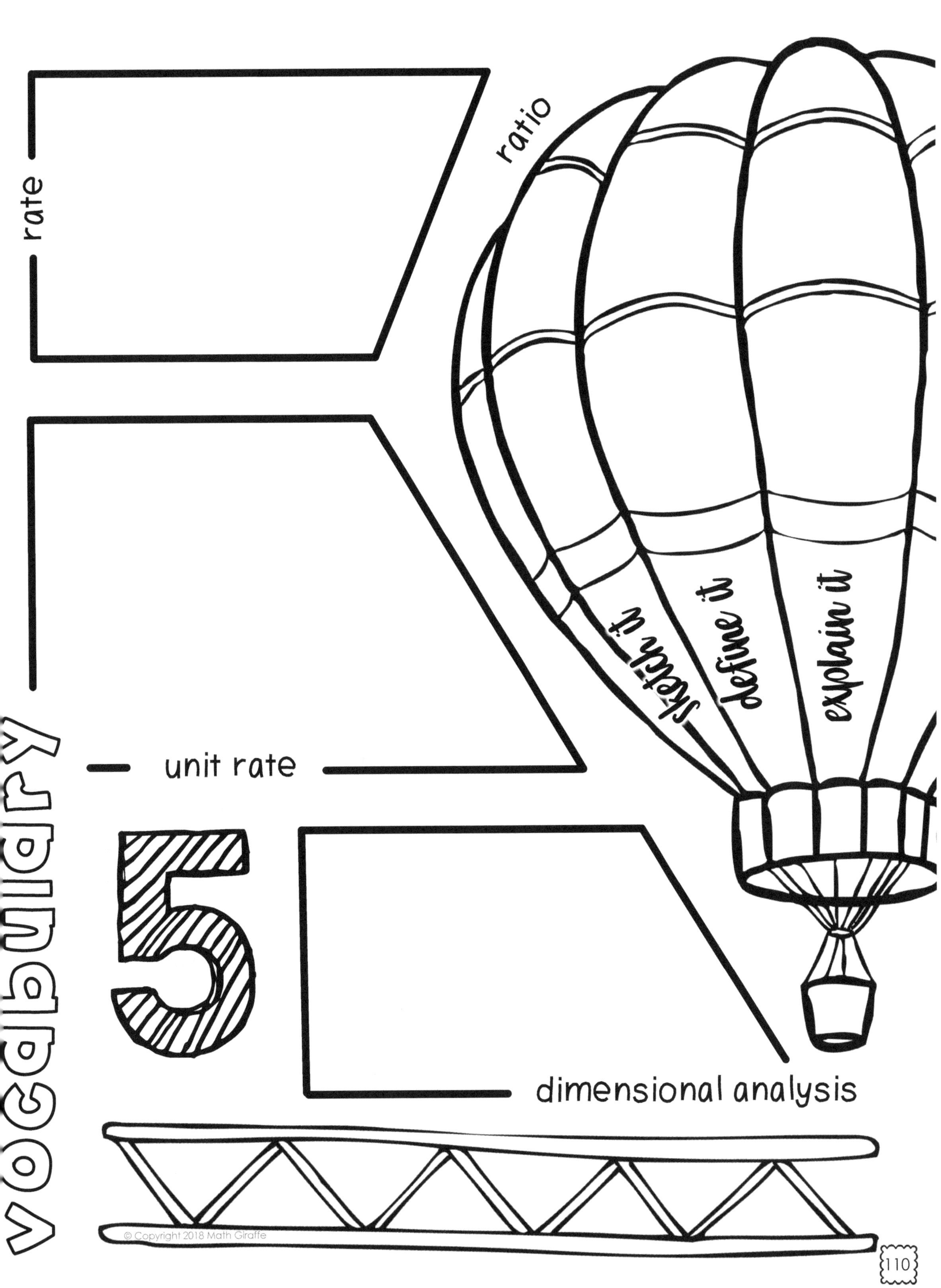

vocabulary
5
rate
ratio
unit rate
dimensional analysis
sketch it
define it
explain it

percent

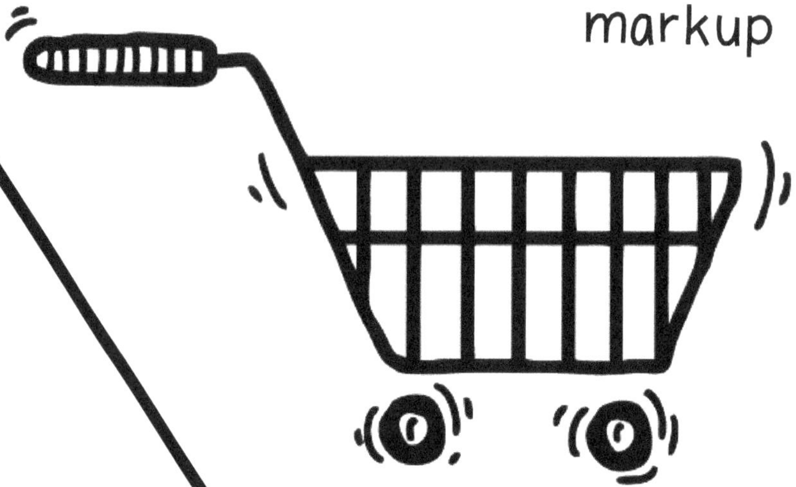

markup

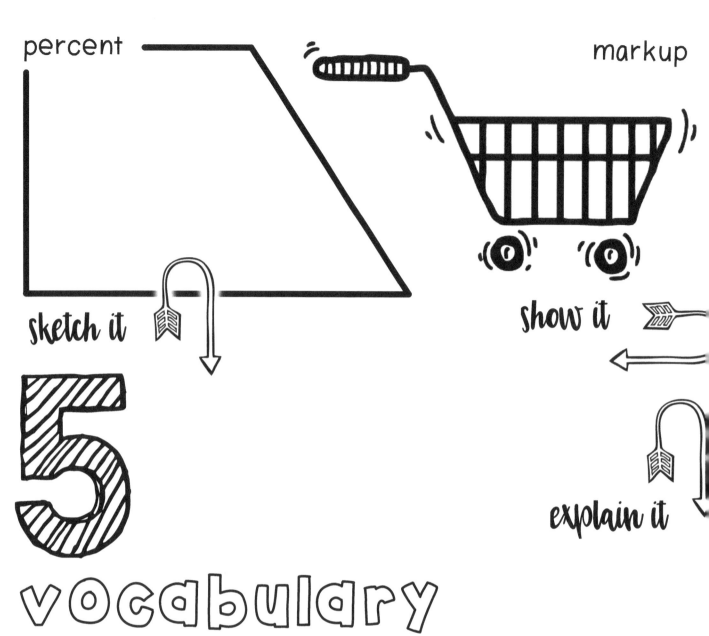

sketch it

show it

5

explain it

discount

show it

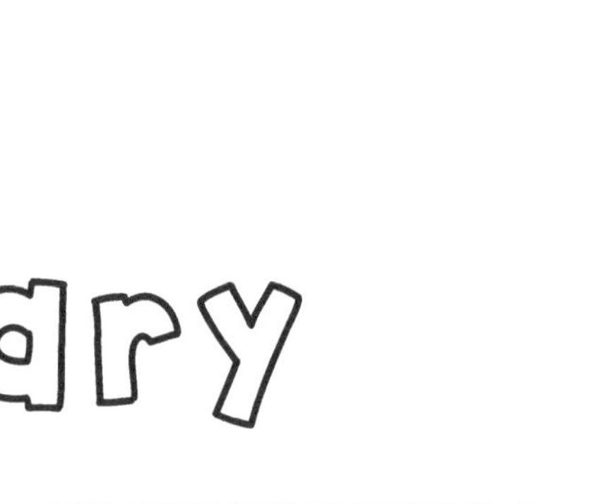

proportion

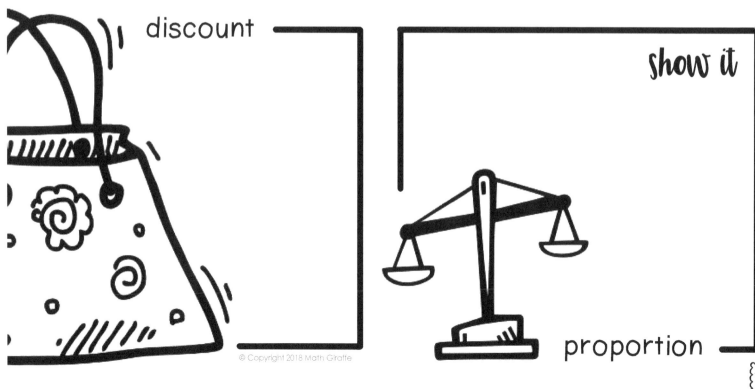

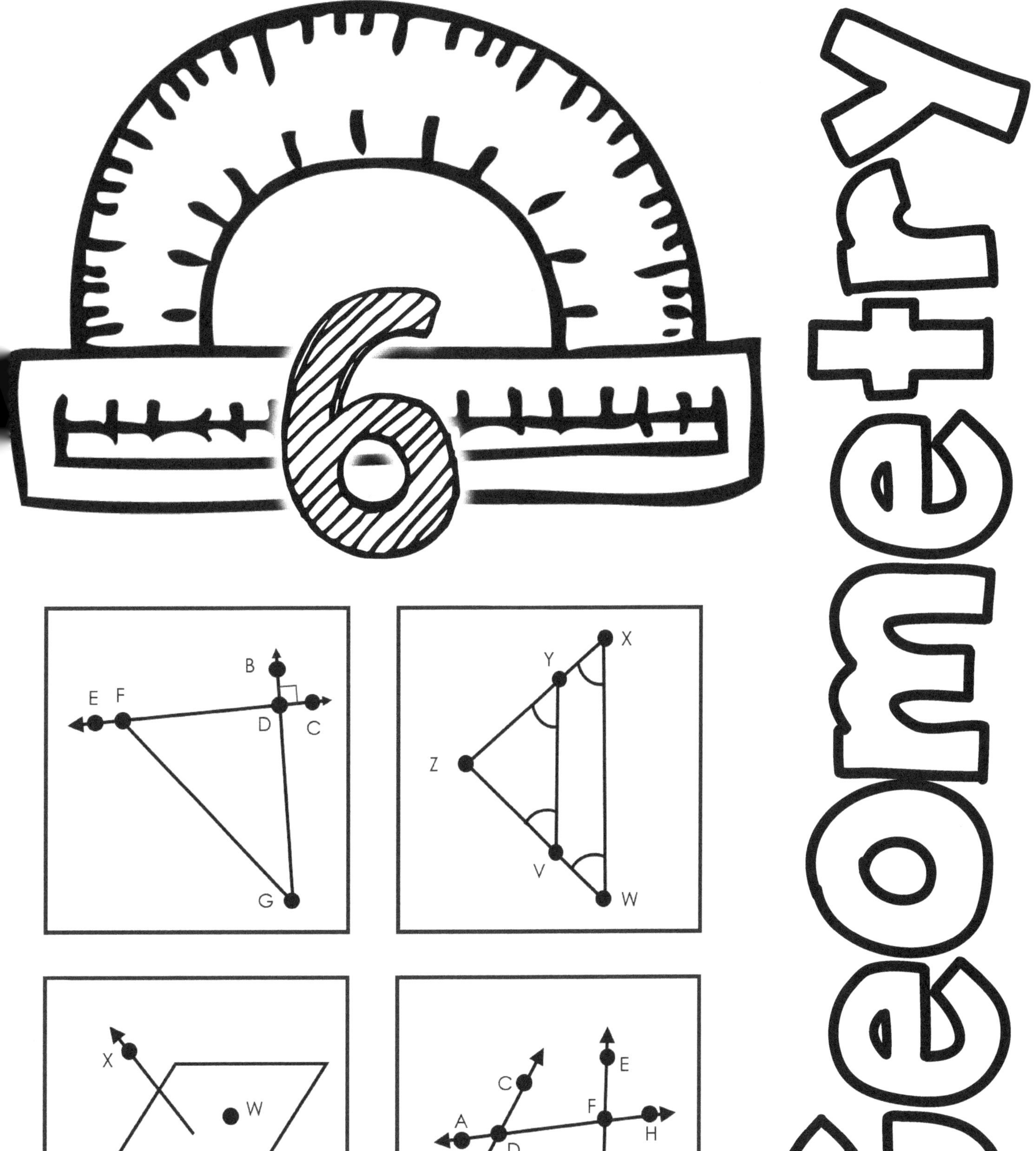

Geometry

point

line

plane

define it

identify it

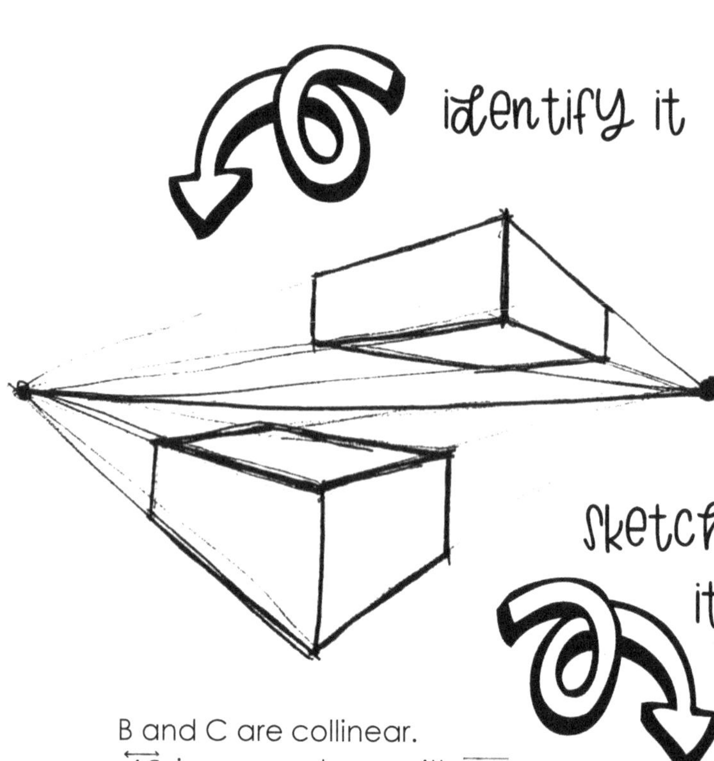

collinear:

coplanar:

sketch it

B and C are collinear.
$\overleftrightarrow{AC}$ is non coplanar with $\overline{MN}$.

points, lines, & planes

the building blocks of geometry

Name:

segment

ray

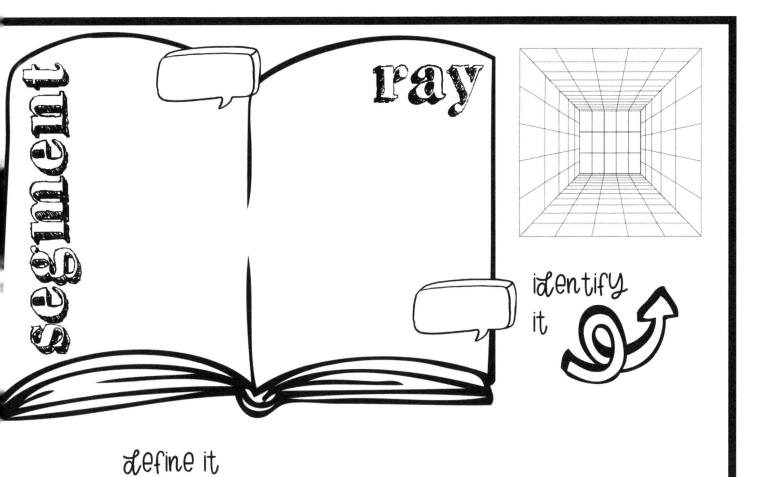

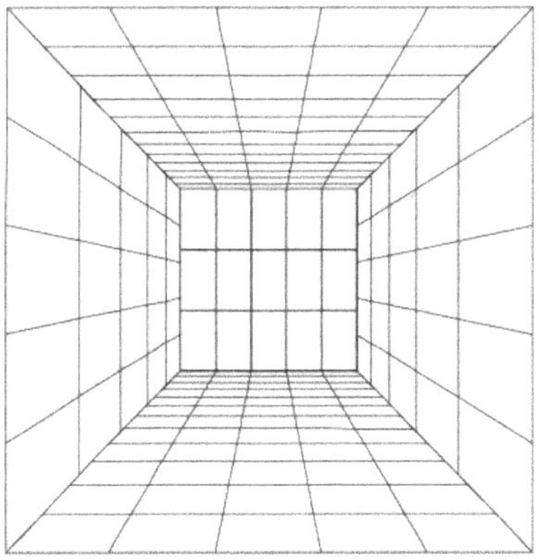

identify it

define it

notation

LABEL & EXPLAIN it

intersecting

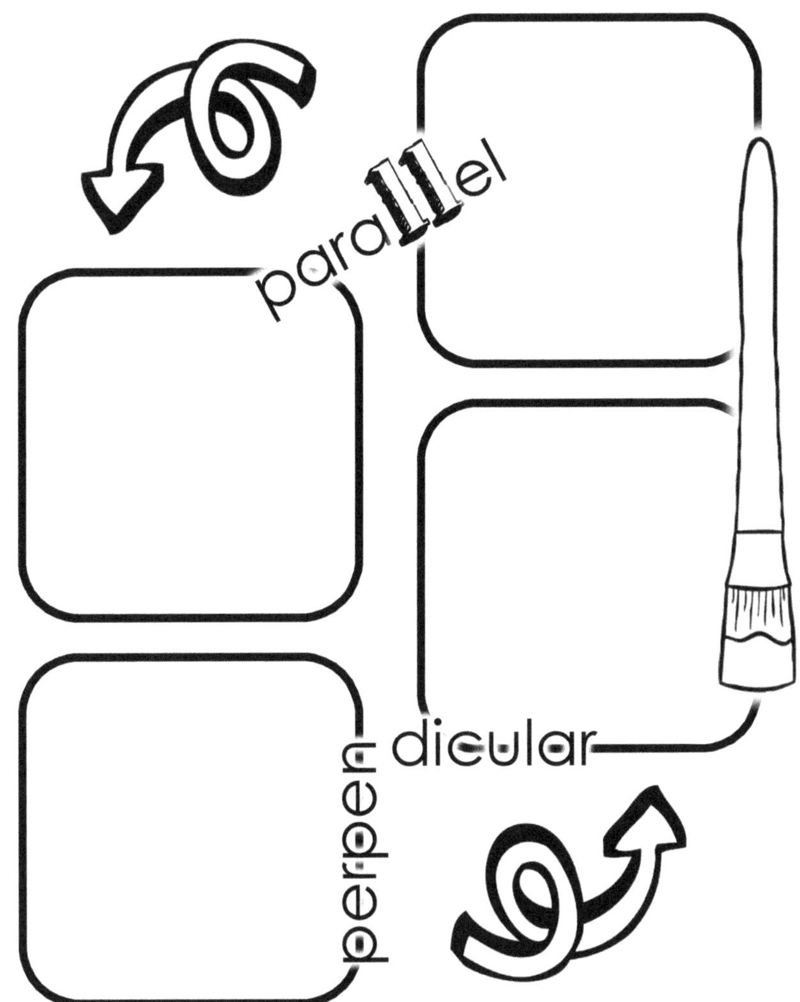

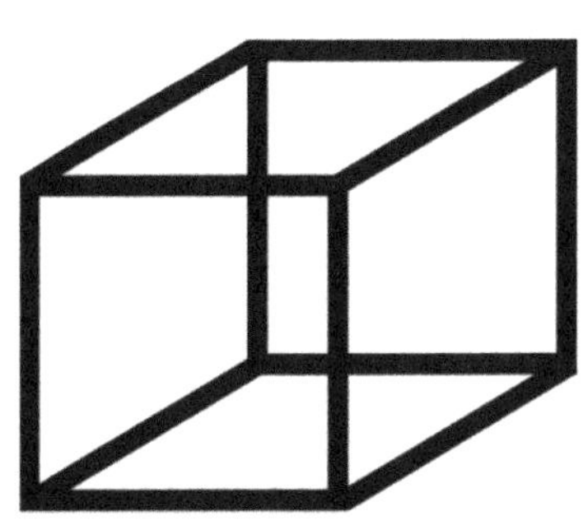

sketch it

Name:

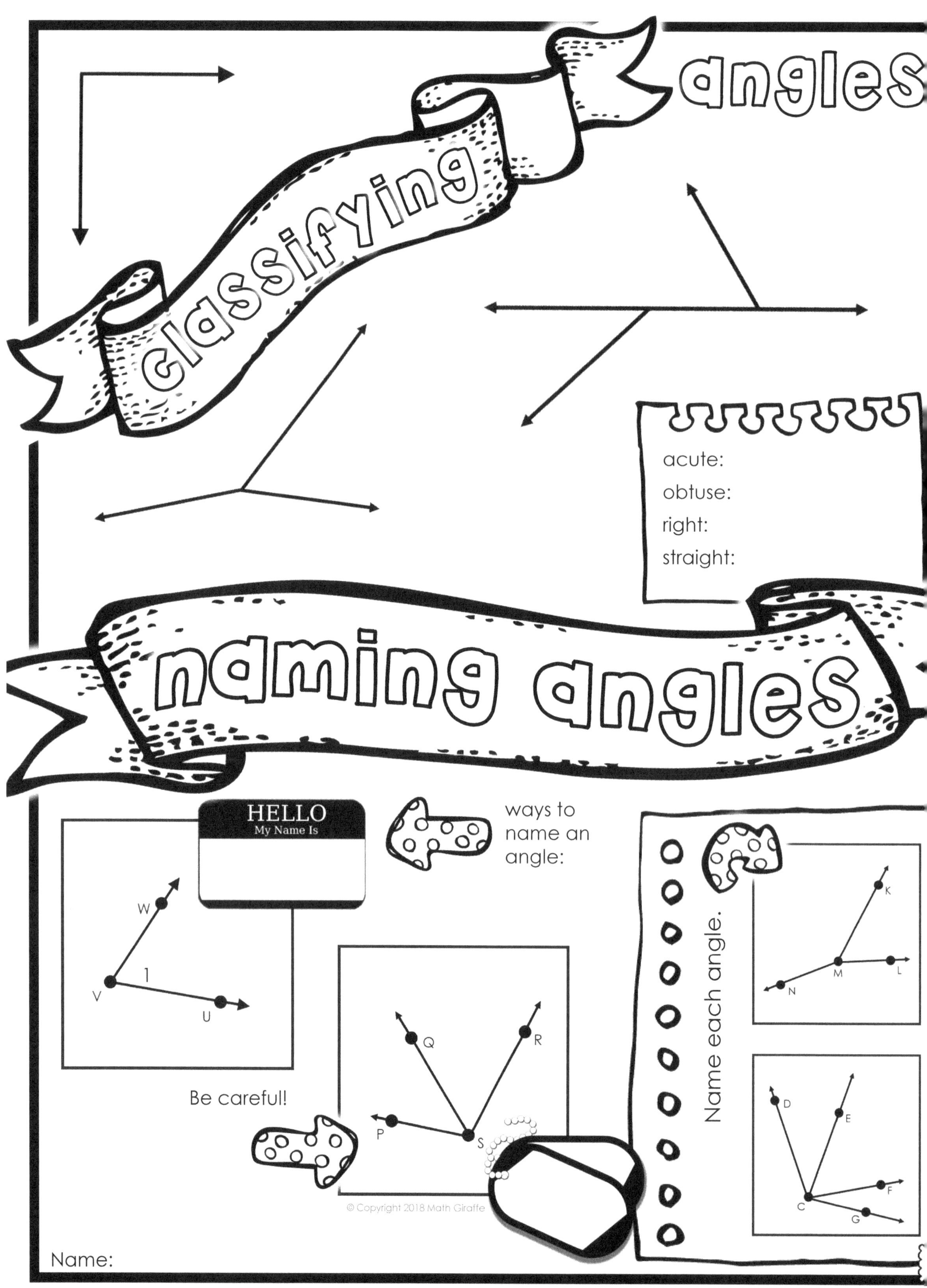
classifying
angles
acute:
obtuse:
right:
straight:
naming angles
HELLO
My Name Is
ways to
name an
angle:
W
V
1
U
Q
R
P
S
Be careful!
Name each angle.
K
M
L
N
D
E
F
C
G
© Copyright 2018 Math Giraffe
Name:

Special pairs of angles
1
2
1
2
2
1
1
2
Name:
supplementary
complementary
vertical
Identify Special pairs
1
2
3
4
5
1
2
3
3
2
4
1
5
6
C
E
A
F
H
D
G
B
A
B
E
D
C
3
2
4
1
5
Sketch a memory trigger

ertical
ngles
meet at the vertex
like
kissing
fish

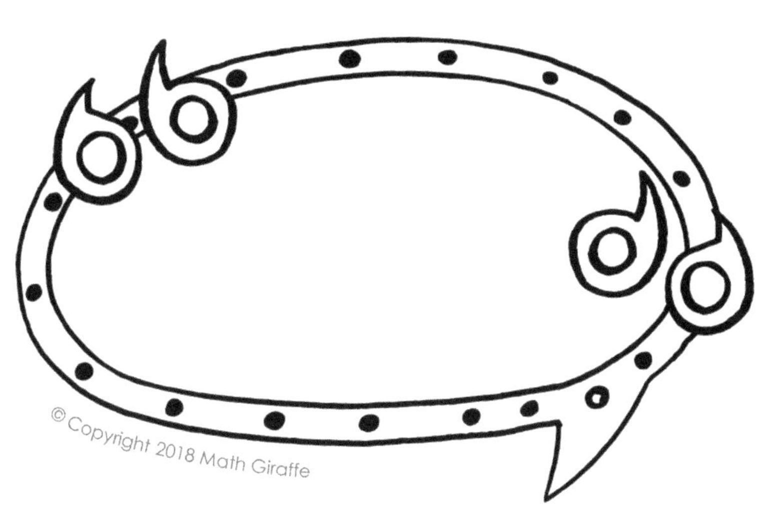

sketch it:
an angle bisector

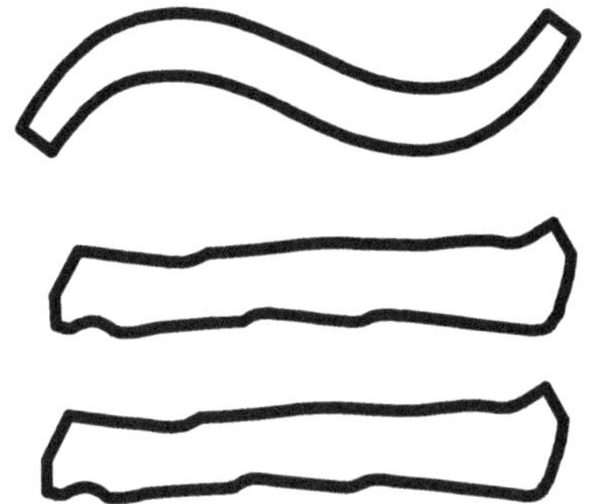

marking congruent parts

writing congruence statements

example

Name:

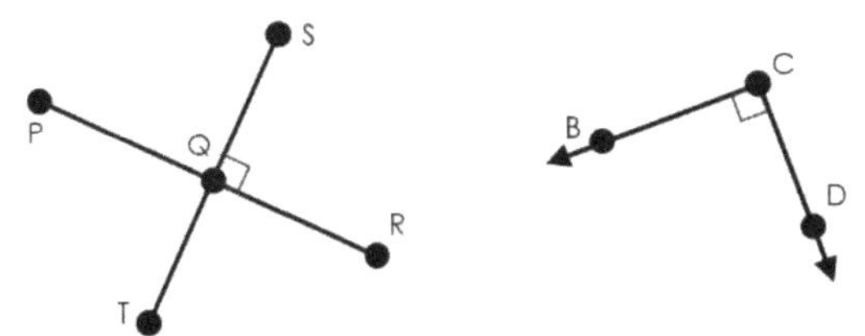

$\overline{PR}$ and $\overline{TS}$ bisect one another. Mark congruent parts and write as many congruence statements as possible.

Name:

CAUTION

≃ vs. ≅

don't slip up

segments

notation

angles

AB ○ CD

$\overline{AB}$ ○ $\overline{CD}$

m<1 ○ m<2

<1 ○ <2

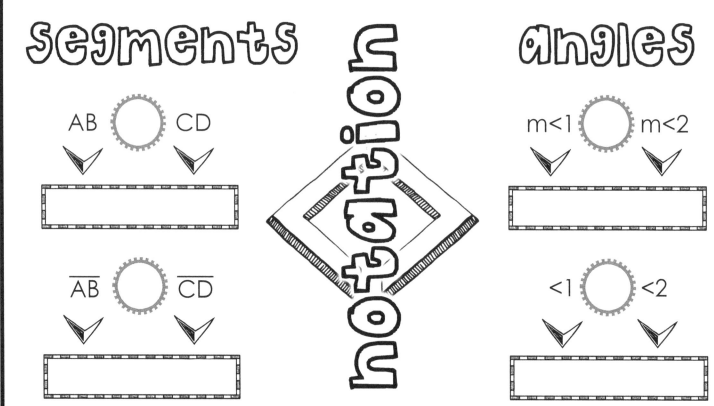

Transversals

A transversal is a line that _______________ at least two other lines in the same plane.

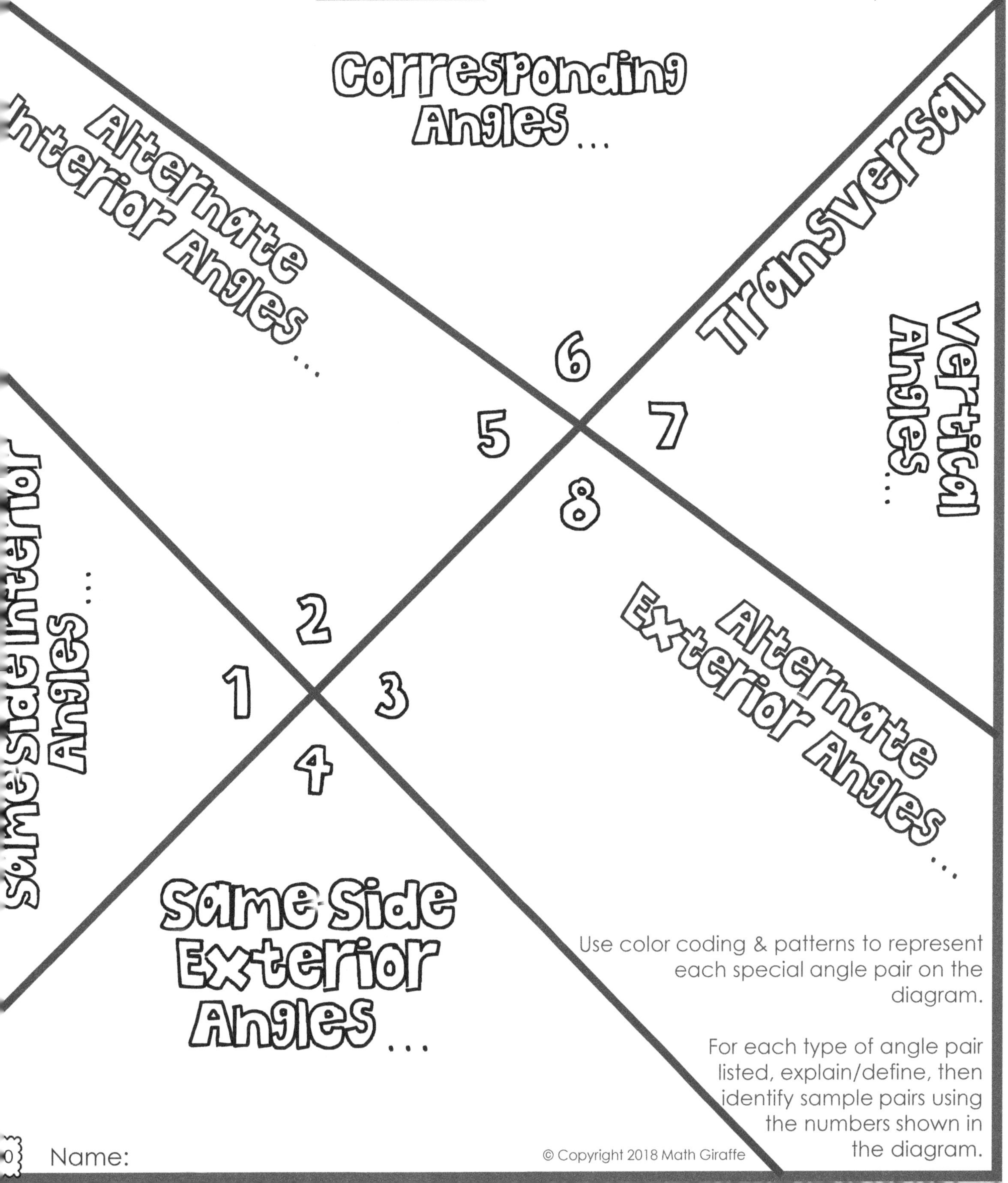

Corresponding Angles ...

Alternate Interior Angles ...

Transversal

6

5

7

8

Vertical Angles ...

Same Side Interior Angles ...

2

1

3

4

Alternate Exterior Angles ...

Same Side Exterior Angles ...

Use color coding & patterns to represent each special angle pair on the diagram.

For each type of angle pair listed, explain/define, then identify sample pairs using the numbers shown in the diagram.

Name:

Parallel

lines intersected by a transversal

Name:

properties / special angle pairs used:

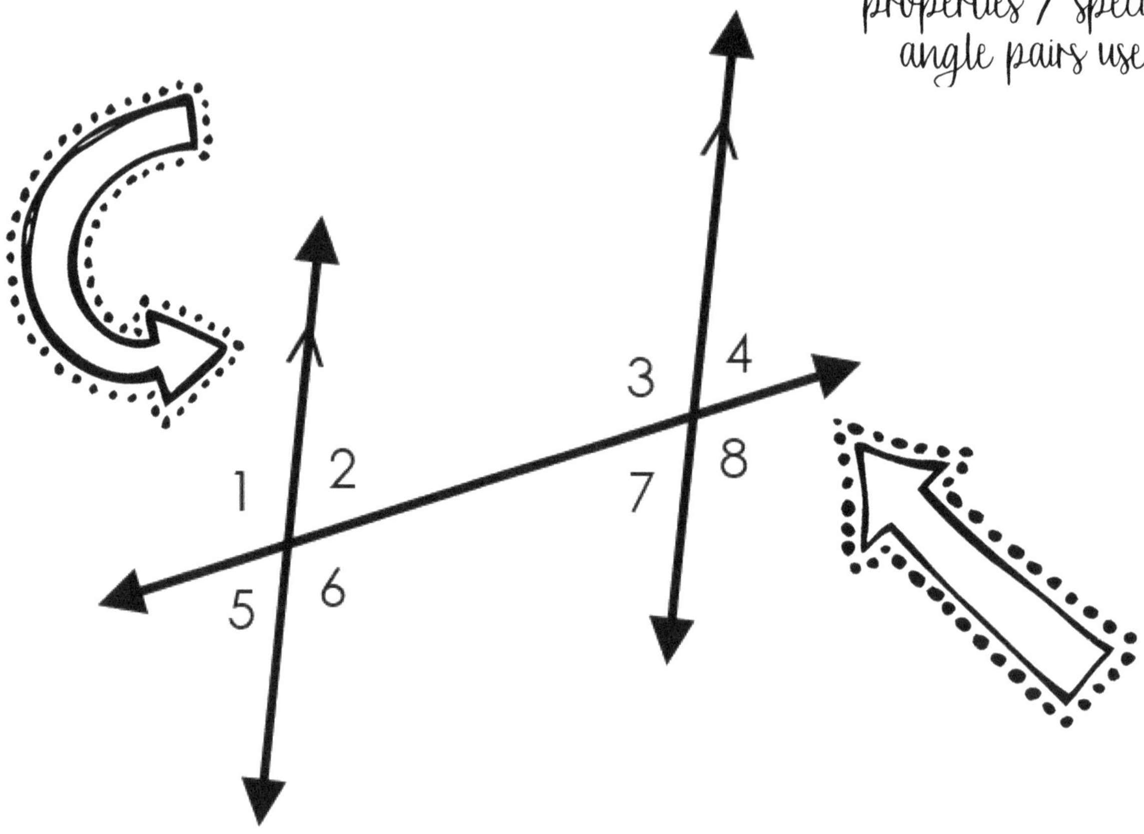

1 2 5 6 3 4 7 8

color code to show congruent angles

Write a BIG list of congruency statements. Then convert each into an equivalence statement using "measure of" notation for each angle.

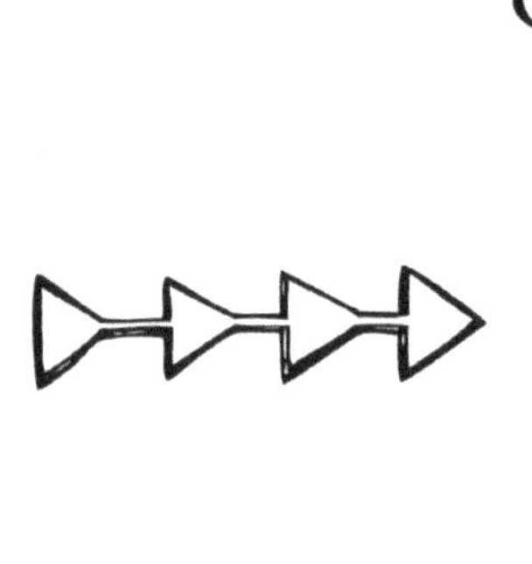

classifying

- ❑ hexagon
- ❑ pentagon
- ❑ triangle
- ❑ quadrilateral
- ❑ octagon
- ❑ heptagon

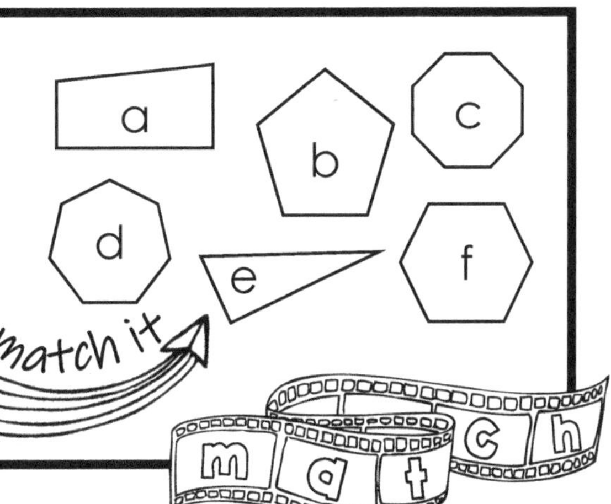

Polygons are classified by…

Congruent figures have…

Write a congruency statement for the triangles below:

Color code corresponding parts to show congruency.

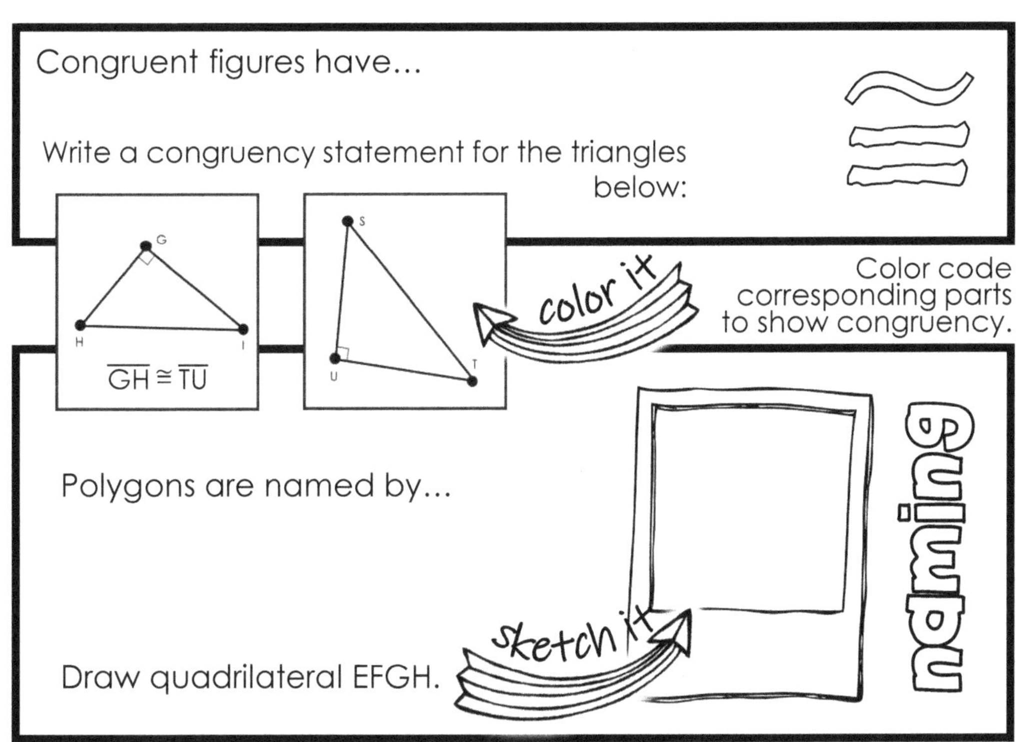

Polygons are named by…

Draw quadrilateral EFGH.

Name:

did you know
sides
An equilateral triangle is always also…
In an isosceles triangle, the base angles (opposite the congruent sides) are always…
both
angles
T
45°
V
45°
U
Name:
classifying triangles

Triangle Sum Theorem

Name:

side note

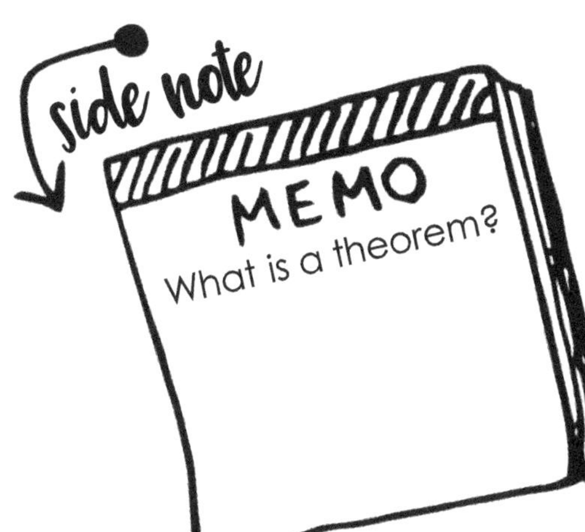

theorem

how to use it

formula & diagram

sketch it

try it

Solve for the unknown angle measure(s) in each triangle.

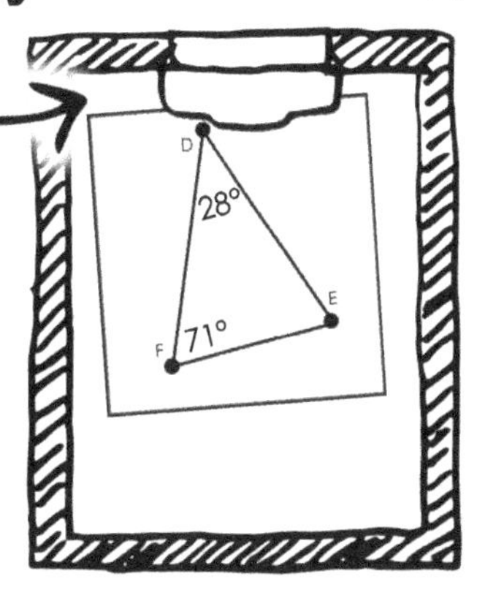

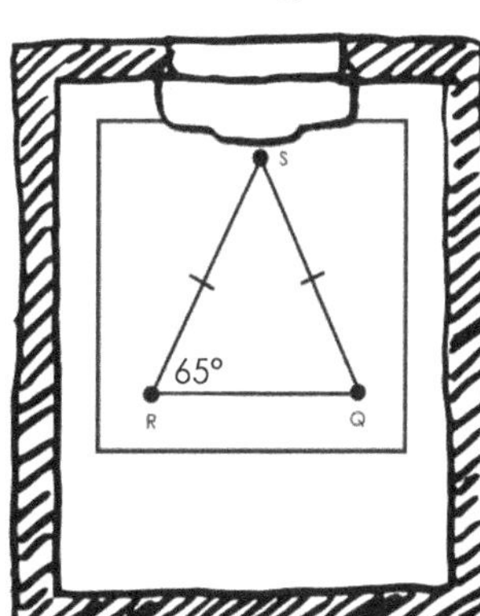

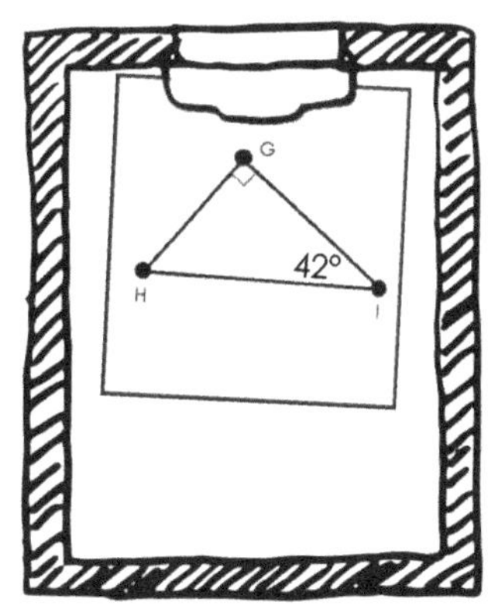

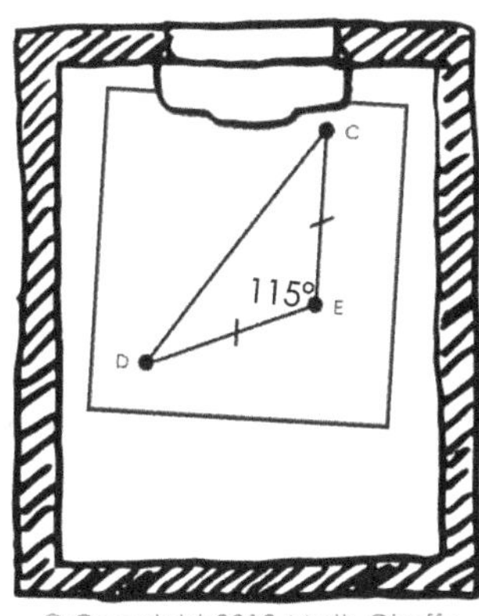

hands on

why it works

Make a large paper triangle. Rip off each vertex. Can you line up the three angles to make a line (or straight angle)? What does this tell you about the number of degrees? Now try with another type of triangle.

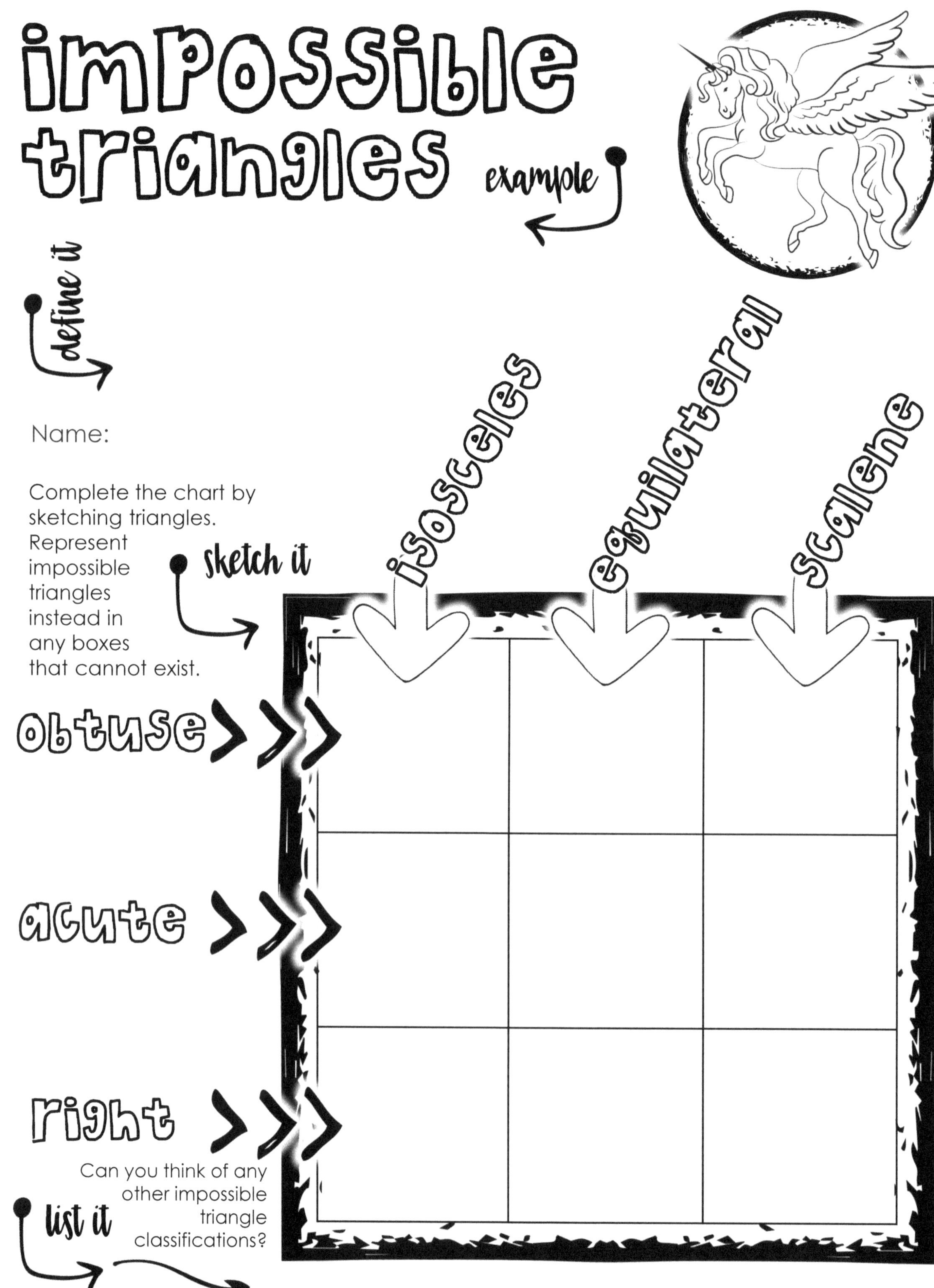
impossible triangles
example
define it
Name:
Complete the chart by sketching triangles. Represent impossible triangles instead in any boxes that cannot exist.
sketch it
isosceles
equilateral
scalene
obtuse
acute
right
Can you think of any other impossible triangle classifications?
list it

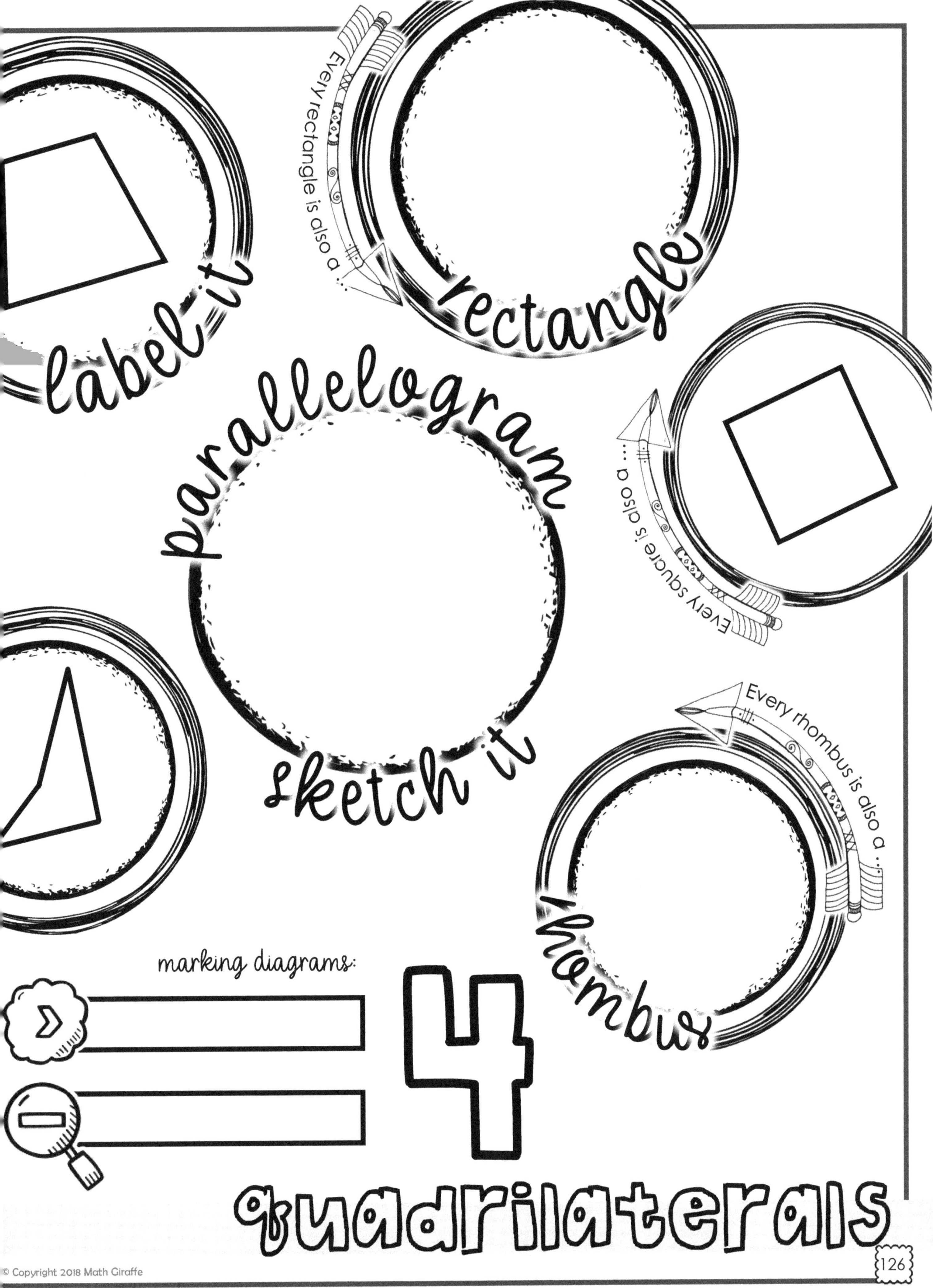

label it
rectangle
Every rectangle is also a ...
parallelogram
sketch it
Every square is also a ...
Every rhombus is also a ...
rhombus
marking diagrams:
4
quadrilaterals

quadrilaterals

definition:

find & label:

- [ ] trapezoid
- [ ] right trapezoid
- [ ] quadrilateral
- [ ] parallelogram
- [ ] square
- [ ] rhombus
- [ ] rectangle

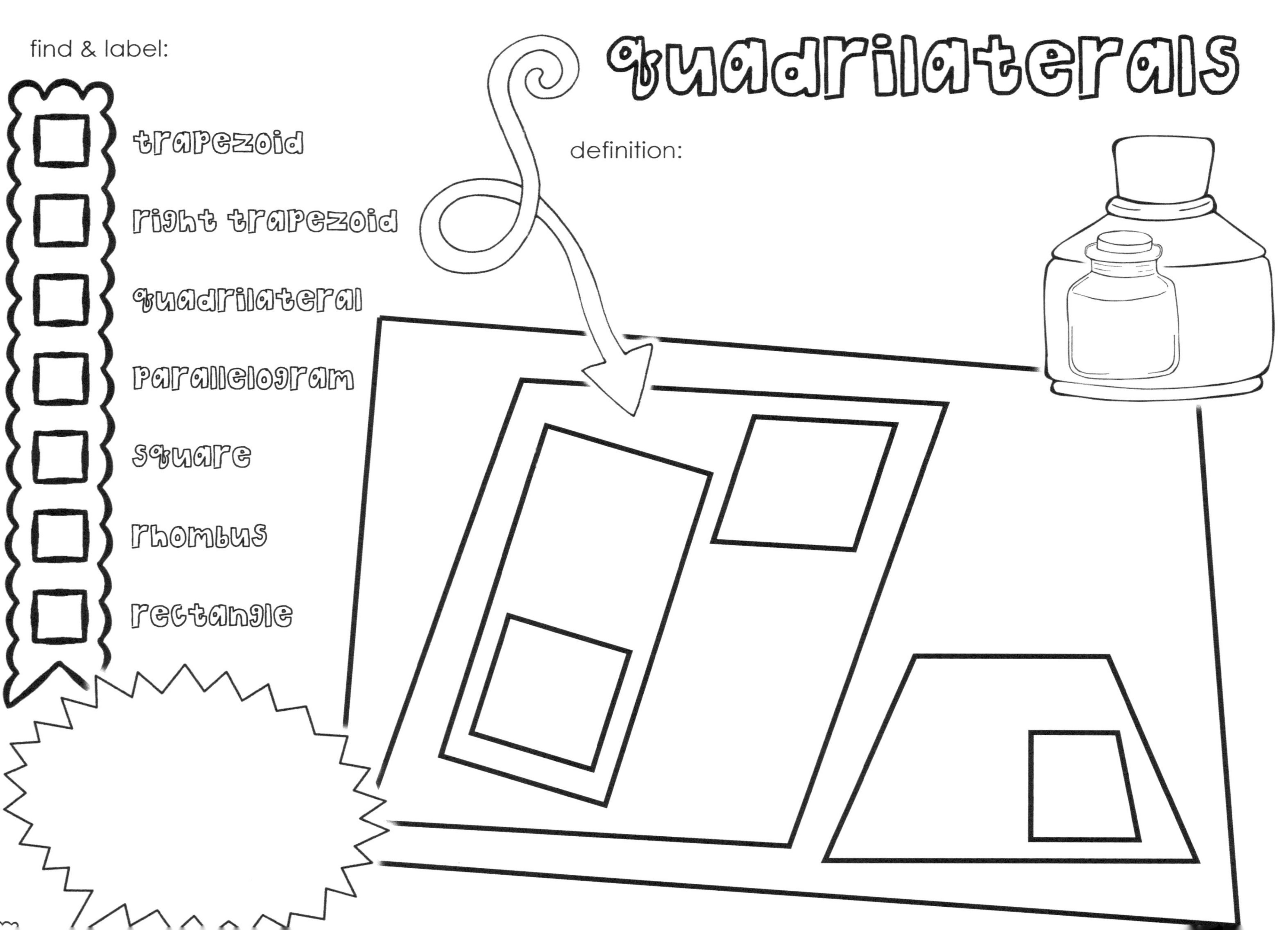

The word "circle" comes from the Latin word "circulus," which means "________________."

PARTS OF A CIRCLE

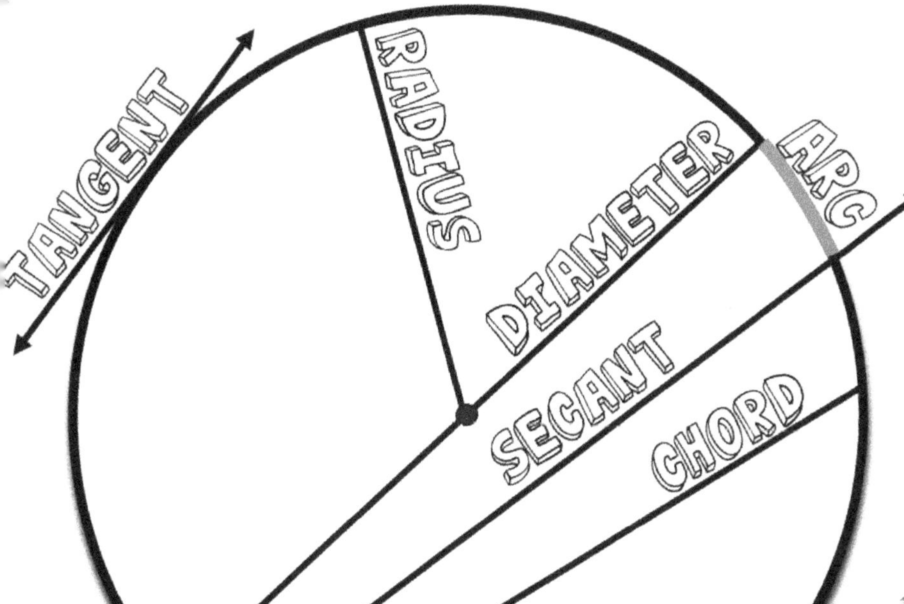

Doodles "To Do" List:

- ❑ Label a "sector."
- ❑ Label a "segment."
- ❑ Draw an arrow to show that "circumference" goes all the way around.
- ❑ Label the "center."
- ❑ Complete the definitions.
- ❑ Fill in each formula.
- ❑ Highlight & embellish key ideas!

CALCULATE

There are _____ degrees in a circle. A sector that is ¼ of the circle would have a central angle measuring...

Name: ________________

____

DEFINITIONS

a line segment that passes through the center and has endpoints on the circumference

a line segment that has 1 endpoint at the center and the other on the circumference

a line segment that has both endpoints on the circumference

a line segment that touches only 1 point on the circumference

a portion of the circumference

a portion defined by a chord and an arc

a portion defined by 2 radii and an arc

a line that intersects the circumference at 2 points

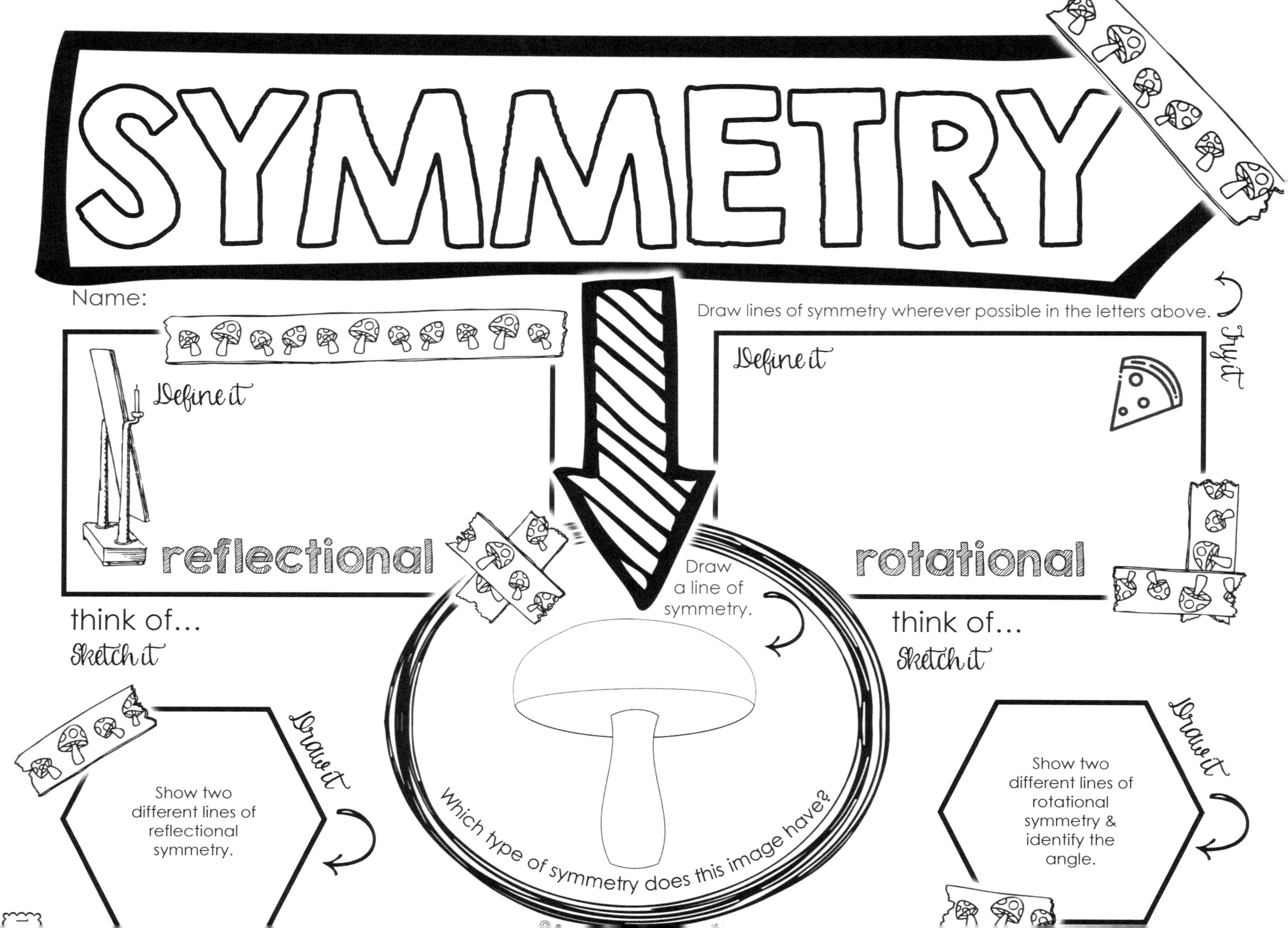
SYMMETRY
Name:
Draw lines of symmetry wherever possible in the letters above.
Try it
Define it
reflectional
think of...
Sketch it
Define it
rotational
think of...
Sketch it
Draw
a line of
symmetry.
Which type of symmetry does this image have?
Draw it
Show two
different lines of
reflectional
symmetry.
Draw it
Show two
different lines of
rotational
symmetry &
identify the
angle.

Name:

A figure and its reflection are...

otation

Reflect the figure across the y-axis.

y

O

x

1

Identify key points in the ________.

2

Create the image such that each point maintains its ________ from the ________ ________.

FLIP

REFLECTION

y

x

Reflect the figure across the line y = x.

y

x

Transformations

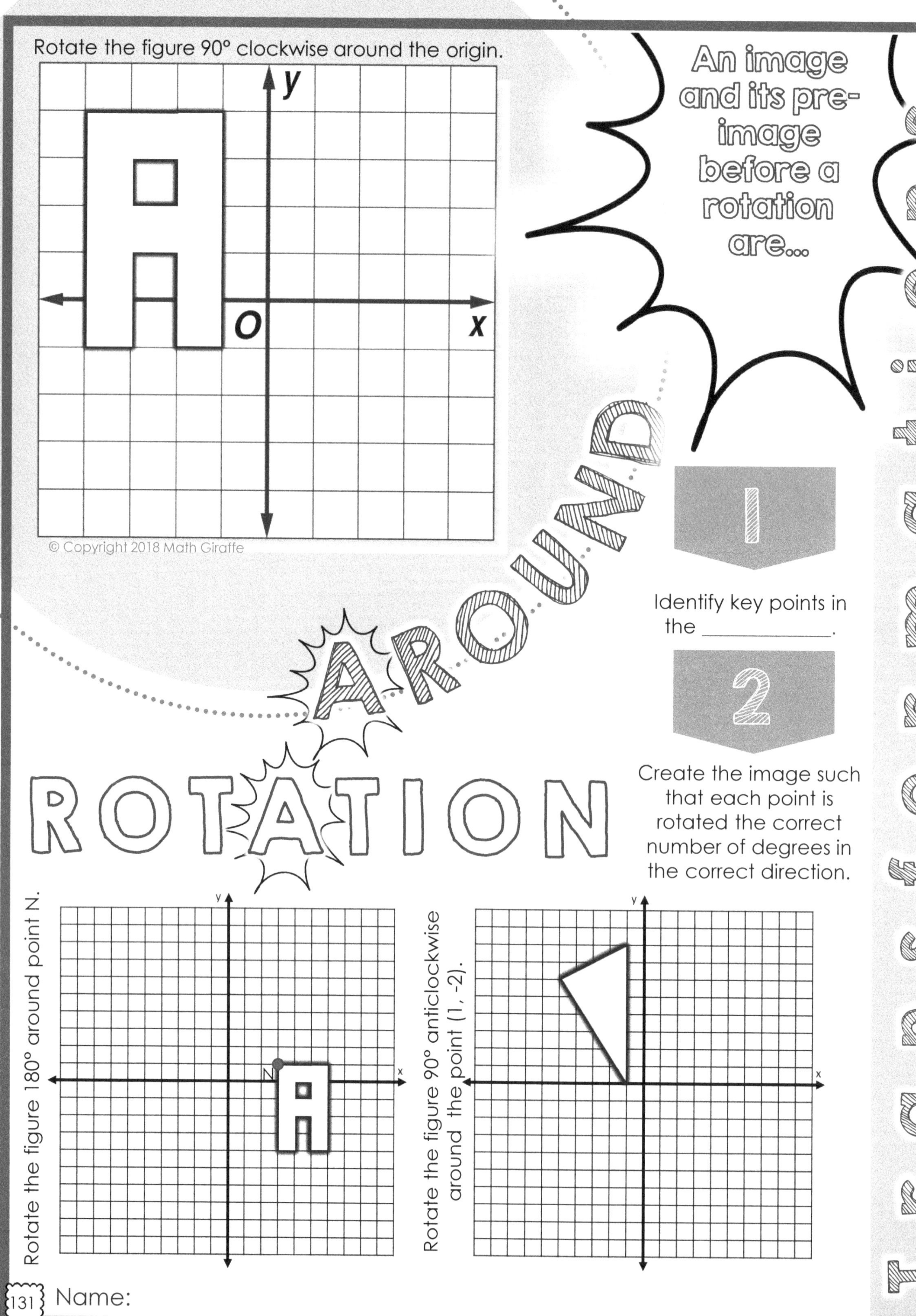
Rotate the figure 90° clockwise around the origin.
y
O
x
© Copyright 2018 Math Giraffe
An image and its pre-image before a rotation are...
Transformations
AROUND
ROTATION
1
Identify key points in the ______________.
2
Create the image such that each point is rotated the correct number of degrees in the correct direction.
Rotate the figure 180° around point N.
N
Rotate the figure 90° anticlockwise around the point (1, -2).
Name:

Name:

Translate the figure up 4 units and left 3 units.

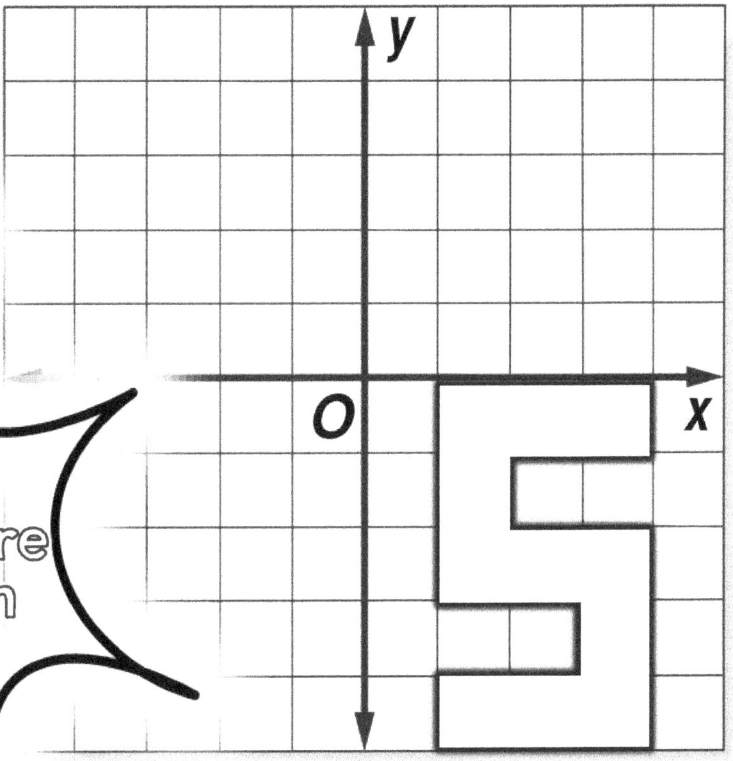

1

Identify key points in the ______.

2

Add or subtract from each coordinate as directed to get the resulting ______.

notation

An image and its pre-image before a translation are...

SLIDE

TRANSLATION

Use the translation (x, y) → (x + 11, y − 7)

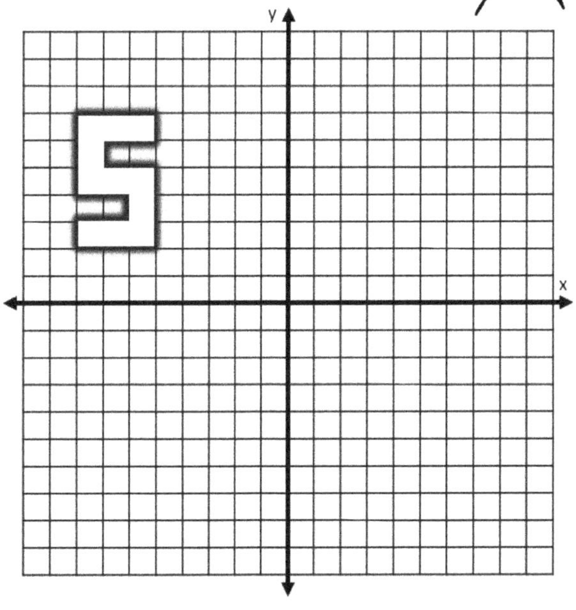

Use the translation (x, y) → (x, y + 5)

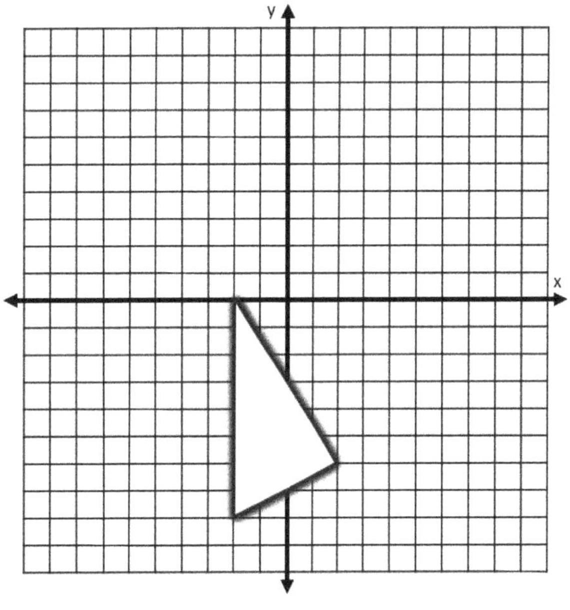

Transformations

Similar Figures
notation
Name:
Say it out loud!
Label and write congruent / equal statements.
K
F
E
baby clothes hanger
J
adult clothes hanger
L
△EFG ~ △JKL
properties
Corresponding angles are ...
Similar figures have the same...
but
Corresponding sides are ...
Try It
1
For each, determine whether the triangles are similar. If they are, write a similarity statement.
a
R
6
5
Q
6.5
S
2
P
2.2
8
T
b
N
10.4
B
3
6
M
A
8
7.8
C
4
L
The third angle pair must be congruent as well because of ...
2
The two sails on a sailboat are similar triangles. The large sail has sides that are 10 m, 26 m, and 24 m. The shortest side of the small sail is 6 m. Find the perimeter of the small sail.
© Copyright 2018 Math Giraffe

Scale drawings & models

272 ft. tall

define it

Notation:

sketch it

Write a...

The model of the tower that sits on a table is a 1 in : 32 ft scale representation of the actual tower. Find its dimensions and draw the model.

______ = ______

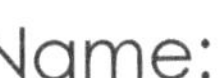

Name:

try it

While traveling on vacation, you fly 2400 miles. On the map, this distance is about 18 inches apart. Find the scale of the map.

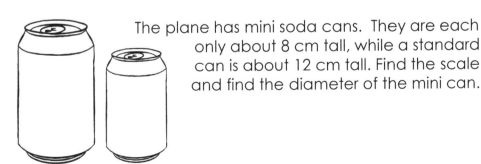

The plane has mini soda cans. They are each only about 8 cm tall, while a standard can is about 12 cm tall. Find the scale and find the diameter of the mini can.

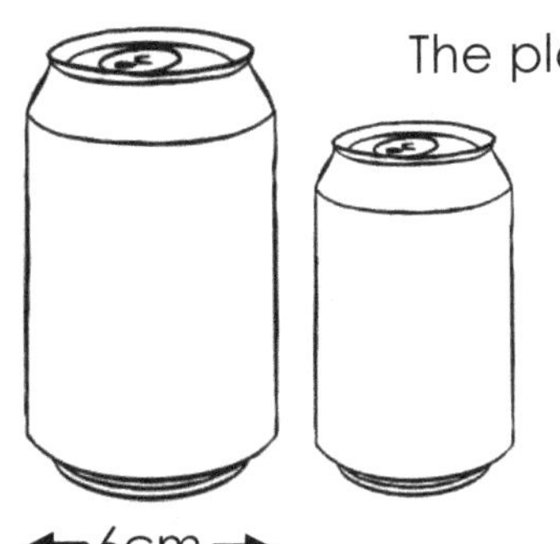

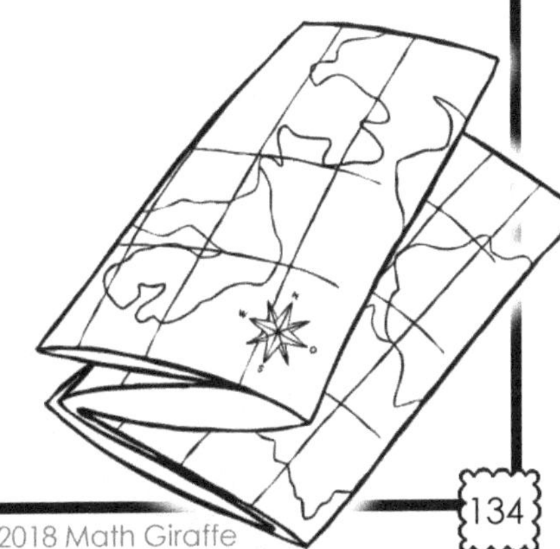

Name:

Use visual representations to find the area and the perimeter in each example.

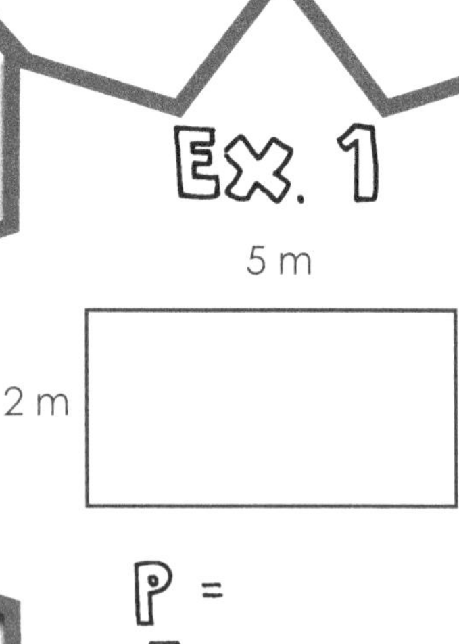

Ex. 1

5 m

2 m

P =

A =

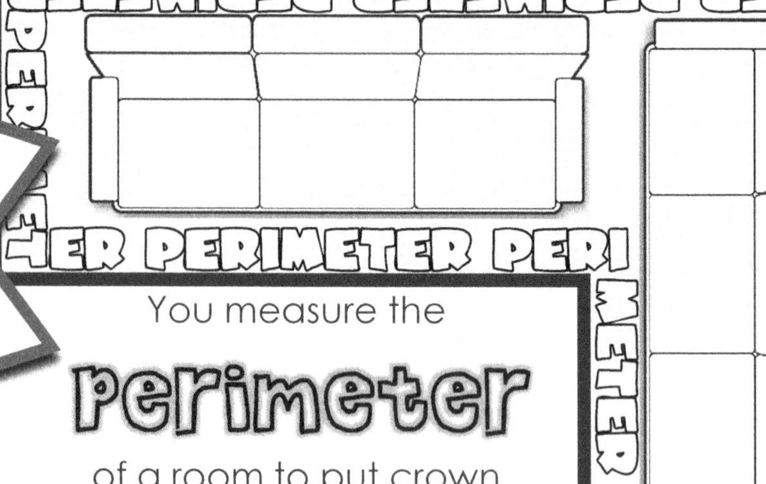

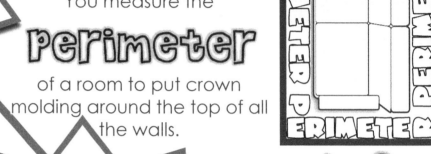

You measure the perimeter of a room to put crown molding around the top of all the walls.

Area is measured in ______ units.

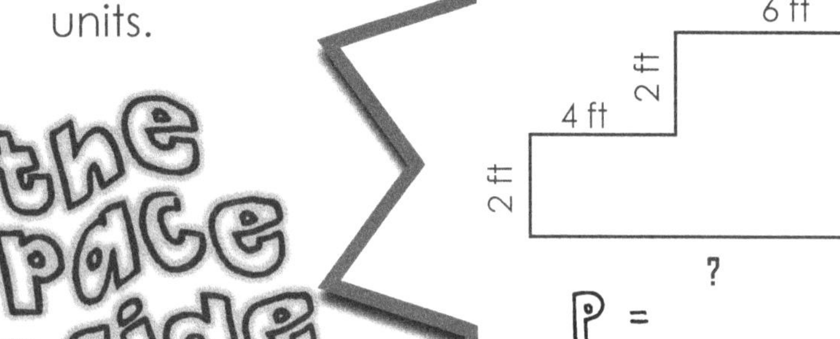

Ex. 2

6 ft

2 ft

4 ft

2 ft

?

?

P =

A =

Perimeter is measured in ______ units.

You measure the  area of a room to install tile on the entire floor.

Ex. 3

3 cm

3 cm

P =

A =

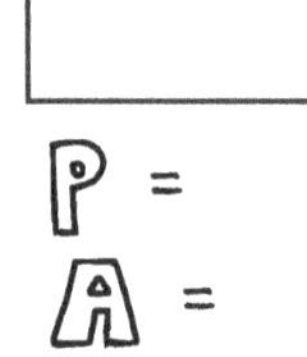

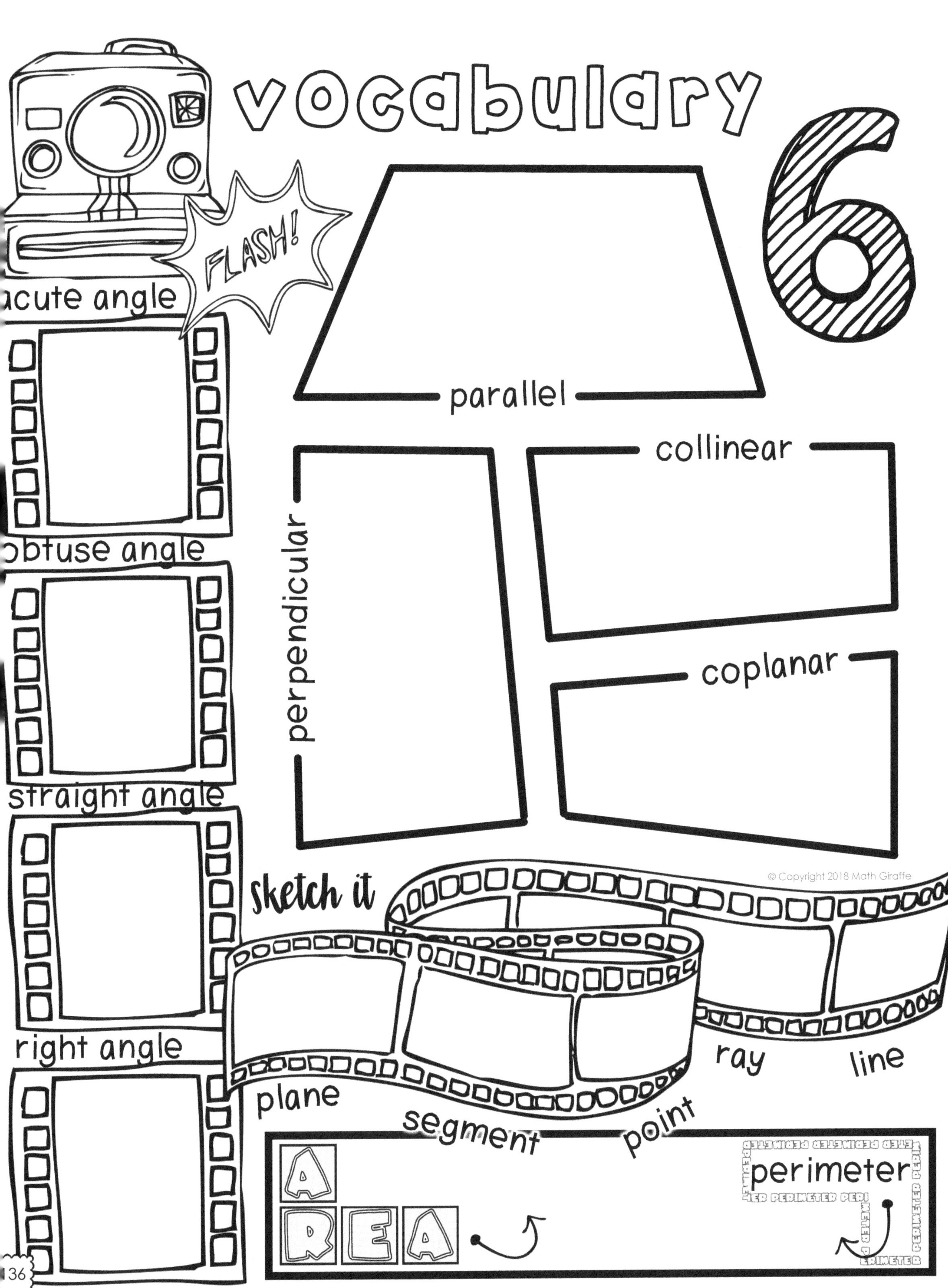
vocabulary
6
FLASH!
acute angle
obtuse angle
straight angle
right angle
parallel
perpendicular
collinear
coplanar
sketch it
plane
segment
point
ray
line
AREA
perimeter

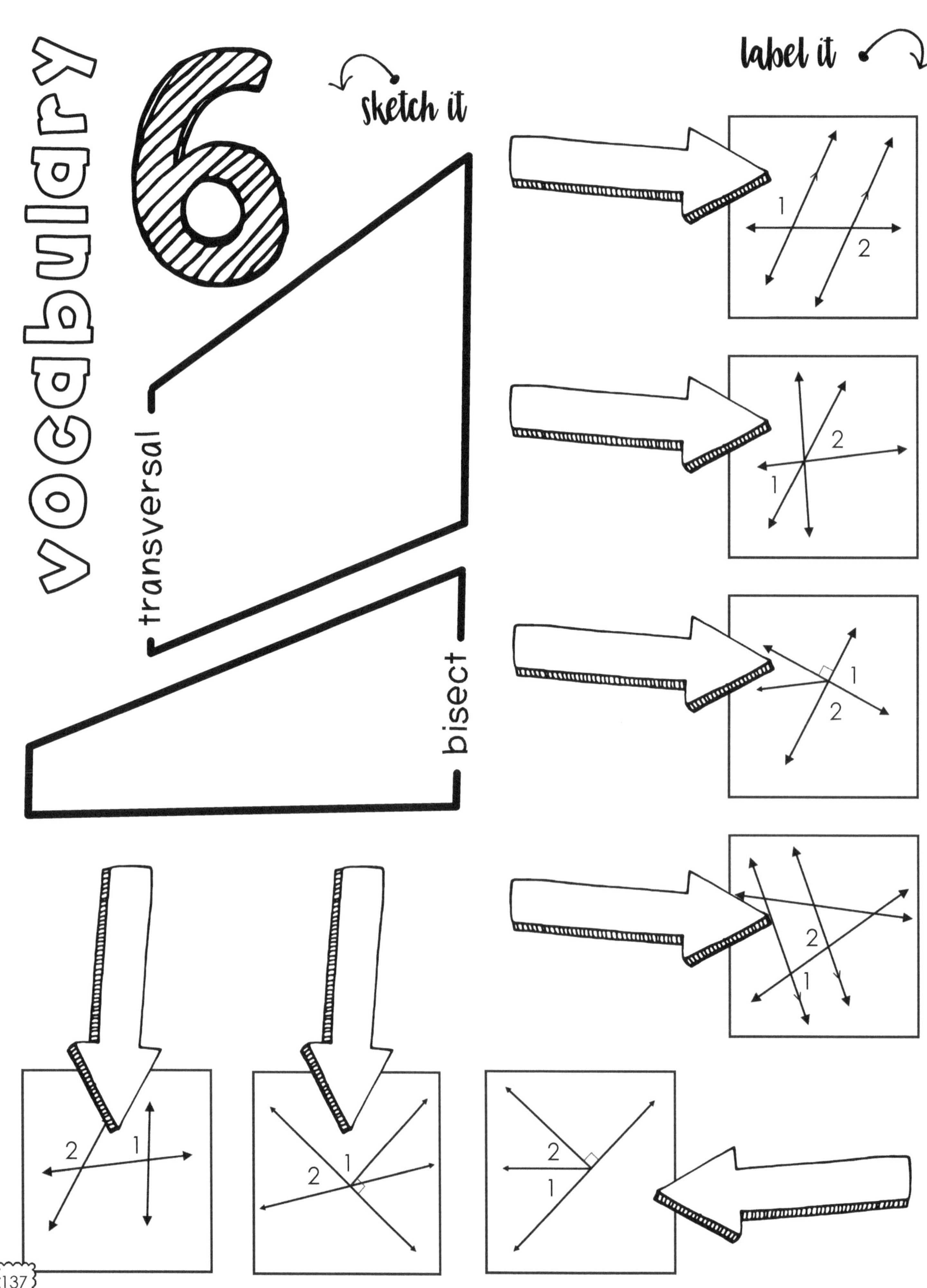

vocabulary
6
sketch it
label it
transversal
bisect
1
2
1
2
1
2
1
2
2
1
2
1
2
1
2
1

vocabulary
polygon
6
regular
by sides
by angles
types of triangles
sketch it
triangle sum theorem
show it
congruent

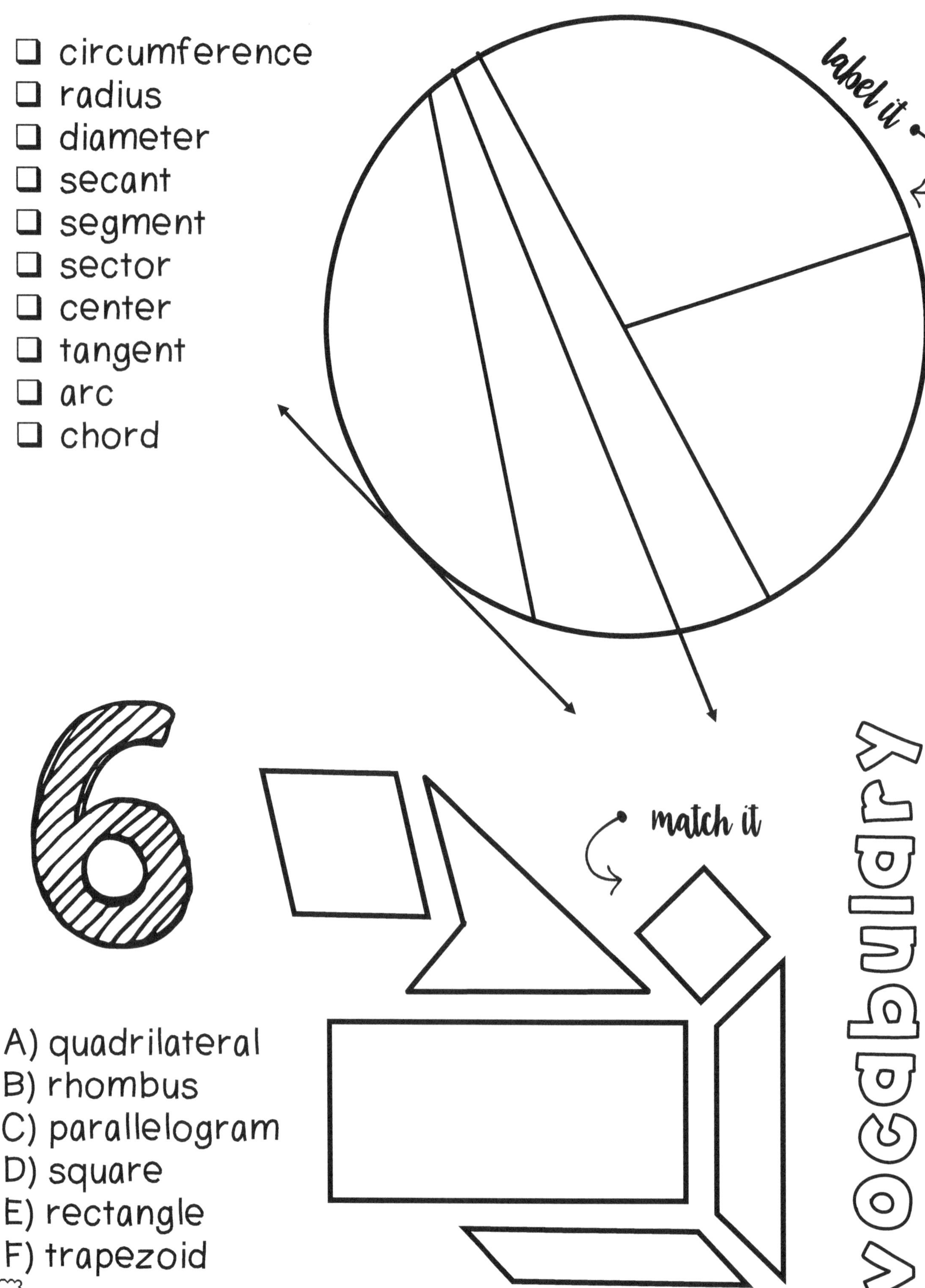
circumference
radius
diameter
secant
segment
sector
center
tangent
arc
chord
label it
6
vocabulary
match it
A) quadrilateral
B) rhombus
C) parallelogram
D) square
E) rectangle
F) trapezoid

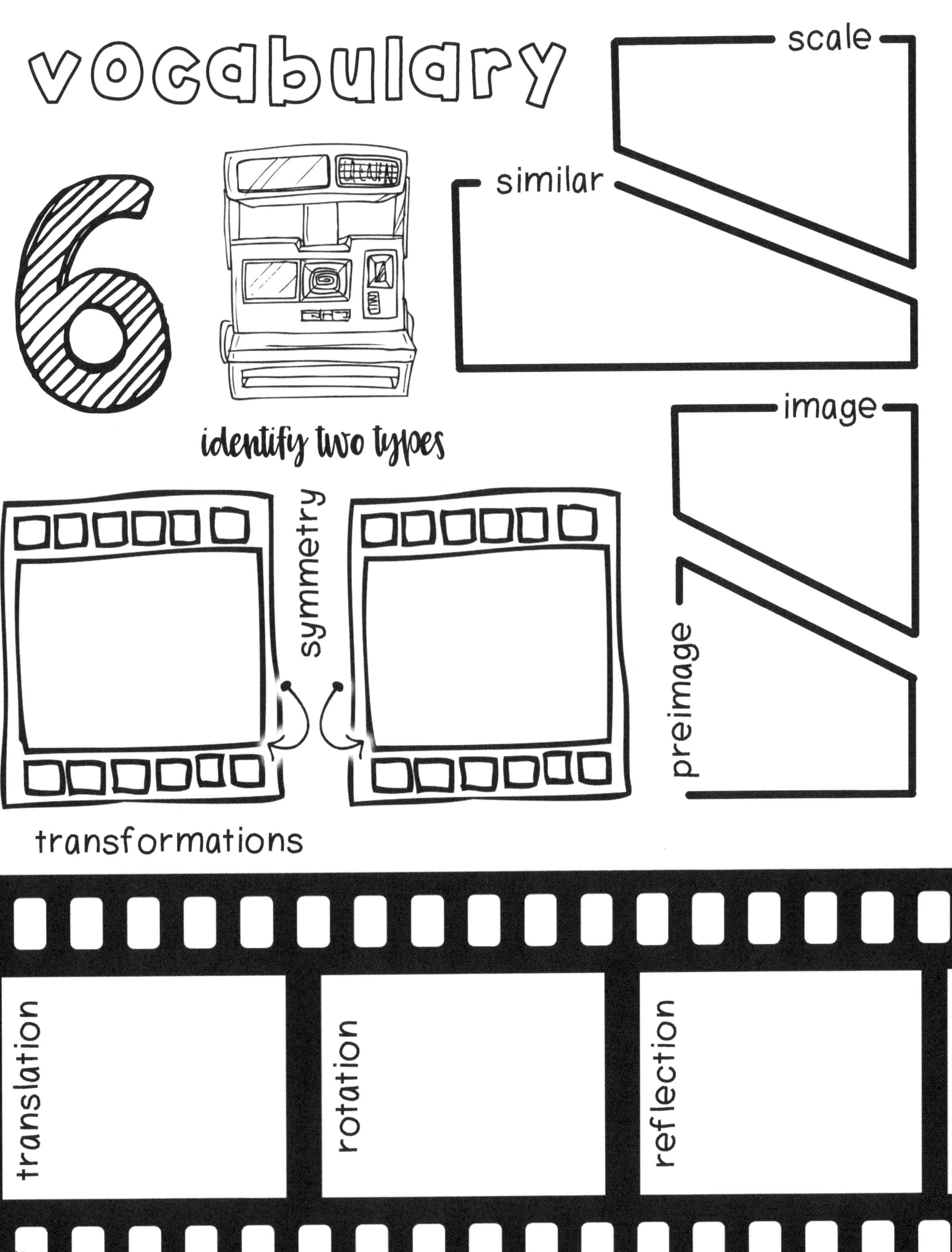
vocabulary
6
scale
similar
identify two types
symmetry
image
preimage
transformations
translation
rotation
reflection

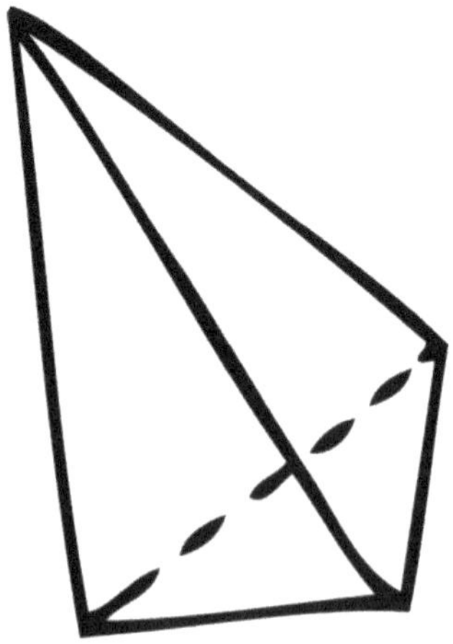

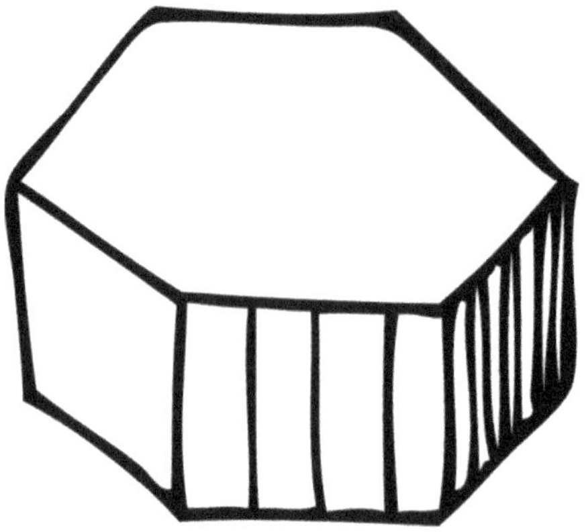

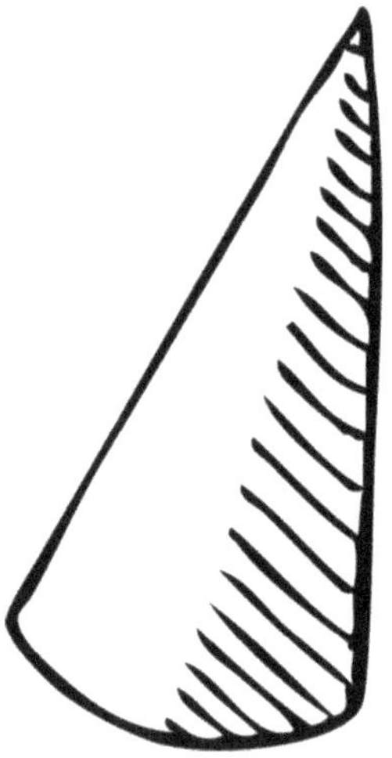

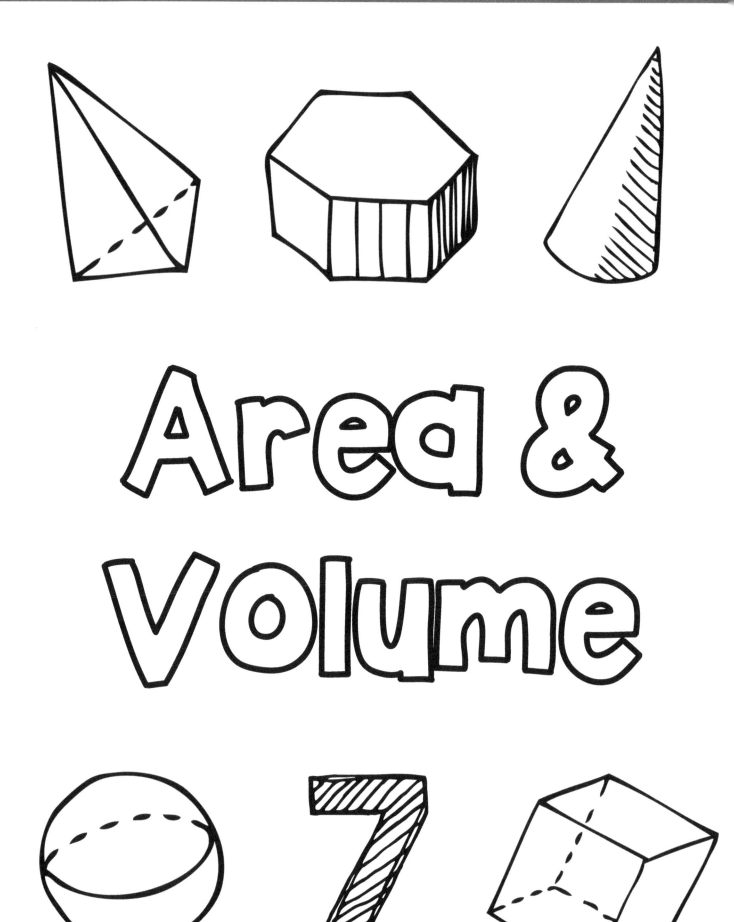

Area & Volume

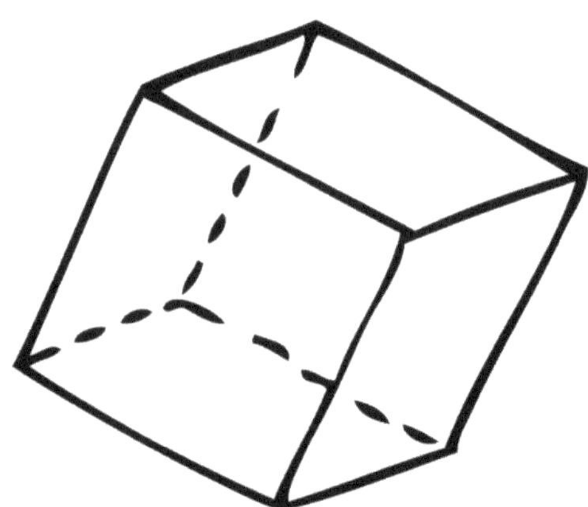

Name:
height
base
The base and the height must meet...
Label the base and height for each quadrilateral.
If the rectangle above has a length of 18 in. and a width of 11 in., what is its area?
formula
area of a parallelogram
label it
try it
Find the area:
3.5 cm
5 cm
What about units?
but...
What is going on here?
why
show it
altitude
Go back up to the shapes in the magnifying glass. Can you represent this reasoning there as well with additional sketches?

areas of composite figures practice

Find the total area for each figure.

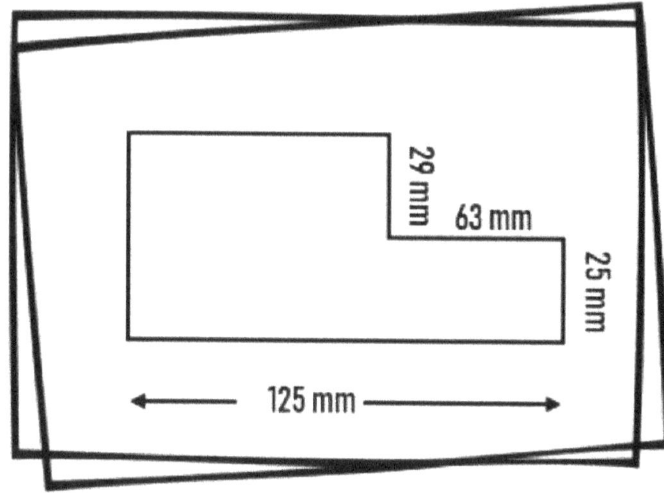

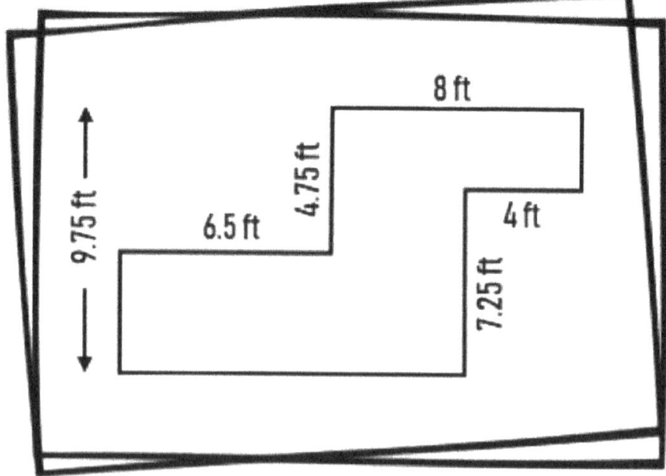

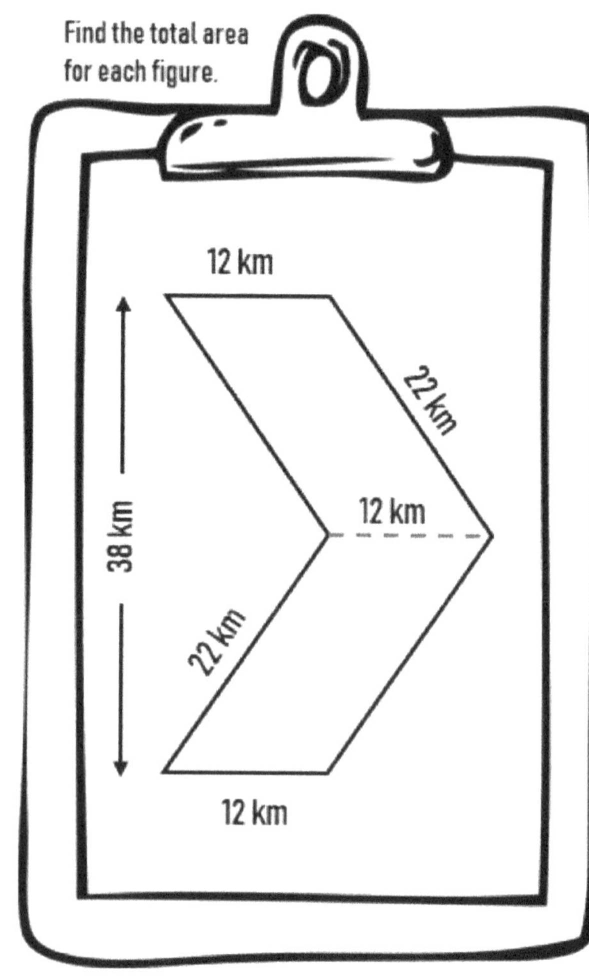

Name:

Find the area of a parallelogram with a base of 5 inches and height of 3 inches with a 1.5 inch square cut out of it.

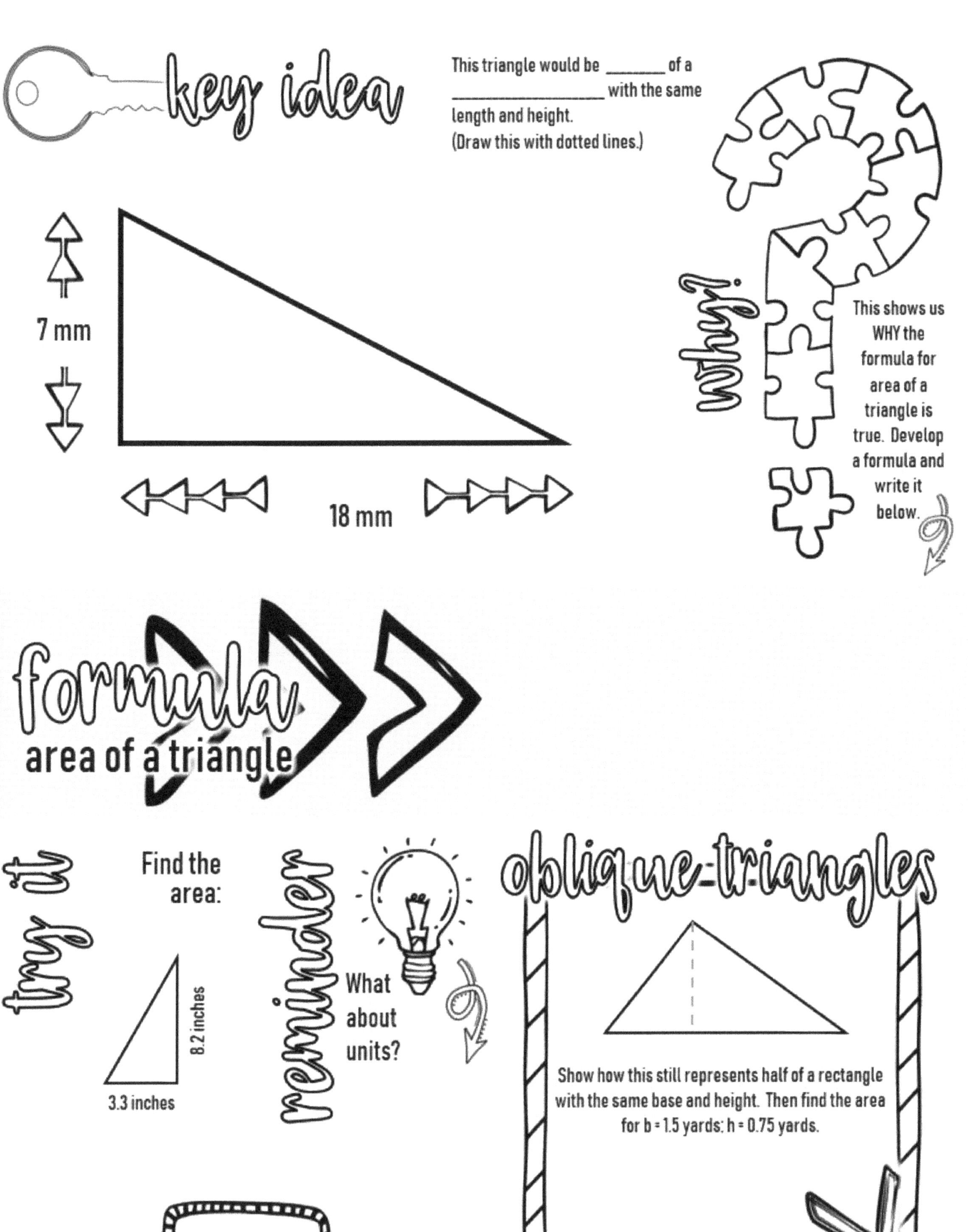

key idea
This triangle would be ________ of a ____________________ with the same length and height.
(Draw this with dotted lines.)
7 mm
18 mm
why?
This shows us WHY the formula for area of a triangle is true. Develop a formula and write it below.
formula
area of a triangle
try it
Find the area:
8.2 inches
3.3 inches
reminder
What about units?
oblique triangles
Show how this still represents half of a rectangle with the same base and height. Then find the area for b = 1.5 yards; h = 0.75 yards.
Does it matter in what order we multiply? (In which step we divide by 2 or multiply by ½?)
Name:

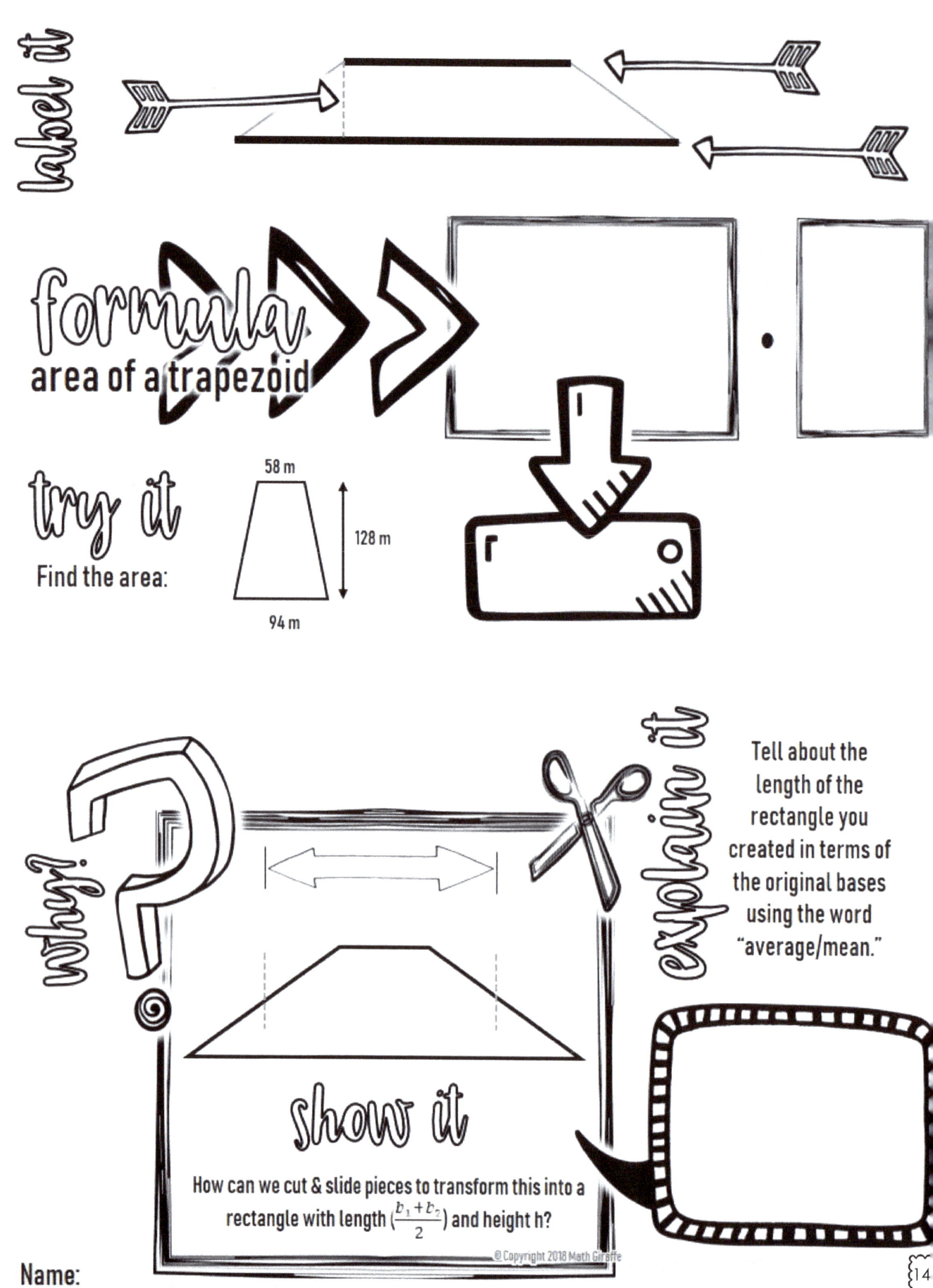
label it
formula
area of a trapezoid
try it
Find the area:
58 m
128 m
94 m
why?
explain it
Tell about the length of the rectangle you created in terms of the original bases using the word "average/mean."
show it
How can we cut & slide pieces to transform this into a rectangle with length ($\frac{b_1+b_2}{2}$) and height h?
Name:

areas of triangles & quadrilaterals

practice

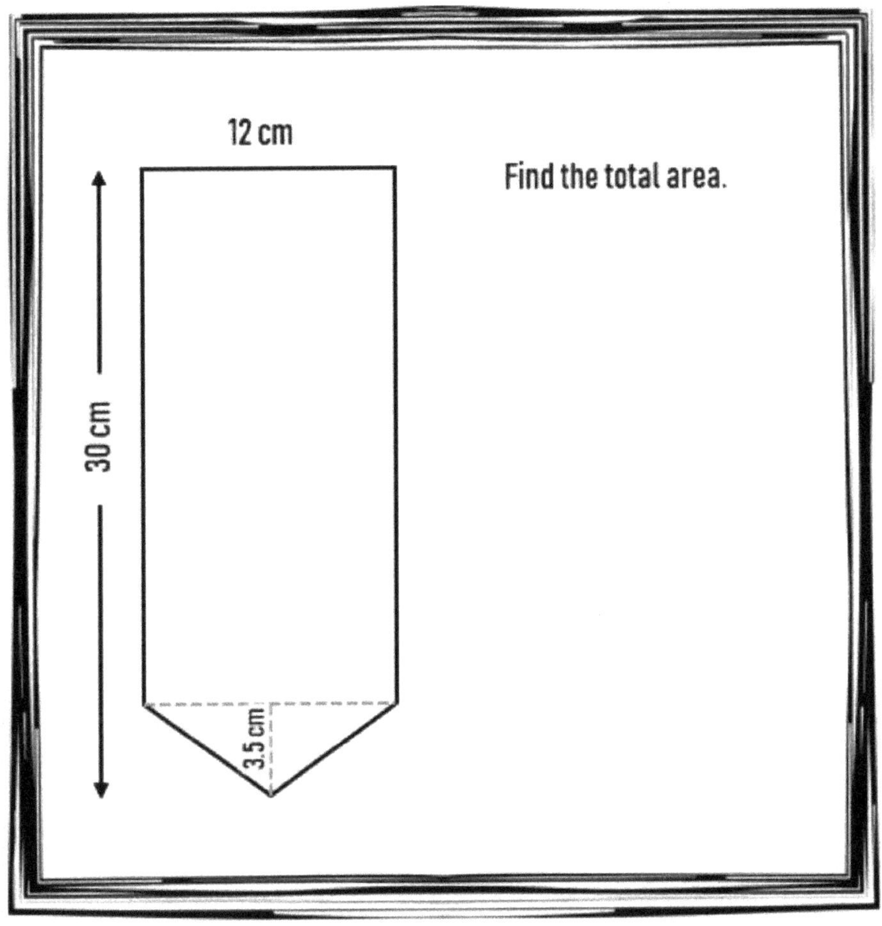

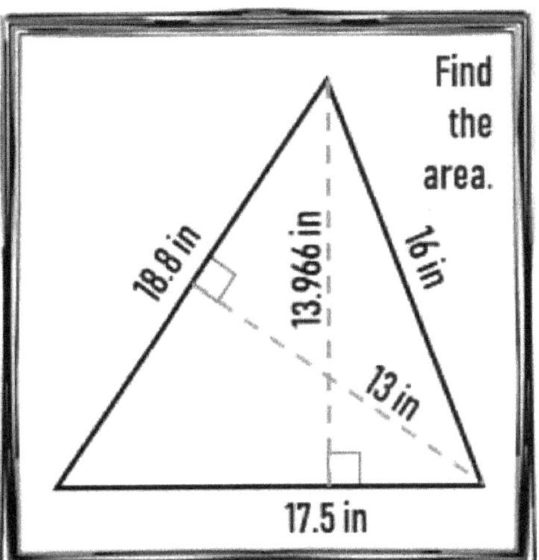

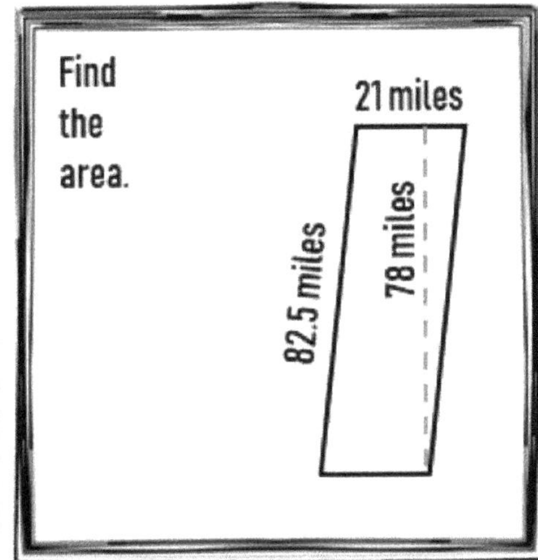

sketch it

Find the area of a trapezoid with bases of 340 feet and 210 feet and a height of 88 INCHES (Be careful about units!)

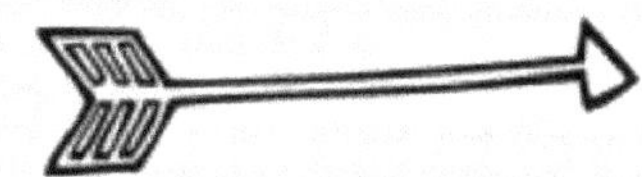

Name:

Name:

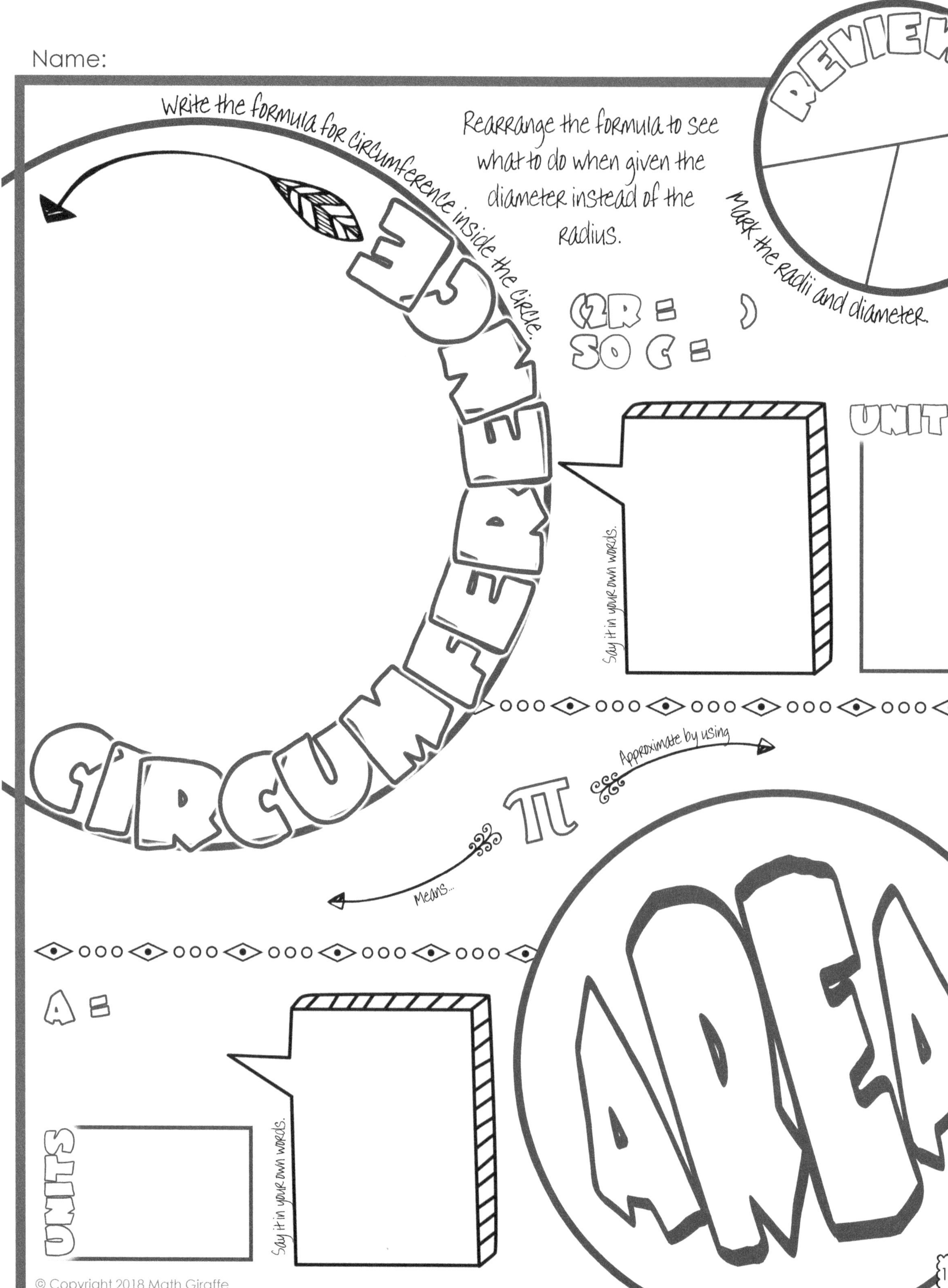

Write the formula for circumference inside the circle.
Rearrange the formula to see what to do when given the diameter instead of the radius.
(2R =)
SO C =
REVIEW
Mark the radii and diameter.
CIRCUMFERENCE
Say it in your own words.
UNIT
Approximate by using
π
Means...
A =
Say it in your own words.
UNITS
AREA

Name:

AREA AND CIRCUMFERENCE

Find the area and circumference of each circle.

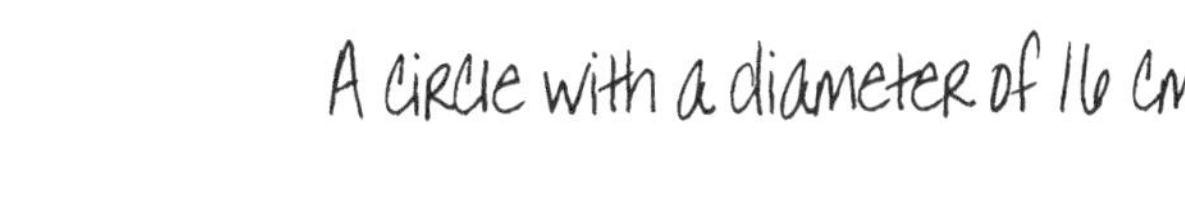

A circle with a radius of 4.8 inches

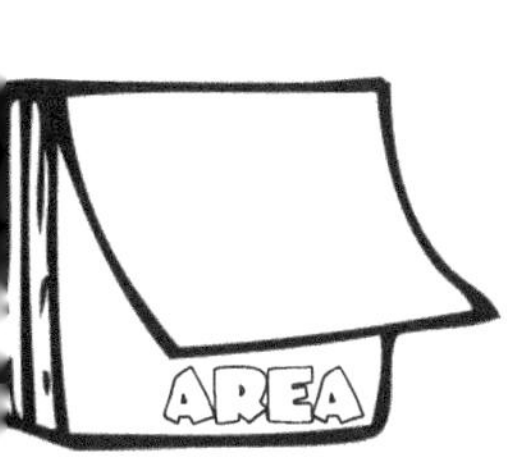

In your own words...

Describe how you'd find the area of the circular donut shape (without including the cut out hole).

Name:

Describe how you'd find the area of square mat in this picture frame (without including the square that is cut out for a photo).

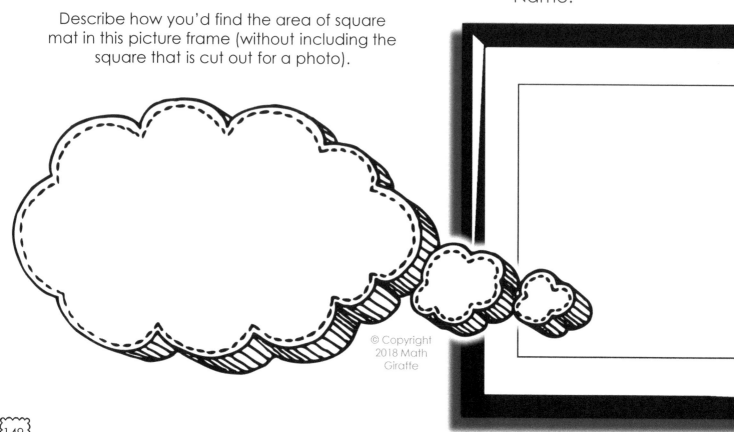

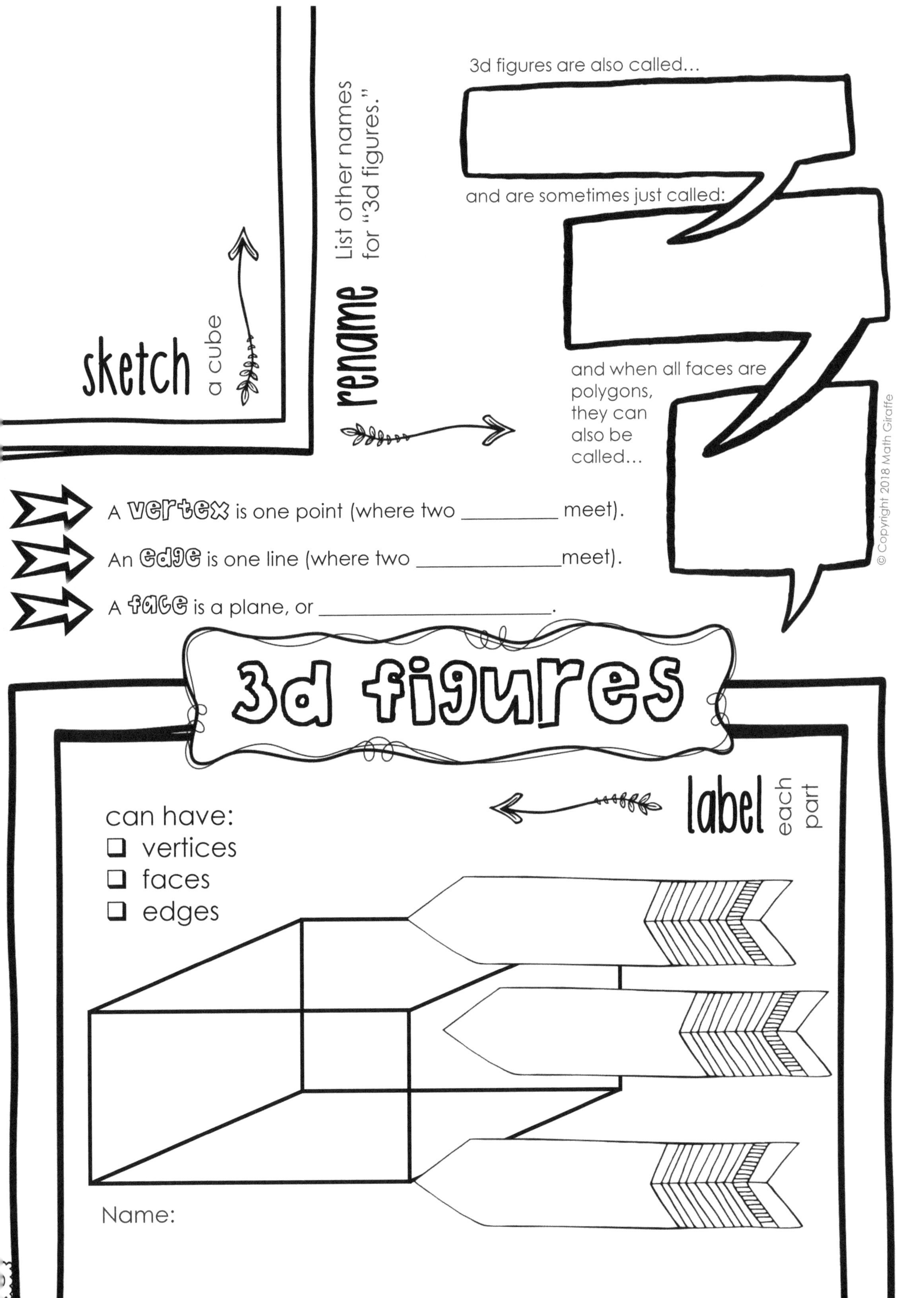
sketch
a cube
rename
List other names for "3d figures."
3d figures are also called…
and are sometimes just called:
and when all faces are polygons, they can also be called…
© Copyright 2018 Math Giraffe
A vertex is one point (where two __________ meet).
An edge is one line (where two _____________meet).
A face is a plane, or _______________________.
3d figures
label
each part
can have:
❑ vertices
❑ faces
❑ edges
Name:

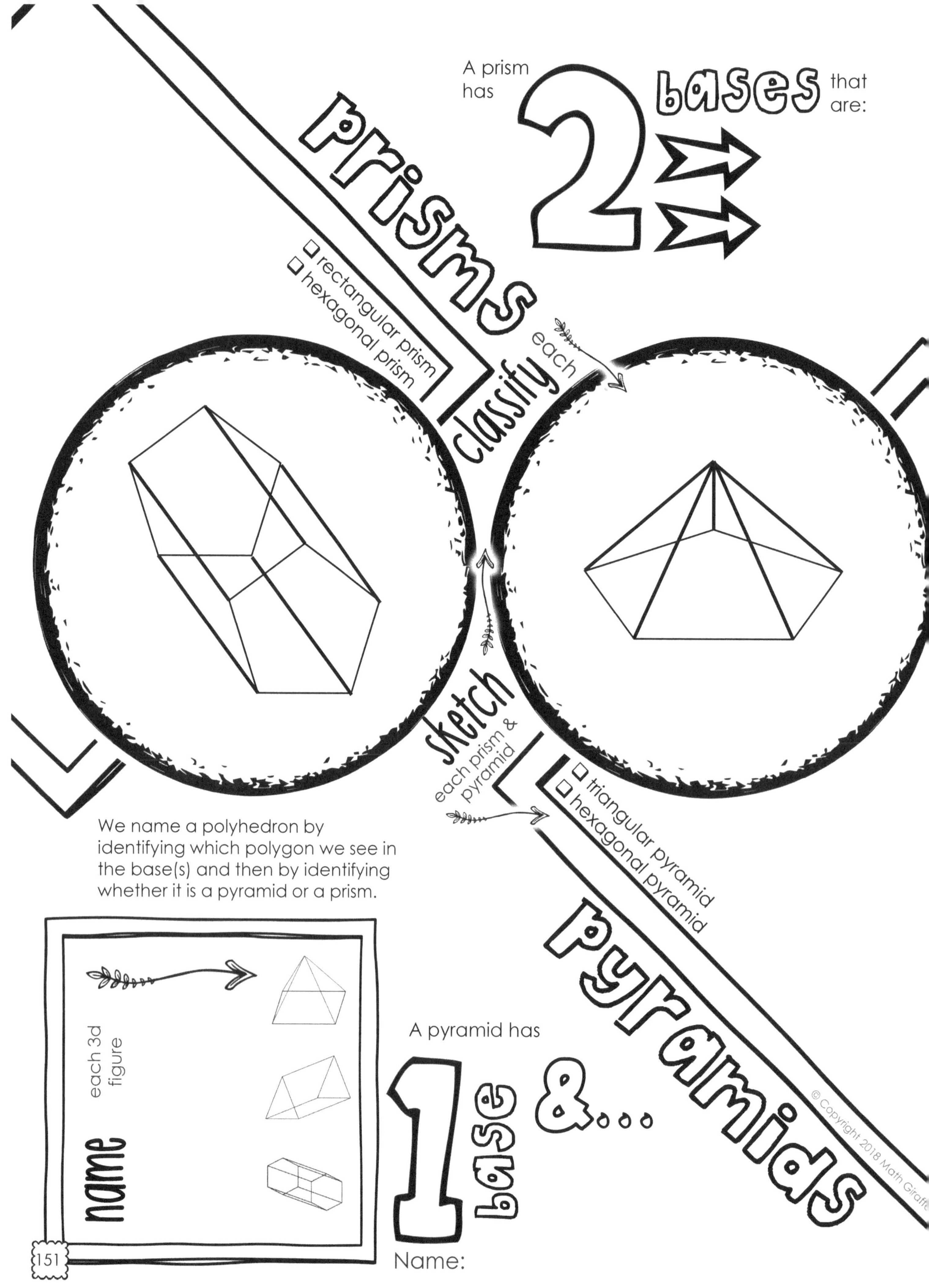
A prism has
2
bases that are:
prisms
rectangular prism
hexagonal prism
classify each
sketch each prism & pyramid
triangular pyramid
hexagonal pyramid
We name a polyhedron by identifying which polygon we see in the base(s) and then by identifying whether it is a pyramid or a prism.
name each 3d figure
A pyramid has
1
base
&...
pyramids
© Copyright 2018 Math Giraffe
151
Name:

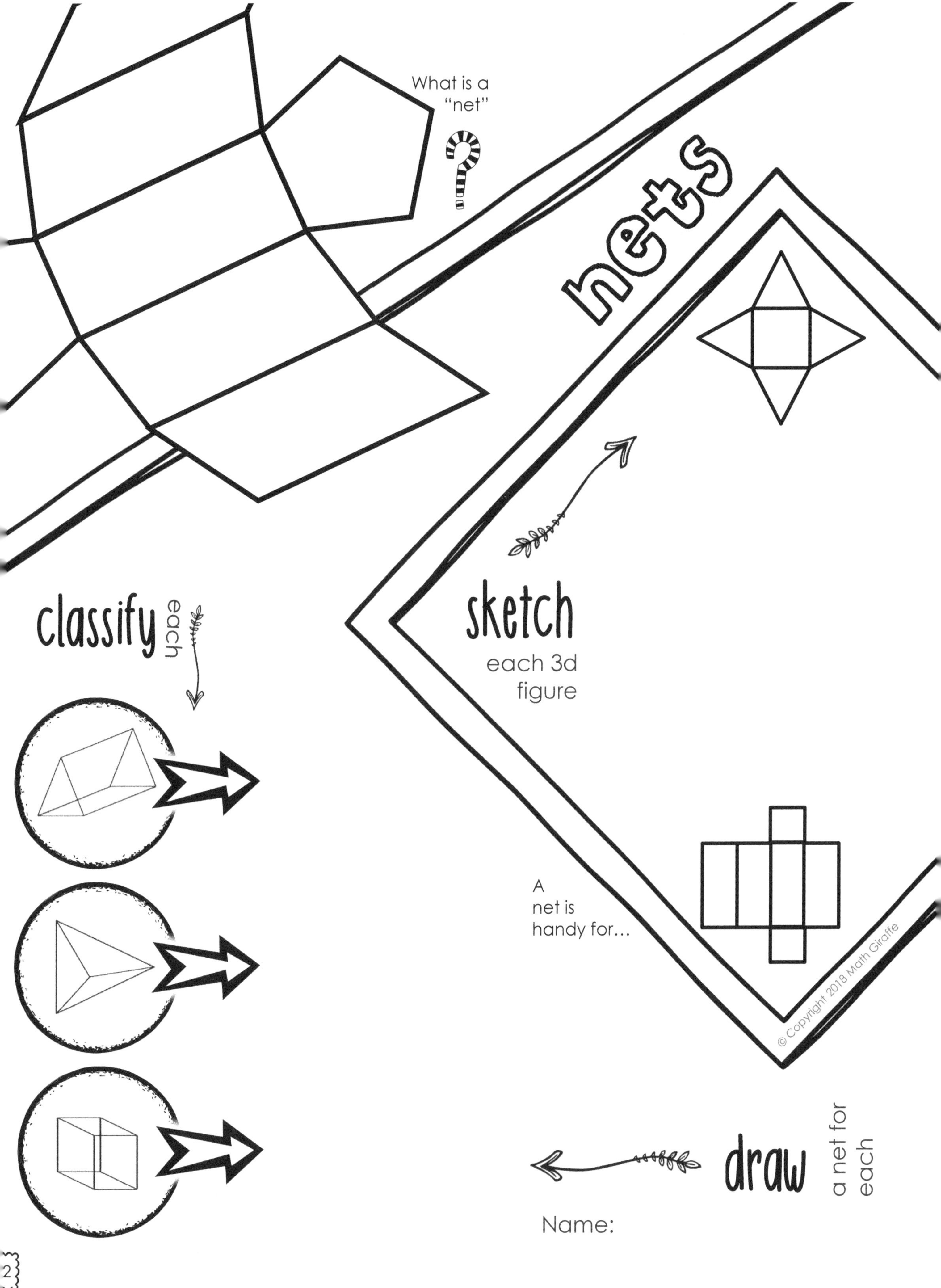
What is a "net"
?
nets
classify each
sketch each 3d figure
A net is handy for...
draw a net for each
Name:
© Copyright 2018 Math Giraffe
2

Name:

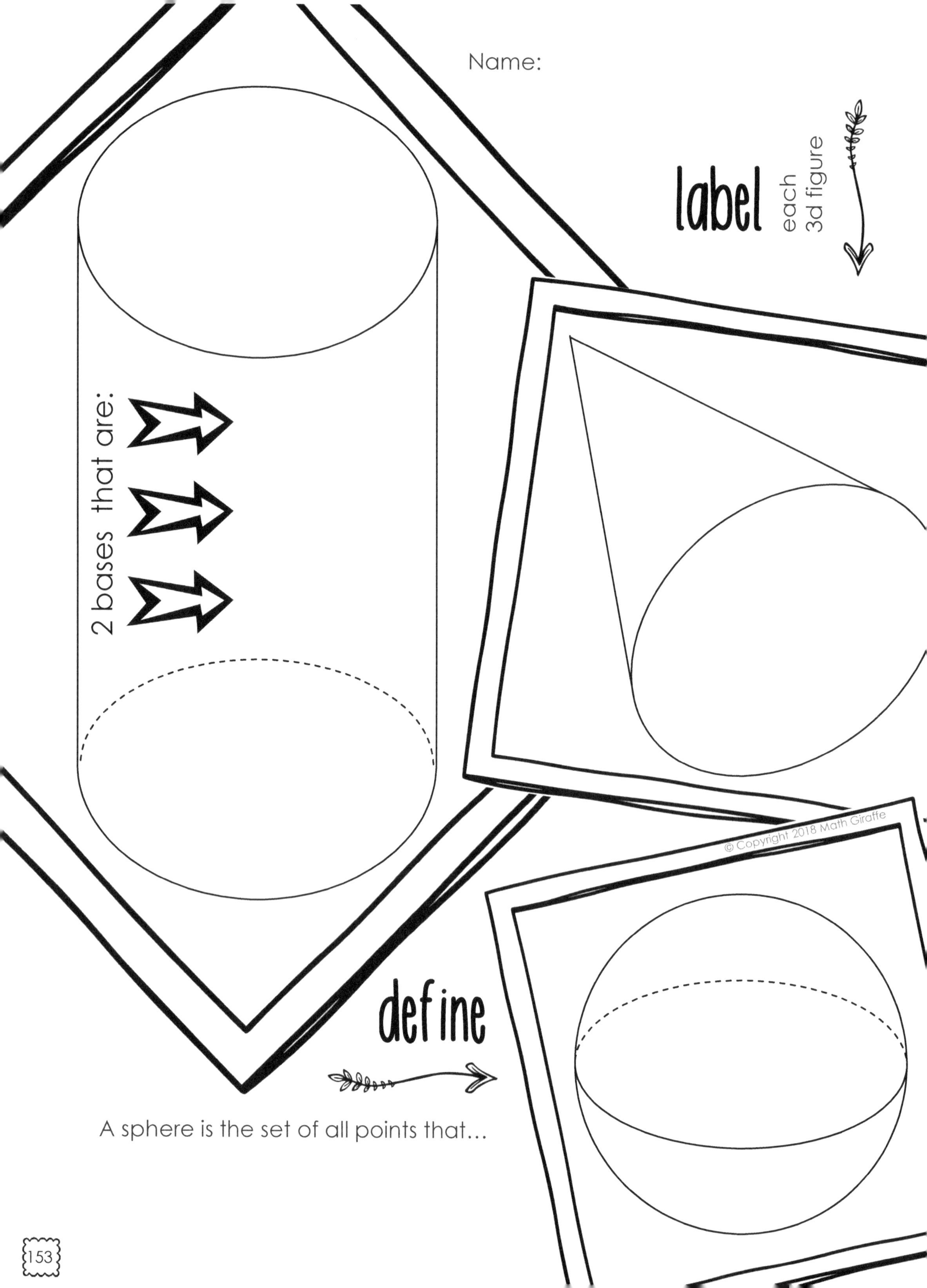

A sphere is the set of all points that...

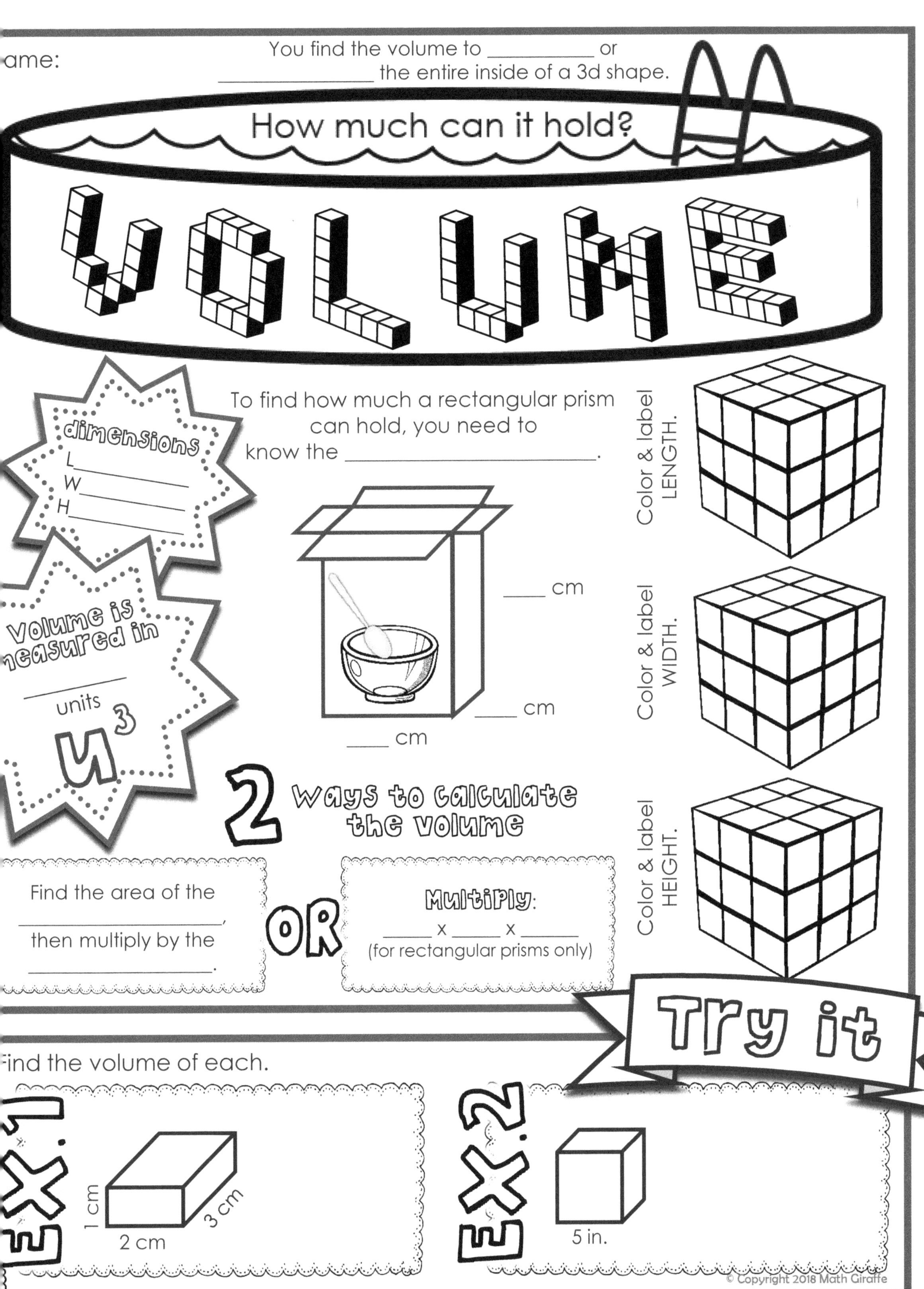
ame:
You find the volume to __________ or
_______________ the entire inside of a 3d shape.
How much can it hold?
VOLUME
dimensions
L________
W________
H________
To find how much a rectangular prism can hold, you need to know the _______________________.
Color & label LENGTH.
Color & label WIDTH.
Color & label HEIGHT.
____ cm
____ cm
____ cm
Volume is measured in
__________ units
u^3
2 ways to calculate the volume
Find the area of the
__________________,
then multiply by the
__________________.
OR
Multiply:
_____ x _____ x ______
(for rectangular prisms only)
Try it
Find the volume of each.
EX.1
1 cm
3 cm
2 cm
EX.2
5 in.

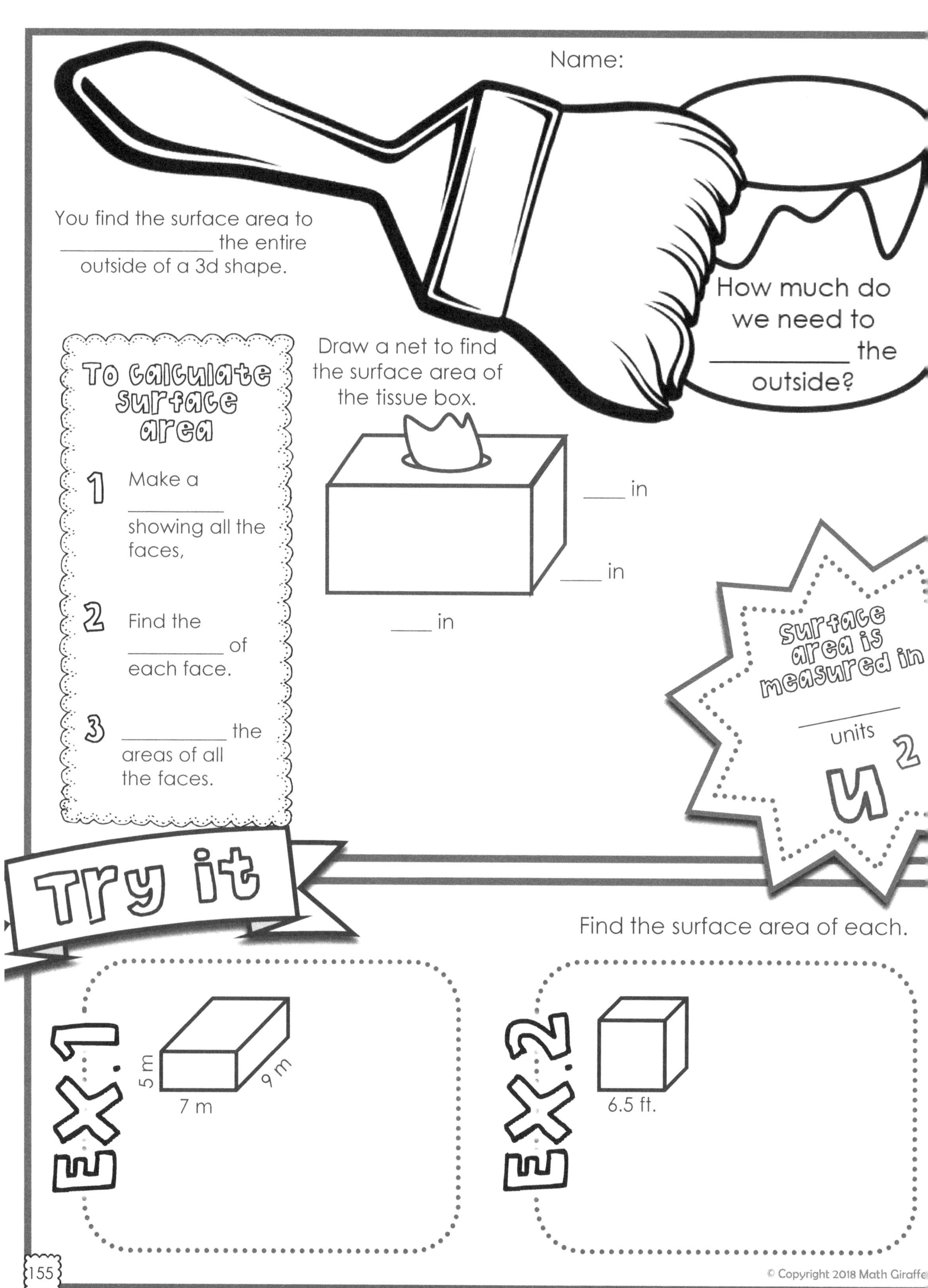
Name:
You find the surface area to _______________ the entire outside of a 3d shape.
How much do we need to __________ the outside?
To calculate surface area
1 Make a __________ showing all the faces,
2 Find the __________ of each face.
3 __________ the areas of all the faces.
Draw a net to find the surface area of the tissue box.
____ in
____ in
____ in
Surface area is measured in __________ units
u 2
Try it
Find the surface area of each.
Ex.1
5 m
9 m
7 m
Ex.2
6.5 ft.
155
© Copyright 2018 Math Giraffe

units of measure

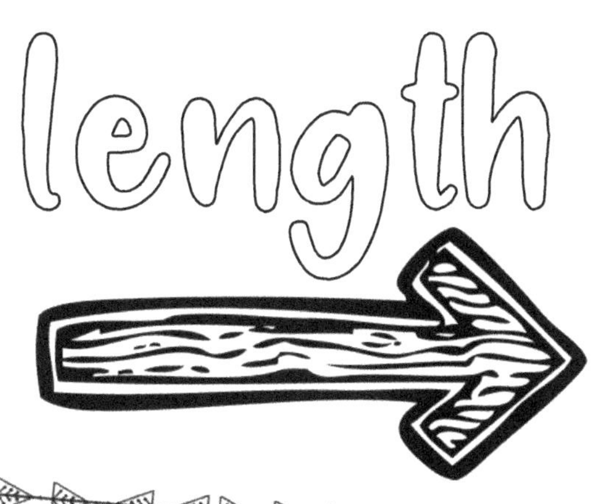

is measured in
_______________ units.

Name:

is measured in
____
units.

is measured in
_______________ units.

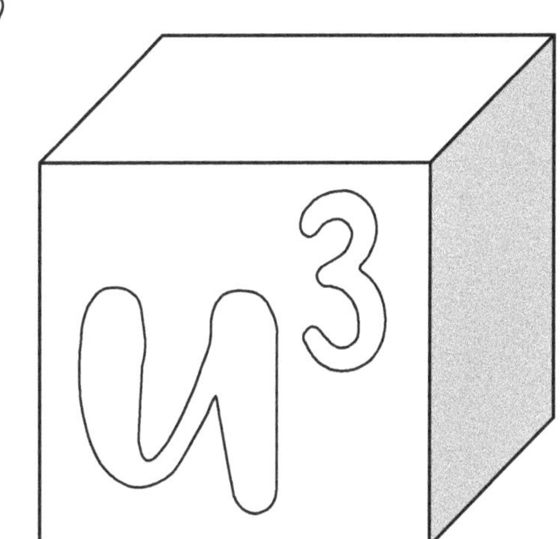

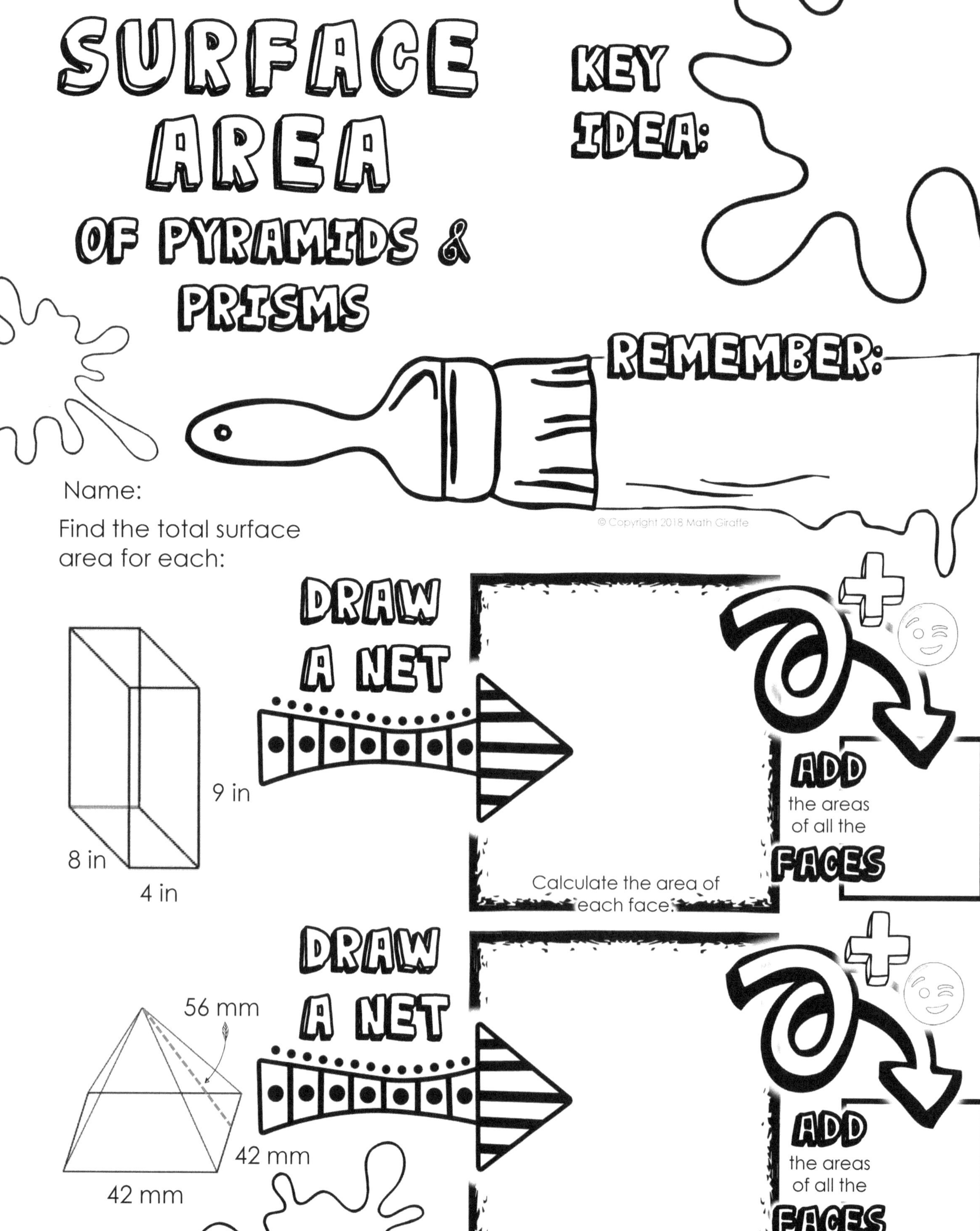
SURFACE AREA OF PYRAMIDS & PRISMS
KEY IDEA:
REMEMBER:
Name:
Find the total surface area for each:
DRAW A NET
9 in
8 in
4 in
Calculate the area of each face.
ADD the areas of all the FACES
DRAW A NET
56 mm
42 mm
42 mm
Calculate the area of each face.
ADD the areas of all the FACES

"LATERAL" SURFACE AREA

means just the "side" surfaces (includes everything EXCEPT the ___________).

Shade / color the faces that would NOT be included in the lateral surface area.

CYLINDER

LABEL IT:

Identify the radius, circumference and the height of the cylinder.

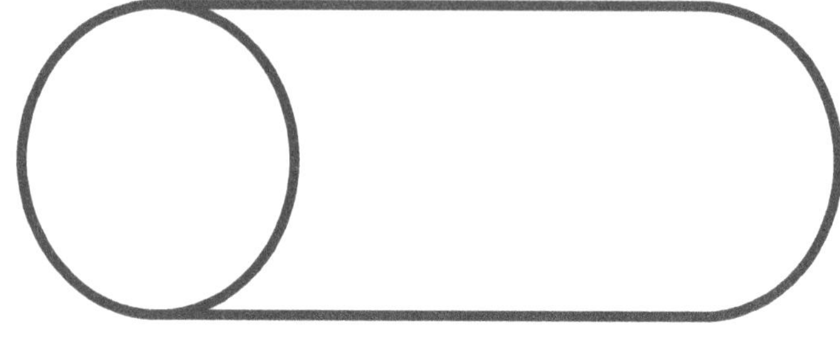

"UNROLL" IT

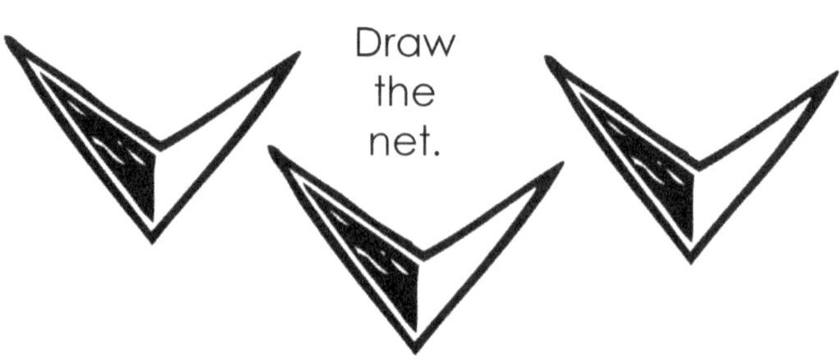

Reminder: Circumference

Transfer your labels for circumference and height EVERYWHERE they apply. Be careful!

Name:

SURFACE AREA

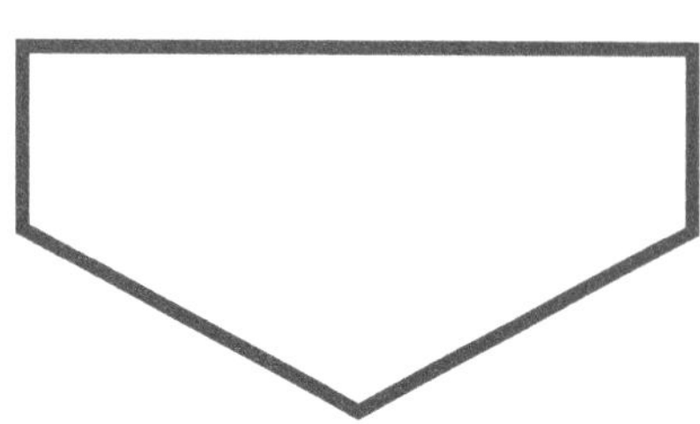

CONE

LABEL IT: Identify the radius, circumference and the slant height of the cone.

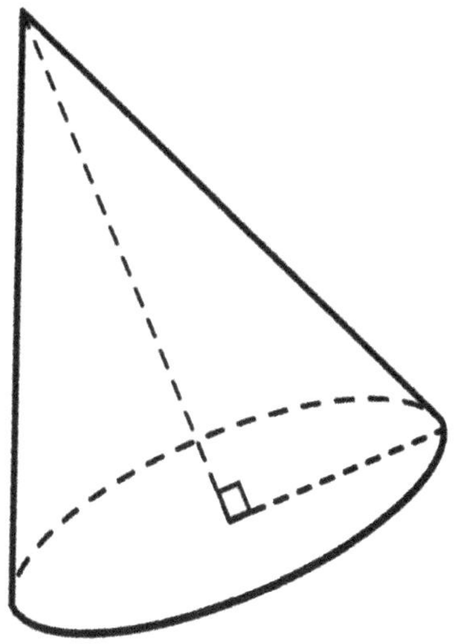

Draw the net.

"UNROLL" IT

SURFACE AREA

S.A.= [ ] + [ ]

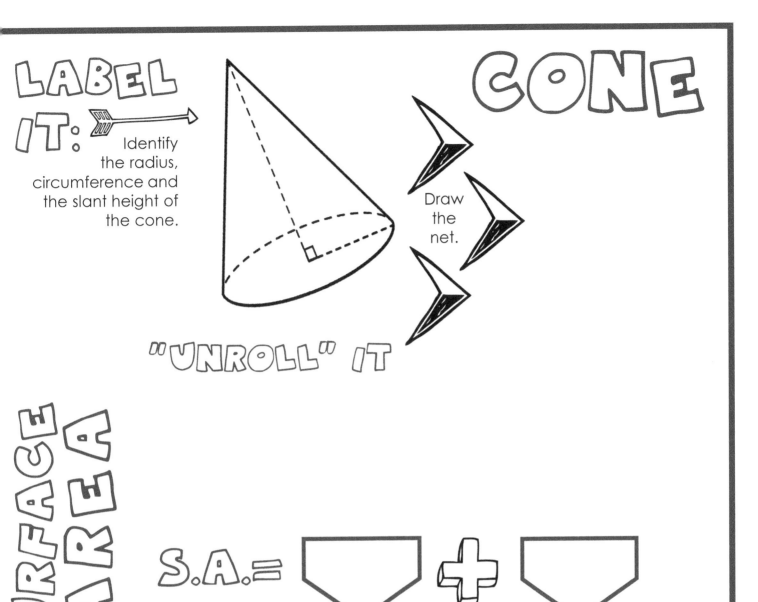

Name:

TRY IT

Name:

Find the surface area of each figure.

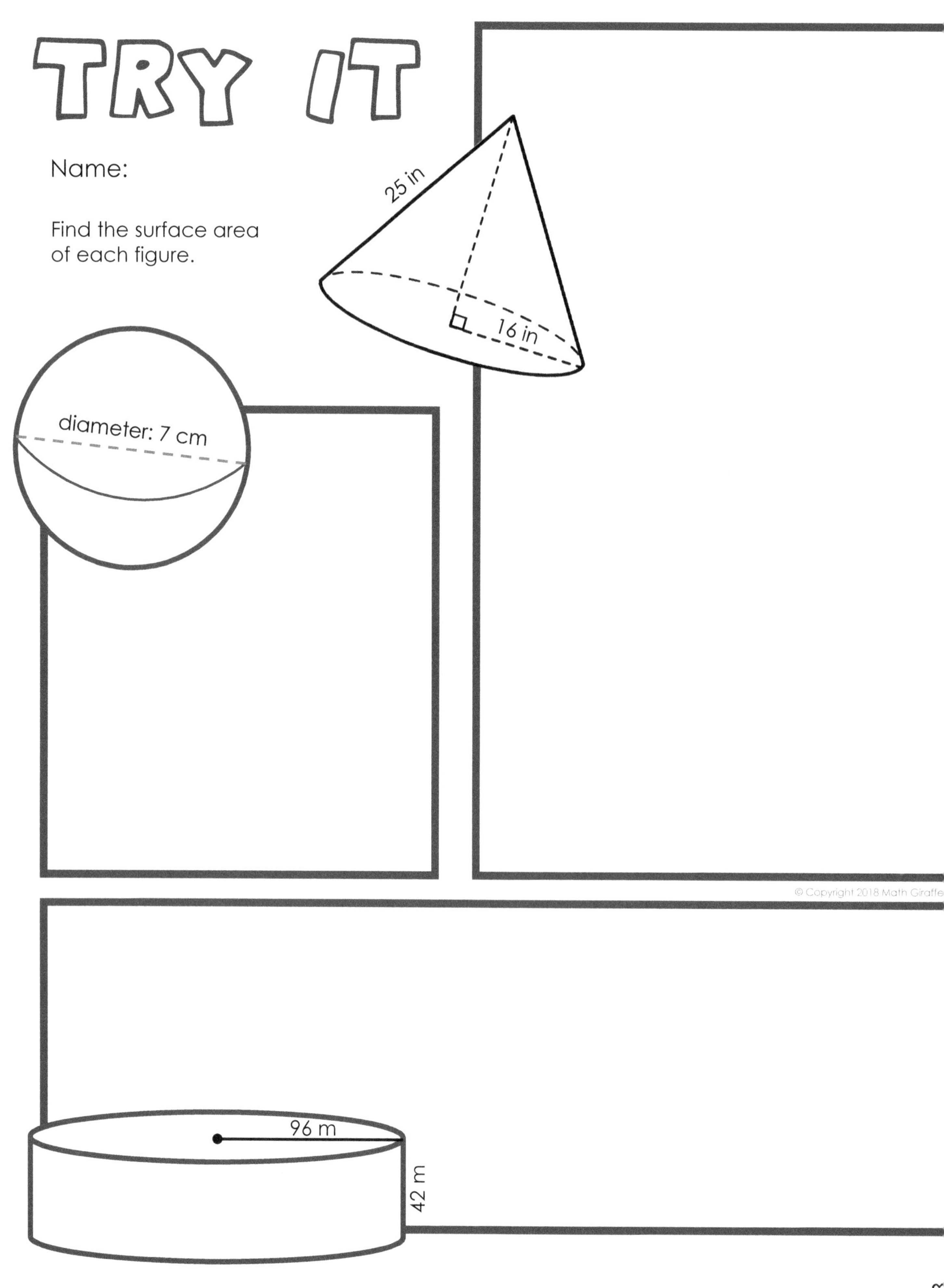

Name:
sketch it
Draw and identify the base: a rectangular prism with dimensions of 4m x 8m x 2.5m:
volume
for prisms & cylinders:
try it
Find the volume of the rectangular prism.
units
height: 18 in
diameter: 15 in
© Copyright 2018 Math Giraffe
162

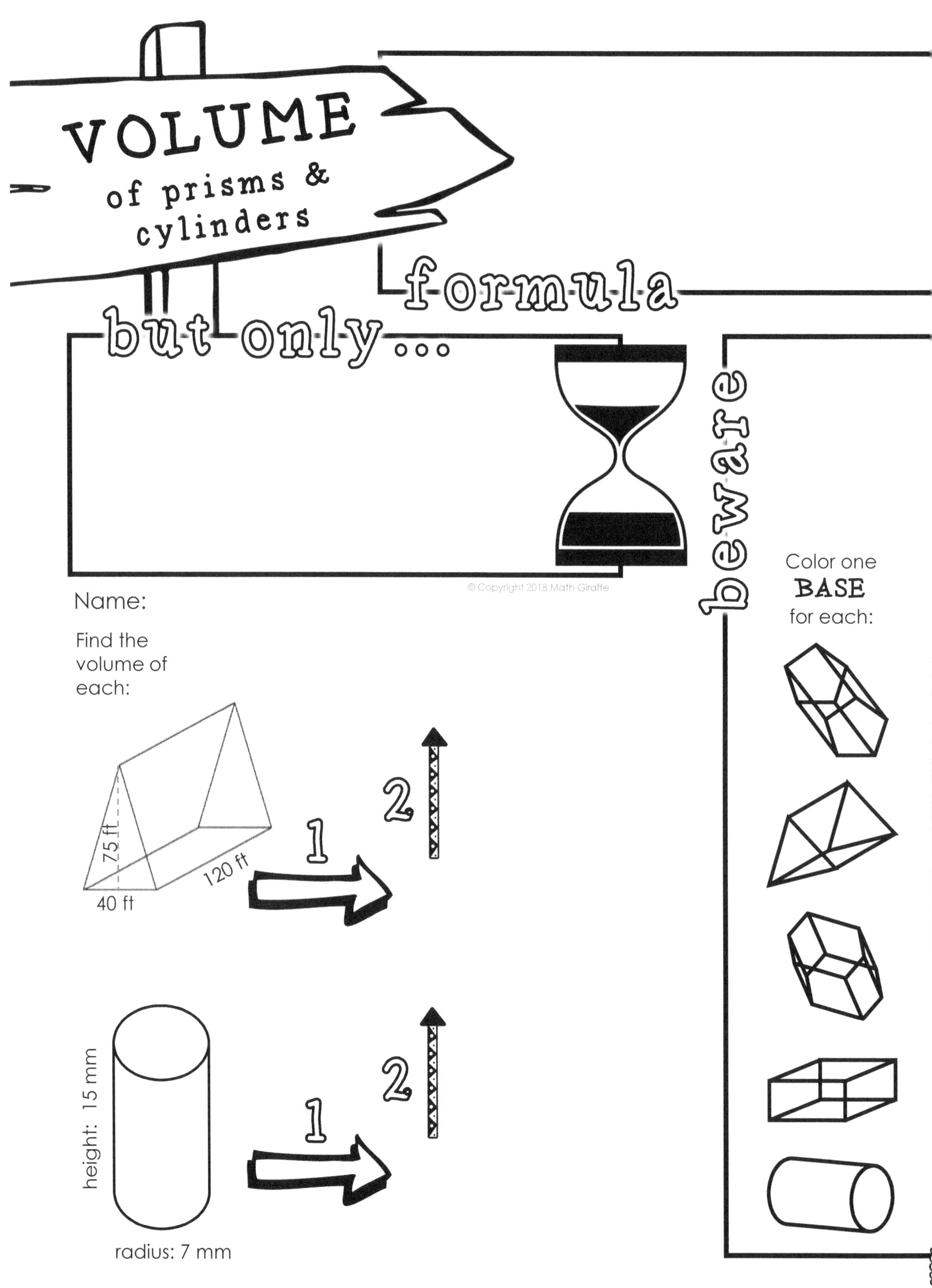
VOLUME
of prisms &
cylinders
formula
but only...
beware
© Copyright 2018 Math Giraffe
Name:
Find the volume of each:
75 ft
120 ft
40 ft
1
2
height: 15 mm
radius: 7 mm
1
2
Color one BASE for each:

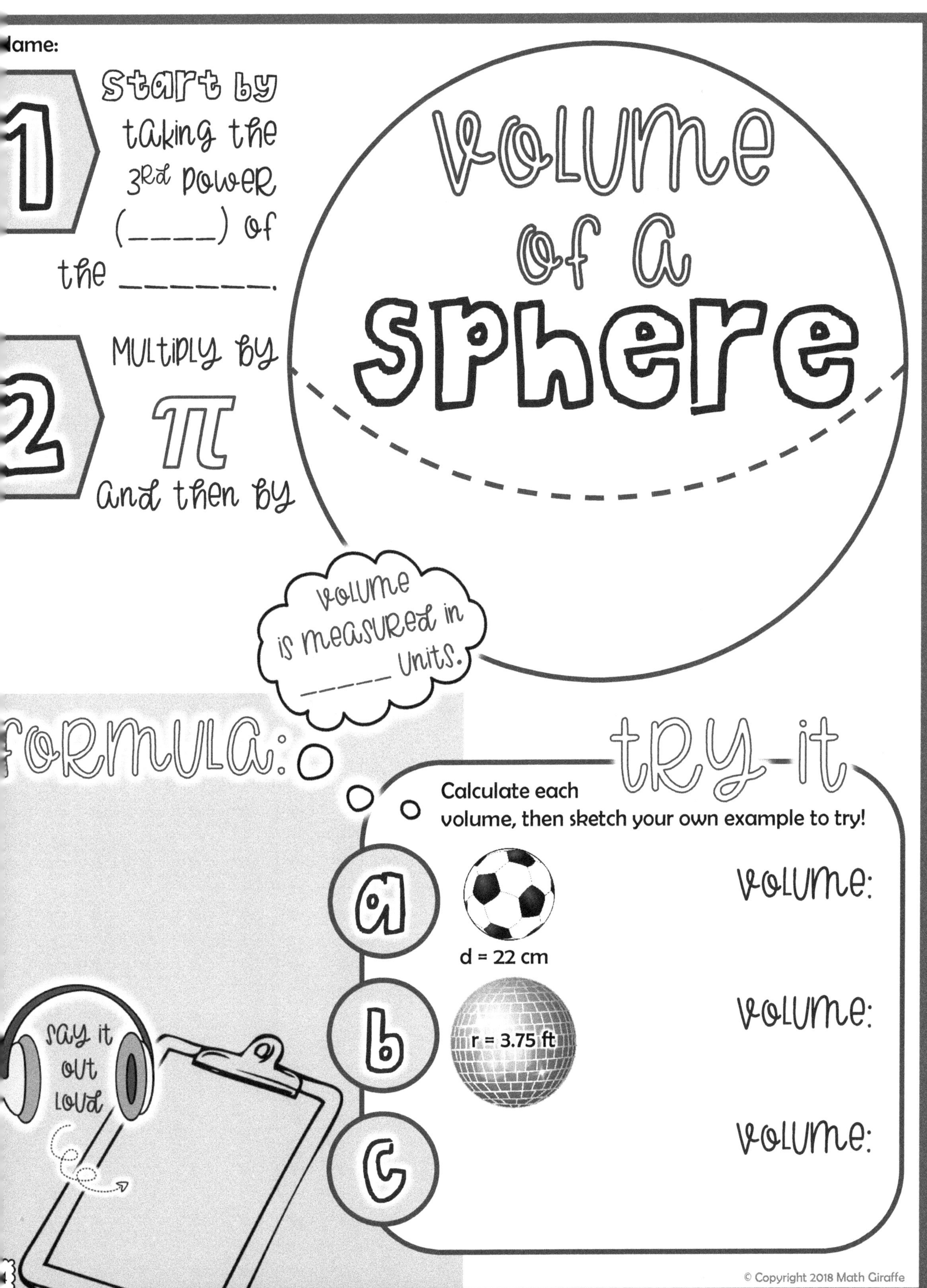
Name:
Volume of a Sphere
1 Start by taking the 3rd power (_____) of the _______.
2 Multiply by π and then by
Volume is measured in _____ units.
Formula:
Try it
Calculate each volume, then sketch your own example to try!
a
d = 22 cm
Volume:
b
r = 3.75 ft
Volume:
c
Volume:
say it out loud

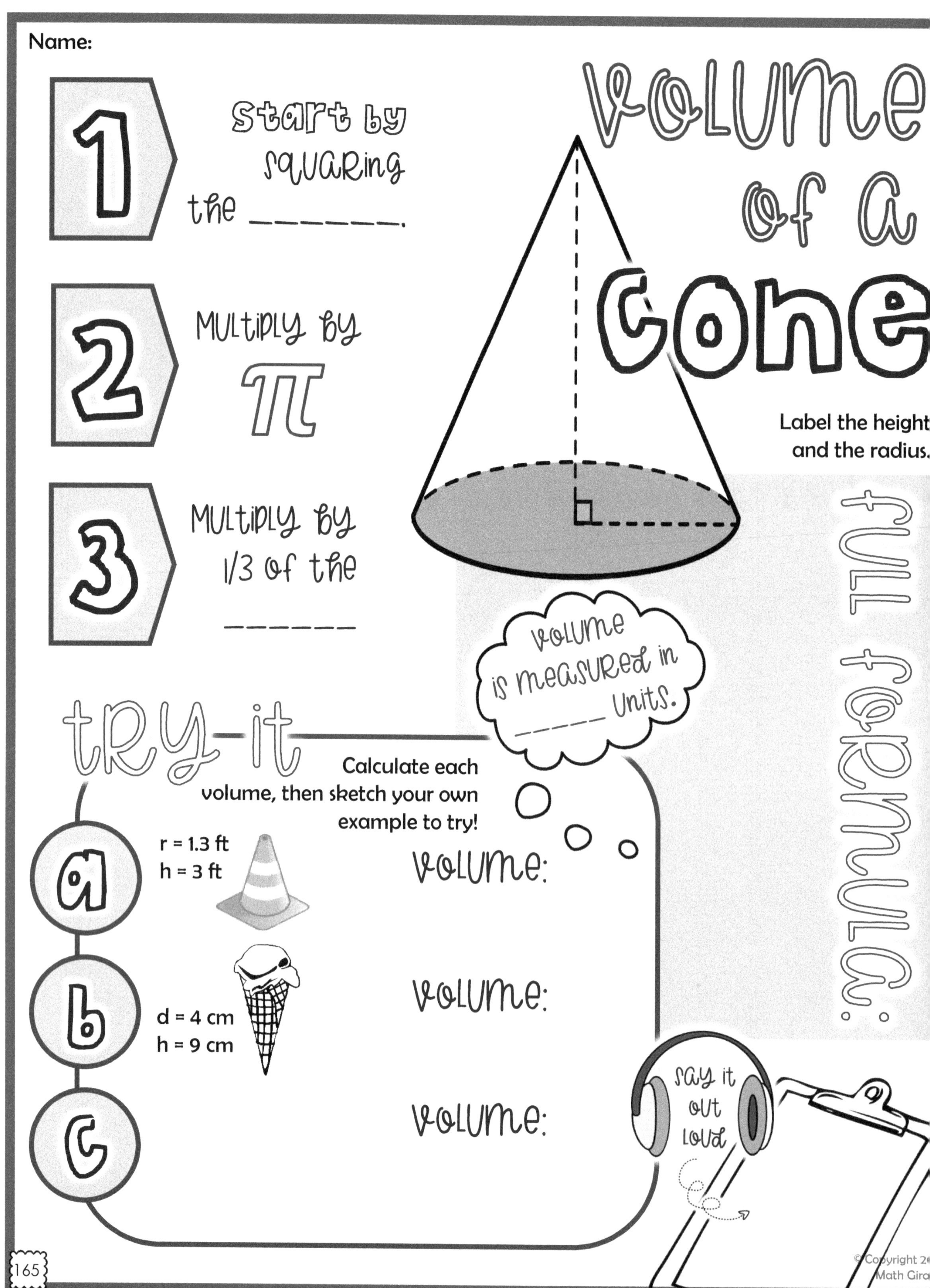
Name:
Volume of a Cone
1
Start by squaring the ________.
2
Multiply by π
3
Multiply by 1/3 of the ________
Label the height and the radius.
Full formula:
Volume is measured in _____ units.
try it
Calculate each volume, then sketch your own example to try!
a
r = 1.3 ft
h = 3 ft
Volume:
b
d = 4 cm
h = 9 cm
Volume:
c
Volume:
Say it out loud

volume of a pyramid

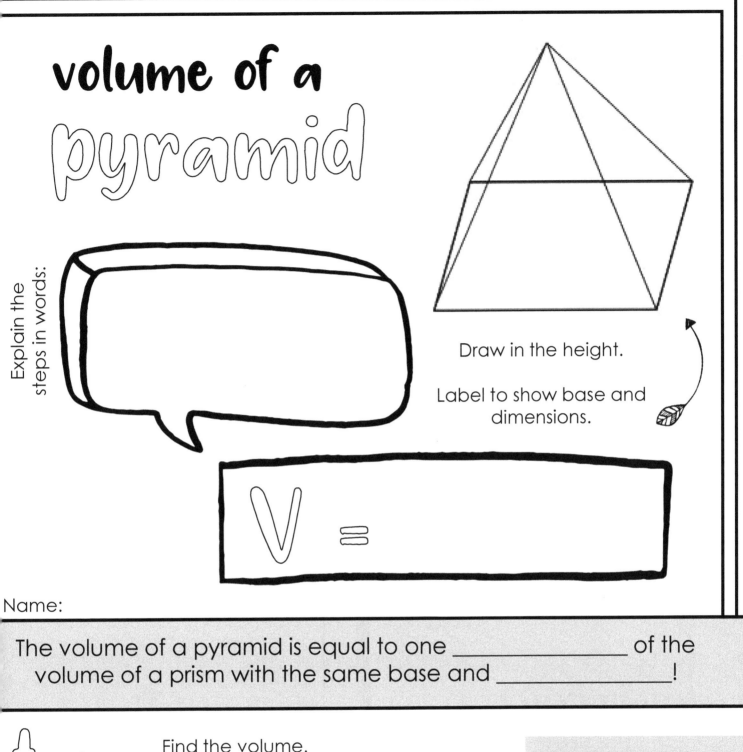

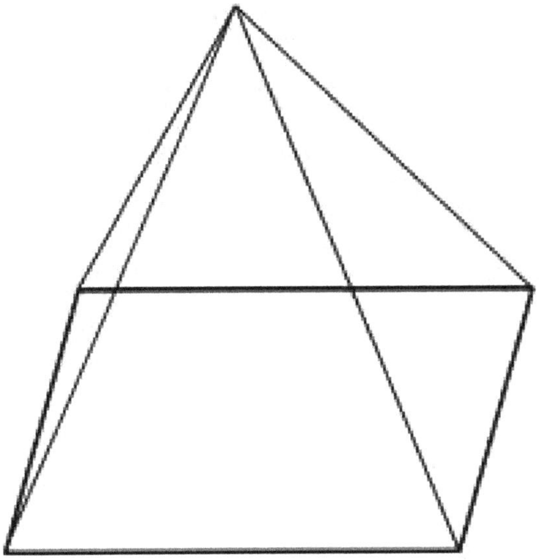

Explain the steps in words:

Draw in the height.

Label to show base and dimensions.

V =

Name:

The volume of a pyramid is equal to one _______________ of the volume of a prism with the same base and _______________!

try it:

Find the volume.

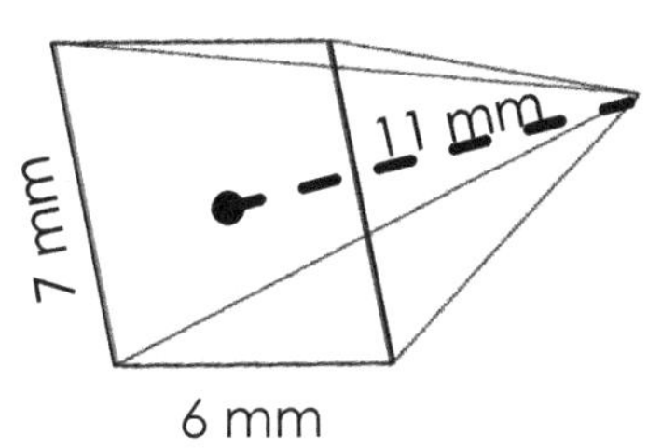

apex

What is the apex of a pyramid?

How do we identify the height/altitude?

Reminder about units:

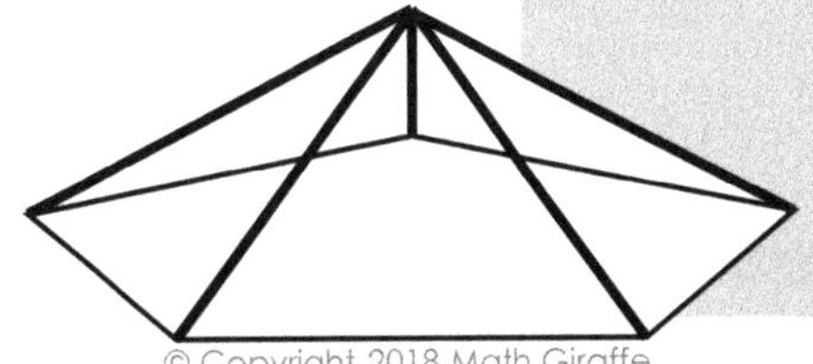

Name:

Composite
3d shapes

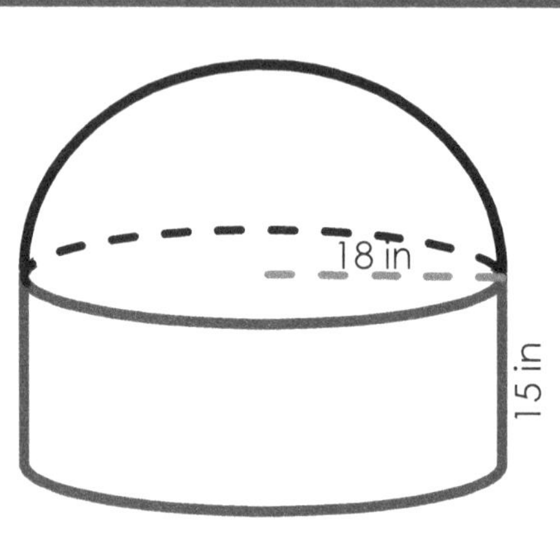

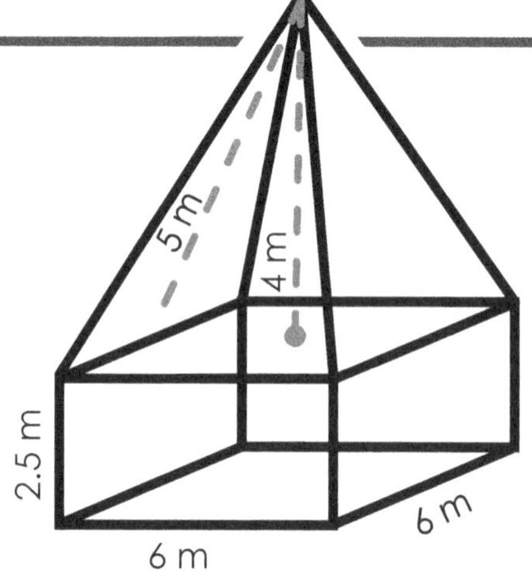

total surface area >>

total surface area >>

total volume >>

total volume >>

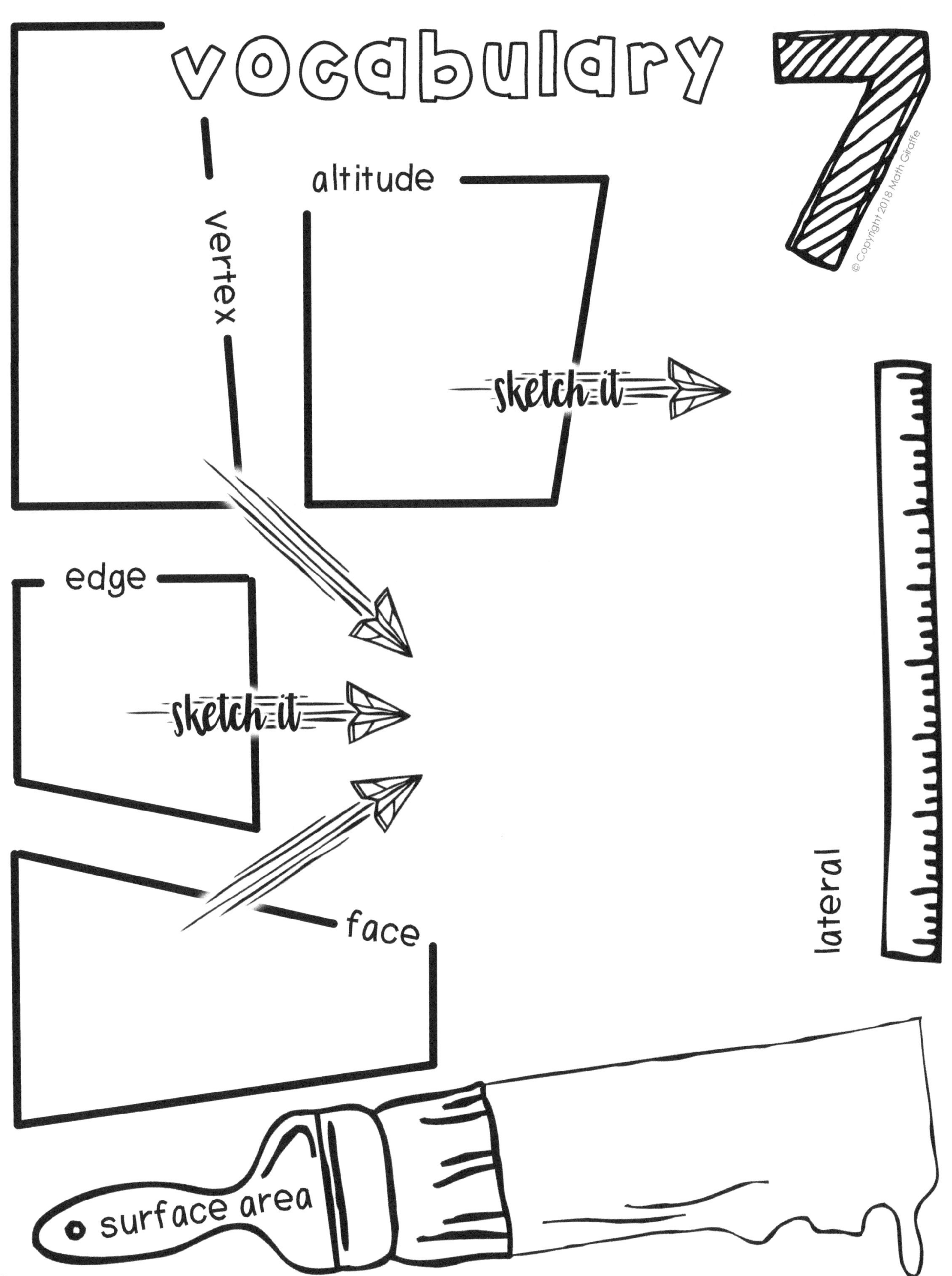
vocabulary 7
vertex
altitude
sketch it
edge
sketch it
face
lateral
surface area
© Copyright 2018 Math Giraffe

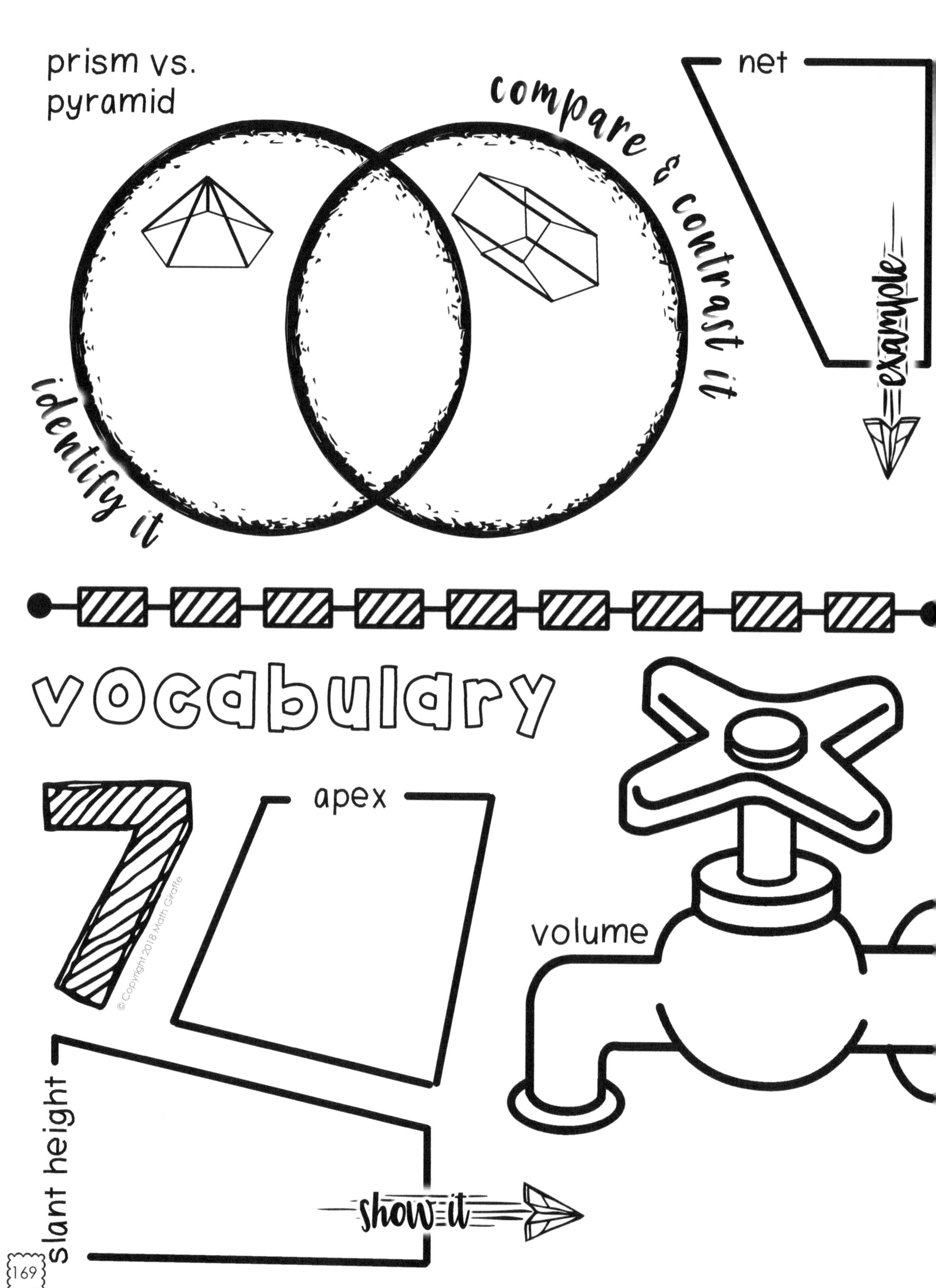
prism vs.
pyramid
compare & contrast it
net
example
identify it
vocabulary
apex
volume
slant height
show it

8
Statistics
& Probability
0
1
2
3
4
A
B
C
D

Describe each measure and how to find it. Then, in the "example" box for each, calculate the measure for the full data set shown (prices of t-shirts on entire page). List and define the measures of central tendency in the middle. Doodle, color, and embellish.

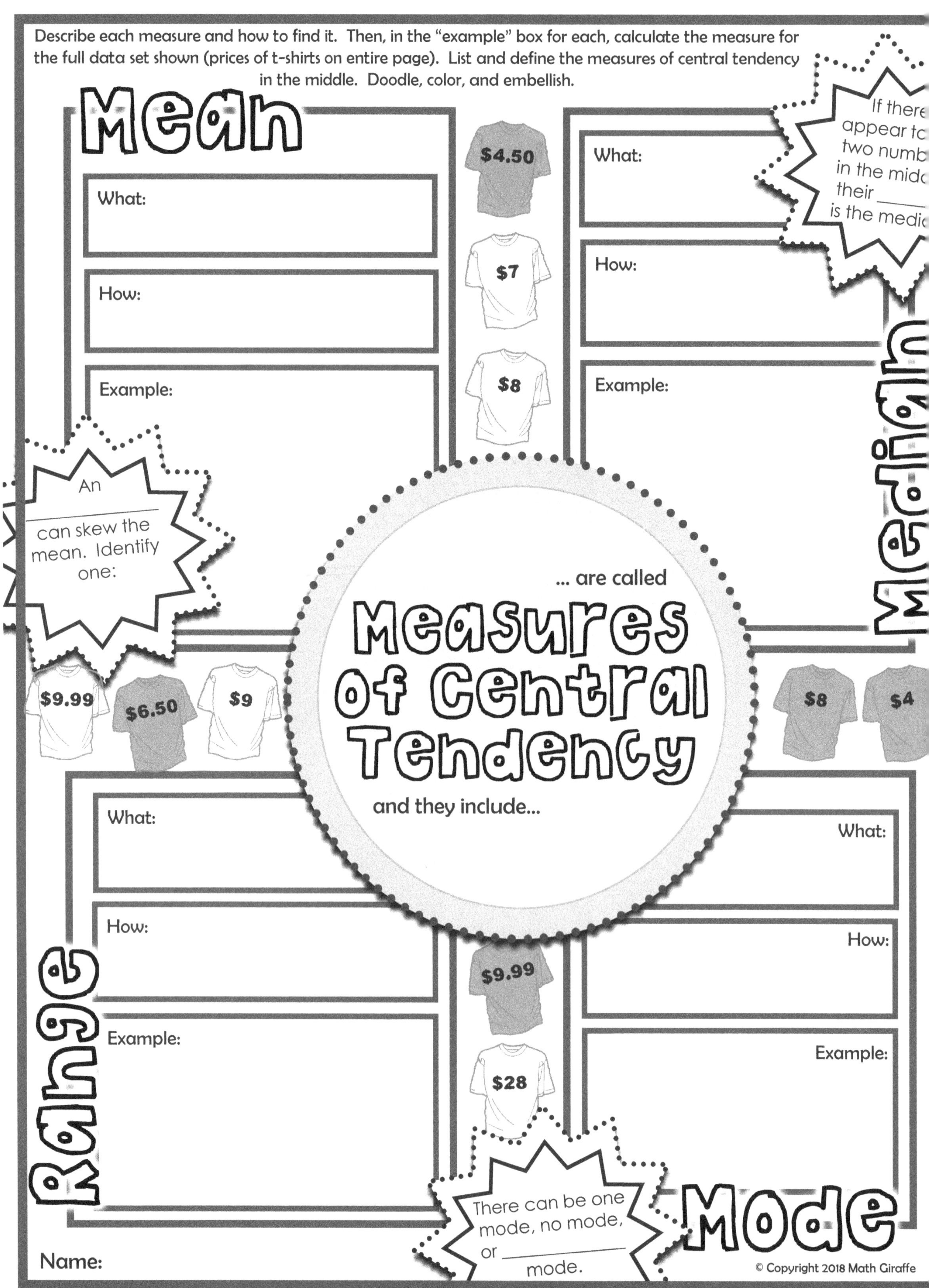

Mean
What:
How:
Example:
$4.50
$7
$8
Median
What:
How:
Example:
If there appear to two numb in the midc their _____ is the medic
An _____ can skew the mean. Identify one:
$9.99
$6.50
$9
... are called
Measures of Central Tendency
and they include...
$8
$4
Range
What:
How:
Example:
$9.99
$28
Mode
What:
How:
Example:
There can be one mode, no mode, or _____ mode.

Name:

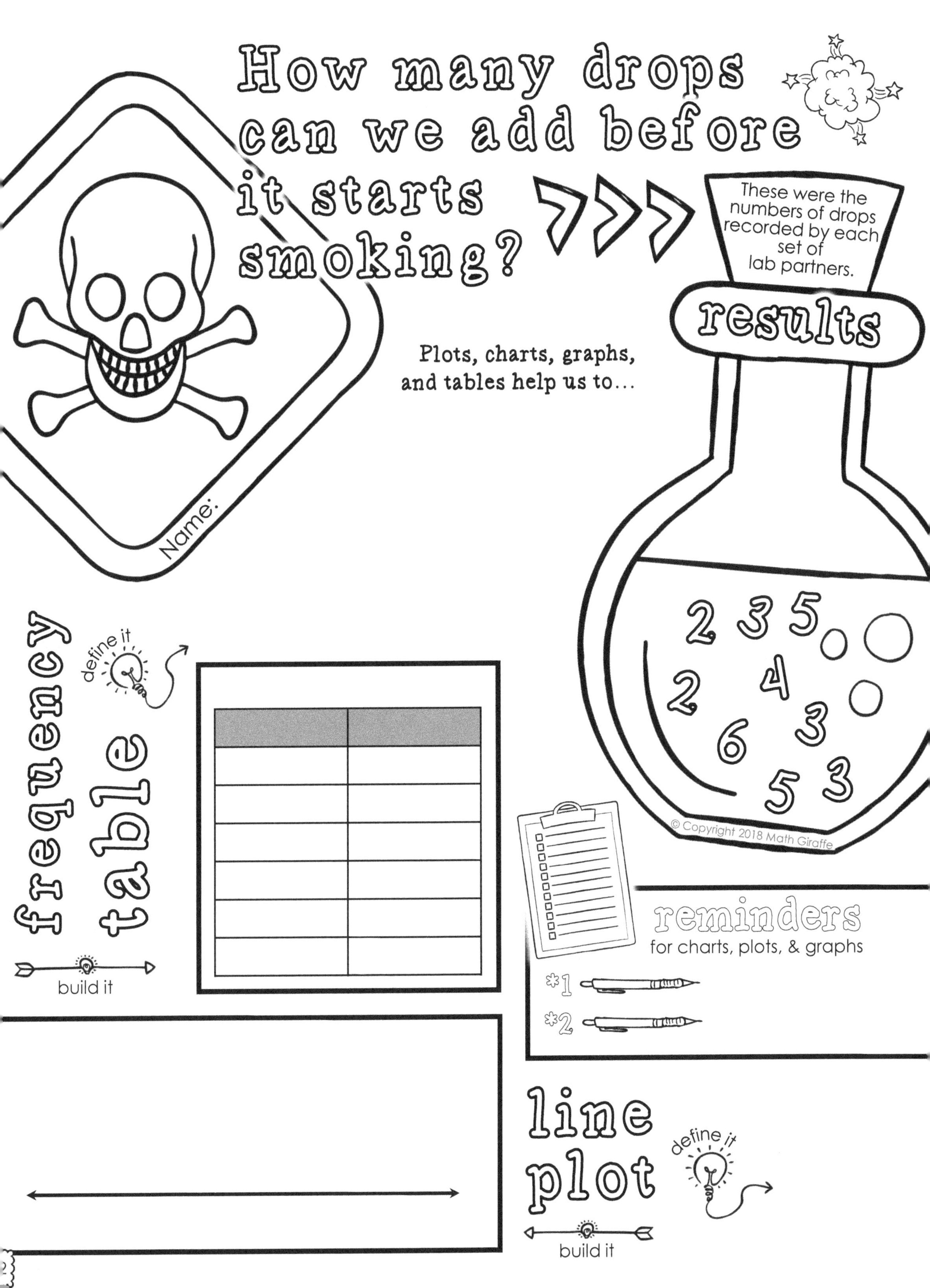

How many drops can we add before it starts smoking?
These were the numbers of drops recorded by each set of lab partners.
results
Plots, charts, graphs, and tables help us to...
Name:
2 3 5
4
2
3
6
5 3
frequency table
define it
build it
reminders
for charts, plots, & graphs
*1
*2
line plot
define it
build it

CREATE IT!

AGE	TSP. SUGAR PER CUP
7	1.2
15	1
35	0.3
18	0.8
18	0.6
29	0.6
44	0.7
53	0
51	0.5

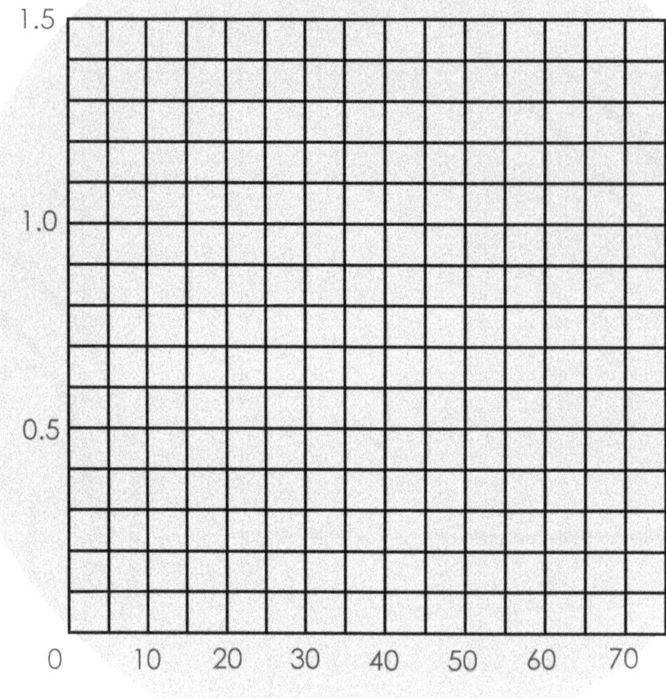

CORRELATION

LABEL IT!

Axes & Title

Name:

TREND LINE OR "LINE OF BEST ______"

Use a __________ to draw a trend line that overlays the graph, with the same number of ___________ on each side of the line.

ANALYZE IT!

1. What type of correlation do you see in the graph above?

2. Which would you expect to be most likely for a 38 year old: 0.2 tsp, 0.5 tsp, or 1.1 tsp of sugar?

3. What type of correlation would you expect between these two variables? Explain why.
x: number of minutes taken to deliver pizza
y: number of stars given for customer service rating

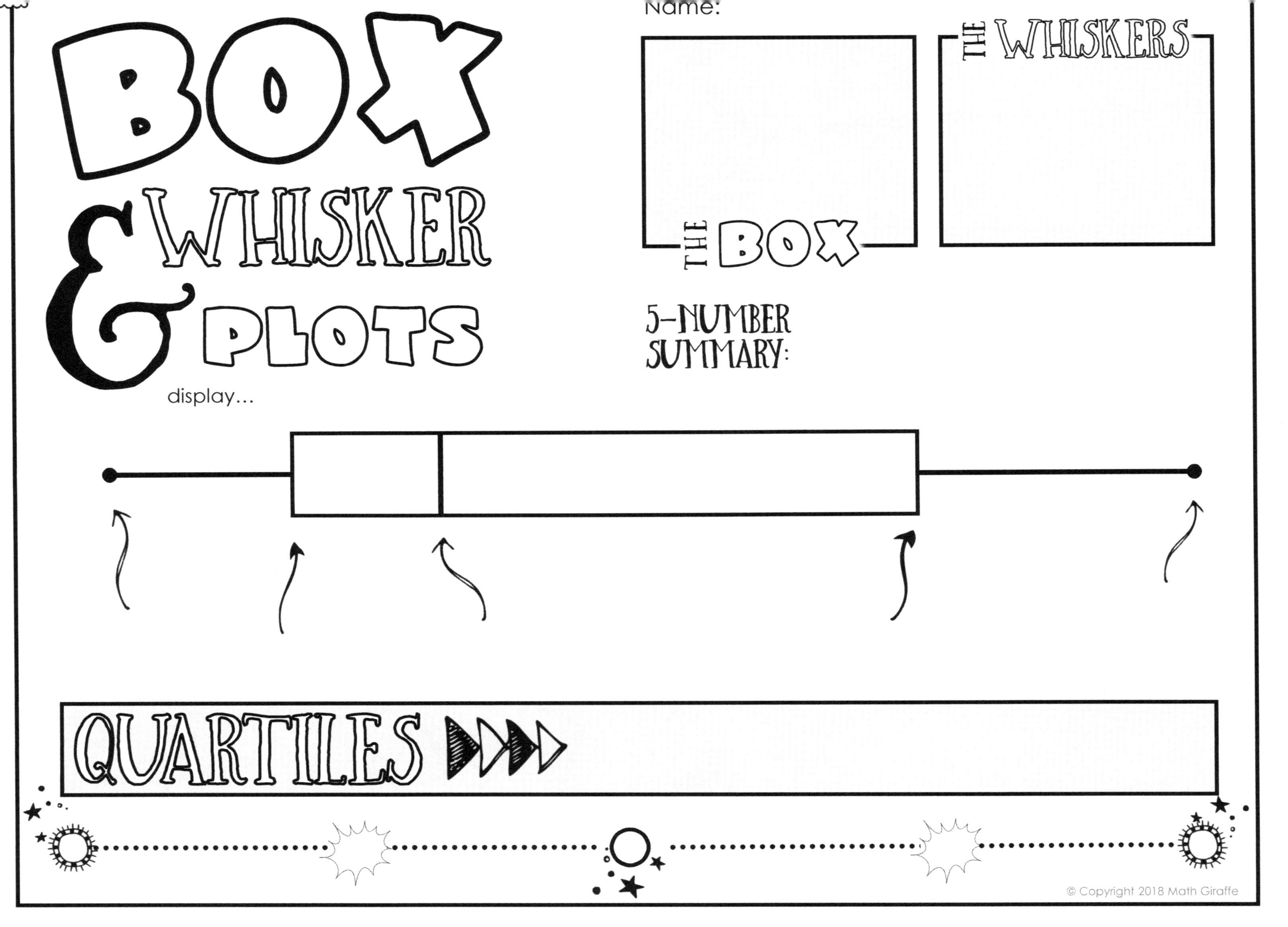
Name:
BOX & WHISKER PLOTS
display...
THE BOX
THE WHISKERS
5-NUMBER SUMMARY:
QUARTILES

BUILD A BOX & WHISKER PLOT

HOUSE PRICES

Create a box and whisker plot to display the data from the table.

House Prices in Stark County (in thousands)	
$278.8	$154
$159	$230
$220	$245
$175.5	$389.9
$169	$89
$179	$122
$230	$456
$289	$269.9
$140	$189
$129	$215
$309	$165

1 Arrange the data in order from ________ to ____________.

Find the median.

2 Identify the lower and the upper ________________.

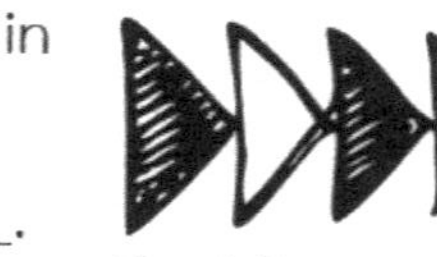

3 Create a number line and mark:

A 

B 

C 

Then draw in the box according to the quartiles

DON'T FORGET!

Number of Hours Calvin Spent Playing Outside Per Week

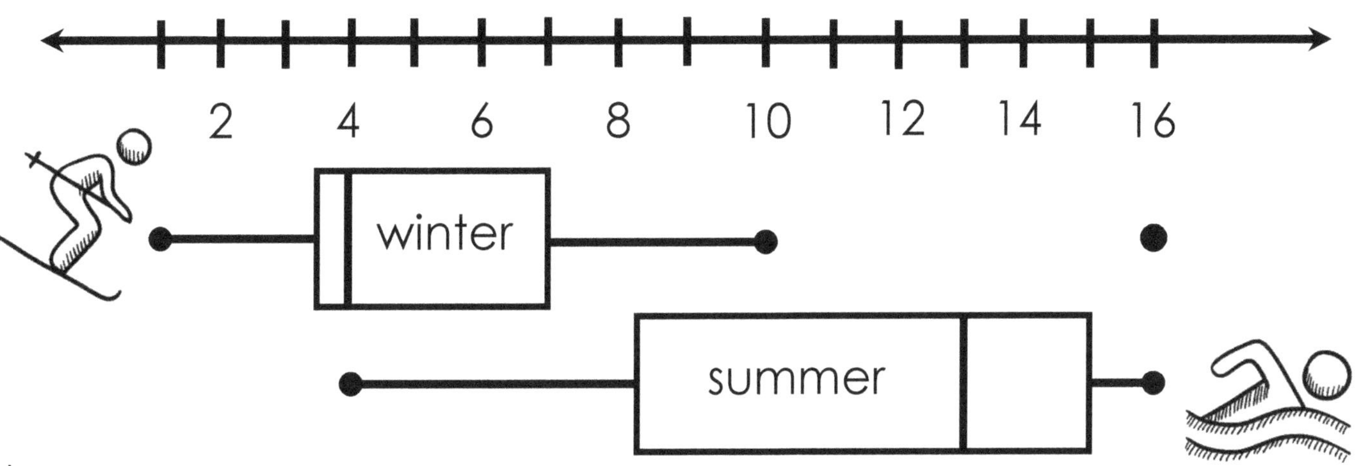

Name:

Explain each of the following and what they mean in terms of the plot above.

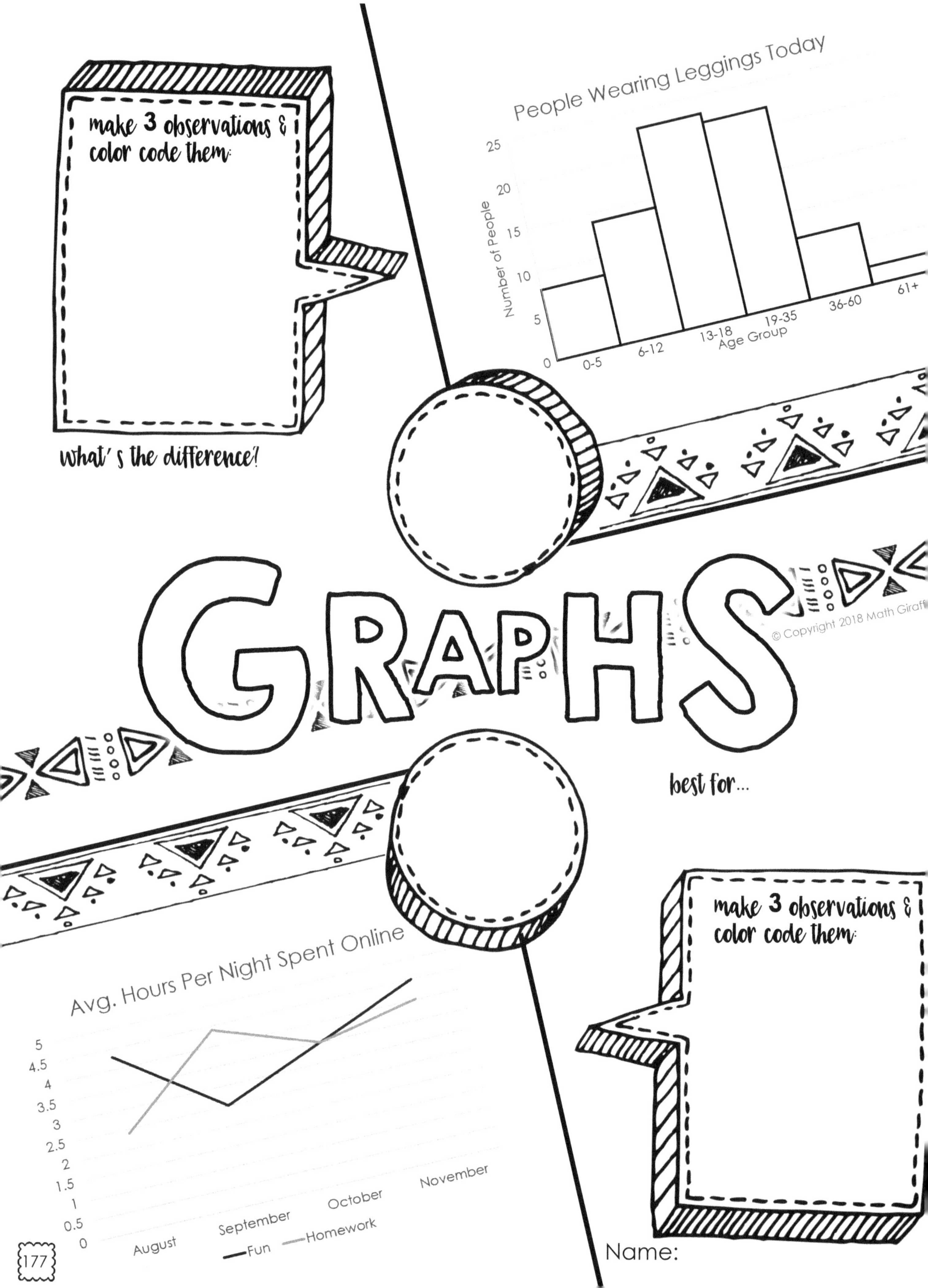
People Wearing Leggings Today
Number of People
25
20
15
10
5
0
0-5
6-12
13-18
19-35
36-60
61+
Age Group
make 3 observations & color code them:
what's the difference?
GRAPHS
best for...
Avg. Hours Per Night Spent Online
5
4.5
4
3.5
3
2.5
2
1.5
1
0.5
0
August
September
October
November
Fun
Homework
make 3 observations & color code them:
Name:

reading GRAPHS & misleading GRAPHS

Name:

Use color to identify what is misleading about each graph. Explain / describe it in complete sentences.

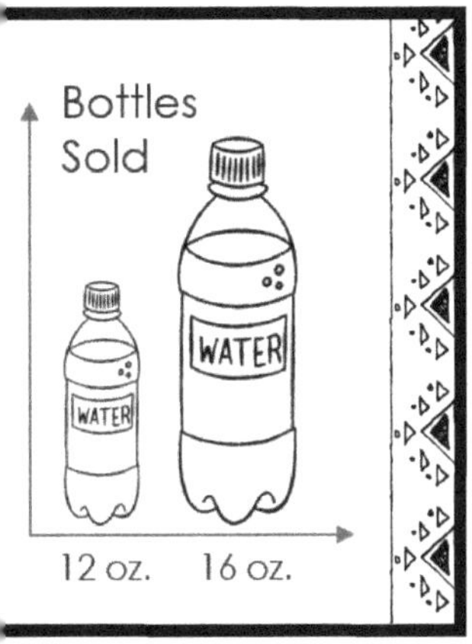

misleading:

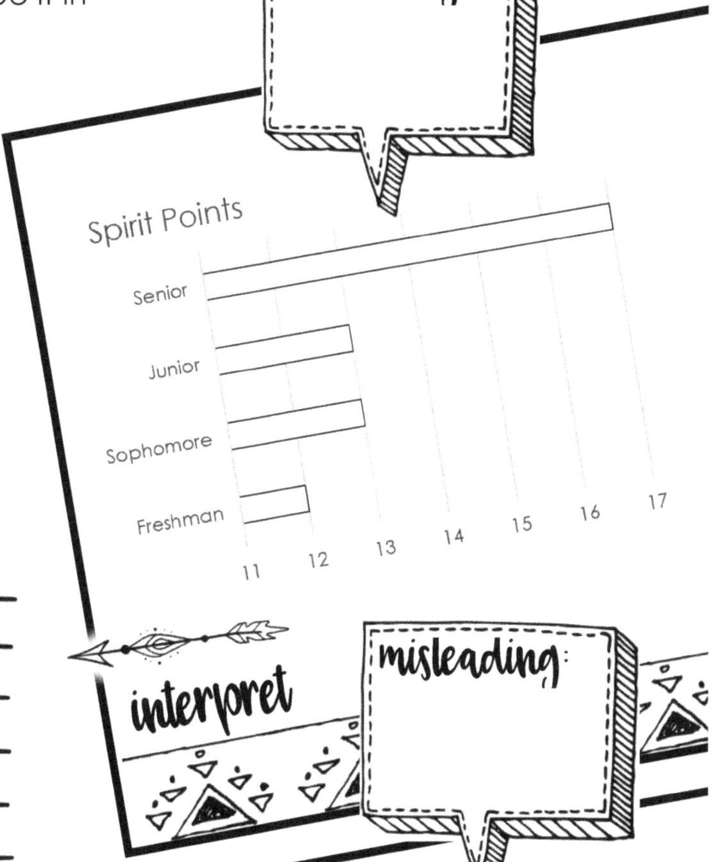

Bottles Sold

WATER

WATER

12 oz. 16 oz.

Spirit Points

Senior	
Junior	
Sophomore	
Freshman	

11 12 13 14 15 16 17

interpret

misleading:

examples of ways a graph can mislead:

interpret

Probability

Name:

Probability (A) = $\dfrac{\text{Number of ways an event can occur}}{\text{Total number of possible outcomes}}$

"The probability of the event A happening is equal to..."

experiment

a scenario that involves ____________ or an uncertain result & can have different __________.

outcome

the ________________________ of a single performance of an experiment

event

a particular outcome of an

____

probability

a measure of ___________ _________________ a particular event is

Example:

A box of donuts contains 6 sprinkled, 3 coconut, and 3 chocolate donuts. If you reach in and pull one out without looking, what is the probability that you get a chocolate donut?

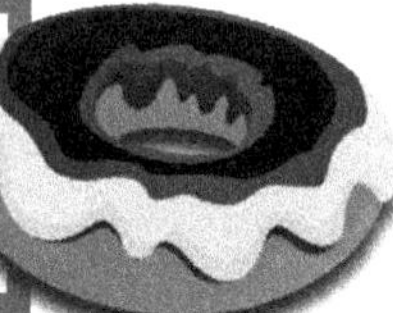

$\frac{3}{12} = \frac{1}{4}$

(1/4 is than h so it's as likely gettin sprink do would b

$0 \leq P \leq 1$

Probability is always between zero and one.

____ = certai

____ = impossibl

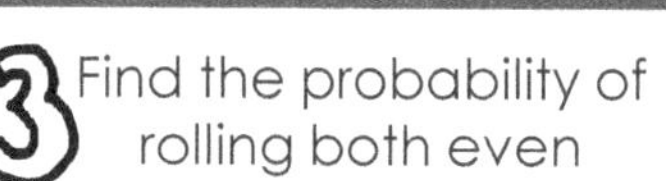

Try it

1 Find the probability of rolling an even number on a standard 6-sided die.

Ways the event can occur:
rolling a 2, a 4, or a 6 (3 ways)

Possible outcomes:
rolling a 1, 2, 3, 4, 5, or 6 (6 total)

Probability:
$\frac{3}{6}$ → Simplify:

2 Find the probability of getting "heads" on three coin flips in a row.

Ways the event can occur:

Possible outcomes:

Probability:

3 Find the probability of rolling both even numbers when you roll two 6-sided dice at once

The

COMPLEMENT

of an event is its opposite.

P (A) = SNOW

P (NOT A) = NO SNOW

P (NOT A) = 1 − P (A)

ODDS
ODDS
in favor
= # of favorable outcomes / # of unfavorable outcomes
ODDS
against
= # of unfavorable outcomes / # of favorable outcomes
The odds of beating level 52 of the video game on the first try is 1/99.
WOW!
That means that only one out of every 100 players passes the level on the first try!

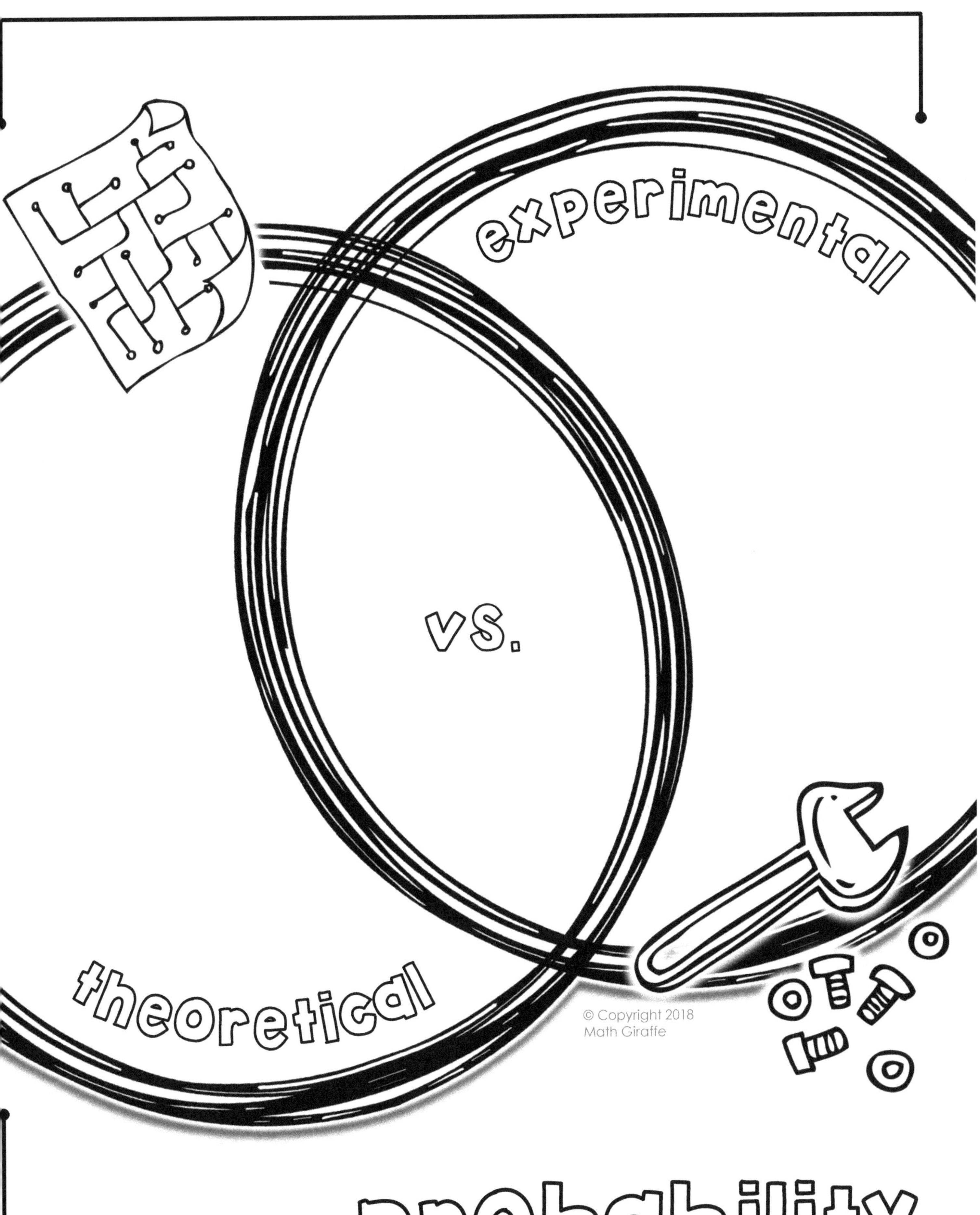

probability

Sample Space:

IN A PERFECT WORLD...

You are about to shake out two candy pieces from a box that has an equal number of red and pink candies. List all possible outcomes.

(sketch a memory trigger for "theoretical" = "perfect world")

theoretical probability

You roll two dice. What is the probability of rolling "doubles"? (two of the same number)

Name:

try it

One friend hates tomatoes. Half of the cups of guacamole have tomato, and half do not. You choose 3 blindly and take them to the table. What is the probability of all 3 having tomato, devastating your friend?

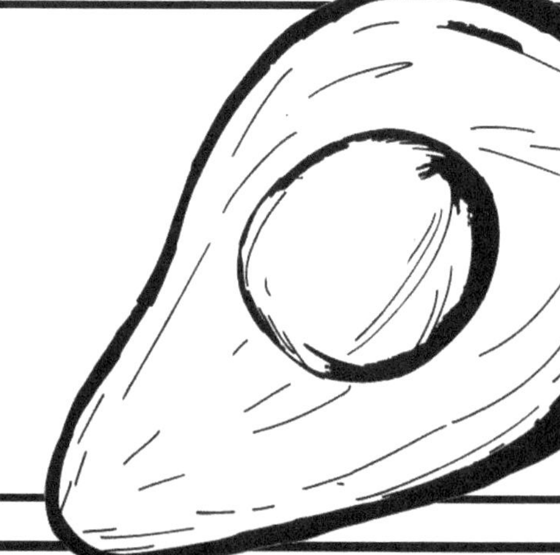

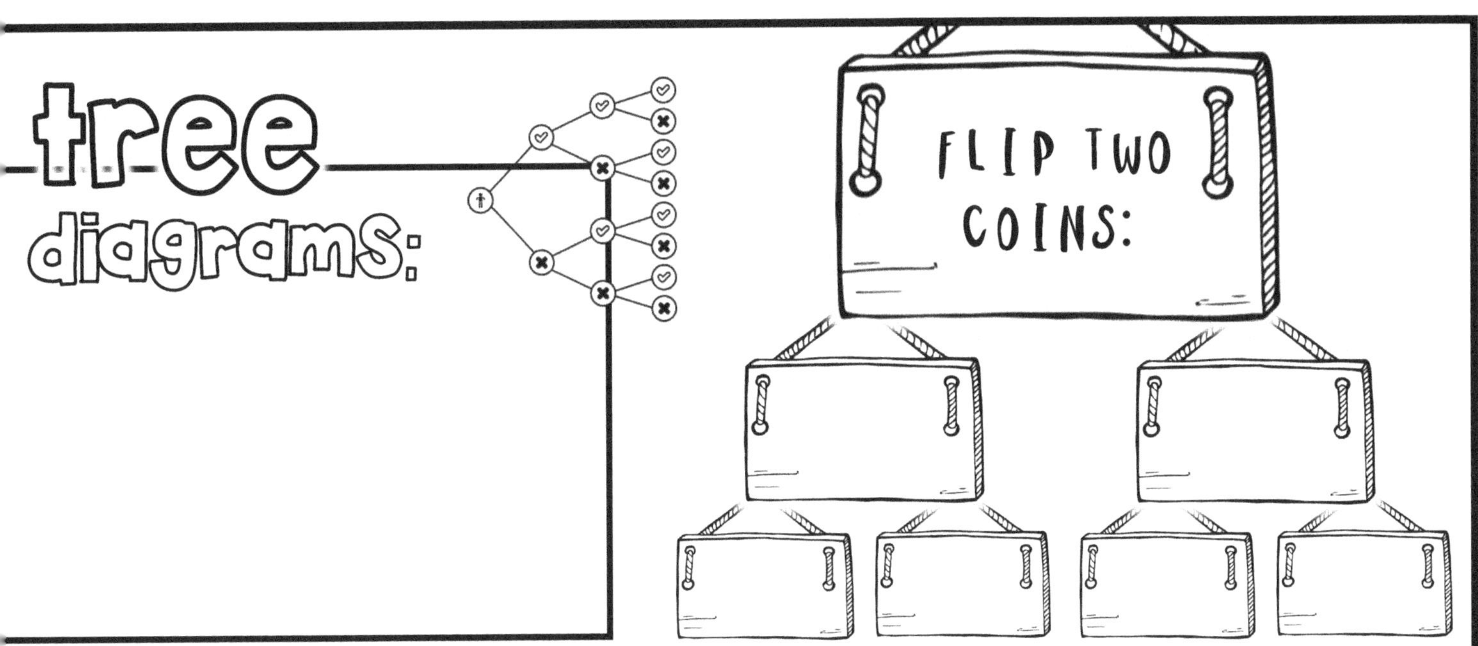

P(event) =

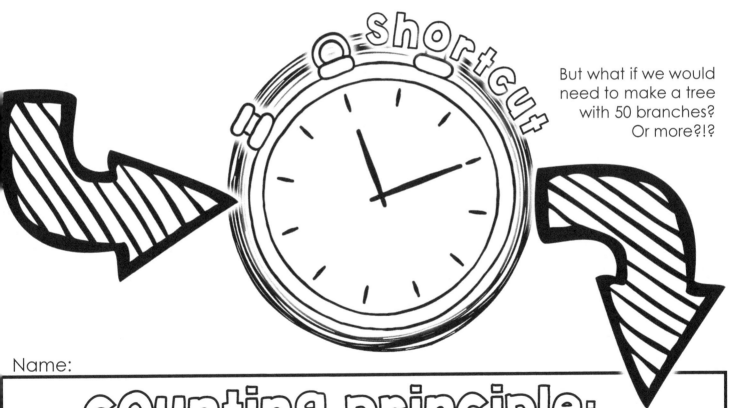

Shortcut

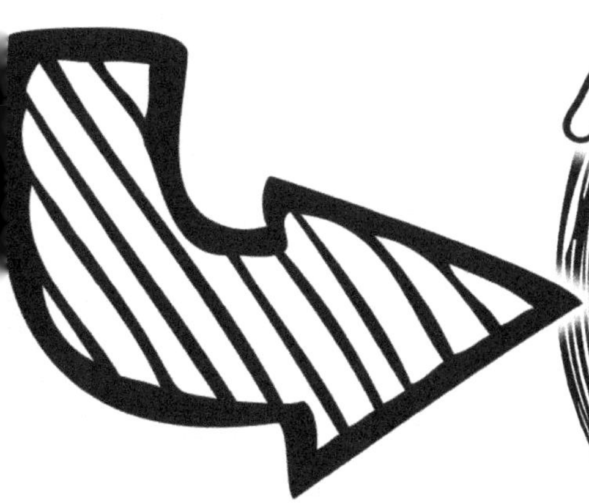

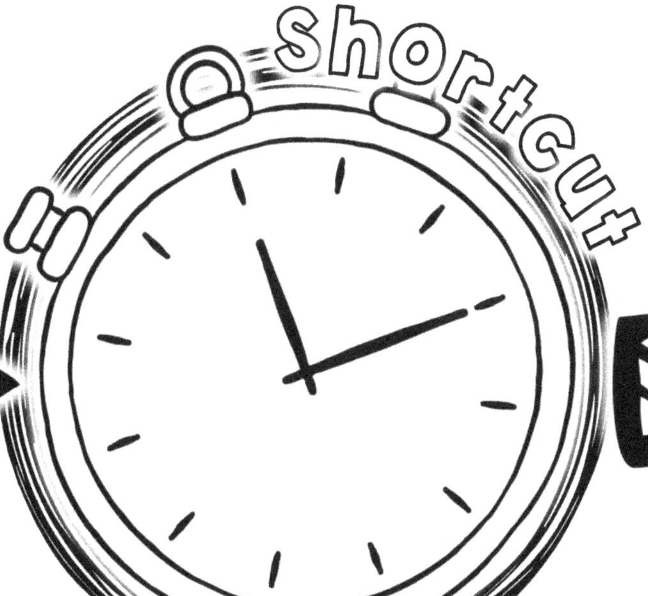

But what if we would need to make a tree with 50 branches? Or more?!?

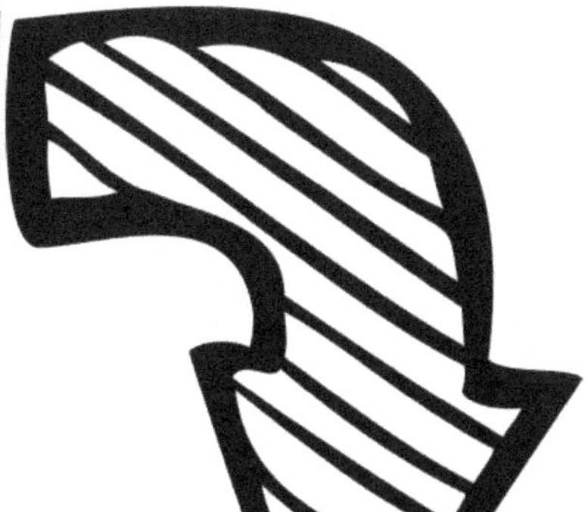

Name:

counting principle:

independent & dependent events

Read carefully, then determine whether the monster you battle DOES depend on where the slide spits you out, or whether if DOES NOT. Write the label "DEPENDENT" or "INDEPENDENT" in the box. Draw a flowchart for each game setup starting with the arrows as the sliding tunnel options.

Name:

A video game includes a sliding tunnel you ride down. Half of the time, it spits you out in a mud pit. The other half of the time, you land in a field. (As the player, you cannot control any of this). In the mud pit, the game chooses to either give you a pig monster or a whirling windmill robot to battle. (each with equal probability of appearing, but randomly chosen by the game). If you land in the field, the game either gives you a troll, a rabbit, or a zombie to battle (each with equal probability of appearing but randomly chosen by the game).

Another level within the same video game includes a different sliding tunnel you ride down. Half of the time, it spits you out in a dungeon. The other half of the time, you land in an old house. (As the player, you cannot control any of this). Then, no matter where you land, the game either gives you a soldier, a tiger, or a giant to battle (each with equal probability of appearing but randomly chosen by the game).

P(A, then B) =

P(A, then B) =

Find the probability of landing in the field and getting a troll to battle in this game level.

Find the probability of getting a monster pig.

What about landing in a field and getting a windmill robot to battle?

Find the probability of landing in the dungeon and getting a tiger to battle in this game level.

Find the probability of getting a soldier to battle.

You use a 4-digit passcode to unlock your phone. You know the numbers you chose are 6, 7, 8, and 9 but you cannot remember what order you chose to put them in. How many possible passcodes would you have to try (max) before it unlocked?

permutations

1st #: 2nd #: 3rd #: 4th #:

choices choices choices choices

ame:

permutation notation

Example: A baseball team has 20 players. How many ways can 9 be selected and arranged in positions on the field? Write out and expand the permutation notation.

TRY IT OUT

P =

You are considering 7 different cities for your road trip. If you have to narrow it down and can only choose 5, how many possible travel itineraries could you form by arranging a sequence of 5 to visit in order?

try it

Use permutation notation and show work.

combinations

- Pepperoni
- Sausage
- Mushrooms
- Olives
- Peppers
- Tomato
- Bacon
- Pineapple

How many possibilities are there for a pizza with two toppings?

notation

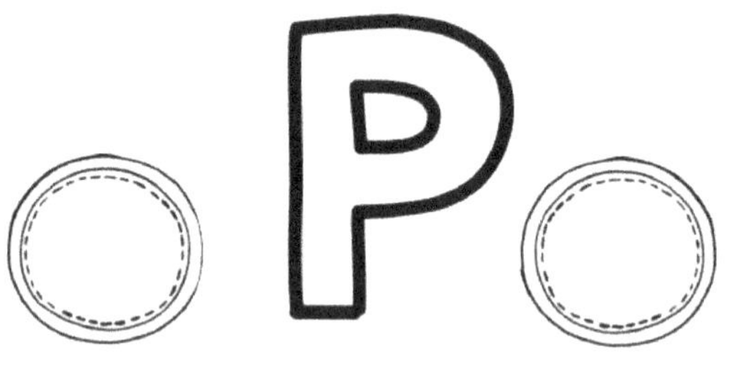

Name:

Use combination notation and show work.

How many possibilities are there for a pizza with three toppings?

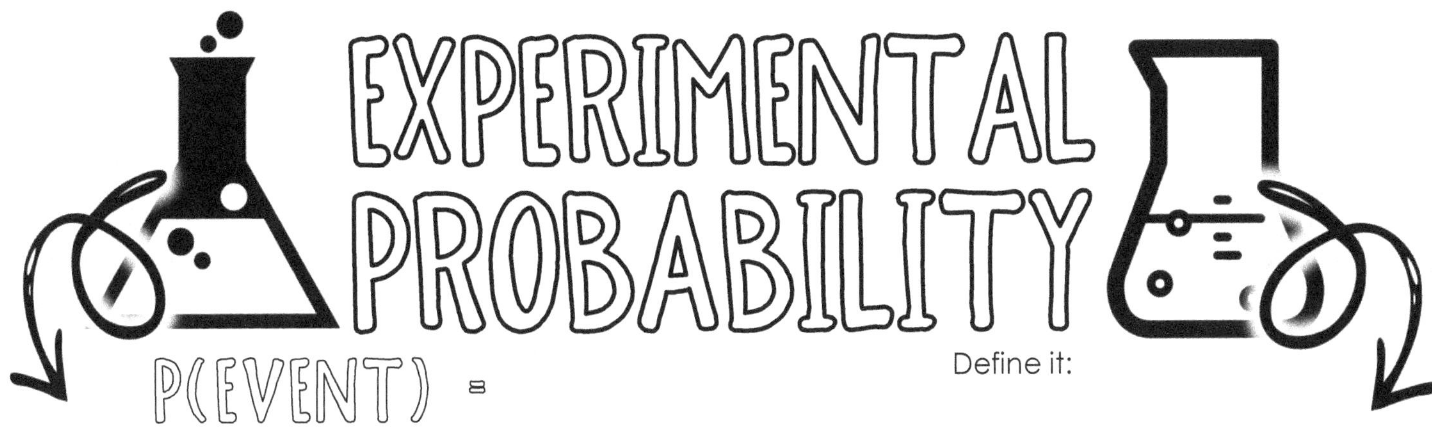

P(EVENT) =

Define it:

Flip a coin 14 times. Record your results. Does the probability of "heads" that you recorded match the theoretical probability of getting "heads"?

RANDOM SAMPLE:

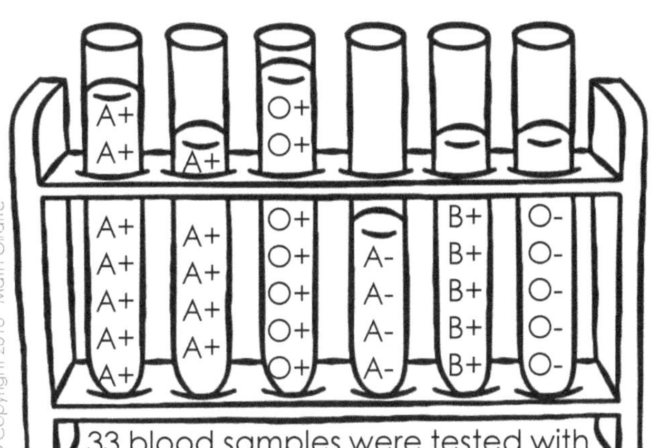

33 blood samples were tested with the above results as blood types. Find each experimental probability as a percent. Round to the nearest tenth of a percent.

P(A+) =

P(O−) =

P(B−) =

*(Note: The theoretical probability of B- is actually 2%. What does that show about this experiment?)

You need to determine what percent of shoppers will boycott a particular grocery store chain for some recent bad publicity.

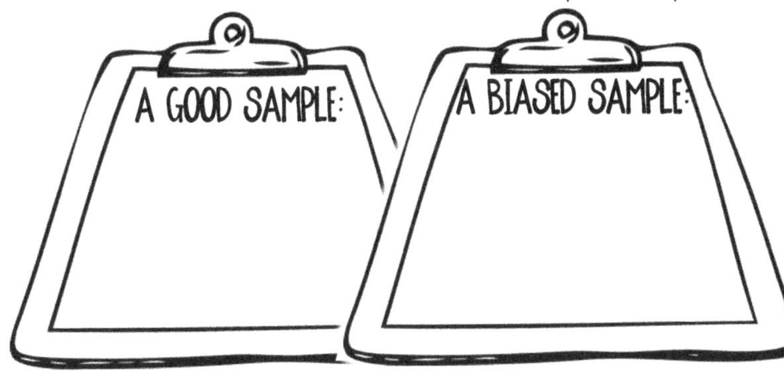

If 12 out of 150 surveyed would boycott, how many people in the total population of 160,000 shoppers might?

Name:

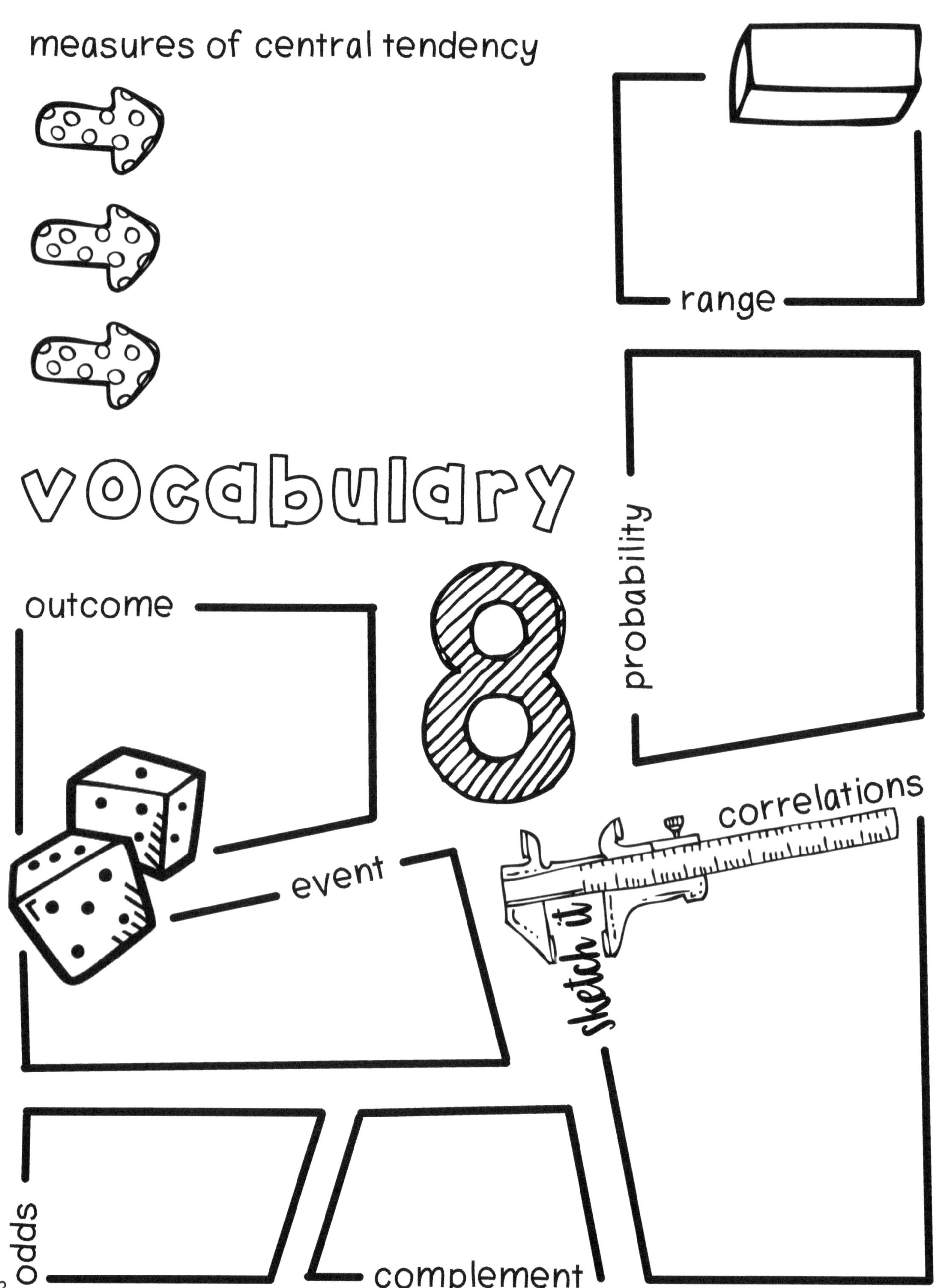
measures of central tendency
range
vocabulary
8
probability
outcome
event
correlations
sketch it
odds
complement

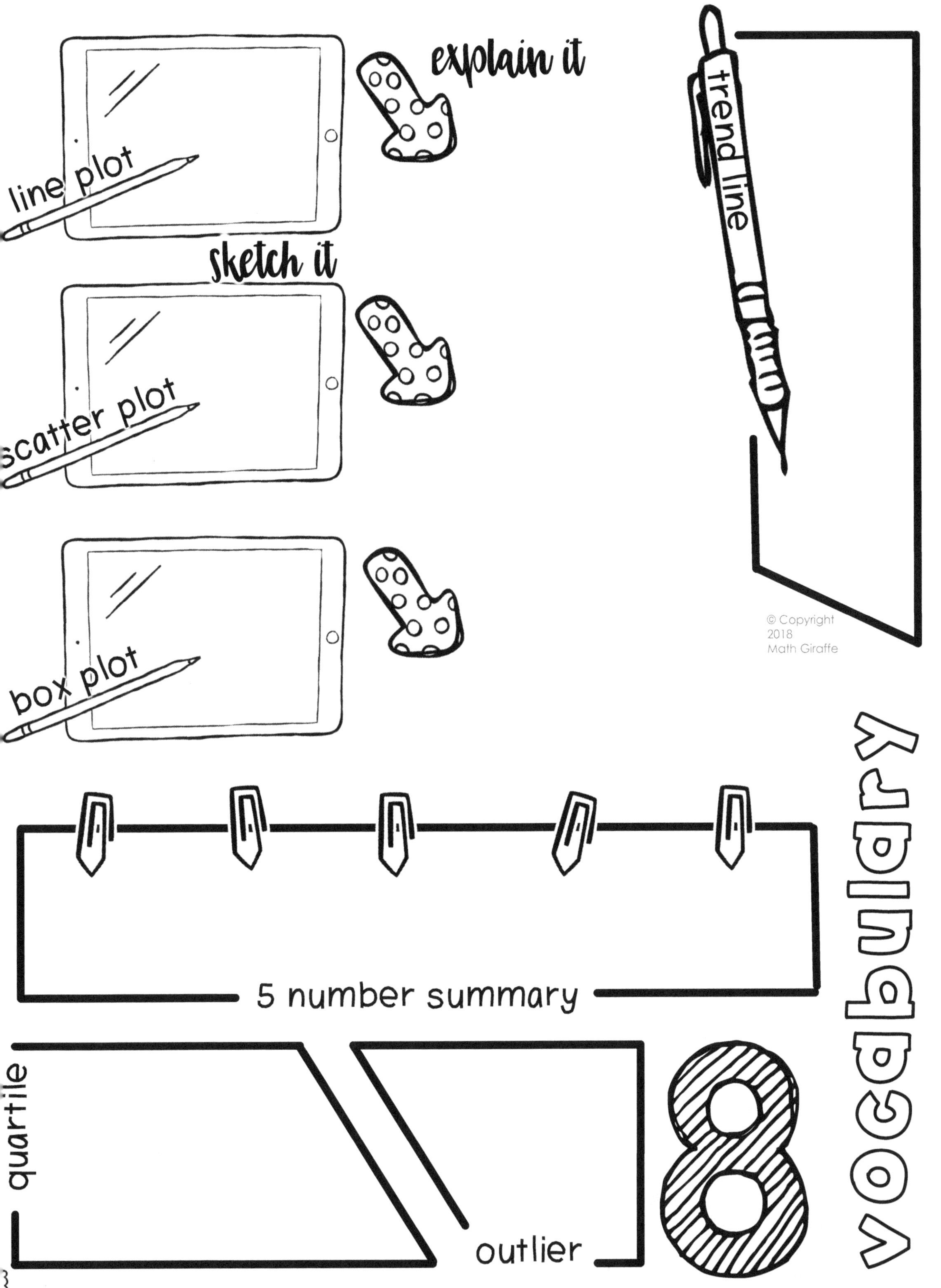

vocabulary
8
explain it
sketch it
line plot
scatter plot
box plot
trend line
5 number summary
quartile
outlier
© Copyright 2018 Math Giraffe

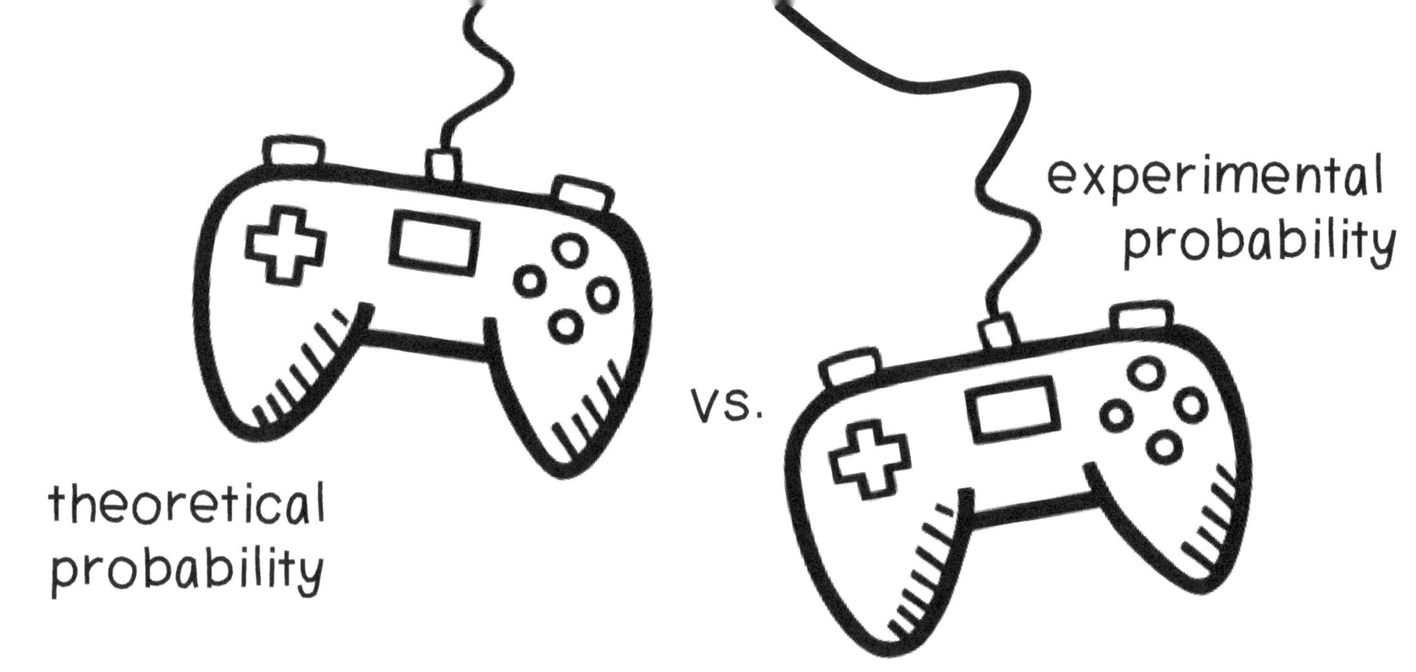

independent events vs. dependent events

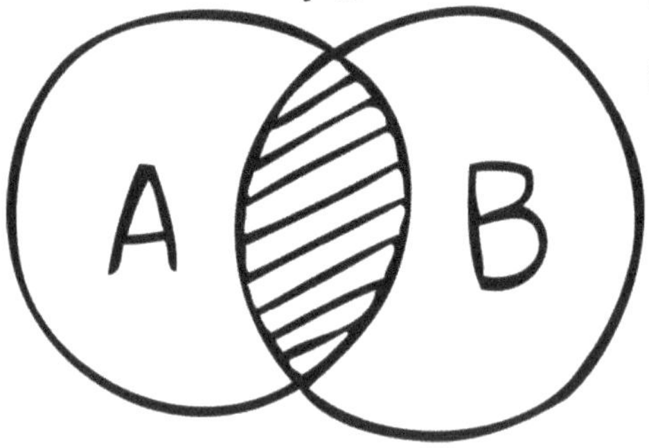

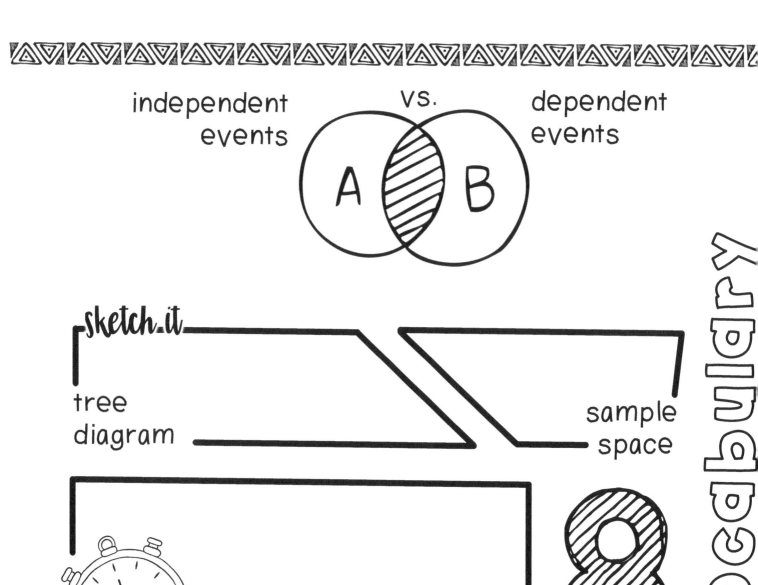

A

B

sketch it

tree diagram

sample space

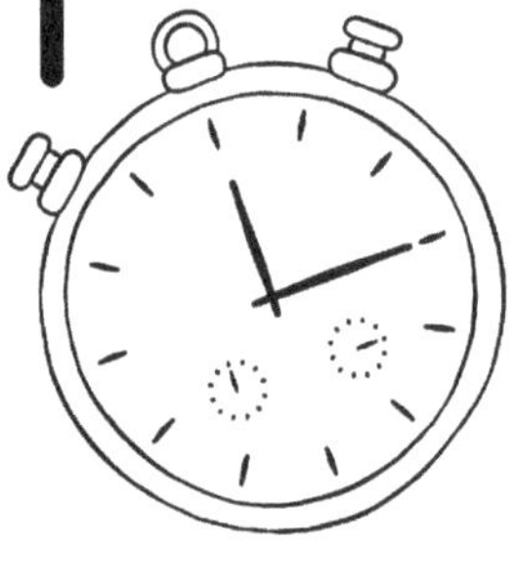

counting principle

8 vocabulary

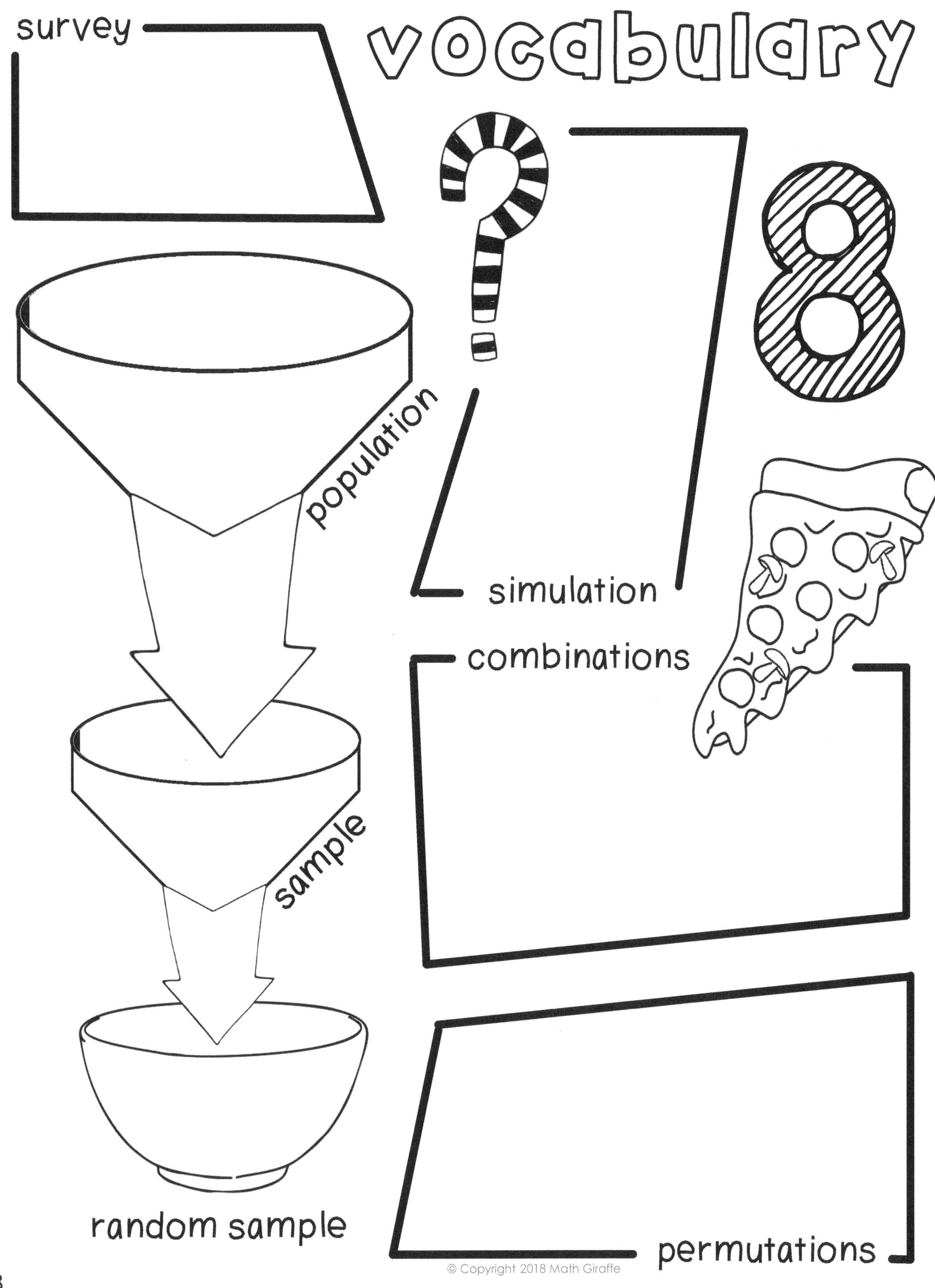
vocabulary
survey
population
sample
random sample
simulation
combinations
permutations

Formulas & Finance

FIRST...

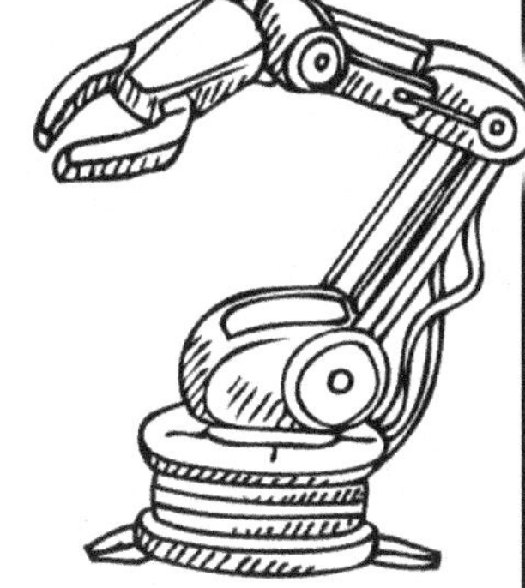

THEN...

2

WHAT IS A FORMULA?

DISTANCE:

PERIMETER:

TEMPERATURE:

TRY IT:

Find the side length of a rectangle with a perimeter of 20 m and a side width of 3 m.

Find the speed of a bike that travels 22 miles in an hour and a half.

Find the temperature in Fahrenheit that is equivalent to 18 degrees Celsius.

Name:

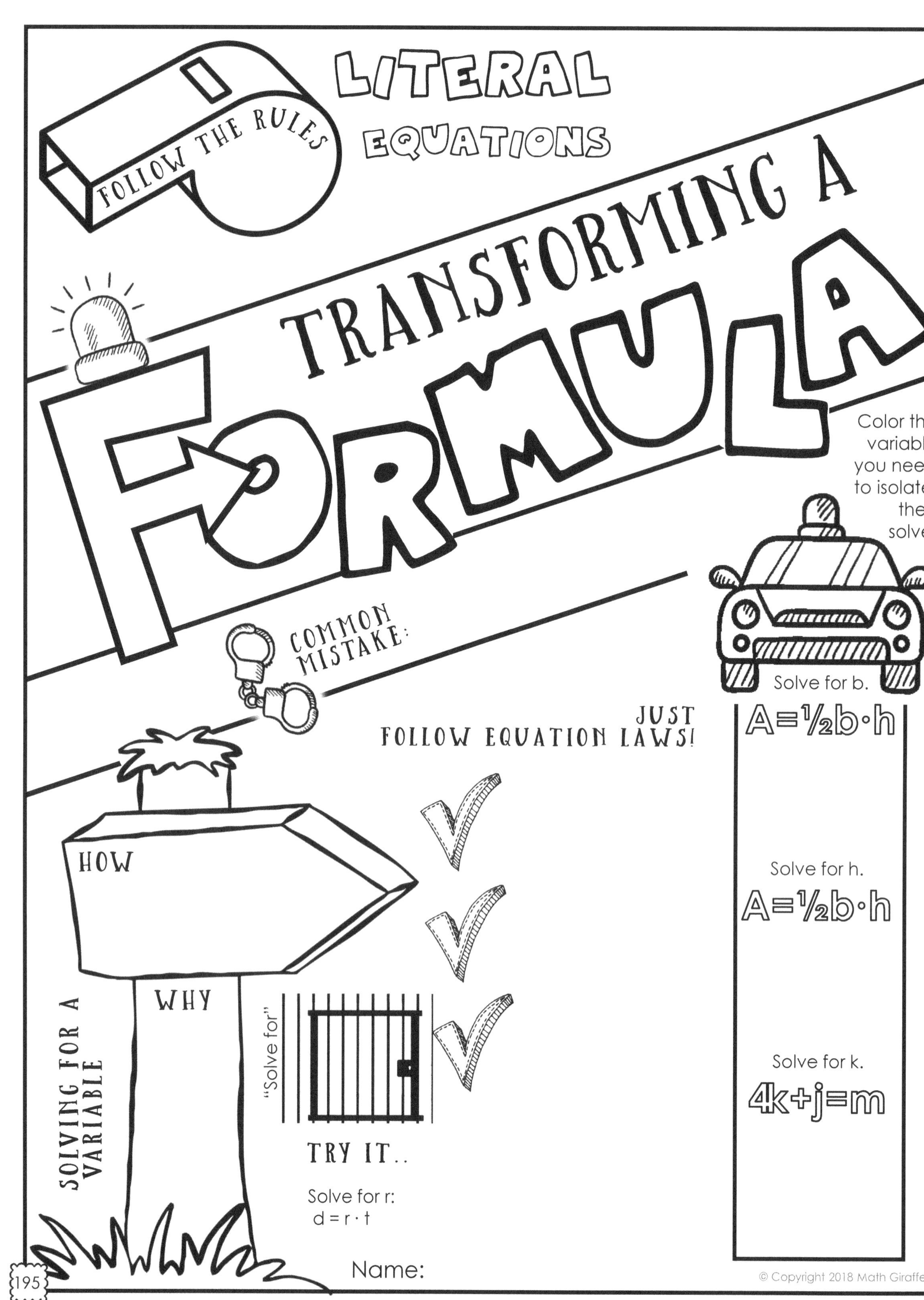
LITERAL
EQUATIONS
FOLLOW THE RULES
TRANSFORMING A
FORMULA
Color the
variable
you need
to isolate
the
solve
COMMON
MISTAKE:
JUST
FOLLOW EQUATION LAWS!
Solve for b.
A=½b·h
HOW
WHY
Solve for h.
A=½b·h
SOLVING FOR A
VARIABLE
"Solve for"
Solve for k.
4k+j=m
TRY IT..
Solve for r:
d = r · t
Name:

writing an equation

strategy

label the steps

Name:

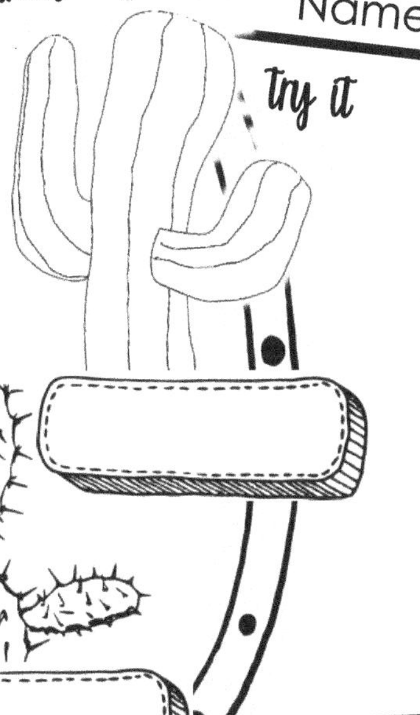

define variables

try it

example

Two cactus plants have a total height of 14 inches. One is 2 inches taller than the other. Find the height of each plant.

practice

define variables, write an equation, & solve

A sweater is $21 cheaper than it's original price once a 20% discount is applied. Find its original price.

Lucy is a florist. She charges $1.75 per rose, plus a delivery fee of $12. If a customer's bill for a bouquet of roses to be delivered is $38.25, how many roses are in the bouquet?

A mini golf facility normally charges each player $8. Their group discount allows large groups to play at a lower rate. If a group of 14 players is charged the same amount that a group of 9 players is charged, what is the group rate?

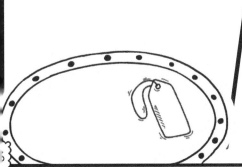

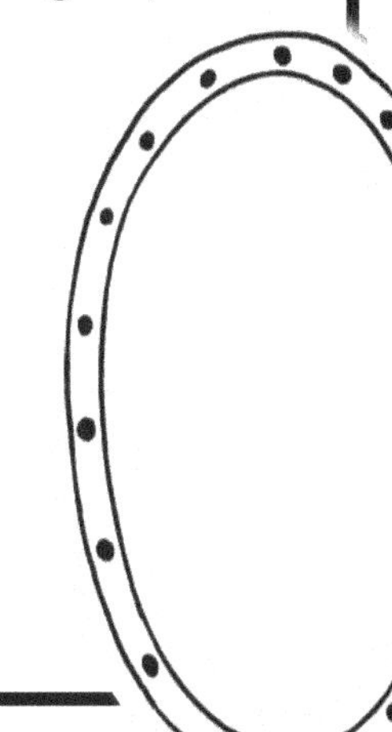

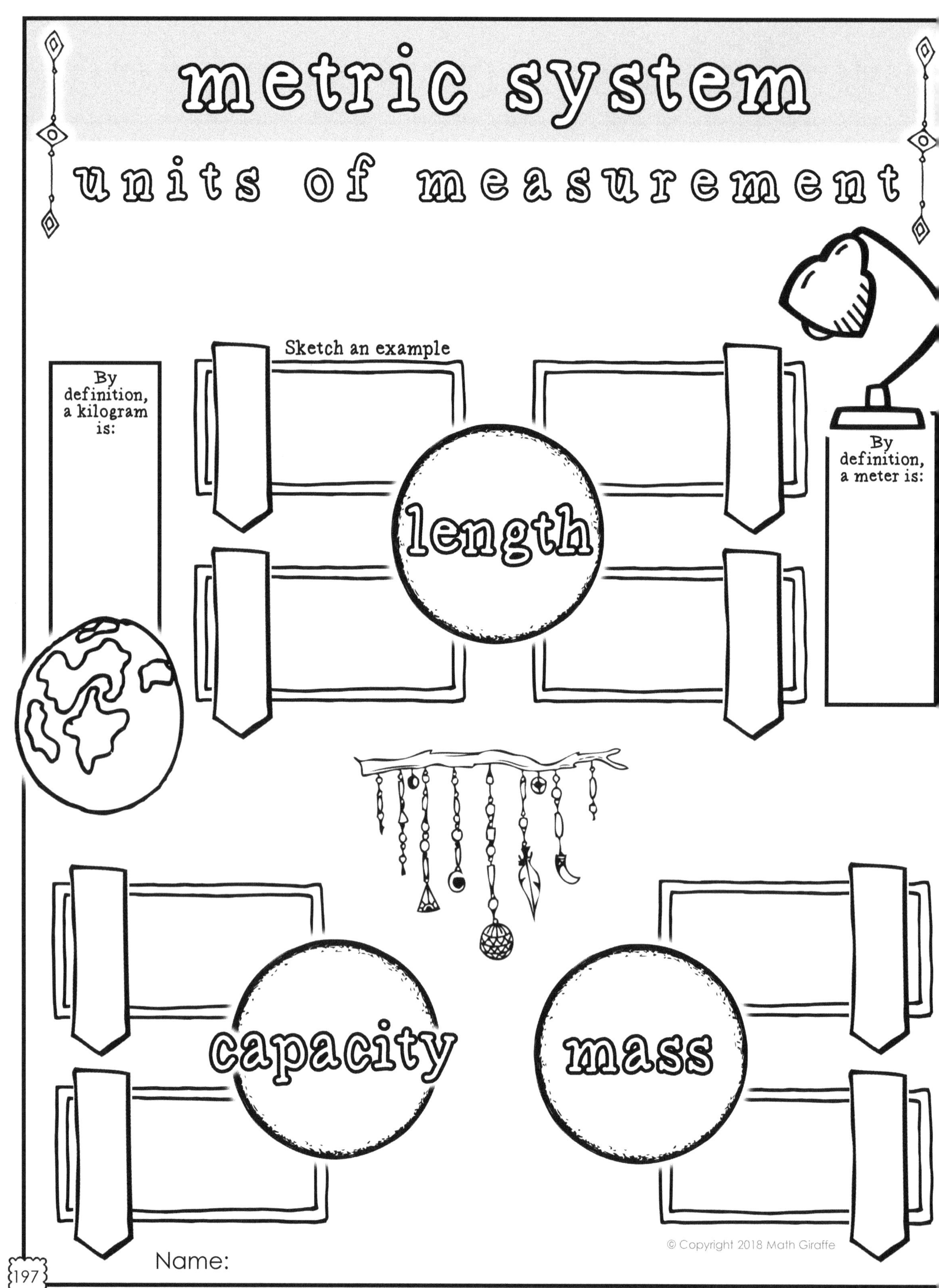
metric system
units of measurement
Sketch an example
By definition, a kilogram is:
By definition, a meter is:
length
capacity
mass
Name:

metric units

understanding & estimating

Color code / match each item with the appropriate unit of measure

- 1 gram
- 1 liter
- 1 kilogram
- 1 meter
- 7 centimeters

Your best guess:

How many mL will the cup hold?

How many kg is the girl?

How many mm long is the weight?

How many cm tall is the bottle?

How many grams is the guitar?

Compare estimates with a friend and talk it through. Who is more accurate?

Name:

metric system
converting units
units
Sketch a memory trigger
Add a curved arrow above to represent shifting TWO steps. What about THREE?
Name:
try it
How many centimeters is equal to 15 meters?
How many kilograms is equal to 416 grams?
Use ______________________ if you are converting from a larger unit to a smaller unit, because there will be ______________ of the smaller units.
Use ______________________ if you are converting from a smaller unit to a larger unit, because there will be ______________ of the larger units.

customary system
units of measurement
capacity
What is the best unit for the capacity of a teacup?
Once you finish creating this reference page, can you determine how many ounces a 18.5 pound backpack weighs?
Name:
What is the best unit for the weight of an elephant?
weight
What is the best unit for the length of a house?
length
© Copyright 2018 Math Giraffe

Name:

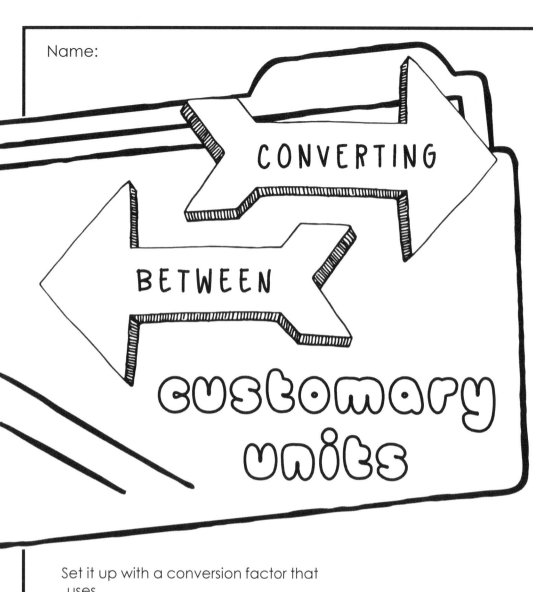

CONVERTING BETWEEN customary units

DIMENSIONAL ANALYSIS:

CONVERSION FACTOR:

Example: Convert 18 pints into gallons.

Set it up with a conversion factor that uses ______________________________

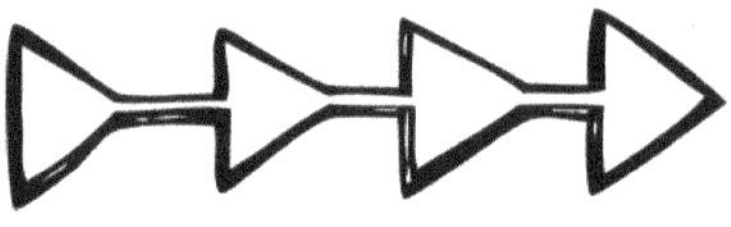

Divide / ___________.

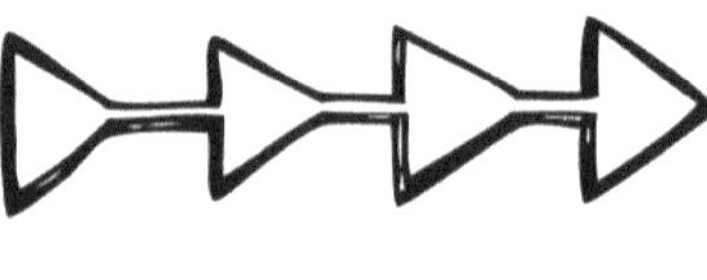

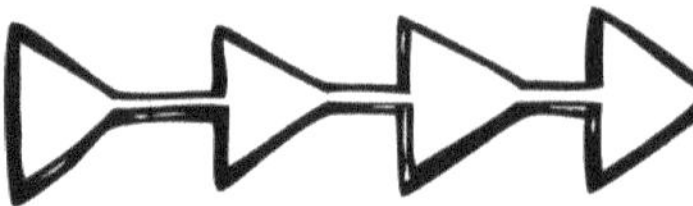

FUN FACT

The abbreviation "lb" comes from

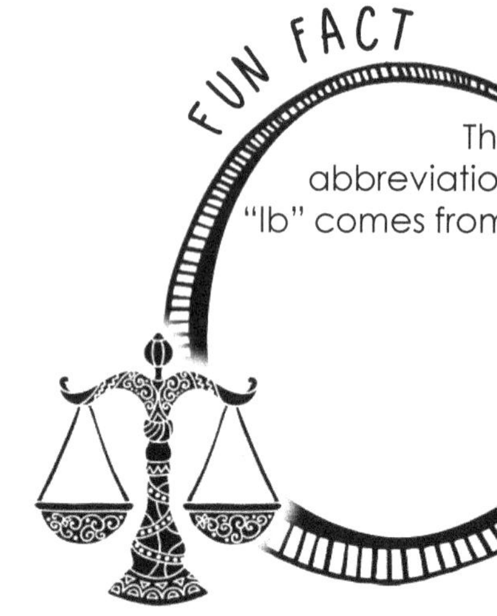

TRY IT

A 60 gallon fish tank costs $120. Another tank that holds 200 quarts has the same price. Which is a better deal by capacity?

How many inches does a track athlete run in the 400 yard race?

Units of Capacity
1 cup
1 pint
1 ounce
1 gallon
1 quart
ame:

DEFINE EACH:
principal
interest rate
interest
© Copyright 2018 Math Giraffe
NAME:
Simple interest
EXPLAIN IT:
MAP OUT HOW IT WORKS:
FILL IT IN:
I = Prt

Simple interest

USING THE FORMULA

I = Prt

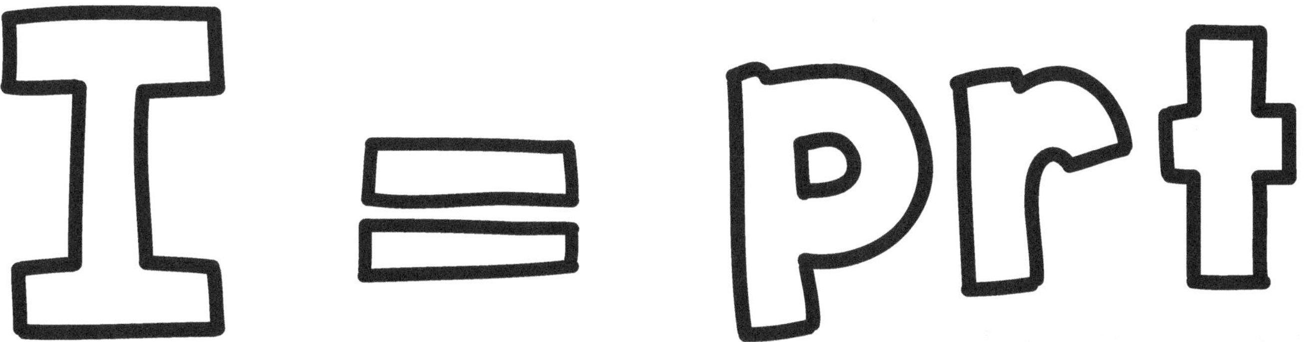

I DEPOSITED $350 INTO A SAVINGS ACCOUNT WITH 3% INTEREST. HOW MUCH WILL BE IN THE ACCOUNT AFTER 2 YEARS?

I PLAN TO PUT $500 PER QUARTER INTO AN ACCOUNT WITH 6% INTEREST. HOW MUCH INTEREST WILL I EARN IN THE FIRST QUARTER?

I'M PUTTING ALL $282.50 I EARNED FROM MY JOB INTO MY BANK ACCOUNT. WHAT INTEREST RATE WOULD GET ME TO $300 BY THE END OF THE YEAR?

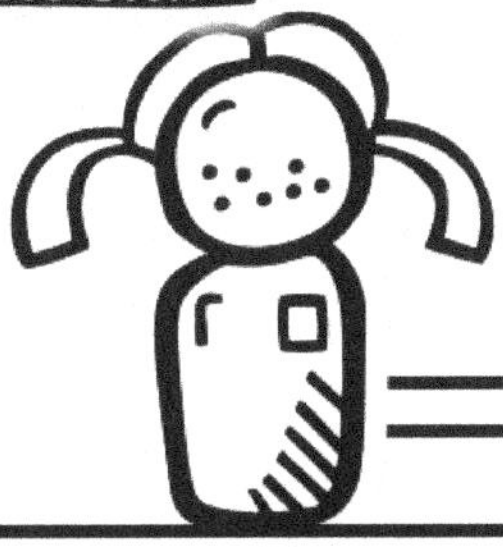

NAME:

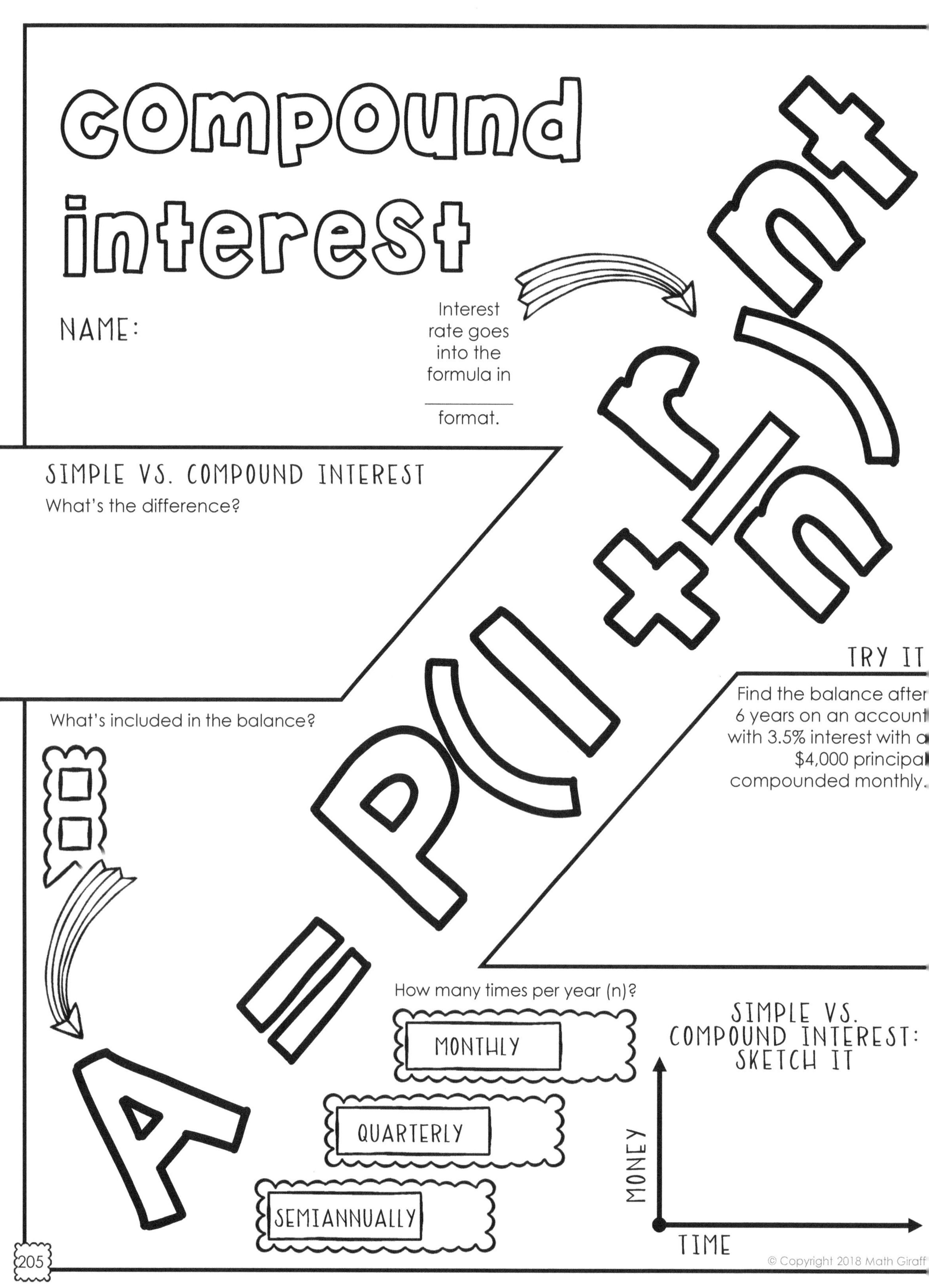
compound interest
NAME:
Interest rate goes into the formula in ________ format.
$A = P(1 + \frac{r}{n})^{nt}$
SIMPLE VS. COMPOUND INTEREST
What's the difference?
What's included in the balance?
TRY IT
Find the balance after 6 years on an account with 3.5% interest with a $4,000 principal compounded monthly.
How many times per year (n)?
MONTHLY
QUARTERLY
SEMIANNUALLY
SIMPLE VS. COMPOUND INTEREST: SKETCH IT
MONEY
TIME

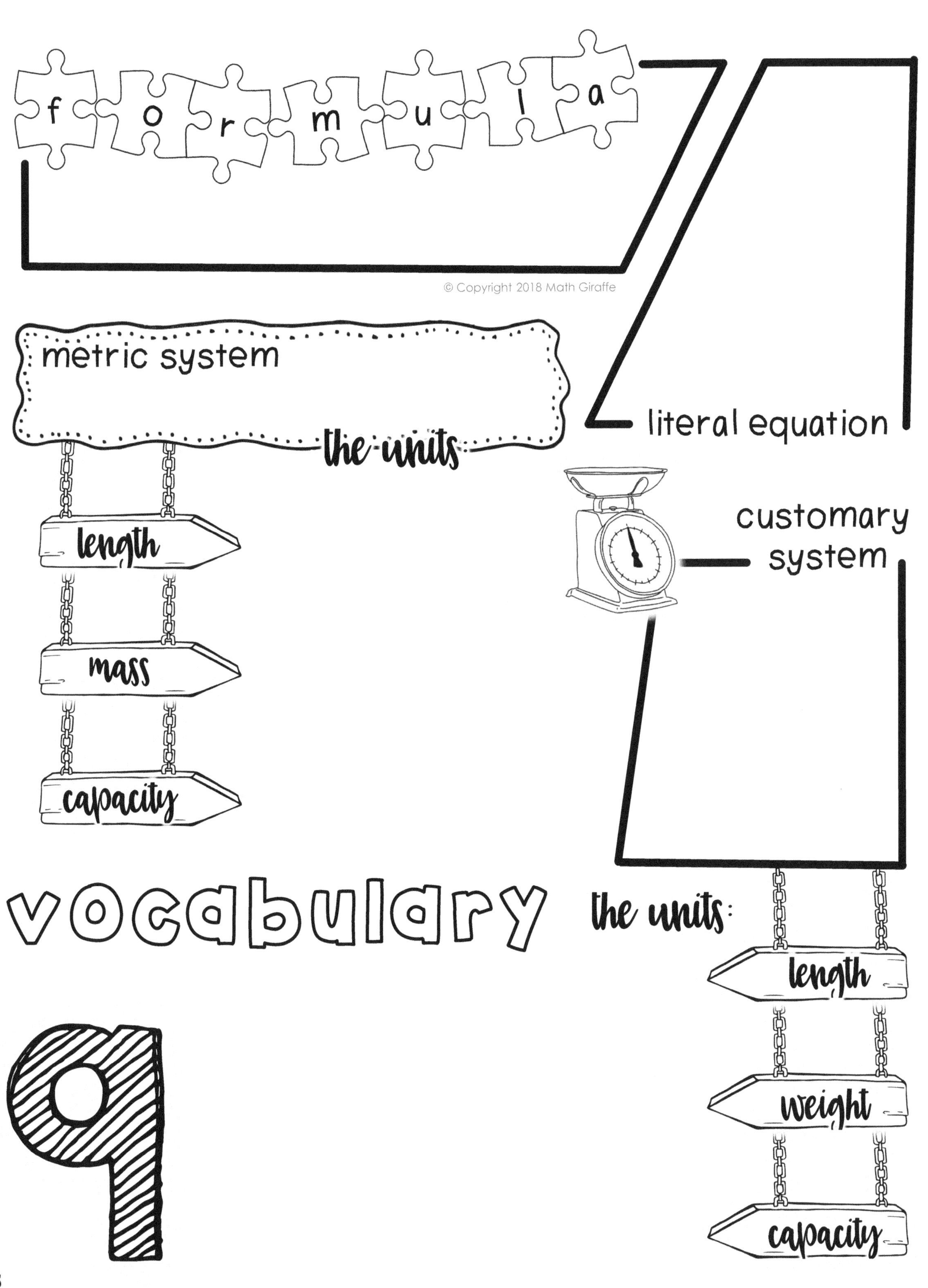
f o r m u l a
literal equation
metric system
the units
length
mass
capacity
customary system
the units:
length
weight
capacity
vocabulary
9

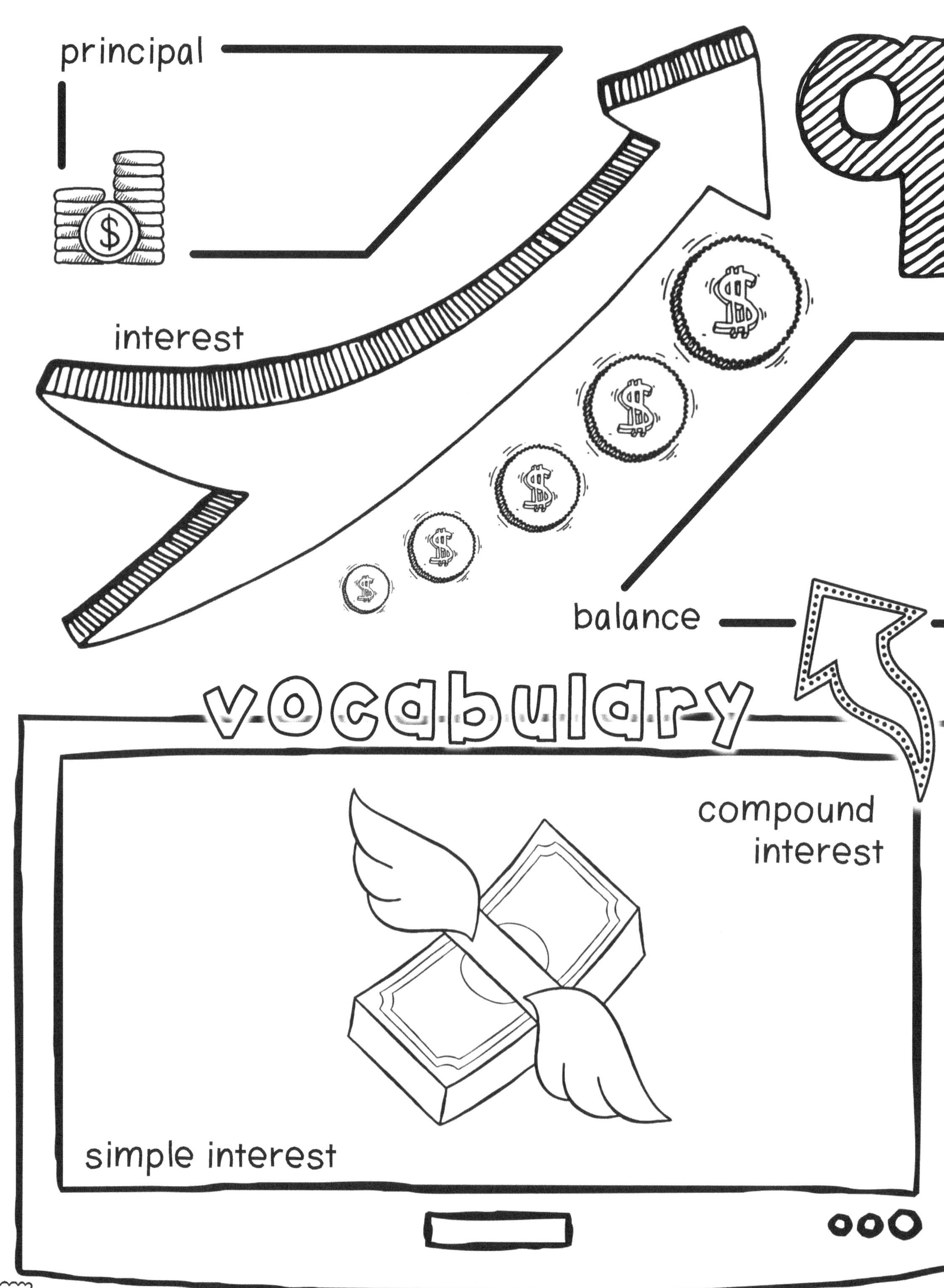
principal
interest
balance
vocabulary
compound interest
simple interest

10
Intro to
Algebra

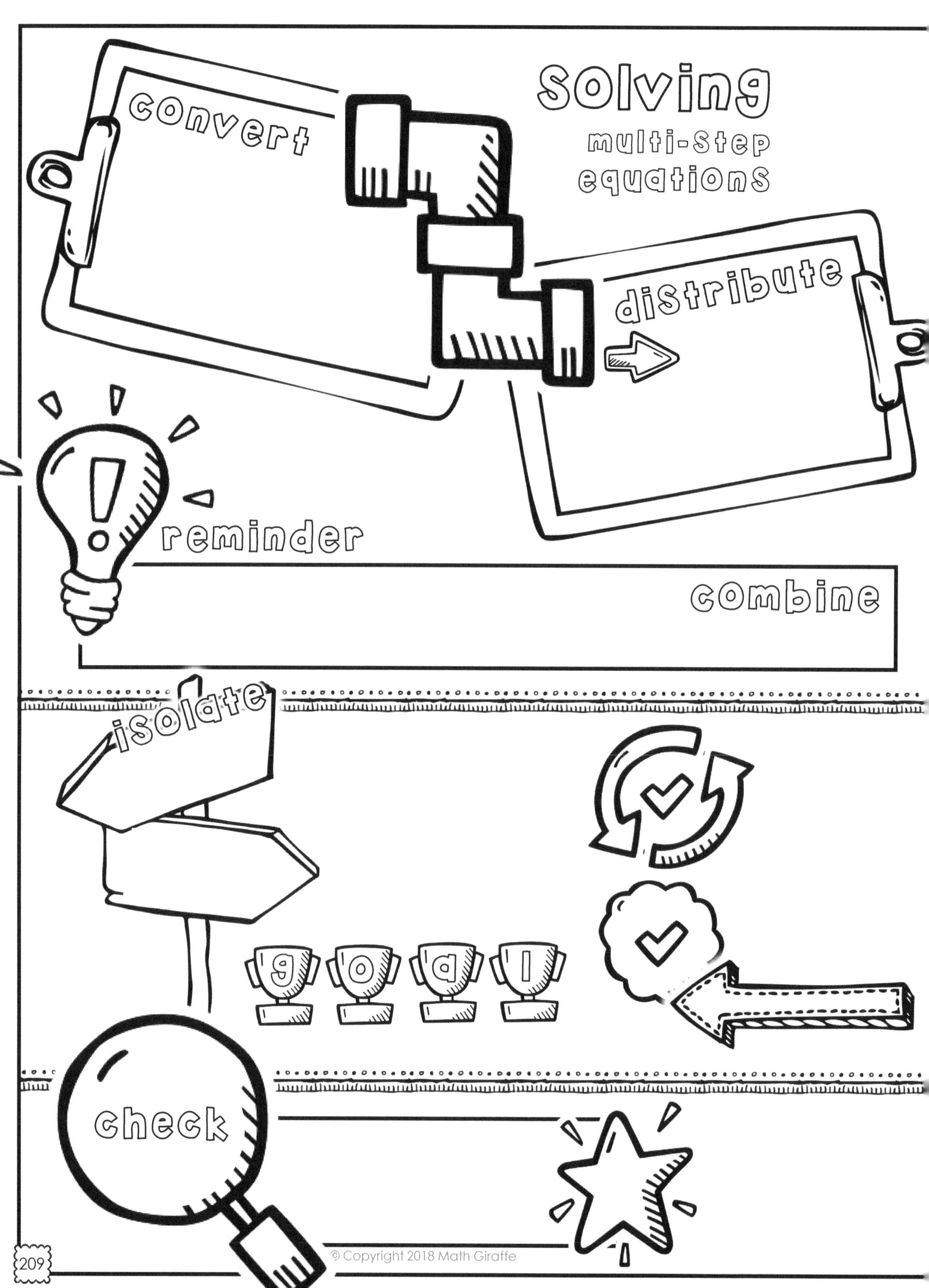
solving
multi-step
equations
convert
distribute
reminder
combine
isolate
g
o
a
l
check

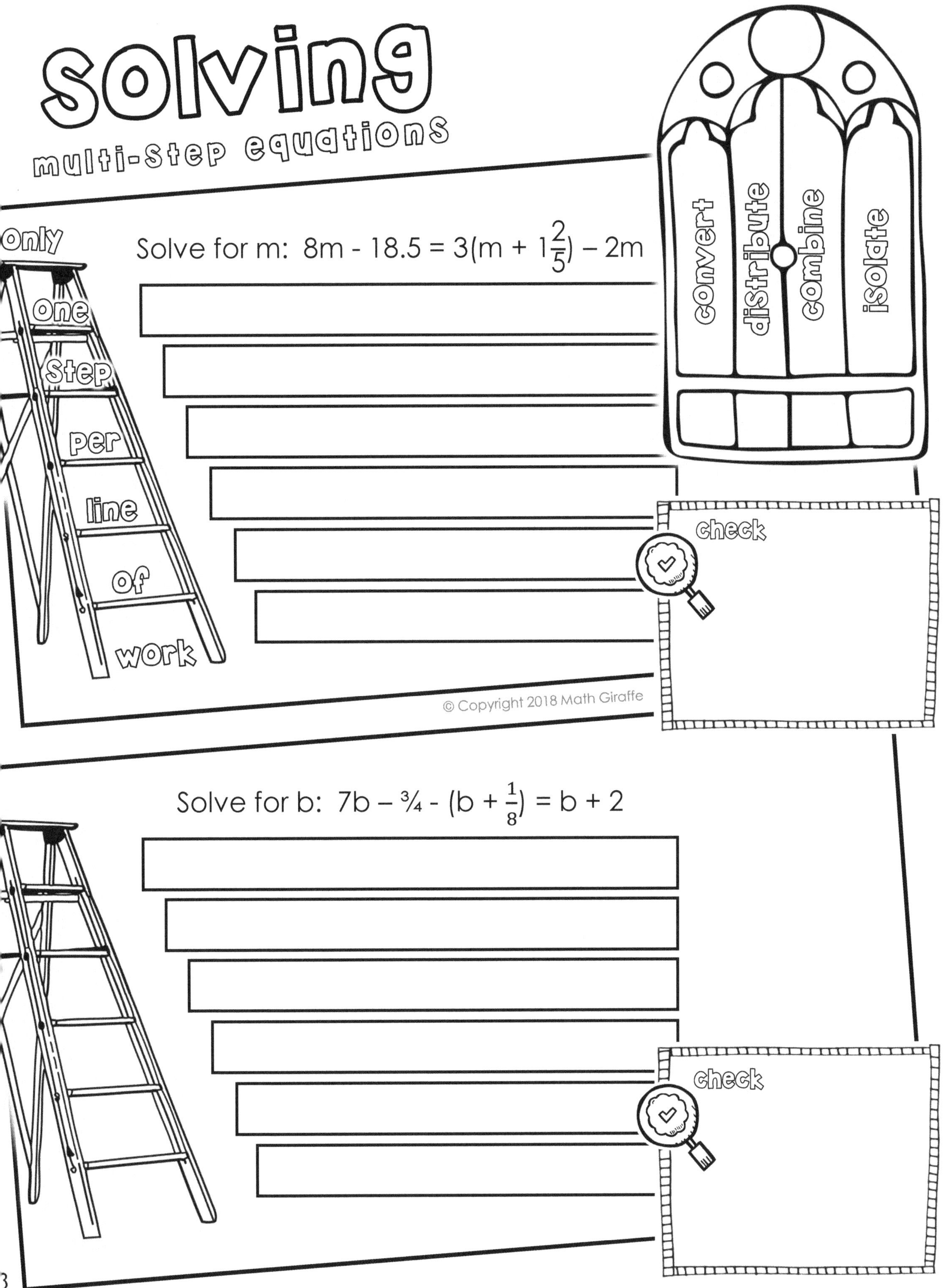
solving
multi-step equations
only
one
step
per
line
of
work
Solve for m: 8m - 18.5 = 3(m + 1$\frac{2}{5}$) – 2m
convert
distribute
combine
isolate
check
Solve for b: 7b – ¾ - (b + $\frac{1}{8}$) = b + 2
check

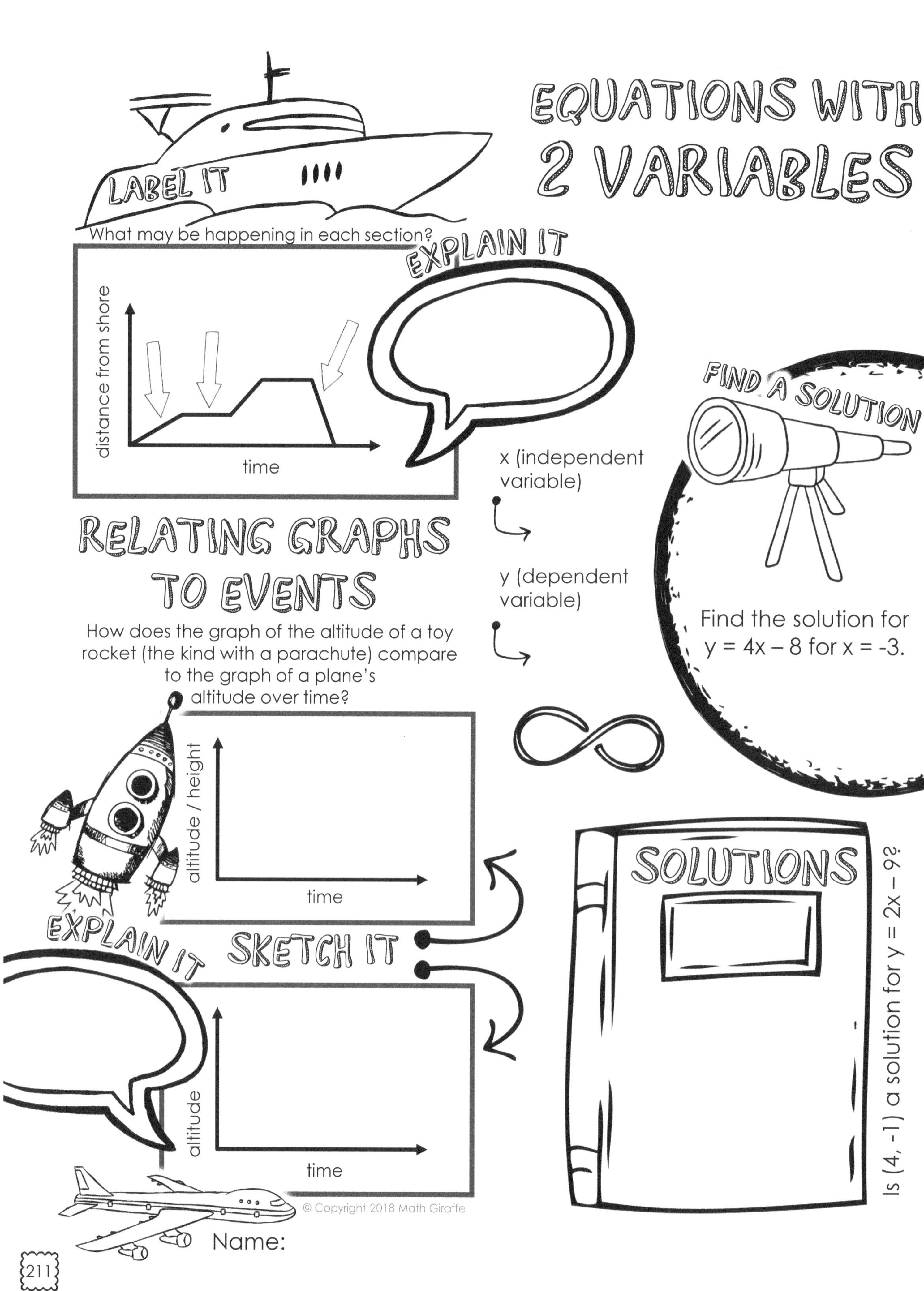
EQUATIONS WITH 2 VARIABLES
LABEL IT
What may be happening in each section?
EXPLAIN IT
distance from shore
time
x (independent variable)
y (dependent variable)
FIND A SOLUTION
Find the solution for y = 4x – 8 for x = -3.
RELATING GRAPHS TO EVENTS
How does the graph of the altitude of a toy rocket (the kind with a parachute) compare to the graph of a plane's altitude over time?
altitude / height
time
EXPLAIN IT
SKETCH IT
altitude
time
SOLUTIONS
Is (4, -1) a solution for y = 2x – 9?
Name:

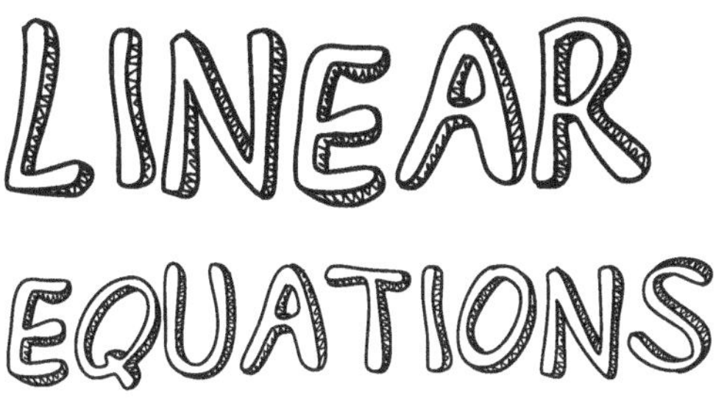

Graph
$y = x - 2$

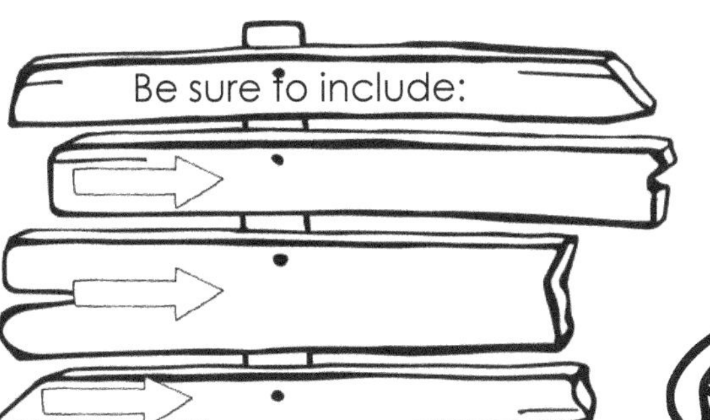

Be sure to include:

x	y
-2	
-1	
0	

2 PLOT THE POINTS

& draw a line through them

Define "linear."

Name:

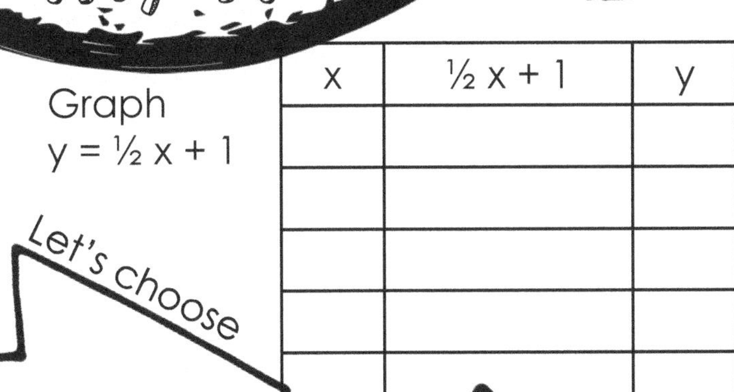

Graph
$y = \frac{1}{2}x + 1$

x	½ x + 1	y

Let's choose

Let's add

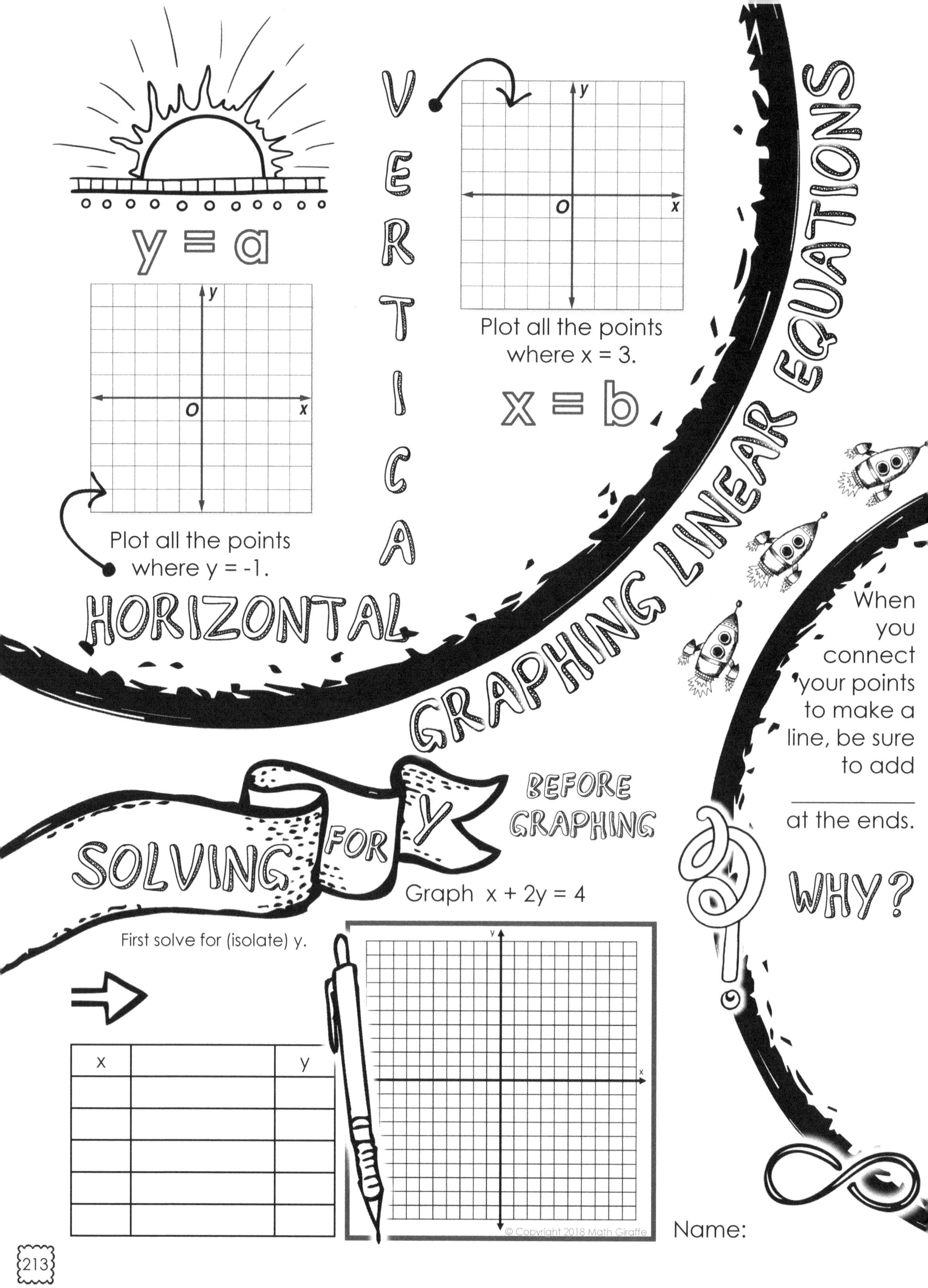

GRAPHING LINEAR EQUATIONS
y = a
Plot all the points where y = -1.
HORIZONTAL
VERTICAL
Plot all the points where x = 3.
x = b
When you connect your points to make a line, be sure to add
at the ends.
WHY?
SOLVING FOR Y BEFORE GRAPHING
Graph x + 2y = 4
First solve for (isolate) y.
x
y
Name:
© Copyright 2018 Math Giraffe

the function machine

____

______

____

____

______

____

A **function** is a ______________ with one or more ______________ where each ______________ has a single ______________.

functions

Each __________ is only allowed to correspond to ONE ________!

This relation is a function because none of the input values (x-values) has more than one different output (y-value).

x	0	1	2	3	4
y	0	3	6	9	12

This relation is NOT a function because at least one of the input values (x-values) has more than one different output (y-value).

x	-3	2	-3	5	8
y	1	5	-1	2	-2

When we put something IN, we can always expect a consistent result to come back OUT.

notation

When dealing with functions, you will see ________ in place of y.

How to say it out loud: "   "

evaluating

To evaluate a function for a particular x-value, just ______________ and then simplify!

Example: If f(x) = 2x + 1, find f(3).

Work: f(3) = 2(3) + 1

Answer: _____

ame:

determining whether a relation is a function

Add circles, arrows, lines, etc. to demonstrate why each relation is/isn't a functio

table

Review the columns. The relation will not be a function if any ______________________ corresponds to more than one different ____________________________.

A

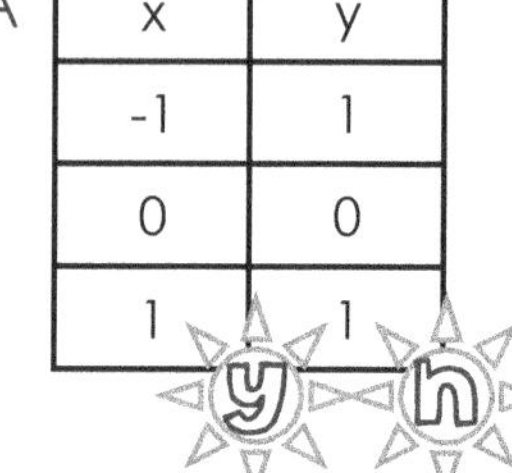

x	y
-1	1
0	0
1	1

B

x	y
3	-6
0	1
3	6

y n

set notation

Review each ordered pair. The relation will not be a function if any ______________________ corresponds to more than one different ____________________________.

C

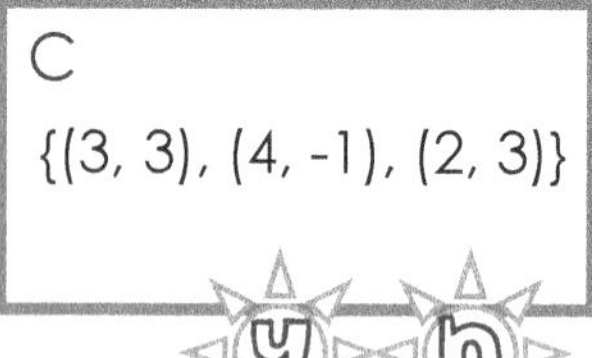

{(3, 3), (4, -1), (2, 3)}

D

{(1, 8), (0, -2), (1, -3

y n

graph

Use the ____.
The relation will not be a function if a vertical line ever ____
____.

E

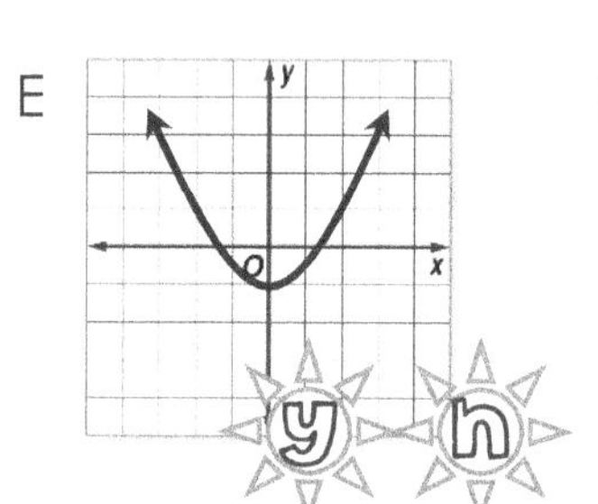

F

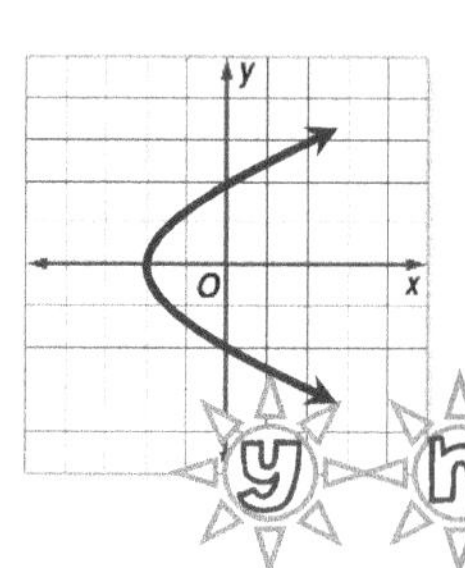

mapping diagram

Review the arrows. The relation will not be a function if any ______________________ maps to more than one different ____________________.

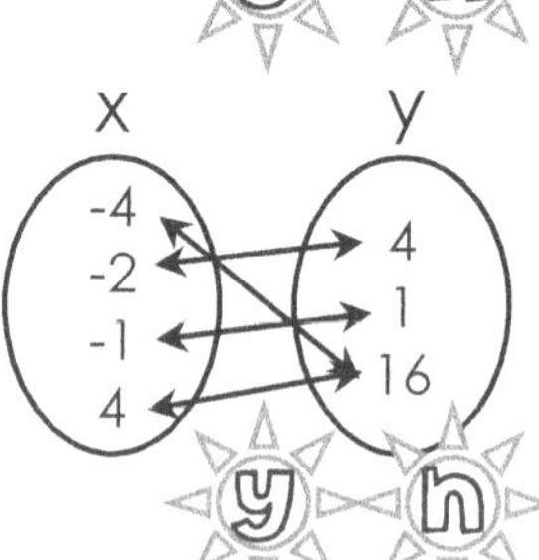

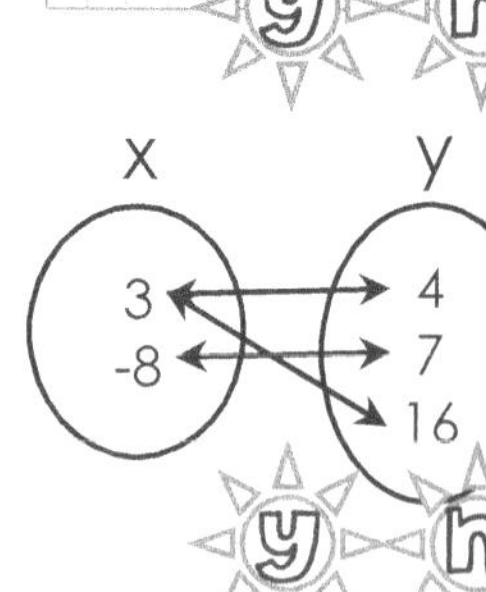

finding domain and range

The domain is the set of all possible ________________.
The range is the set of all possible ________________.

Identify the domain and range of the relations in the "table," "set notation," and "graph' examples above.

	Domain	Range
A		
B		
C		
D		
E		
F		

Try it

Create a mapping diagram and a graph that each represent functions.

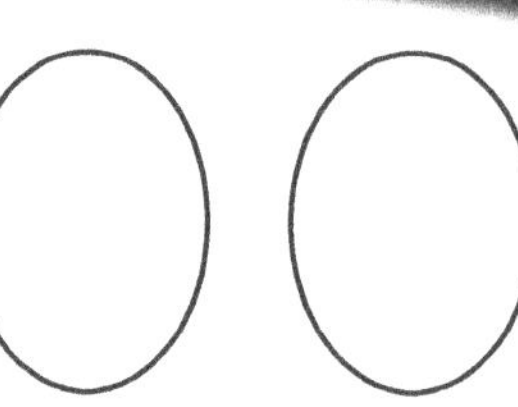

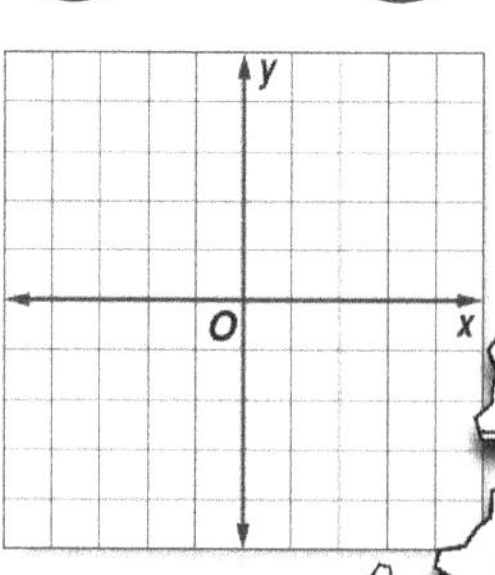

ame:

Slope (vertical change over horizontal change) is represented by the letter "m."

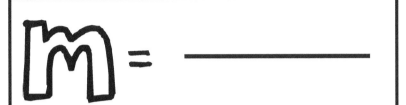

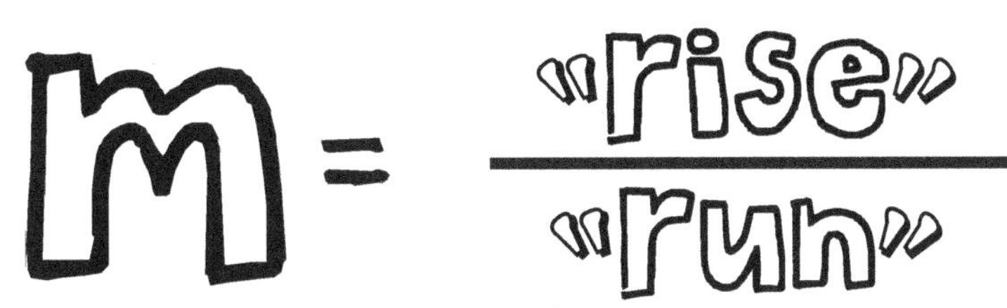

The slope of a line can be determined from a table, by __________ units on a coordinate plane, or by ____________ coordinates.

SLOPE

Find the slope between the two points.

1. (3, -2) and (4, 4)

2. (6, 0) and (-8, -1)

The slope of a horizontal line is ______.

The slope of a vertical line is ______________.

Remember:

UP and RIGHT are __________ movements;

DOWN and LEFT are __________ movements.

Plot a line that starts at the origin and has a slope of -3. Label it "a."

Plot a line that starts at (0, 4) and has a slope of $-\frac{3}{4}$. Label it "b."

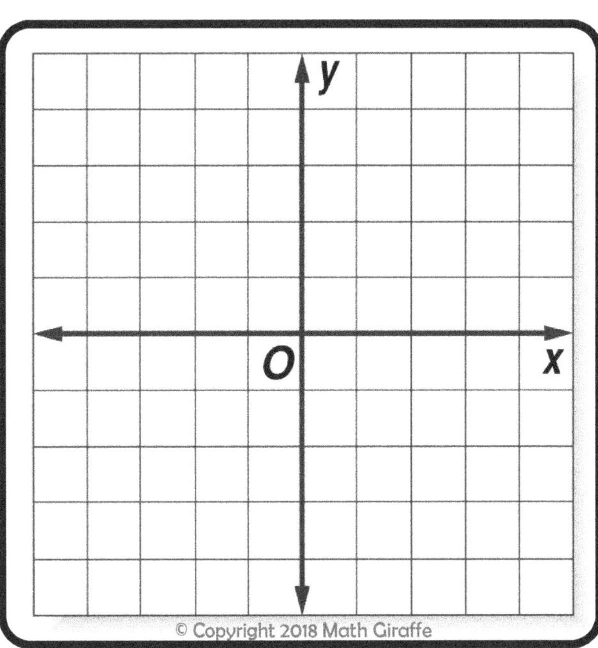

Steeper slopes have greater ______________.

STEEP

Name:

try it

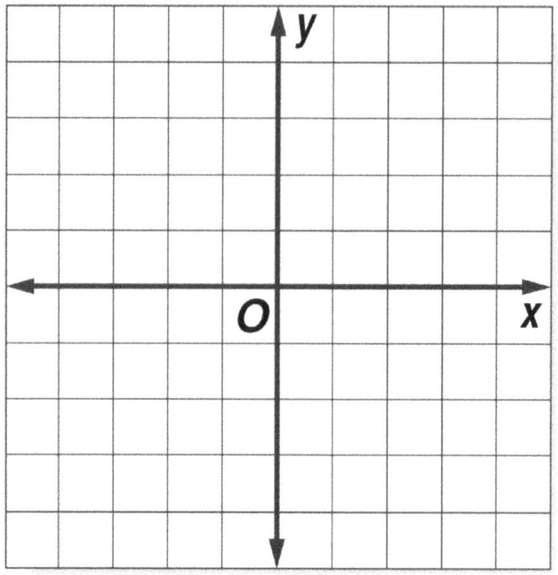

y

O x

Graph four different lines, all with different negative slopes. Show each slope and compare steepness.

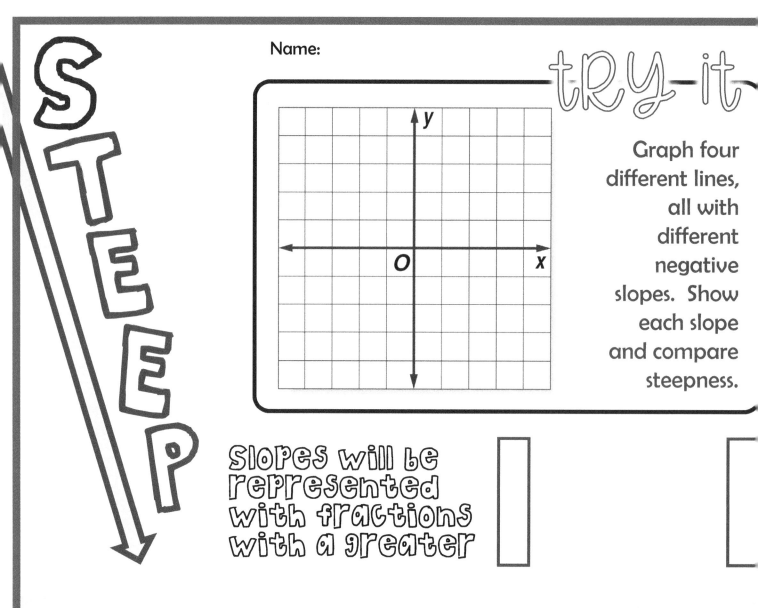

Slopes will be represented with fractions with a greater

Sketch a sample (or a few) of each type of slope. Add a skier if you want! It may help you remember the direction and whether the values are increasing or decreasing

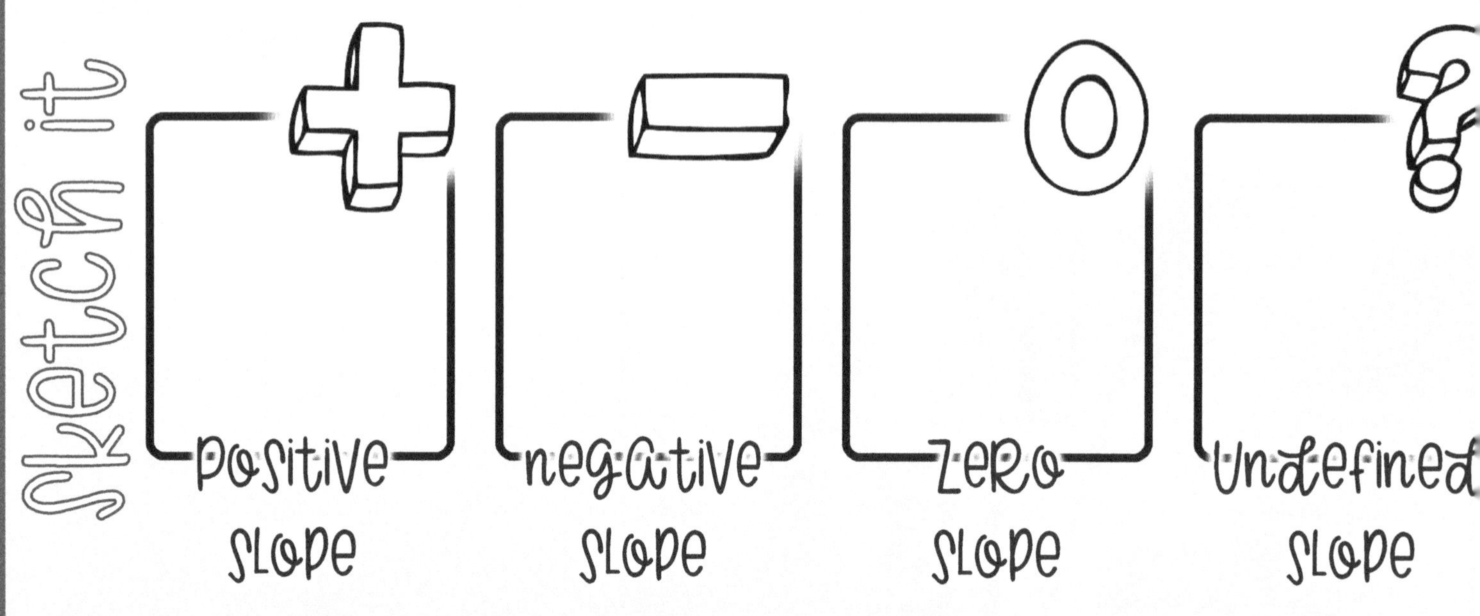

order from steepest to least steep: 1/3, 3, 3/2, 3/4

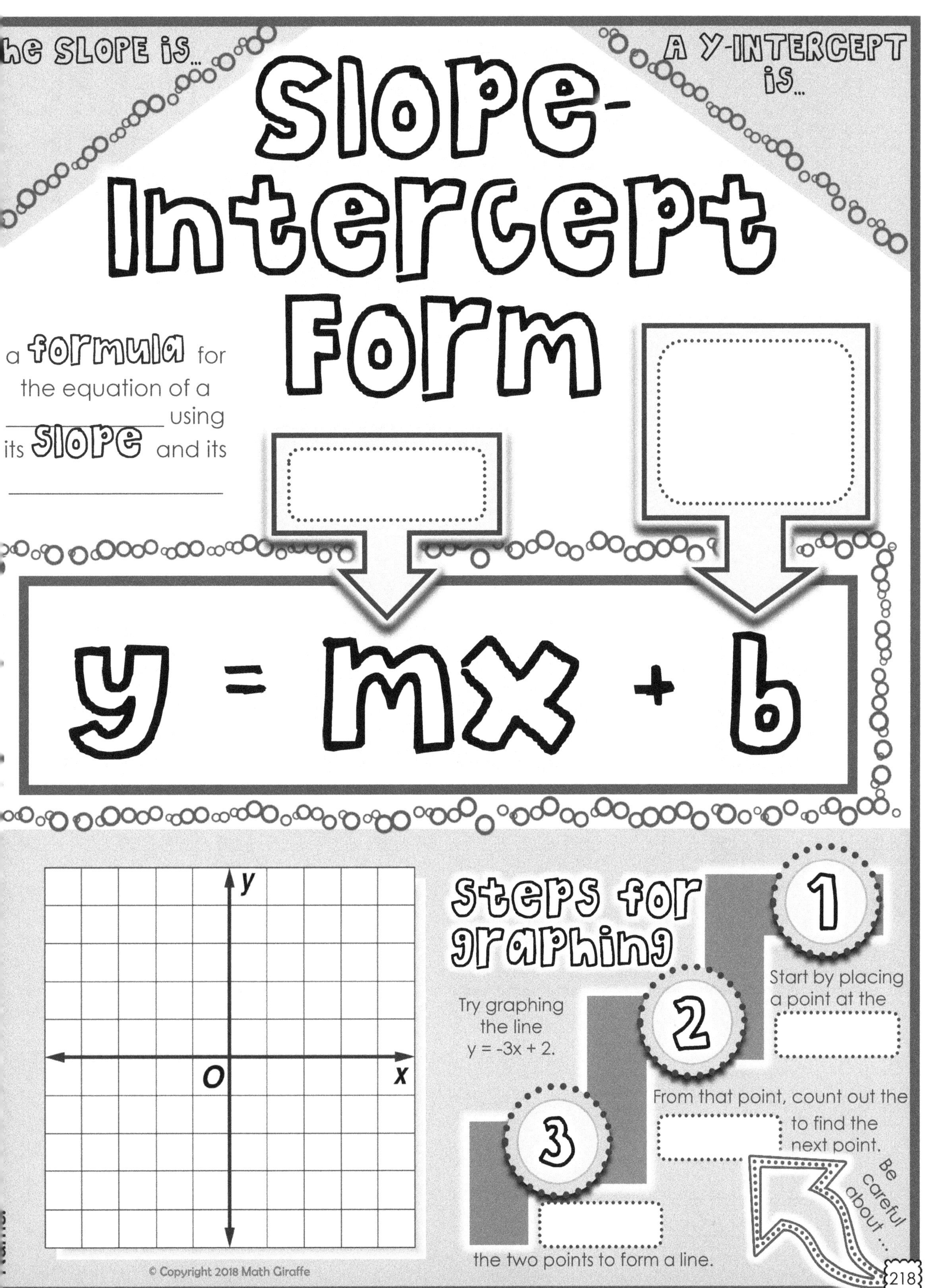
he SLOPE is...
A Y-INTERCEPT is...
Slope-Intercept Form
a formula for the equation of a ______ using its slope and its ______
y = mx + b
y
O
x
Steps for graphing
1
Start by placing a point at the
Try graphing the line y = -3x + 2.
2
From that point, count out the to find the next point.
3
Be careful about...
the two points to form a line.
© Copyright 2018 Math Giraffe
218

practice

Using Slope-Intercept Form

Write an equation for each line in slope-intercept form.

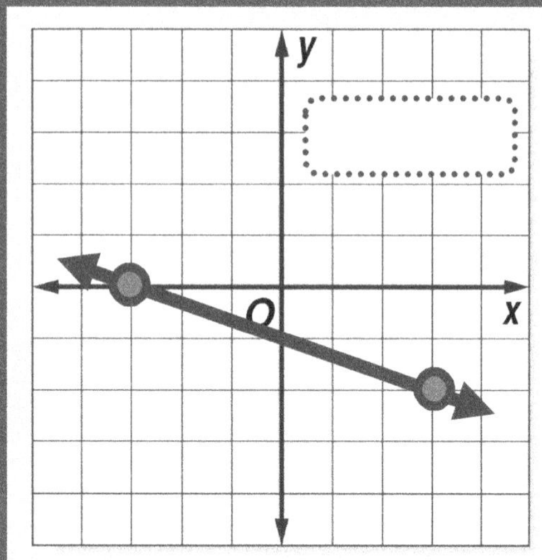

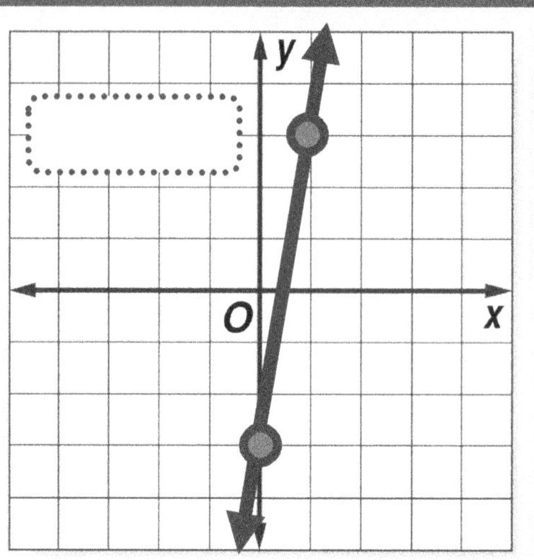

Graph each line.

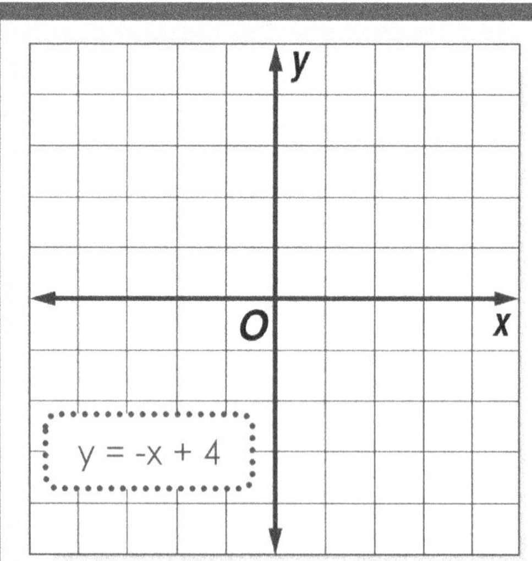

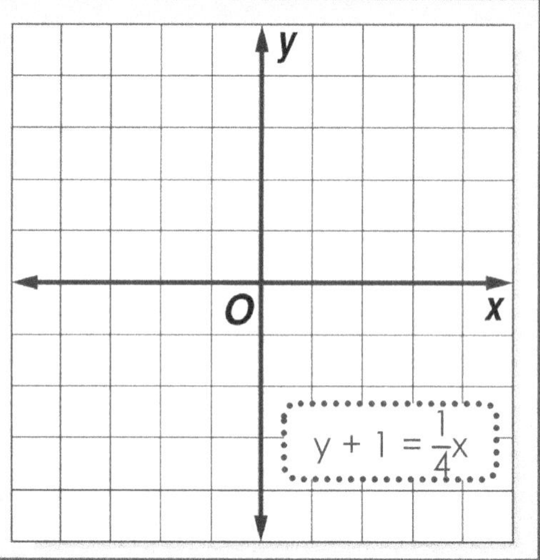

Identify the slope and y-intercept for each.

A. $4y - 2x = 8$

B. $7 = y - x$

C.

x	-2	2	6
y	5	-3	-11

D. the line that passes through (3, 3) and (-6, 0)

Name:

Special Cases

$y = 2x$

$y = 4$

Extra Work

$3y + 5 = 6x$

Systems of Equations

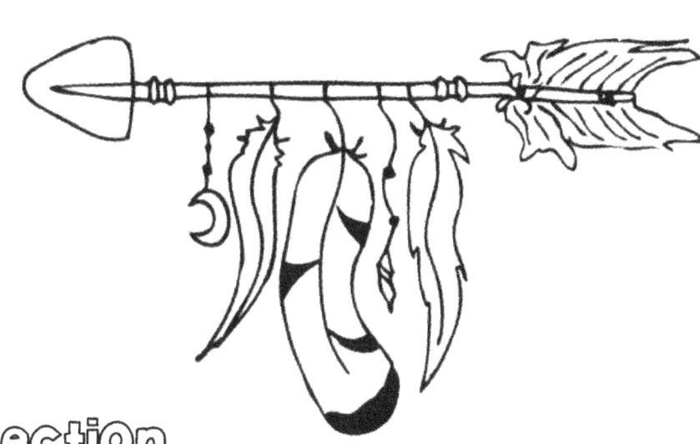

System of linear equations:

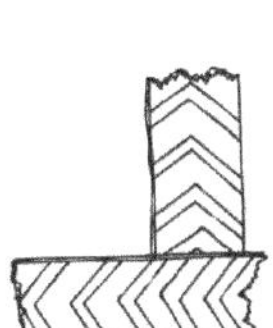

Solution for a system of equations

y = -2x - 1

y = x - 4

x

y

O

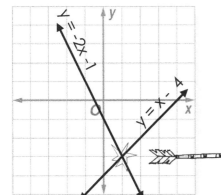

intersection point:

Solve each system by graphing.

Name:

reminder

$y = -x + 2$
$y = 4x - 3$

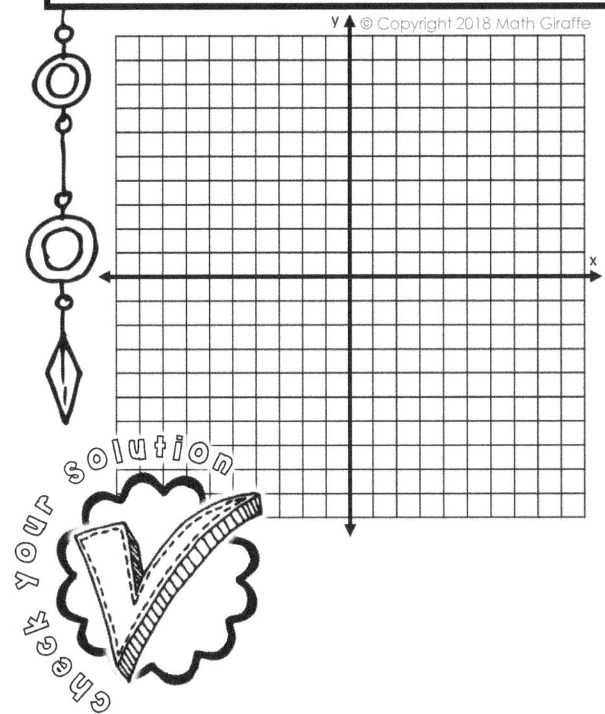

$y = 3x - 3$
$1 = x + y$

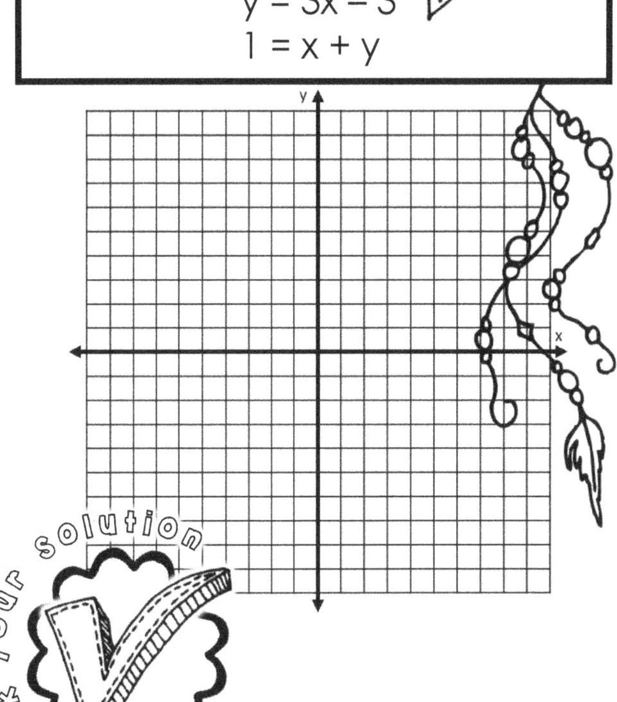

PROBLEM

Name:

Claire is eight years older than Trevon.
In three years, she will be twice his age.
Figure out how old Claire and Trevon are.

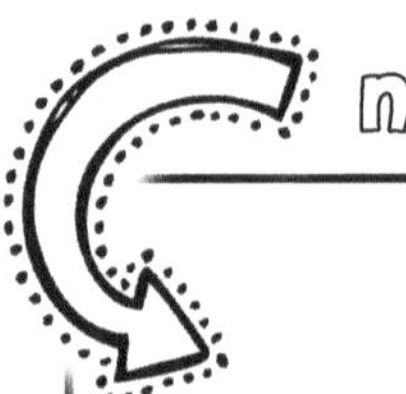

now

in 3 years

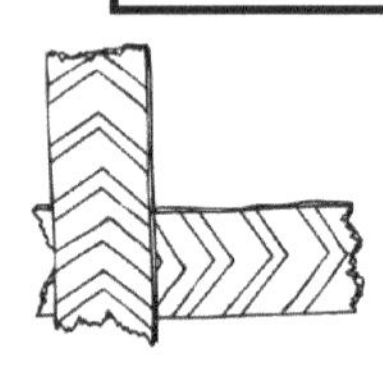

2 separate equations

Systems of Equations

The solution will be the...

Substitution

$$C = T + 8$$
$$C = 2T + 3$$

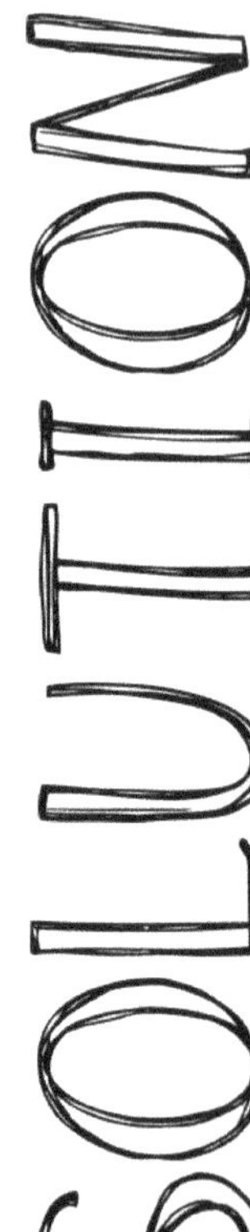

Special Systems of equations

$2x + y = 2$
$y = -2x$

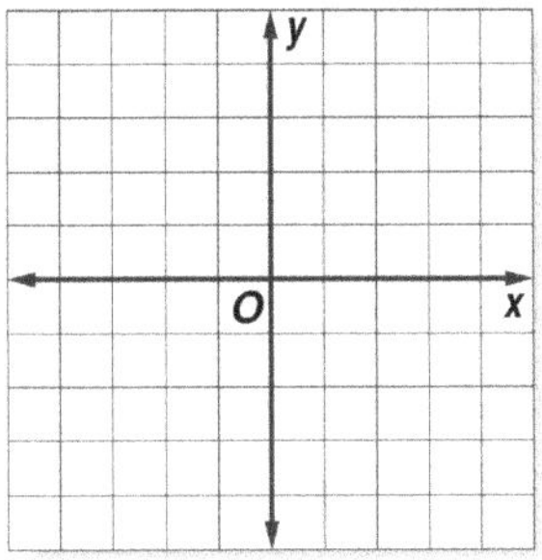

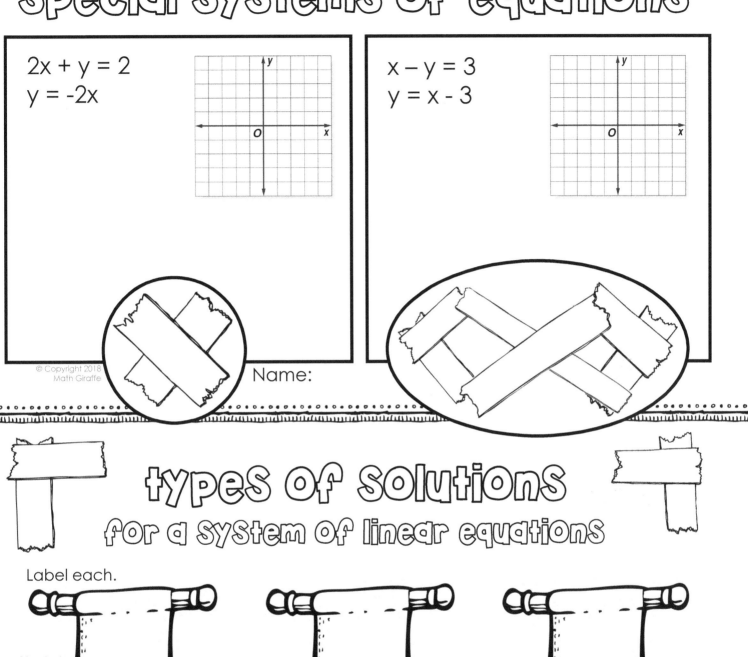

$x - y = 3$
$y = x - 3$

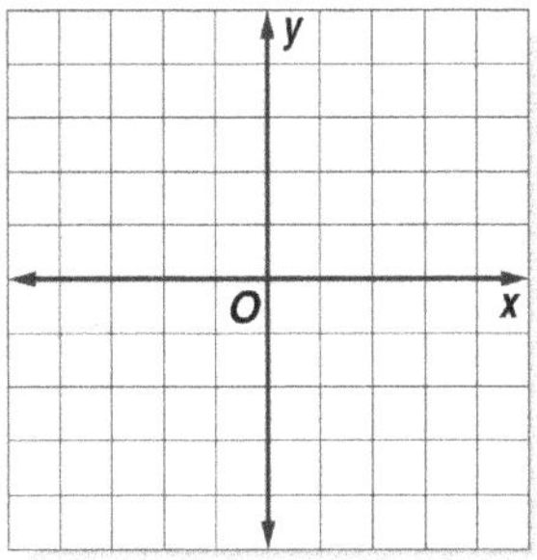

Name:

types of solutions

for a system of linear equations

Label each.

Sketch each.

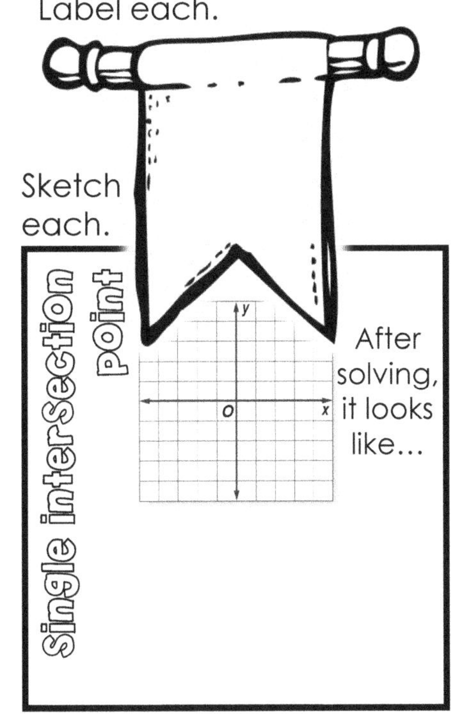

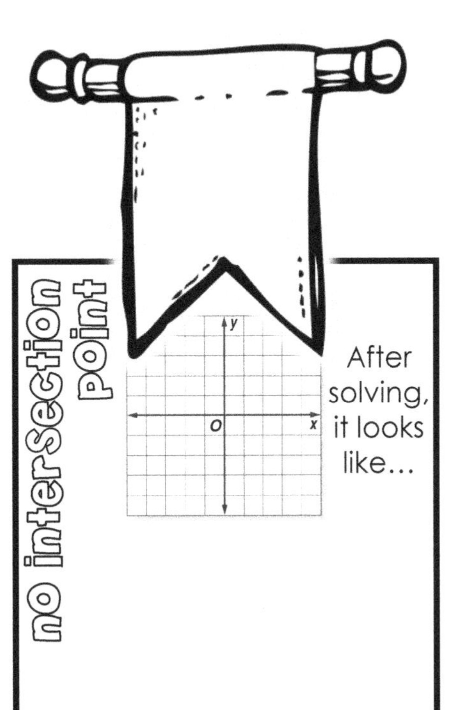

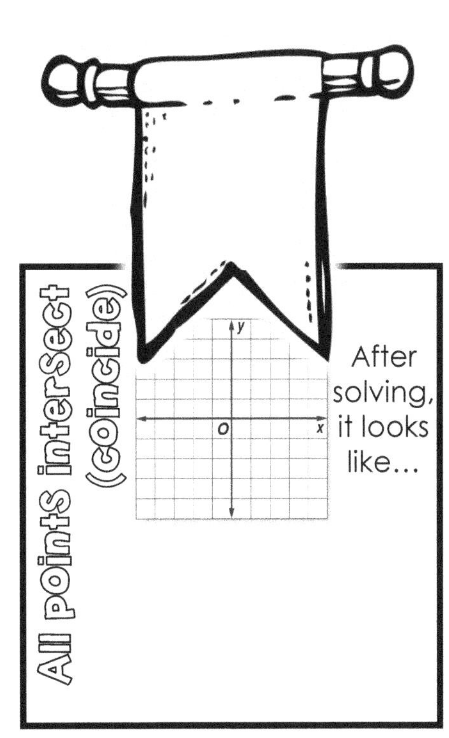

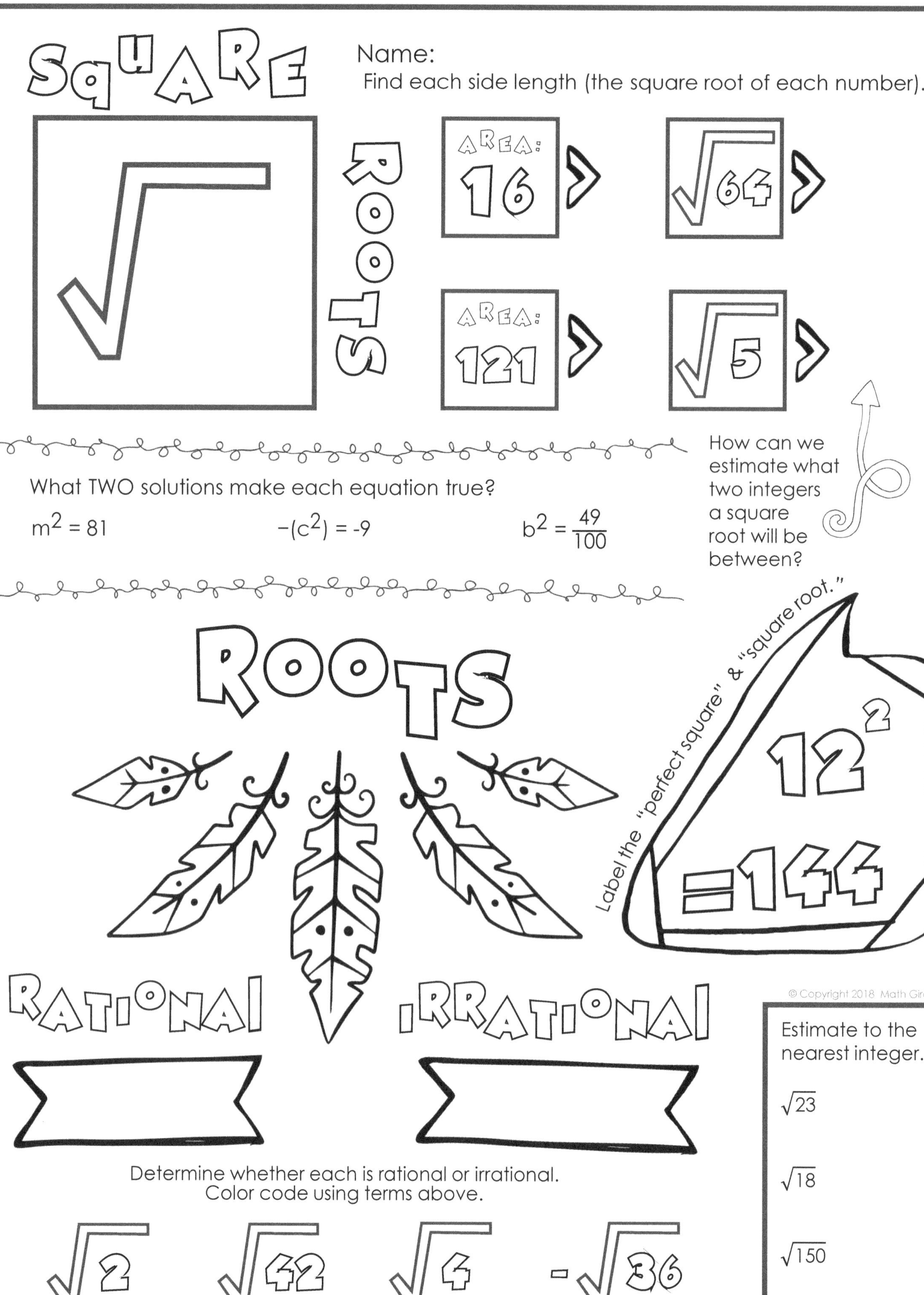
Square
Name:
Find each side length (the square root of each number).
Roots
AREA:
16
64
AREA:
121
5
How can we estimate what two integers a square root will be between?
What TWO solutions make each equation true?
$m^2 = 81$
$-(c^2) = -9$
$b^2 = \frac{49}{100}$
Roots
Label the "perfect square" & "square root."
12^2
=144
Rational
Irrational
Determine whether each is rational or irrational.
Color code using terms above.
$\sqrt{2}$
$\sqrt{42}$
$\sqrt{4}$
$-\sqrt{36}$
Estimate to the nearest integer.
$\sqrt{23}$
$\sqrt{18}$
$\sqrt{150}$
© Copyright 2018 Math Giraff

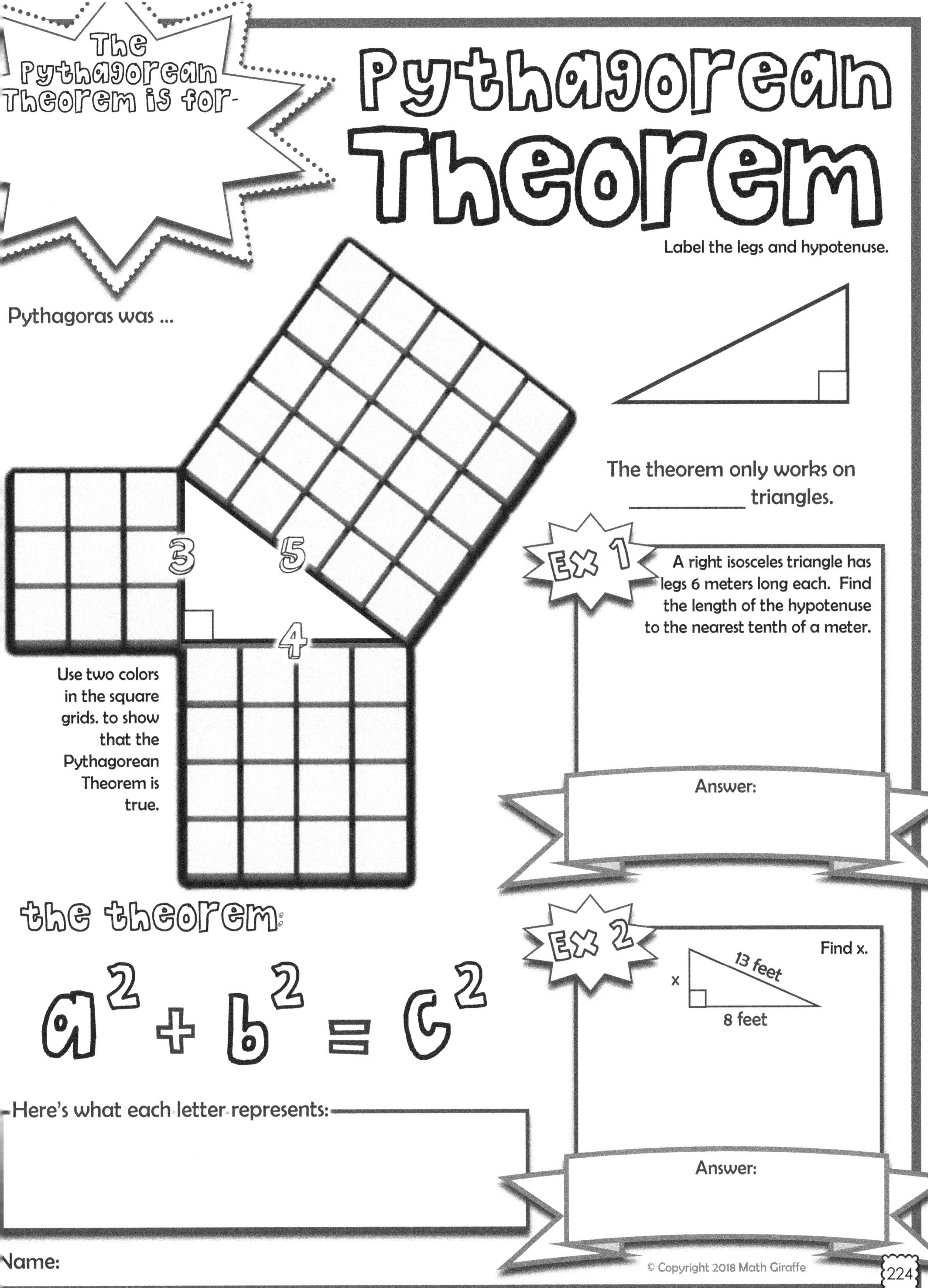

The Pythagorean Theorem is for-
Pythagorean Theorem
Label the legs and hypotenuse.
Pythagoras was ...
The theorem only works on ___________ triangles.
3
5
4
Use two colors in the square grids. to show that the Pythagorean Theorem is true.
Ex 1
A right isosceles triangle has legs 6 meters long each. Find the length of the hypotenuse to the nearest tenth of a meter.
Answer:
the theorem:
a² + b² = c²
Ex 2
Find x.
13 feet
x
8 feet
Here's what each letter represents:
Answer:
Name:

Try It

Find the length of the diagonal

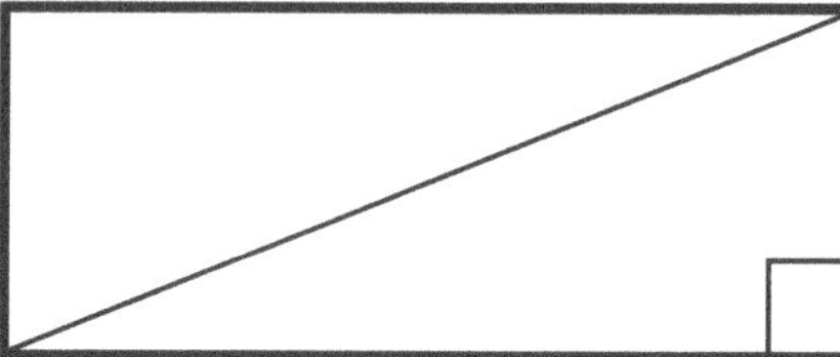

Is this a right triangle?

Side lengths: 8 cm, 10 cm, 16 cm

The theorem works both ways!

1. If a triangle is a right triangle, then ____________________.
2. If ____________________, then the triangle is a right triangle.

History

Although Pythagoras is credited with the first proof of the Pythagorean Theorem (used in Euclidean Geometry), it is believed that Babylonian, Mesopotamian, Chinese, and Indian mathematicians understood the concept before his time. There are many proofs of the Pythagorean theorem, including both algebraic and geometric proofs.

Find the distance between the points (4, -3) and (-2, 1) on the coordinate plane.

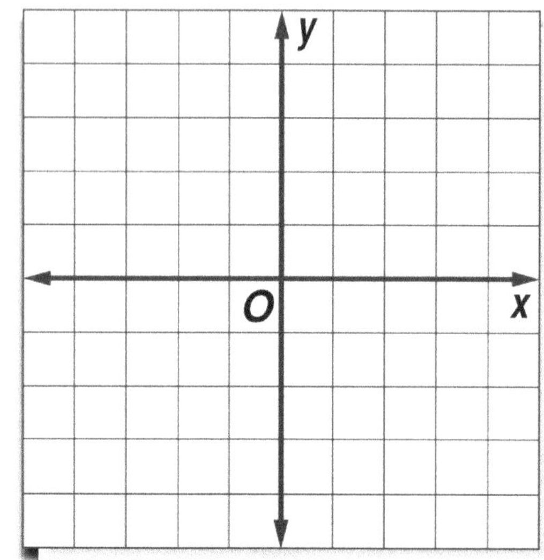

Name:

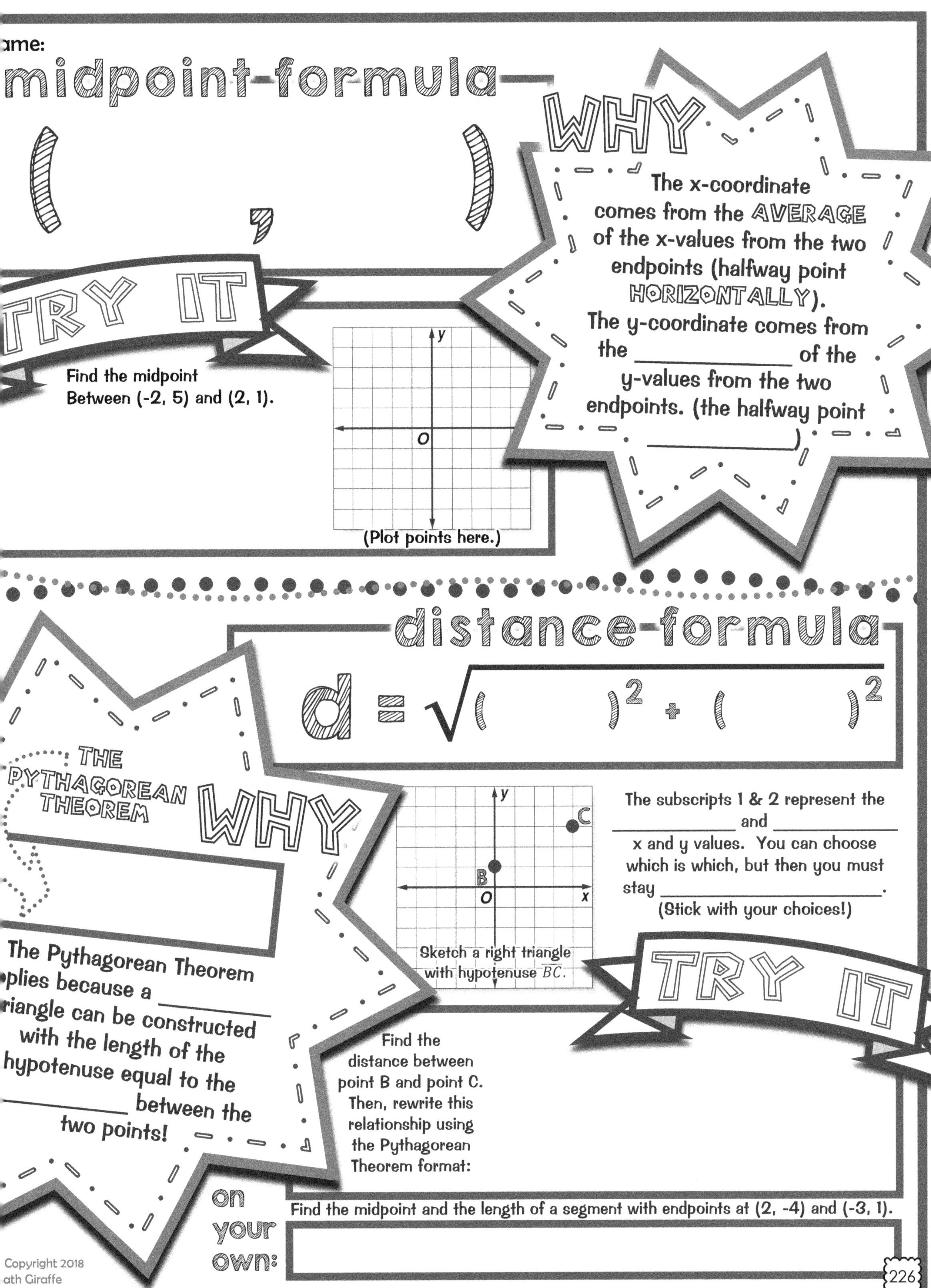
Name:
midpoint formula
(,)
WHY
The x-coordinate comes from the AVERAGE of the x-values from the two endpoints (halfway point HORIZONTALLY).
The y-coordinate comes from the ____________ of the y-values from the two endpoints. (the halfway point ____________)
TRY IT
Find the midpoint Between (-2, 5) and (2, 1).
y
O
(Plot points here.)
distance formula
d = √( )² + ( )²
THE PYTHAGOREAN THEOREM
WHY
The Pythagorean Theorem applies because a ____________ triangle can be constructed with the length of the hypotenuse equal to the ____________ between the two points!
y
C
B
O
x
Sketch a right triangle with hypotenuse BC.
The subscripts 1 & 2 represent the ____________ and ____________ x and y values. You can choose which is which, but then you must stay ____________.
(Stick with your choices!)
TRY IT
Find the distance between point B and point C. Then, rewrite this relationship using the Pythagorean Theorem format:
on your own:
Find the midpoint and the length of a segment with endpoints at (2, -4) and (-3, 1).

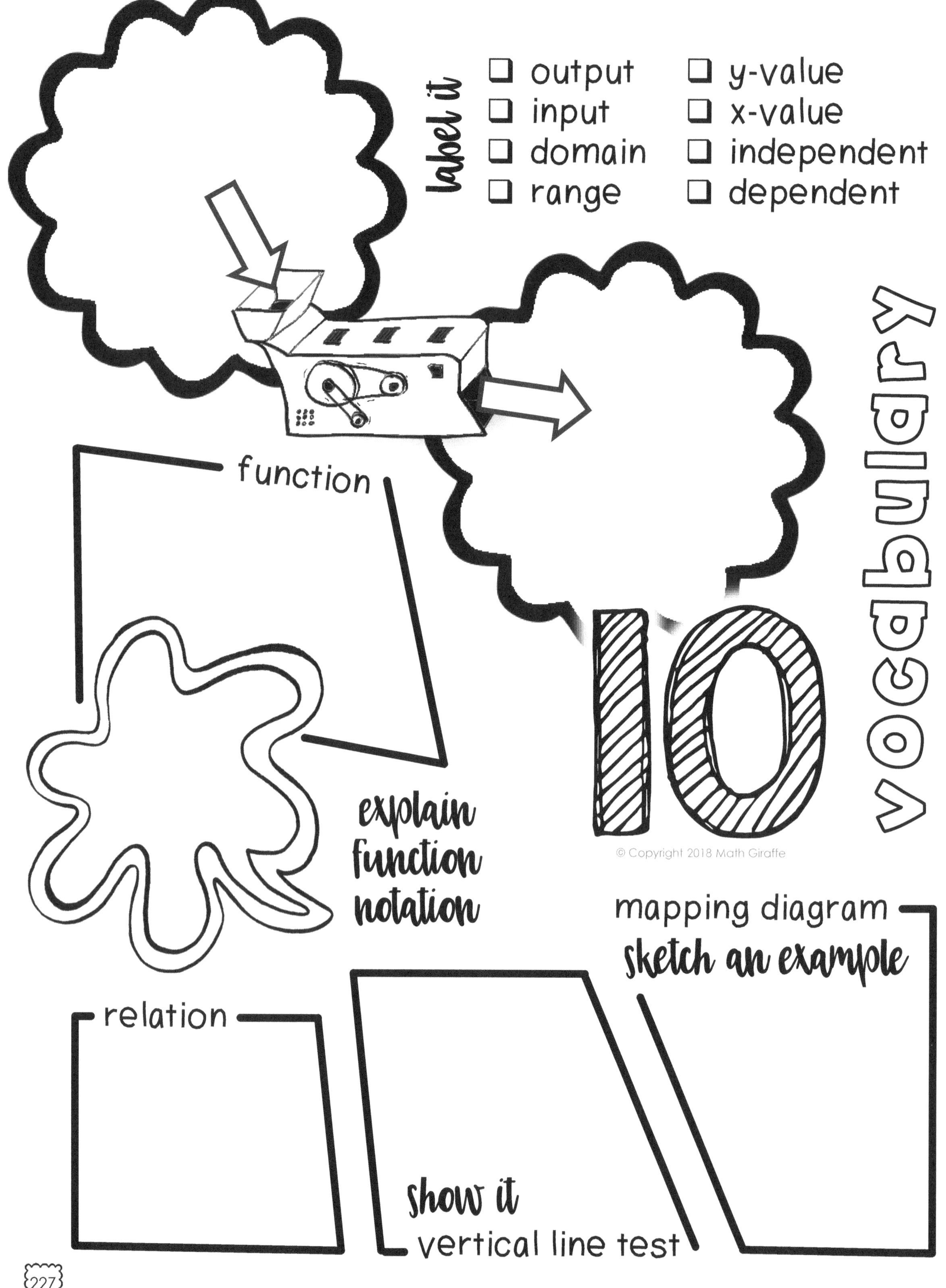
label it
output
input
domain
range
y-value
x-value
independent
dependent
function
10
vocabulary
explain
function
notation
© Copyright 2018 Math Giraffe
mapping diagram
sketch an example
relation
show it
vertical line test

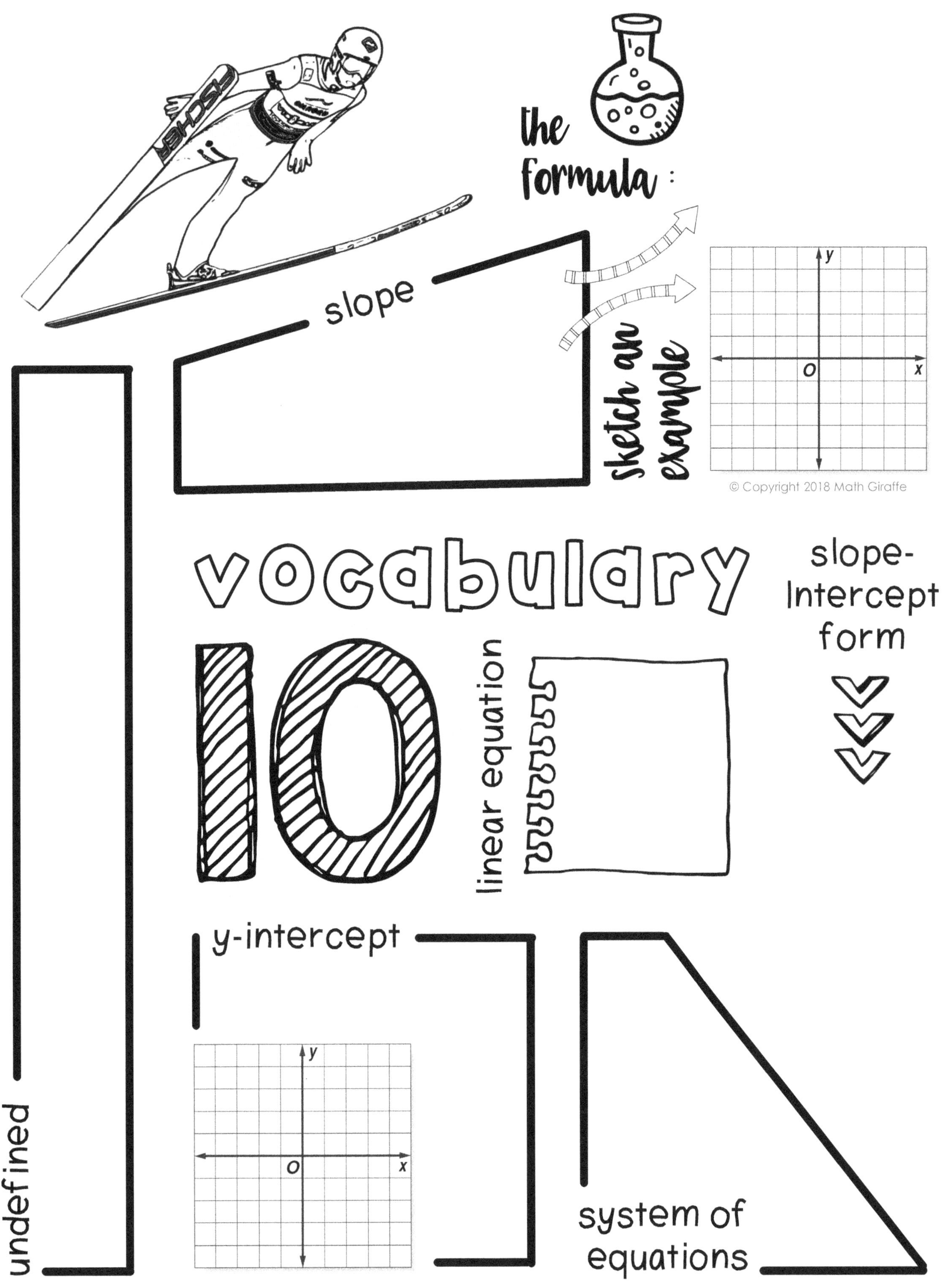
the formula:
slope
sketch an example
y
O
x
© Copyright 2018 Math Giraffe
vocabulary
10
slope-Intercept form
linear equation
y-intercept
y
O
x
undefined
system of equations

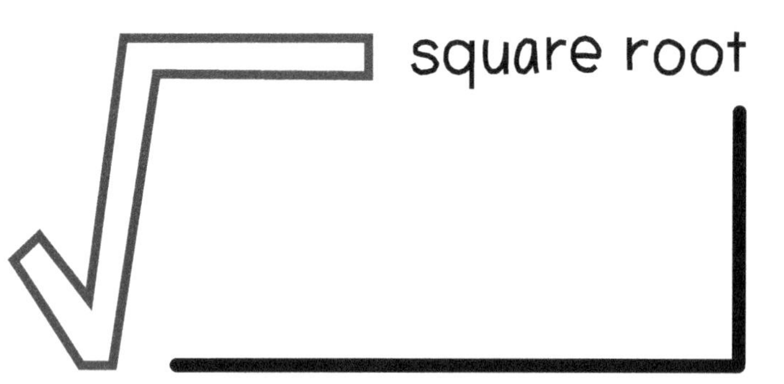

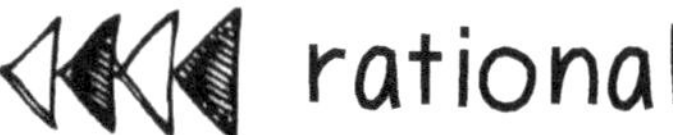

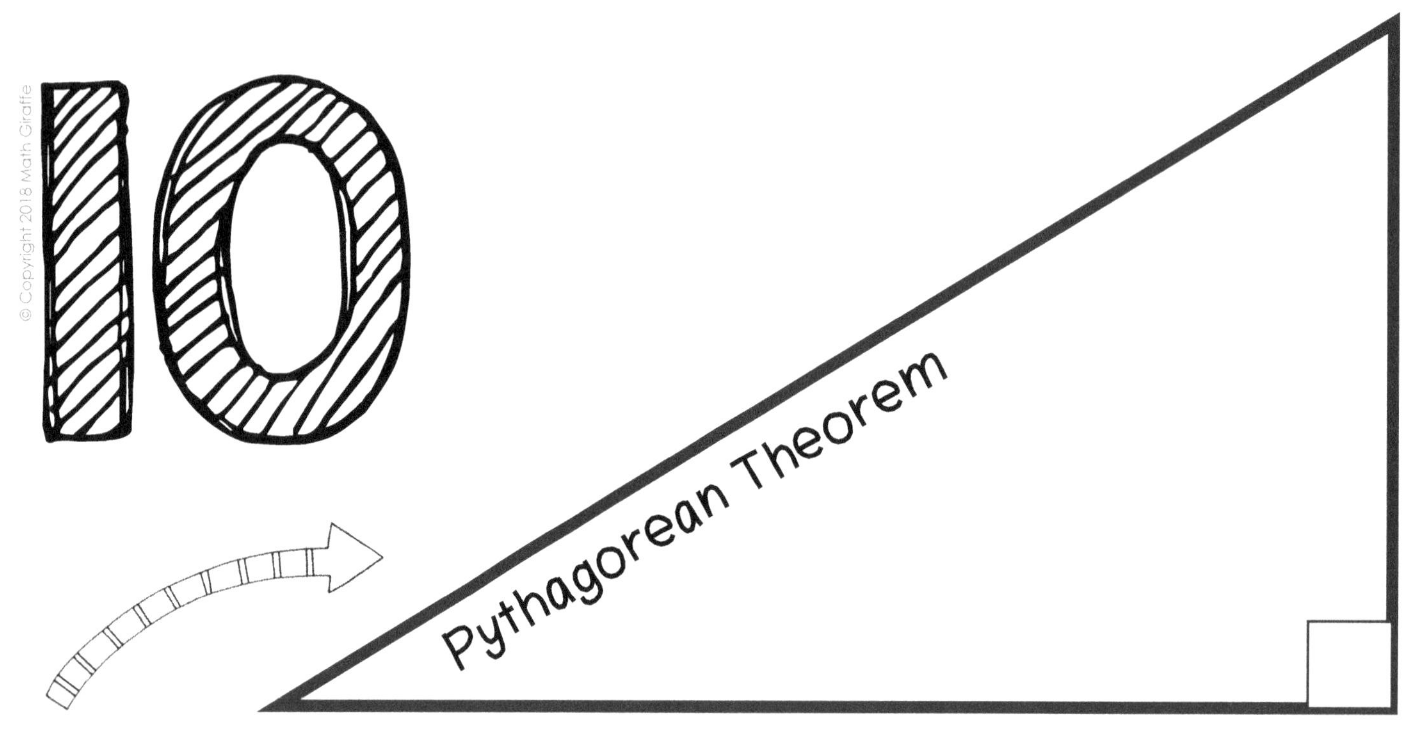

label it

- ❑ legs
- ❑ hypotenuse

vocabulary

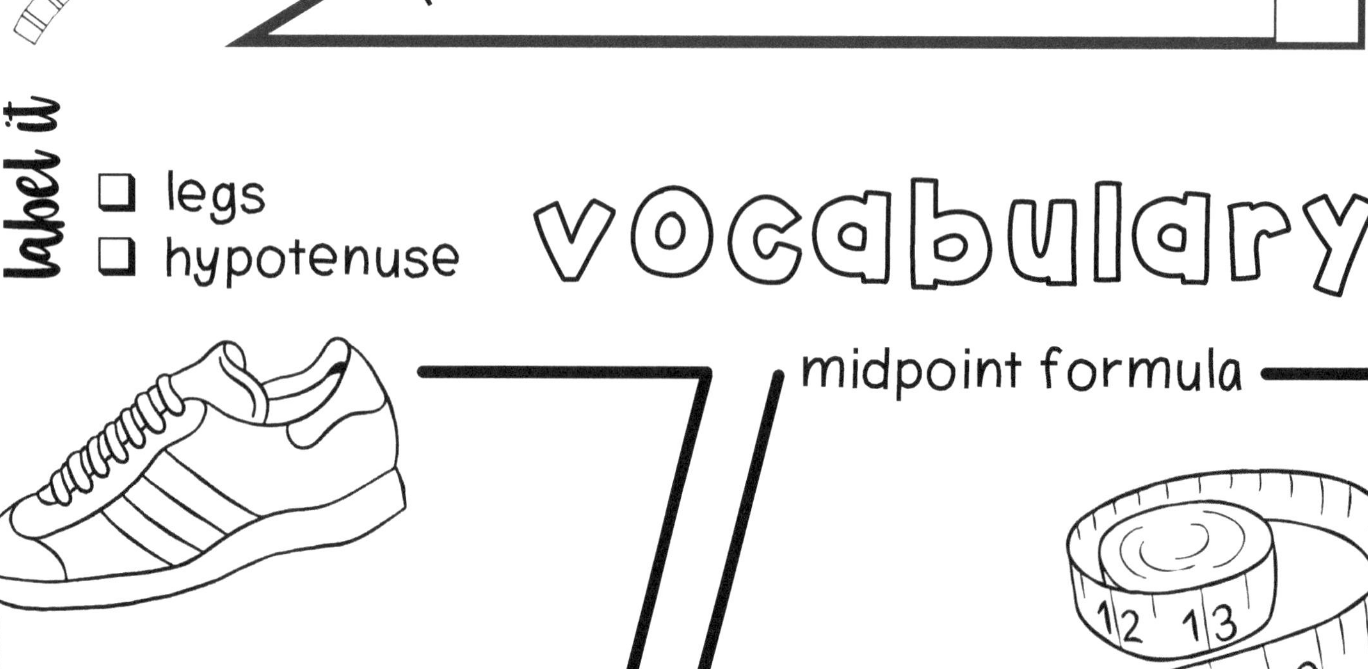

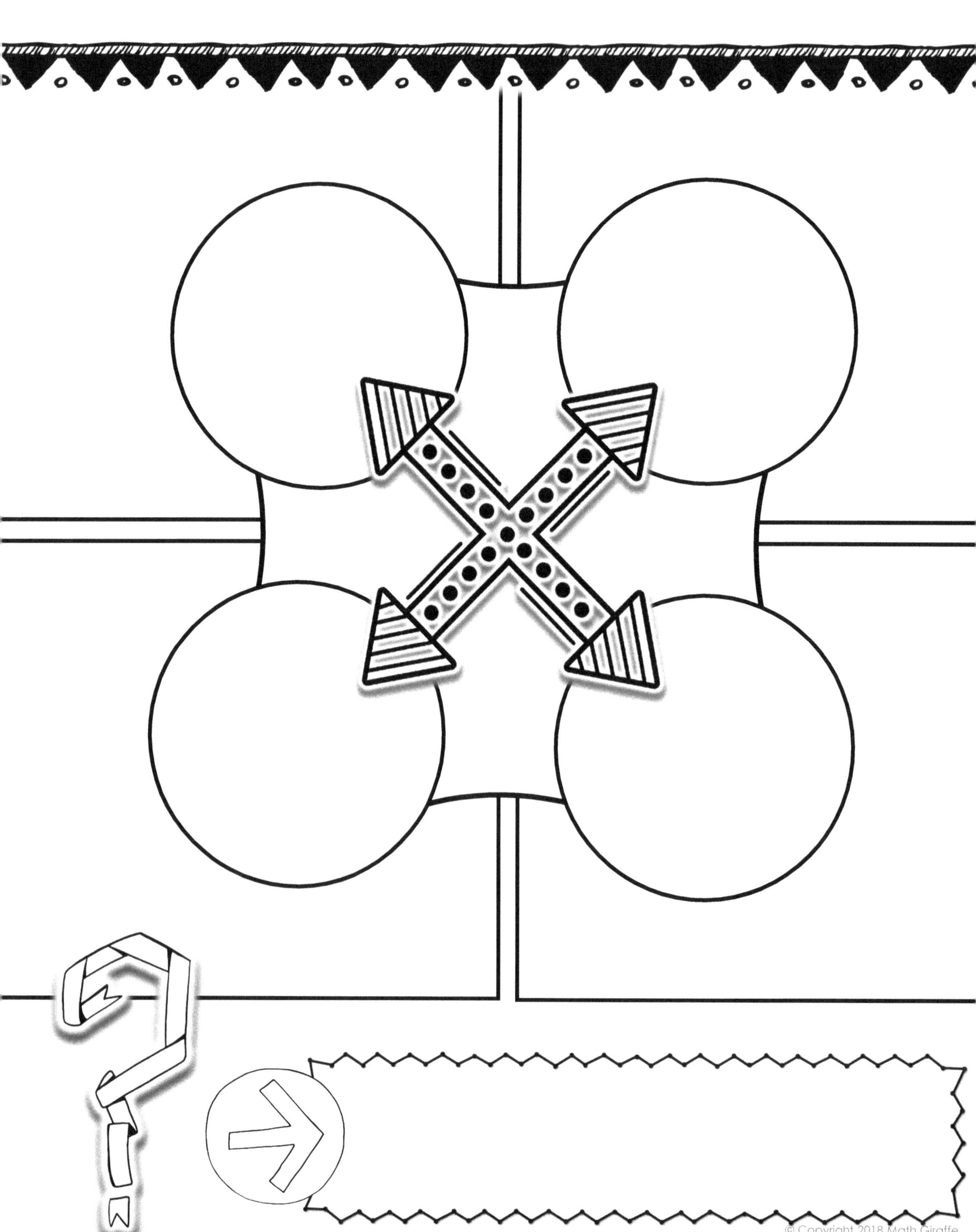

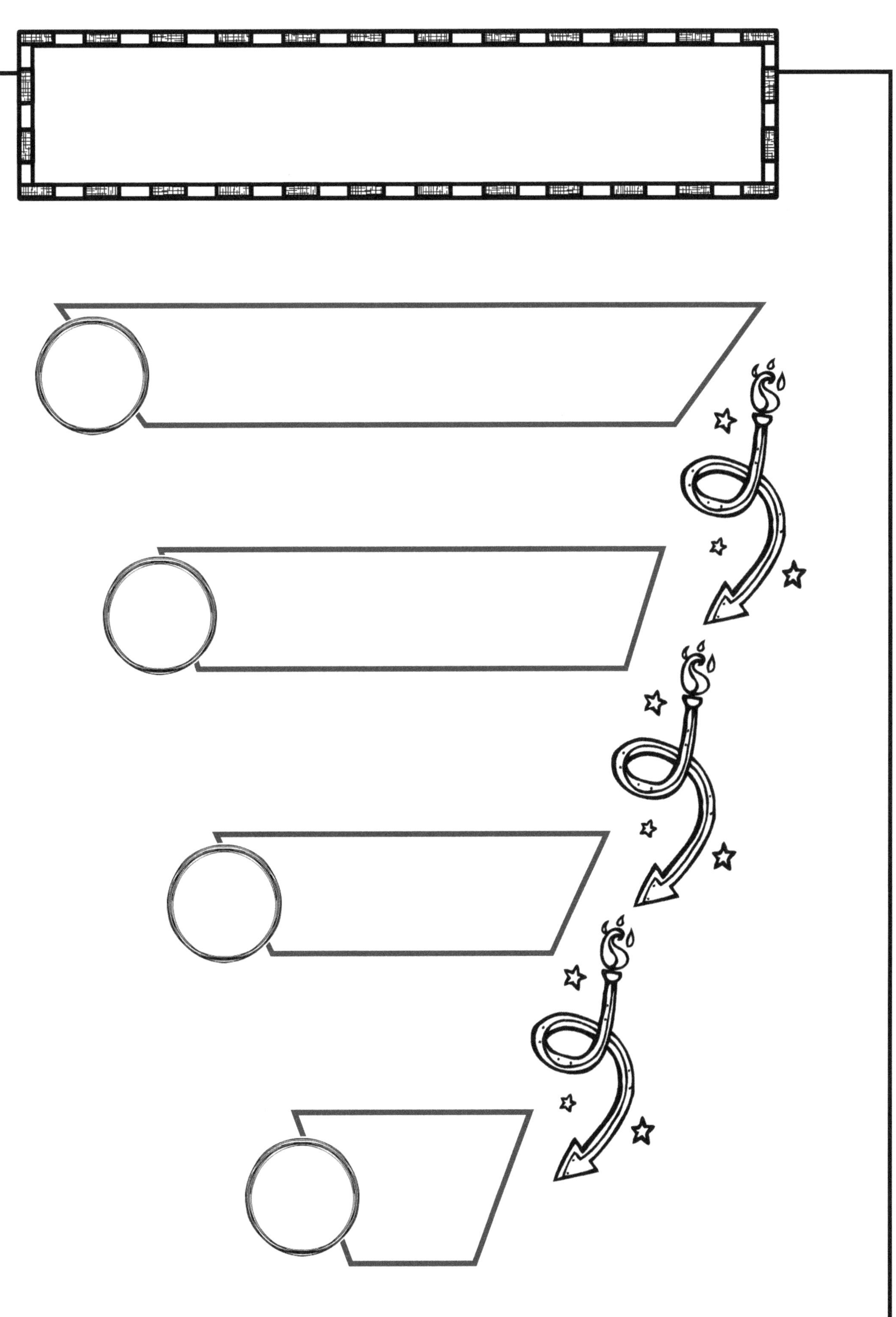

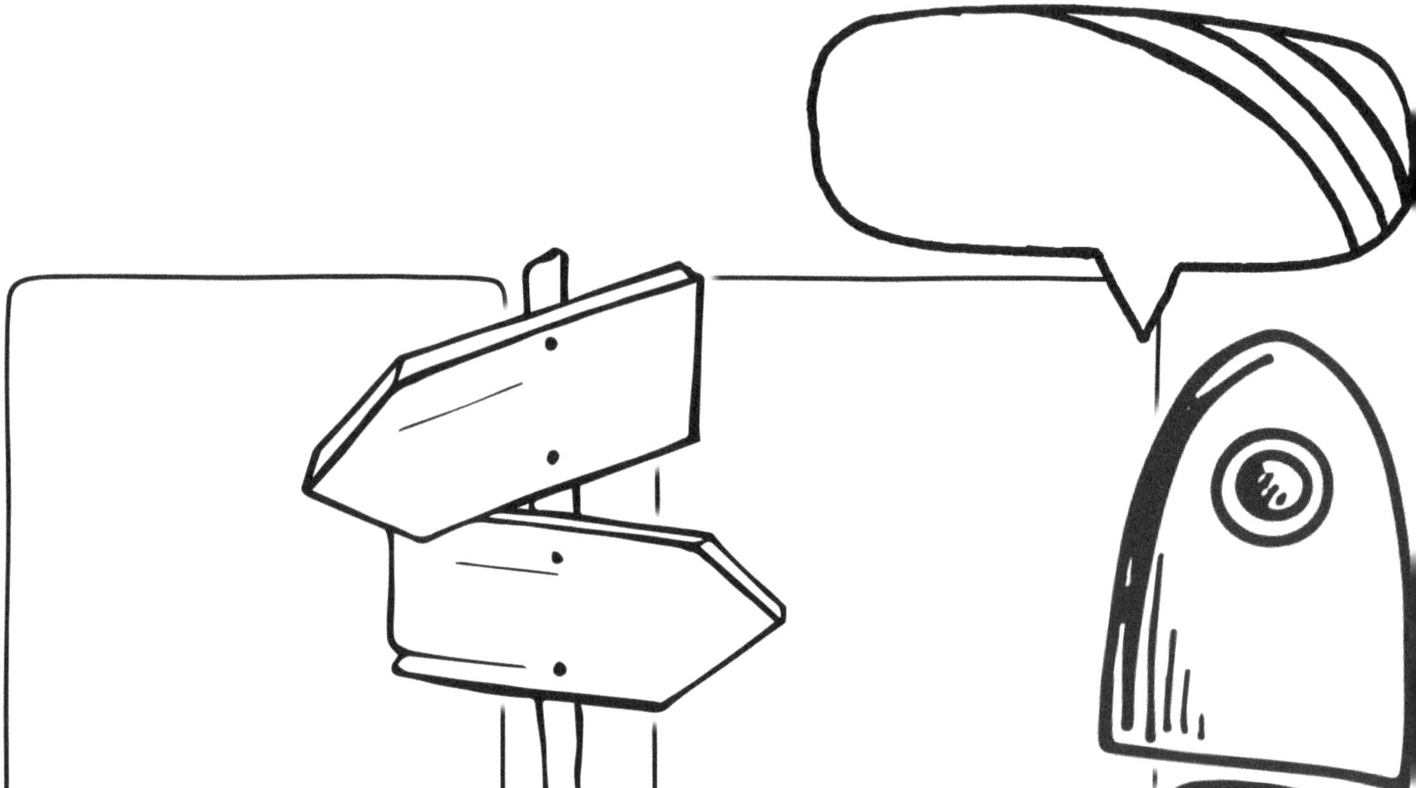

ENGAGE your brain!
- doodle, sketch, color -
TO DO
1 Write left-brain activities inside the left hemisphere of the brain, and right-brain activities in the right hemisphere.
2 Choose 9 of the brain benefits to write on the lines.
3 Label the corpus callosum.
Our core subjects are mainly _LEFT_ brained. When we can add color, doodles, and artistic embellishment, we incorporate the _RIGHT_ brain. This communication between both sides of the brain at the same time (across the ___CORPUS___ ___CALLOSUM___) activates our brains more fully to help us to maximize:
(Students choose 9 of the following to feature on their page):
Concentration
Memory
Relaxation
Focus
Learning
Problem solving
Energy
Retention
Creativity
Alertness
Attention
Coordination
Engagement
left hemisphere
right hemisphere
callosum
corpus
cross-lateral
communication
Right brain examples: coloring painting singing making up a new dance
Left brain examples: doing math problems reading writing sentences
Balance your use of both hemispheres! →
balance
The corpus callosum is...
a bundle of neural fibers that runs down the center of the brain (between the left and right hemispheres)
visual memory triggers
stick in our brains! (When we color in the text of a vocabulary word with a visual analogy or we doodle around a word or idea, we remember it better and can picture it in our mind later on!)
LEFT: words, logic, numbers, reasoning, reading, writing
RIGHT: color, music, art, visual, spatial, creative, design
Name: Teacher Guide / Answer Key
© Copyright 2016 Math Giraffe

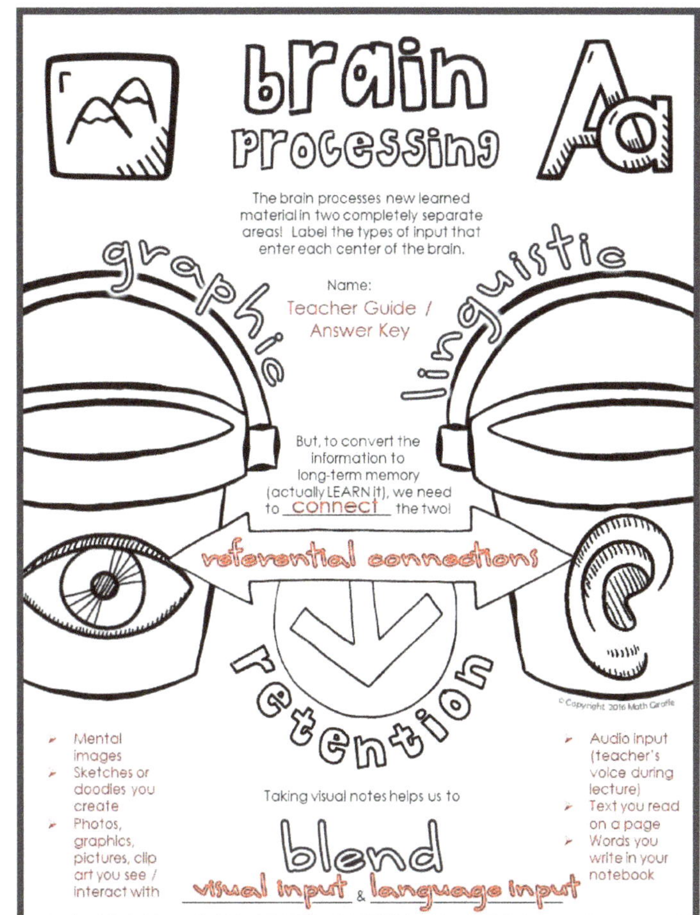
brain processing
The brain processes new learned material in two completely separate areas! Label the types of input that enter each center of the brain.
Name:
Teacher Guide / Answer Key
graphic
linguistic
But, to convert the information to long-term memory (actually LEARN it), we need to connect the two!
referential connections
retention
Mental images
Sketches or doodles you create
Photos, graphics, pictures, clip art you see / interact with
Audio input (teacher's voice during lecture)
Text you read on a page
Words you write in your notebook
Taking visual notes helps us to
blend
visual input & language input
© Copyright 2016 Math Giraffe

ENGAGE your brain!
- doodle, sketch, color -
TO DO
1 Write left-brain activities inside the left hemisphere of the brain, and right-brain activities in the right hemisphere.
2 Choose 9 of the brain benefits to write on the lines.
3 Label the corpus callosum.
Our core subjects are mainly LEFT brained. When we can add color, doodles, and embellishment, we incorporate the RIGHT brain. This communication between both sides of the brain at the same time (across the CORPUS CALLOSUM) activates our brains more fully to help us to maximize:
CONCENTRATION
NEW LEARNING
ATTENTION
PROBLEM
MEMORY
CREATIVITY
FOCUS
RELAXATION
RETENTION
left hemisphere
right hemisphere
CALLOSUM
CORPUS
left
right
balance
The corpus callosum is... a BUNDLE of neural FIBERS that runs between the 2 HEMISPHERES of the BRAIN
visual memory triggers
STICK in our BRAINS
LEFT: words, logic, numbers, reasoning, reading, writing
Name:
© Copyright 2016 Math Giraffe

brain processing
The brain processes new learned material in two completely separate areas! Label the types of input that enter each center of the brain.
Name:
graphic
linguistic
WHY take doodle NOTES
VISUALS
WORDS
written text, auditory input, read info
But, to convert the information to long-term memory (actually LEARN it), we need to CONNECT the two!
REFERENTIAL CONNECTIONS
memory boost
combining leads to
retention
graphic analogies & visual memory triggers
doodles & sketches (even coloring!)
what the teacher says
what you read in the book
what you write
Taking visual notes helps us to
blend
image input & language input
© Copyright 2016 Math Giraffe

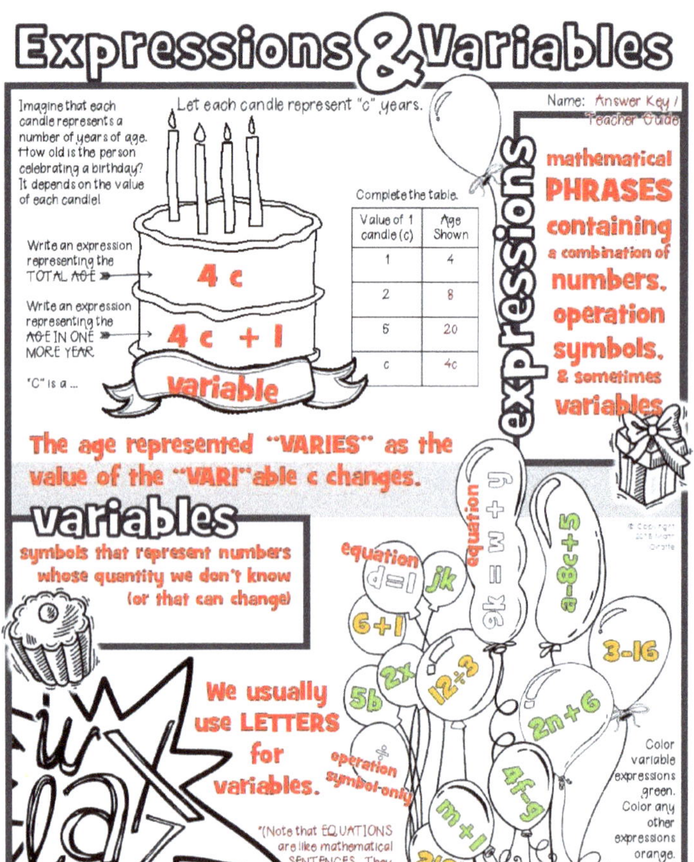

Expressions & Variables
Imagine that each candle represents a number of years of age. How old is the person celebrating a birthday? It depends on the value of each candle!
Let each candle represent "c" years.
Name: Answer Key / Teacher Guide
Complete the table.
Value of 1 candle (c) | Age Shown
1 | 4
2 | 8
5 | 20
c | 4c
Write an expression representing the TOTAL AGE
4c
Write an expression representing the AGE IN ONE MORE YEAR
4c + 1
"C" is a ...
variable
expressions
mathematical PHRASES containing a combination of numbers, operation symbols, & sometimes variables
The age represented "VARIES" as the value of the "VARI"able c changes.
variables
symbols that represent numbers whose quantity we don't know (or that can change)
We usually use LETTERS for variables.
equation
d/l = 1
jk
6+1
5b
2x
12÷3
3-16
2n+6
4f-g
m+1
2(8)
9k = m + 9
a-8c+5
operation symbol-only
Color variable expressions green. Color any other expressions orange. Are any NOT expressions at all? Why? Label.
*(Note that EQUATIONS are like mathematical SENTENCES. They contain the "IS" in "is equal to.")

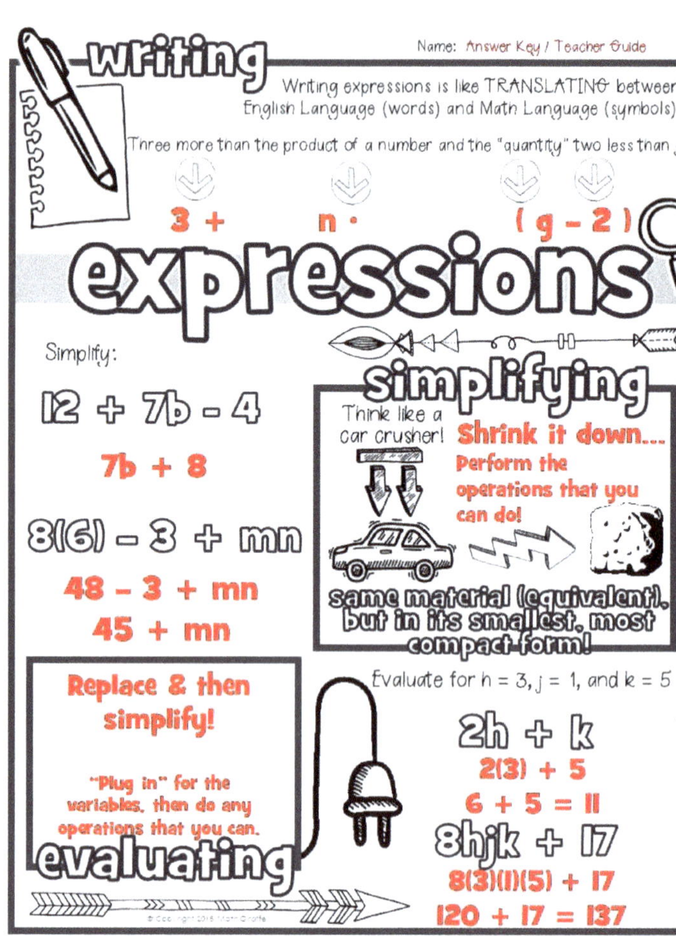

writing
Name: Answer Key / Teacher Guide
Writing expressions is like TRANSLATING between English Language (words) and Math Language (symbols)
Three more than the product of a number and the "quantity" two less than
3 +
n ·
(g - 2)
expressions
Simplify:
12 + 7b - 4
7b + 8
8(6) - 3 + mn
48 - 3 + mn
45 + mn
simplifying
Think like a car crusher!
Shrink it down...
Perform the operations that you can do!
same material (equivalent), but in its smallest, most compact form!
Replace & then simplify!
"Plug in" for the variables, then do any operations that you can.
evaluating
Evaluate for h = 3, j = 1, and k = 5
2h + k
2(3) + 5
6 + 5 = 11
8hjk + 17
8(3)(1)(5) + 17
120 + 17 = 137

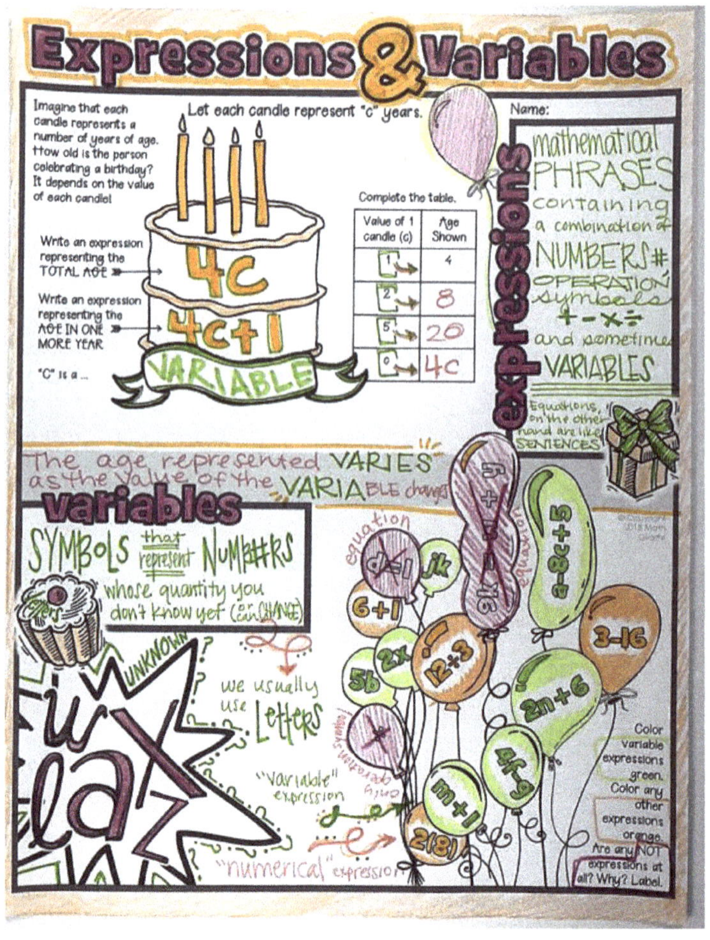

Expressions & Variables
Imagine that each candle represents a number of years of age. How old is the person celebrating a birthday? It depends on the value of each candle!
Let each candle represent "c" years.
Name:
Complete the table.
Value of 1 candle (c) | Age Shown
1 | 4
2 | 8
5 | 20
c | 4c
Write an expression representing the TOTAL AGE
4c
Write an expression representing the AGE IN ONE MORE YEAR
4c+1
"C" is a ...
VARIABLE
expressions
mathematical PHRASES containing a combination of NUMBERS # OPERATION symbols + - × ÷ and sometimes VARIABLES
The age represented VARIES as the value of the VARIABLE changes
variables
SYMBOLS that represent NUMBERS whose quantity you don't know yet (or can CHANGE)
UNKNOWN
we usually use Letters
"variable" expression
"numerical" expression
Color variable expressions green. Color any other expressions orange. Are any NOT expressions at all? Why? Label.

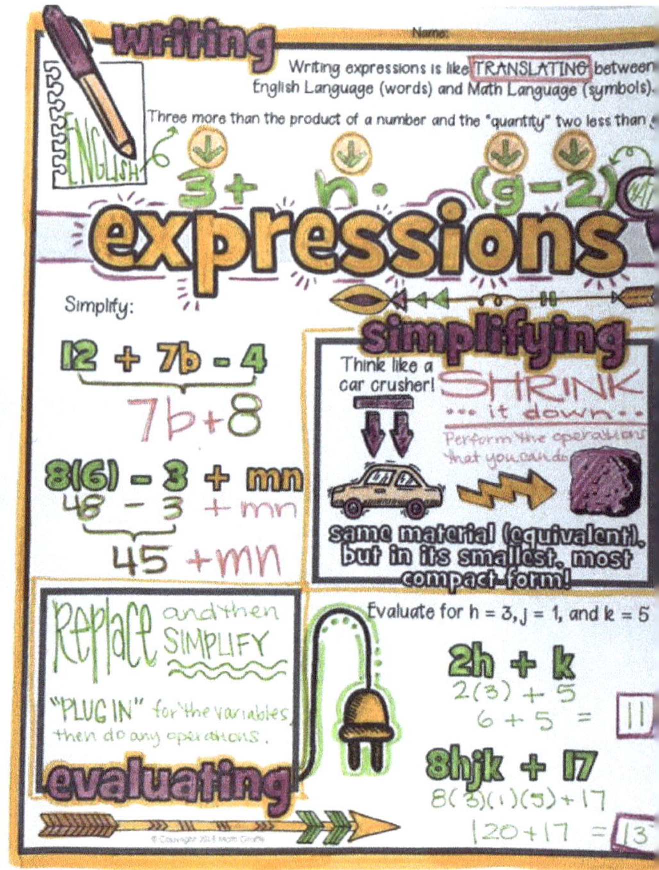

writing
Name:
Writing expressions is like TRANSLATING between English Language (words) and Math Language (symbols).
Three more than the product of a number and the "quantity" two less than
ENGLISH
3+
n ·
(g-2)
expressions
Simplify:
12 + 7b - 4
7b+8
8(6) - 3 + mn
48 - 3 + mn
45 + mn
simplifying
Think like a car crusher!
SHRINK it down
Perform the operations that you can do
same material (equivalent), but in its smallest, most compact form!
Replace and then SIMPLIFY
"PLUG IN" for the variables then do any operations.
evaluating
Evaluate for h = 3, j = 1, and k = 5
2h + k
2(3) + 5
6 + 5 = 11
8hjk + 17
8(3)(1)(5)+17
120+17 = 137

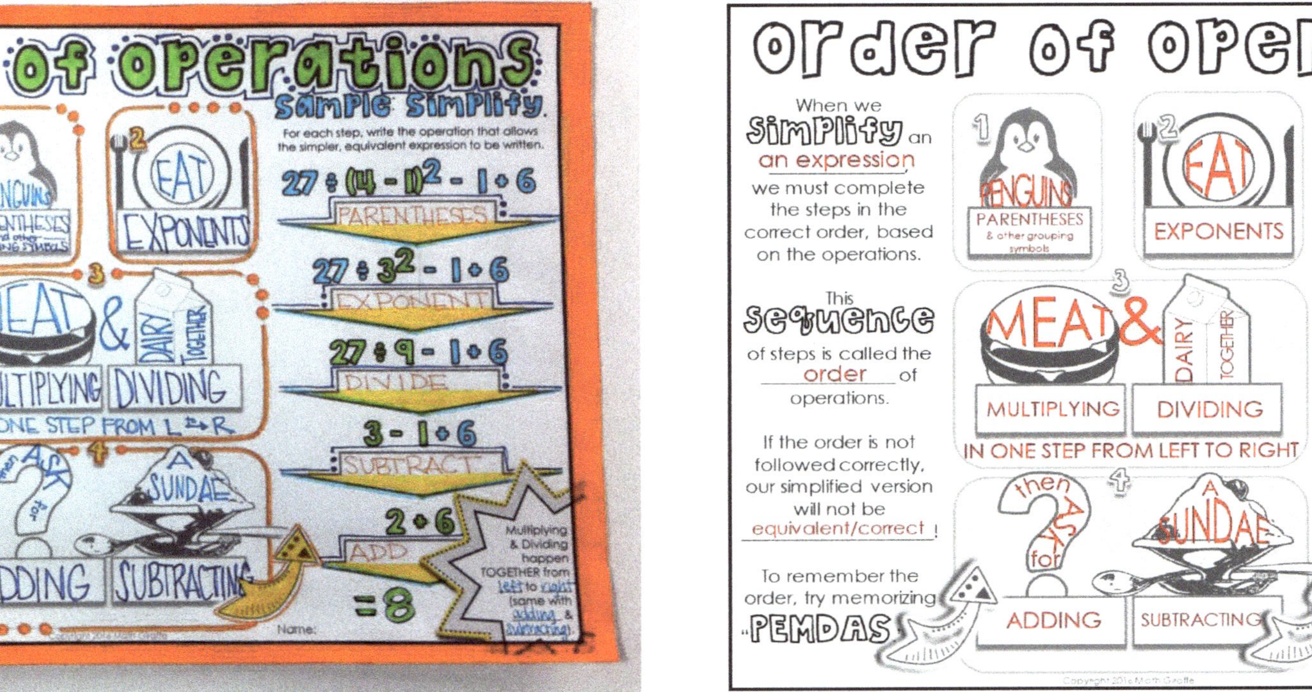

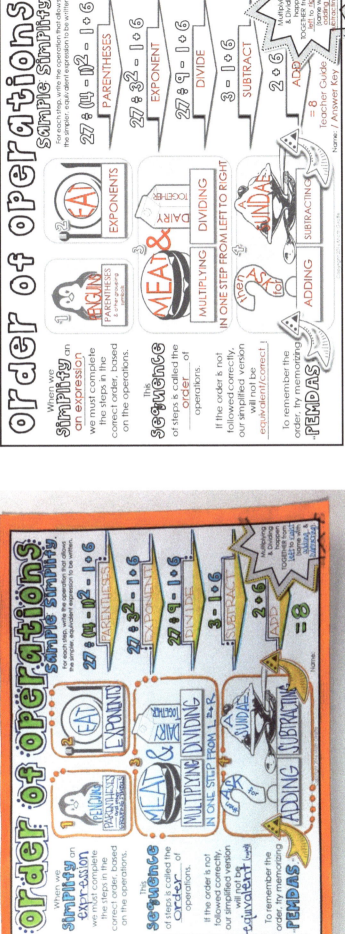

order of operations
Sample: Simplify.
For each step, write the operation that allows the simpler, equivalent expression to be written.
When we simplify an expression we must complete the steps in the correct order, based on the operations.
This sequence of steps is called the order of operations.
If the order is not followed correctly, our simplified version will not be equivalent/correct!
To remember the order, try memorizing "PEMDAS".
1 PENGUINS PARENTHESES & other grouping symbols
2 EAT EXPONENTS
3 MEAT & DAIRY TOGETHER
MULTIPLYING DIVIDING
IN ONE STEP FROM LEFT TO RIGHT
4 then ASK for A SUNDAE
ADDING SUBTRACTING
27 ÷ (4 − 1)2 − 1 + 6
PARENTHESES
27 ÷ 32 − 1 + 6
EXPONENT
27 ÷ 9 − 1 + 6
DIVIDE
3 − 1 + 6
SUBTRACT
2 + 6
ADD
= 8
Multiplying & Dividing happen TOGETHER from left to right (same with adding & subtracting).
Name: Teacher Guide / Answer Key

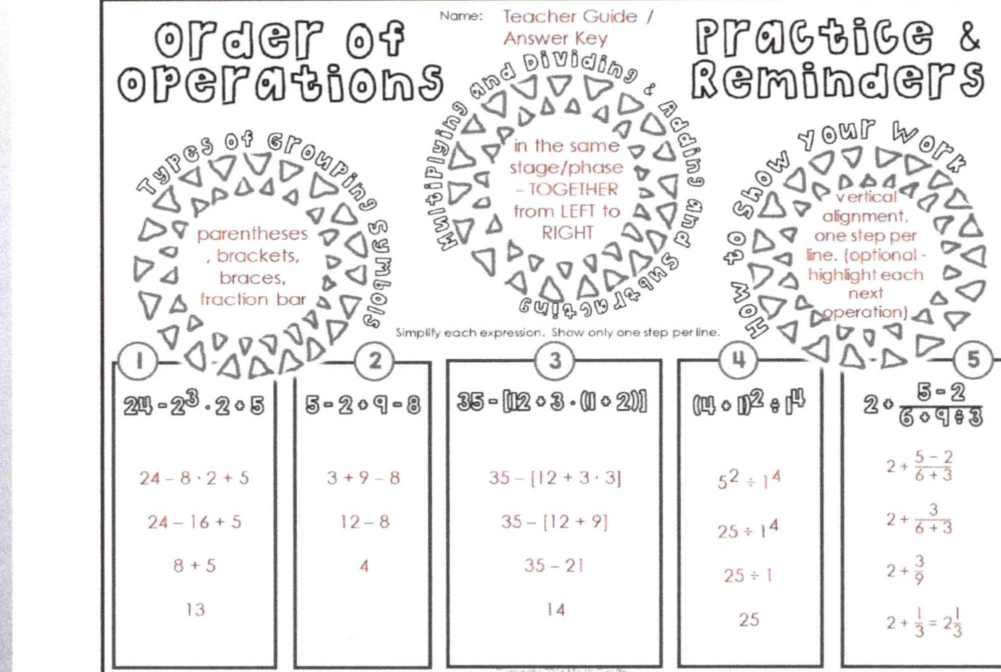

order of operations Practice & Reminders
Name: Teacher Guide / Answer Key
Types of Grouping Symbols
parentheses, brackets, braces, fraction bar
Multiplying and Dividing & Adding and Subtracting
in the same stage/phase – TOGETHER from LEFT to RIGHT
How to Show your work
vertical alignment, one step per line. (optional - highlight each next operation)
Simplify each expression. Show only one step per line.
1 24 − 2³ · 2 + 5
24 − 8 · 2 + 5
24 − 16 + 5
8 + 5
13
2 5 − 2 + 9 − 8
3 + 9 − 8
12 − 8
4
3 35 − [12 + 3 · (1 + 2)]
35 − [12 + 3 · 3]
35 − [12 + 9]
35 − 21
14
4 (4 + 1)² ÷ 1⁴
5² ÷ 1⁴
25 ÷ 1⁴
25 ÷ 1
25
5 2 + (5 − 2)/(6 + 9 ÷ 3)
2 + (5 − 2)/(6 + 3)
2 + 3/(6 + 3)
2 + 3/9
2 + 1/3 = 2 1/3

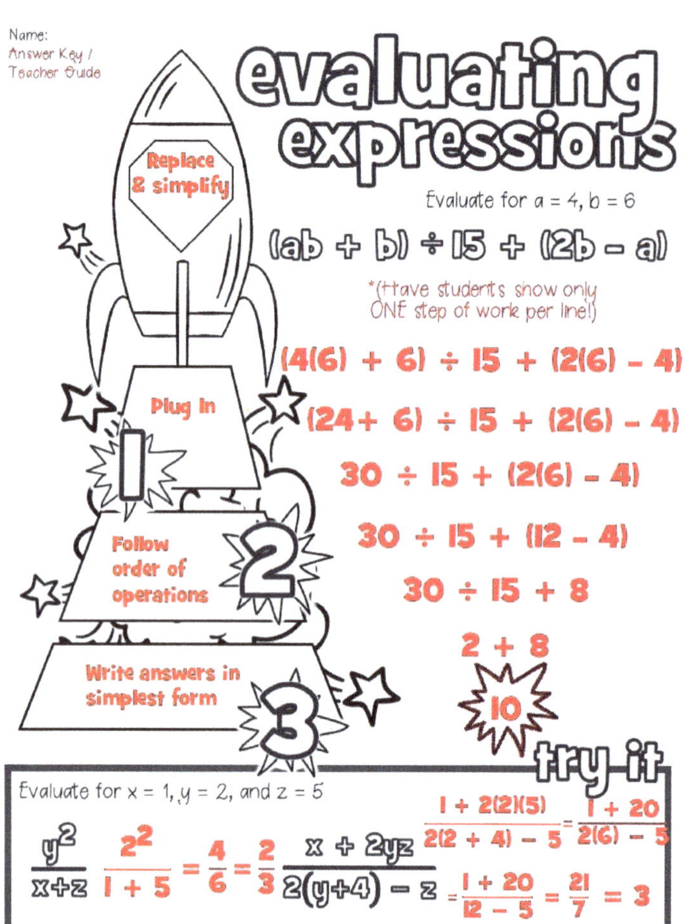
Name:
Answer Key / Teacher Guide
evaluating expressions
Replace & simplify
Evaluate for a = 4, b = 6
(ab + b) ÷ 15 + (2b − a)
*(Have students show only ONE step of work per line!)
(4(6) + 6) ÷ 15 + (2(6) − 4)
(24 + 6) ÷ 15 + (2(6) − 4)
30 ÷ 15 + (2(6) − 4)
30 ÷ 15 + (12 − 4)
30 ÷ 15 + 8
2 + 8
10
Plug in
1
Follow order of operations
2
Write answers in simplest form
3
try it
Evaluate for x = 1, y = 2, and z = 5

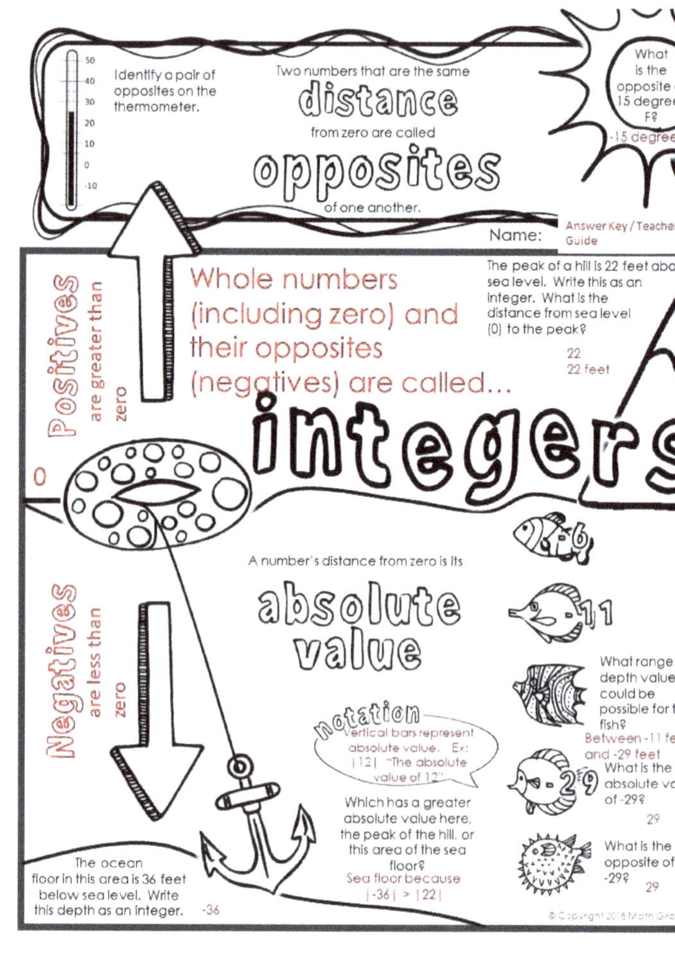
Identify a pair of opposites on the thermometer.
Two numbers that are the same
distance
from zero are called
opposites
of one another.
What is the opposite of 15 degrees F?
-15 degrees
Name:
Answer Key / Teacher Guide
Positives are greater than zero
Whole numbers (including zero) and their opposites (negatives) are called...
integers
The peak of a hill is 22 feet above sea level. Write this as an integer. What is the distance from sea level (0) to the peak?
22
22 feet
0
A number's distance from zero is its
absolute value
Negatives are less than zero
notation
Vertical bars represent absolute value. Ex: |12| "The absolute value of 12"
Which has a greater absolute value here, the peak of the hill, or this area of the sea floor?
Sea floor because |-36| > |22|
The ocean floor in this area is 36 feet below sea level. Write this depth as an integer.
-36
What range depth value could be possible for this fish?
Between -11 feet and -29 feet
What is the absolute value of -29?
29
What is the opposite of -29?
29

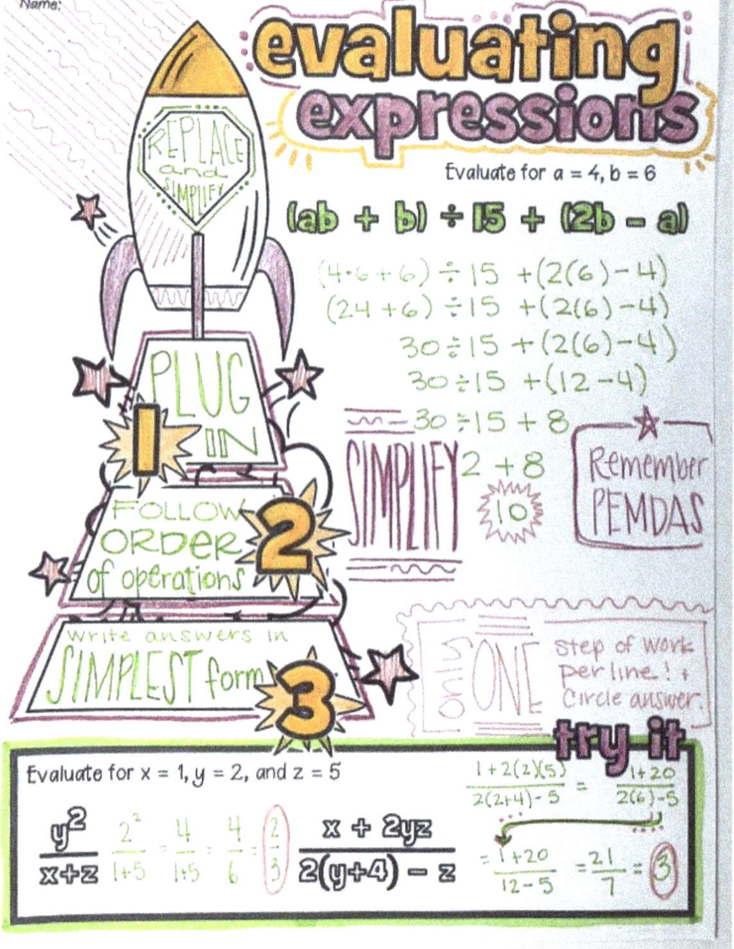
Name:
evaluating expressions
REPLACE and SIMPLIFY
Evaluate for a = 4, b = 6
(ab + b) ÷ 15 + (2b − a)
(4·6 + 6) ÷ 15 + (2(6) − 4)
(24 + 6) ÷ 15 + (2(6) − 4)
30 ÷ 15 + (2(6) − 4)
30 ÷ 15 + (12 − 4)
30 ÷ 15 + 8
SIMPLIFY
2 + 8
10
Remember PEMDAS
PLUG IN
1
FOLLOW ORDER of operations
2
Write answers in SIMPLEST form
3
Only ONE step of work per line! Circle answer.
try it
Evaluate for x = 1, y = 2, and z = 5

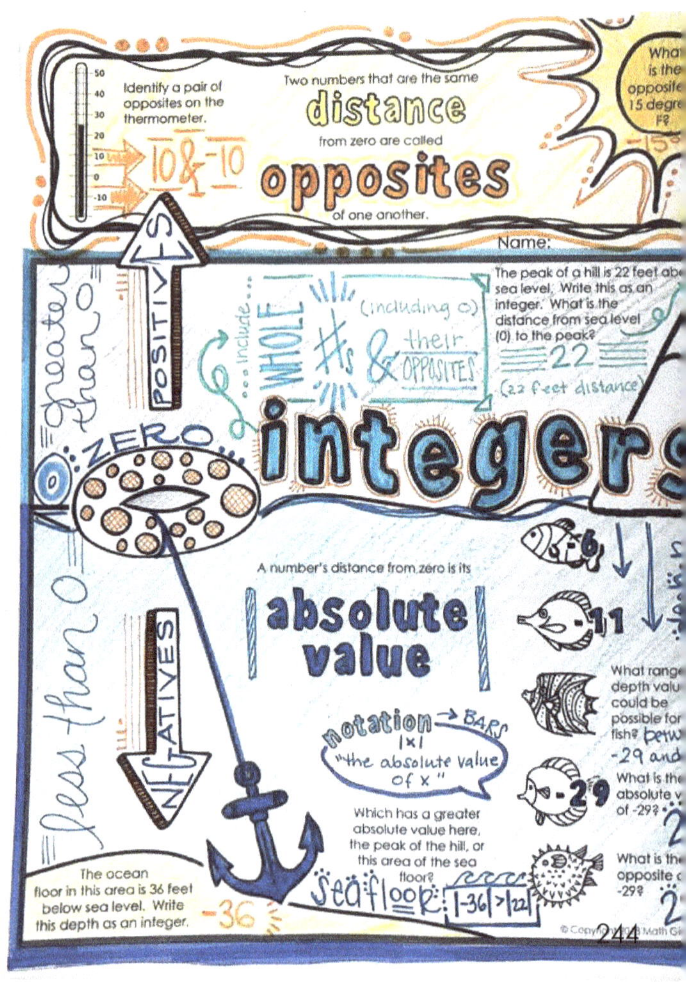
Identify a pair of opposites on the thermometer.
10 & -10
Two numbers that are the same
distance
from zero are called
opposites
of one another.
Name:
POSITIVES
greater than 0
...include... WHOLE #s (including 0) & their OPPOSITES
integers
ZERO
0
The peak of a hill is 22 feet above sea level. Write this as an integer. What is the distance from sea level (0) to the peak?
22
(22 feet distance)
A number's distance from zero is its
absolute value
NEGATIVES
less than 0
notation → BARS
|x|
"the absolute value of x"
Which has a greater absolute value here, the peak of the hill, or this area of the sea floor?
seafloor |-36| > |22|
The ocean floor in this area is 36 feet below sea level. Write this depth as an integer.
-36
What is the absolute value of -29?
What is the opposite of -29?

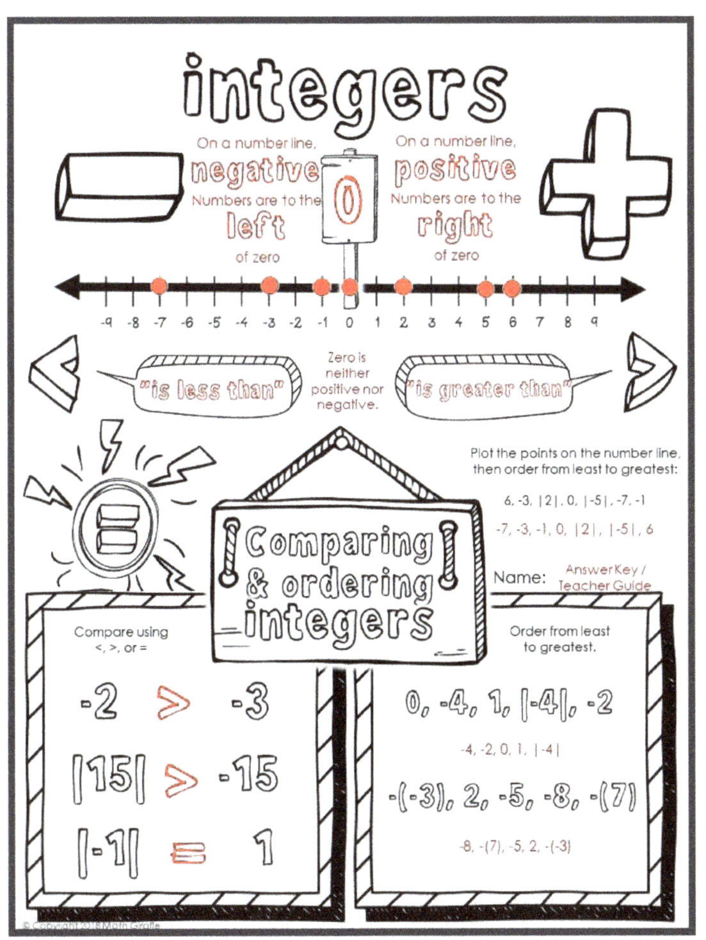
integers
On a number line, negative Numbers are to the left of zero
0
On a number line, positive Numbers are to the right of zero
-9 -8 -7 -6 -5 -4 -3 -2 -1 0 1 2 3 4 5 6 7 8 9
"is less than"
Zero is neither positive nor negative.
"is greater than"
Plot the points on the number line, then order from least to greatest:
6, -3, |2|, 0, |-5|, -7, -1
-7, -3, -1, 0, |2|, |-5|, 6
Comparing & ordering integers
Name: Answer Key / Teacher Guide
Compare using <, >, or =
-2 > -3
|15| > -15
|-1| = 1
Order from least to greatest.
0, -4, 1, |-4|, -2
-4, -2, 0, 1, |-4|
-(-3), 2, -5, -8, -(7)
-8, -(7), -5, 2, -(-3)
© Copyright 2016 Math Giraffe

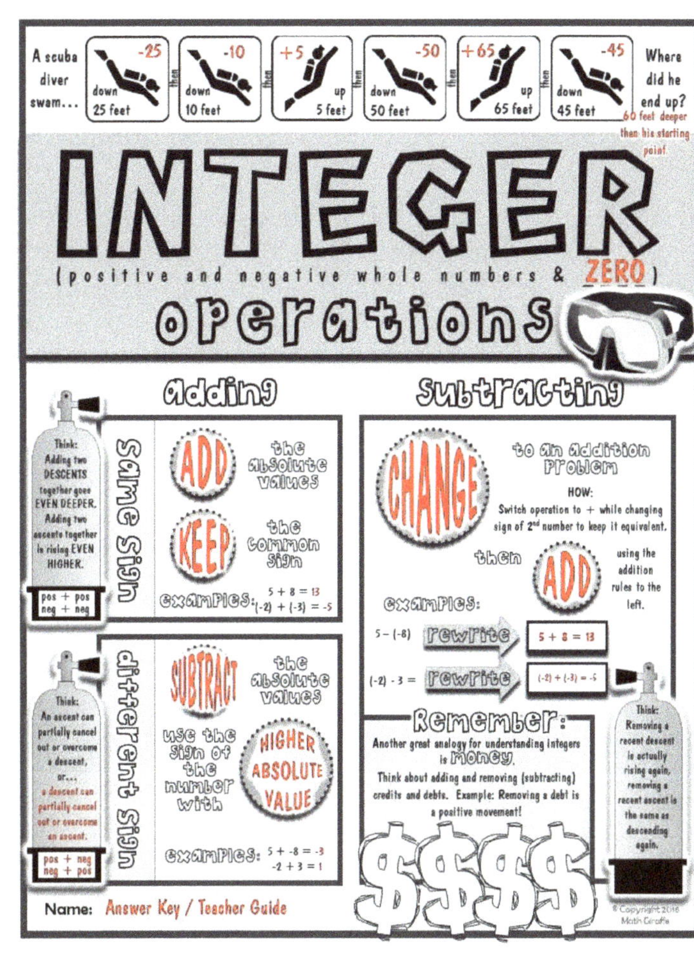
A scuba diver swam...
-25 down 25 feet
then
-10 down 10 feet
then
+5 up 5 feet
then
-50 down 50 feet
then
+65 up 65 feet
then
-45 down 45 feet
Where did he end up? 60 feet deeper than his starting point.
INTEGER
(positive and negative whole numbers & ZERO)
operations
adding
subtracting
Think: Adding two DESCENTS together goes EVEN DEEPER. Adding two ascents together is rising EVEN HIGHER.
pos + pos
neg + neg
same sign
ADD the absolute values
KEEP the common sign
examples: 5 + 8 = 13
(-2) + (-3) = -5
CHANGE to an addition problem
HOW: Switch operation to + while changing sign of 2nd number to keep it equivalent.
then ADD using the addition rules to the left.
examples:
5 - (-8) rewrite 5 + 8 = 13
(-2) - 3 = rewrite (-2) + (-3) = -5
Think: An ascent can partially cancel out or overcome a descent, or... a descent can partially cancel out or overcome an ascent.
pos + neg
neg + pos
different sign
SUBTRACT the absolute values
use the sign of the number with HIGHER ABSOLUTE VALUE
examples: 5 + -8 = -3
-2 + 3 = 1
Remember:
Another great analogy for understanding integers is money.
Think about adding and removing (subtracting) credits and debts. Example: Removing a debt is a positive movement!
Think: Removing a recent descent is actually rising again, removing a recent ascent is the same as descending again.
$$$$
Name: Answer Key / Teacher Guide
© Copyright 2016 Math Giraffe

integers
NEGATIVE
POSITIVE
0
-9 -8 -7 -6 -5 -4 -3 -2 -1 0 1 2 3 4 5 6 7 8 9
"is Less than"
Zero is neither positive nor negative
"is greater than"
Plot the points on the number line, then order from least to greatest:
least → greatest
6, -3, |2|, 0, |-5|, -7, -1
-7, -3, -1, 0, |2|, |-5|, 6
Comparing & ordering integers
Name:
Compare using <, >, or =
-2 > -3
|15| > -15
|-1| = 1
Order from least to greatest.
0, -4, 1, |-4|, -2
4
-4, -2, 0, 1, |-4|
-(-3), 2, -5, -8, -(7)
3
-8, -7, -5, 2, -(-3)
-7

A scuba diver swam...
-25 down 25 feet
-10 down 10 feet
+5 up 5 feet
-50 down 50 feet
+65 up 65 feet
-45 down 45 feet
Where did he end up?
60 ft deeper than start pt.
INTEGER
(positive and negative whole numbers & ZERO)
operations
to SUBTRACT is to Add the OPPOSITE!
adding
subtracting
same sign
ADD the absolute values
KEEP the common sign
examples: 5 + 8 = 13
(-2) + (-3) = -5
CHANGE to an addition problem
HOW: Switch operation to + while changing sign of 2nd number to keep it equivalent.
then ADD using the addition rules to the left.
examples:
5 - (-8) rewrite 5 + 8 = 13
(-2) - 3 = rewrite -2 + -3 = -5
different sign
SUBTRACT the absolute values
use the sign of the number with GREATER ABSOLUTE VALUE
VICE VERSA
pos + neg
neg + pos
examples: 5 + -8 = -3
-2 + 3 = 1
Remember:
Another great analogy for understanding integers is money.
Think about adding and removing (subtracting) credits and debts. Example: Removing a debt is a positive movement!
$$$$
Name:
© Copyright 2016 Math Giraffe

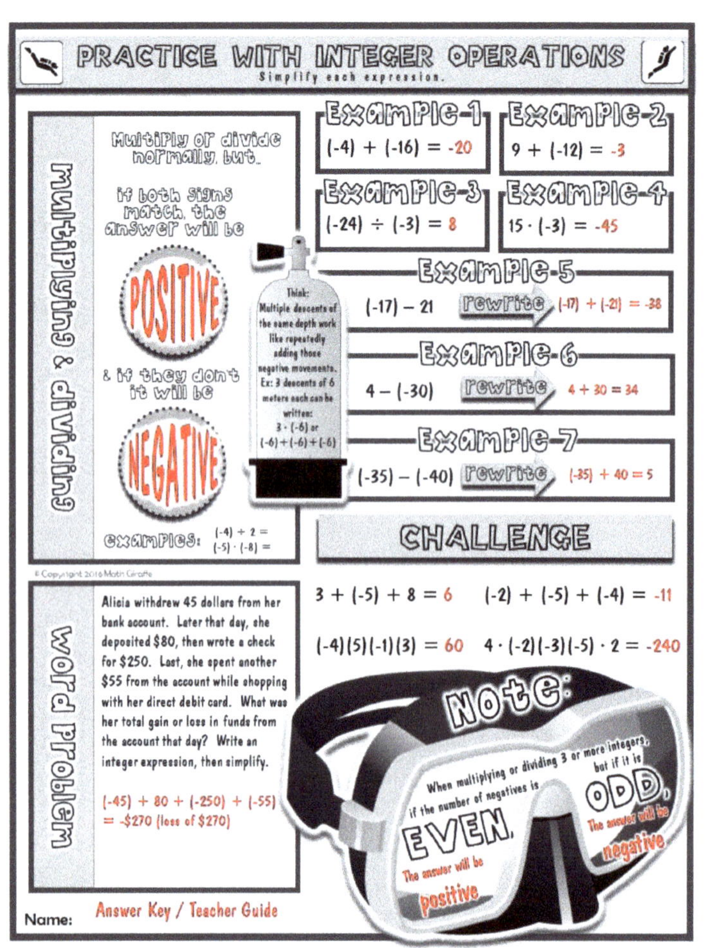
PRACTICE WITH INTEGER OPERATIONS
Simplify each expression.
multiplying & dividing
Multiply or divide normally, but...
if both signs match, the answer will be
POSITIVE
& if they don't it will be
NEGATIVE
examples: (-4) ÷ 2 =
(-5) · (-8) =
Think: Multiple descents of the same depth work like repeatedly adding those negative movements. Ex: 3 descents of 6 meters each can be written: 3 · (-6) or (-6) + (-6) + (-6)
Example-1
(-4) + (-16) = -20
Example-2
9 + (-12) = -3
Example-3
(-24) ÷ (-3) = 8
Example-4
15 · (-3) = -45
Example-5
(-17) – 21 rewrite (-17) + (-21) = -38
Example-6
4 – (-30) rewrite 4 + 30 = 34
Example-7
(-35) – (-40) rewrite (-35) + 40 = 5
CHALLENGE
3 + (-5) + 8 = 6
(-2) + (-5) + (-4) = -11
(-4)(5)(-1)(3) = 60
4 · (-2)(-3)(-5) · 2 = -240
word problem
Alicia withdrew 45 dollars from her bank account. Later that day, she deposited $80, then wrote a check for $250. Last, she spent another $55 from the account while shopping with her direct debit card. What was her total gain or loss in funds from the account that day? Write an integer expression, then simplify.
(-45) + 80 + (-250) + (-55) = -$270 (loss of $270)
Note:
When multiplying or dividing 3 or more integers, if the number of negatives is EVEN, The answer will be positive
but if it is ODD, The answer will be negative
Name: Answer Key / Teacher Guide
© Copyright 2016 Math Giraffe

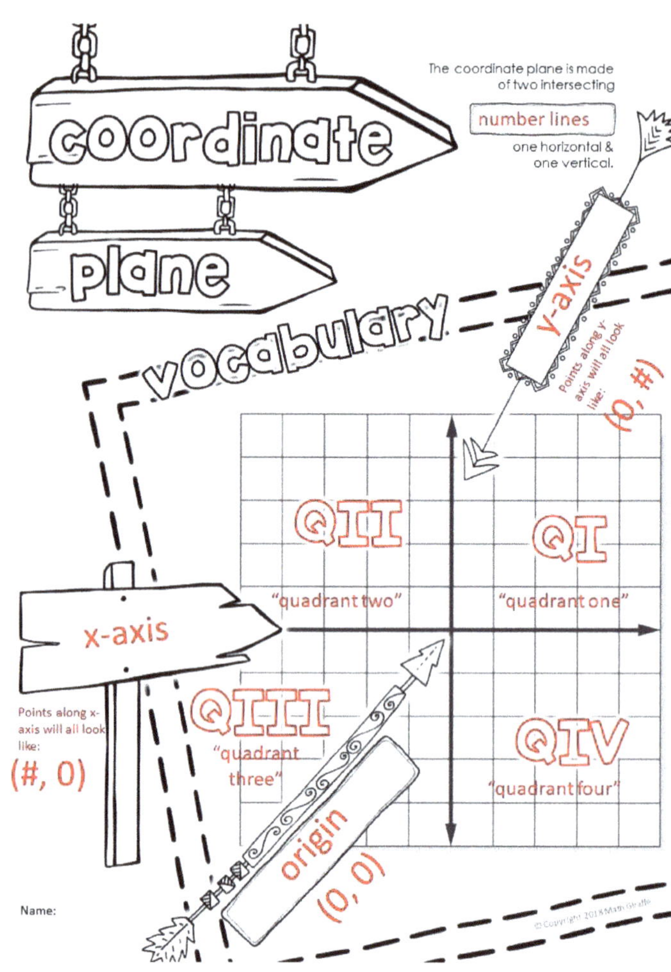
coordinate
plane
The coordinate plane is made of two intersecting
number lines
one horizontal & one vertical.
vocabulary
y-axis
Points along y-axis will all look like:
(0, #)
x-axis
Points along x-axis will all look like:
(#, 0)
QII
"quadrant two"
QI
"quadrant one"
QIII
"quadrant three"
QIV
"quadrant four"
origin
(0, 0)
Name:
© Copyright 2016 Math Giraffe

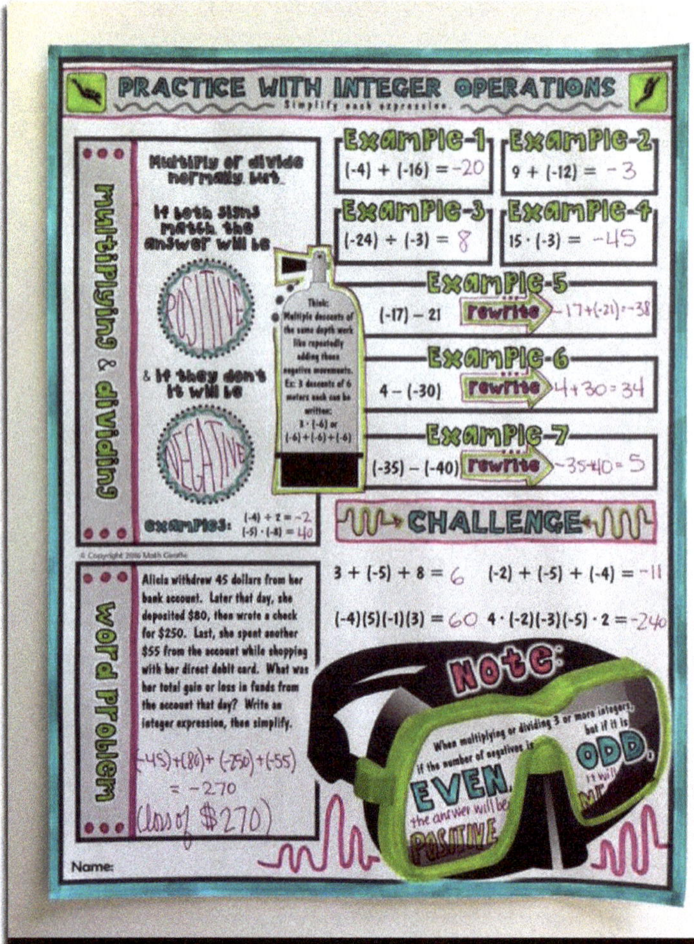
PRACTICE WITH INTEGER OPERATIONS
Simplify each expression.
multiplying & dividing
Multiply or divide normally, but...
if both signs match, the answer will be
POSITIVE
& if they don't it will be
NEGATIVE
examples: (-4) ÷ 2 = -2
(-5) · (-8) = 40
Example-1
(-4) + (-16) = -20
Example-2
9 + (-12) = -3
Example-3
(-24) ÷ (-3) = 8
Example-4
15 · (-3) = -45
Example-5
(-17) – 21 rewrite -17+(-21) = -38
Example-6
4 – (-30) rewrite 4+30 = 34
Example-7
(-35) – (-40) rewrite -35+40 = 5
CHALLENGE
3 + (-5) + 8 = 6
(-2) + (-5) + (-4) = -11
(-4)(5)(-1)(3) = 60
4 · (-2)(-3)(-5) · 2 = -240
word problem
(-45)+(80)+(-250)+(-55) = -270 (loss of $270)
Note:
EVEN
the answer will be POSITIVE
ODD
Name:

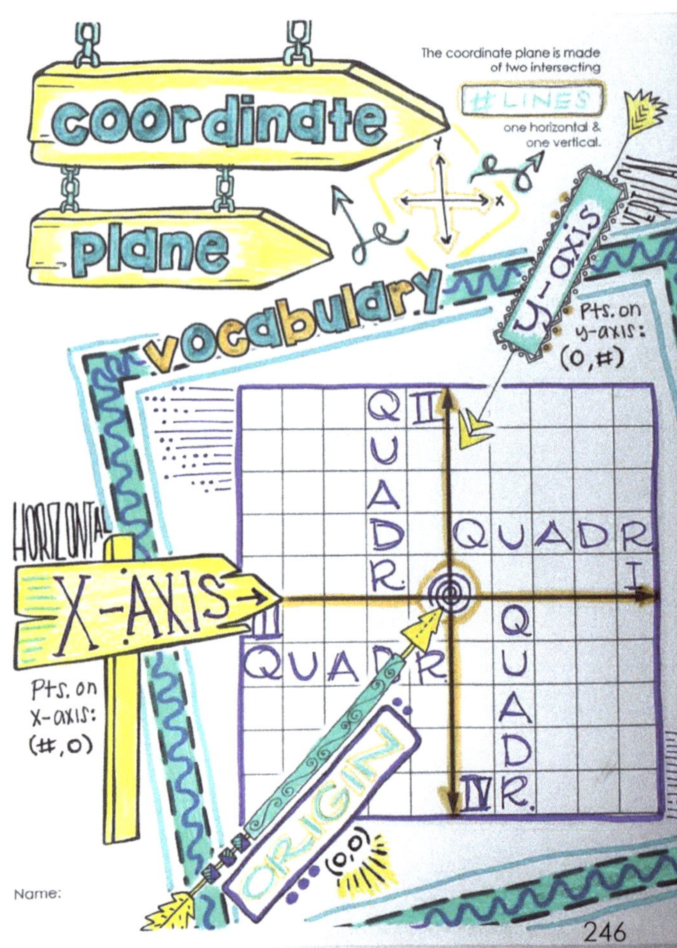
coordinate
plane
The coordinate plane is made of two intersecting
LINES
one horizontal & one vertical.
vocabulary
y-axis
VERTICAL
Pts. on y-axis: (0, #)
HORIZONTAL
X-AXIS
Pts. on X-axis: (#, 0)
QUADR II
QUADR I
QUADR III
QUADR IV
ORIGIN
(0,0)
Name:

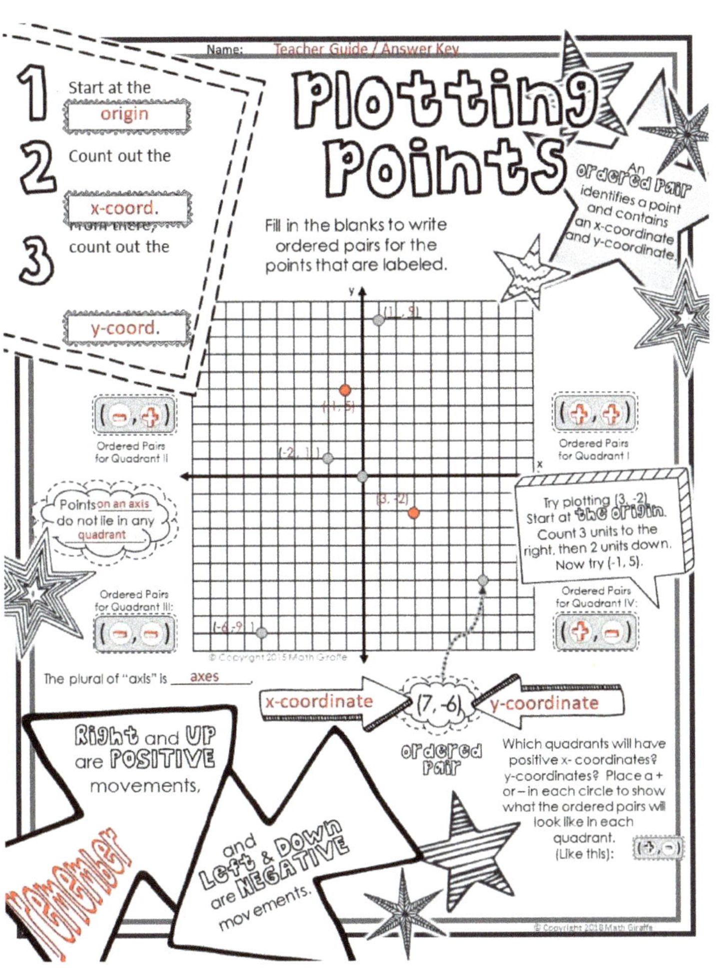

Name: Teacher Guide / Answer Key
Plotting Points
1 Start at the origin
2 Count out the x-coord.
3 count out the y-coord.
An ordered pair identifies a point and contains an x-coordinate and y-coordinate.
Fill in the blanks to write ordered pairs for the points that are labeled.
Ordered Pairs for Quadrant II
Ordered Pairs for Quadrant I
Points on an axis do not lie in any quadrant
Try plotting (3, -2). Start at the origin. Count 3 units to the right, then 2 units down. Now try (-1, 5).
Ordered Pairs for Quadrant III:
Ordered Pairs for Quadrant IV:
The plural of "axis" is axes
x-coordinate
(7, -6)
y-coordinate
ordered pair
Right and UP are POSITIVE movements,
and Left & Down are NEGATIVE movements.
Remember
Which quadrants will have positive x- coordinates? y-coordinates? Place a + or – in each circle to show what the ordered pairs will look like in each quadrant. (Like this):

Properties of Addition & Multiplication
These number properties, or rules, allow us to re-write an expression in a different, but equivalent form.
"EQUALS"
"can be rewritten as"
"commute" means... to travel back and forth, to move
In this case, they move back and forth to change ORDER
commutative property (change order / move back and forth)
1
Commutative Property of Addition
a + b = b + a
Commutative Property of Multiplication
ab = ba
EXAMPLE
7 + 6 = 6 + 7
f · 144 = 144 f
Name: Answer Key / Teacher Guide

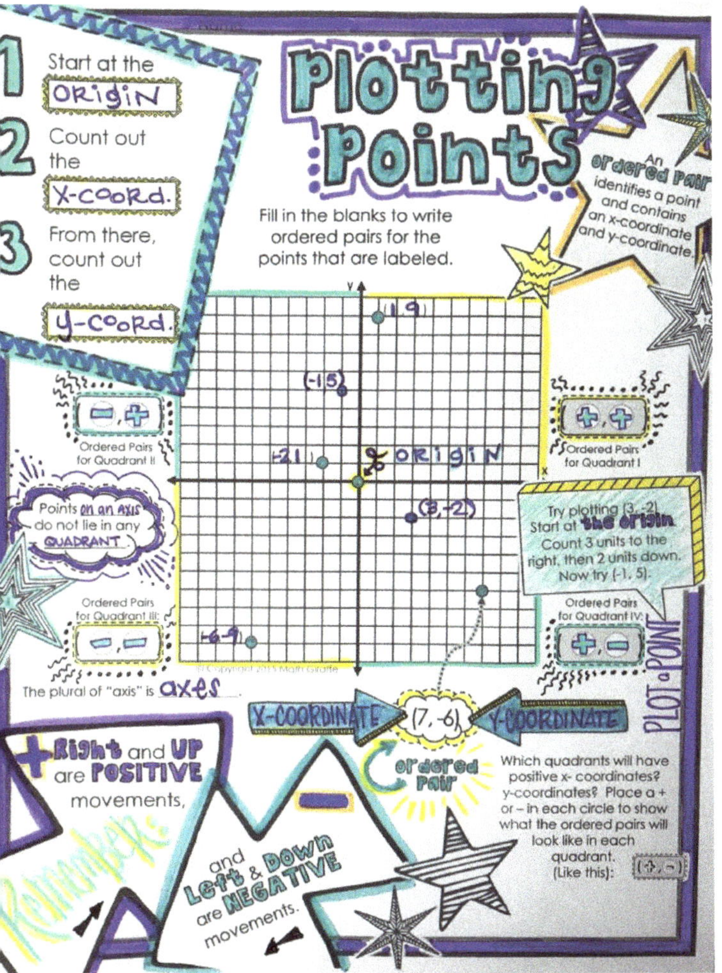

Plotting Points
1 Start at the ORIGIN
2 Count out the X-COORD.
3 From there, count out the Y-COORD.
An ordered pair identifies a point and contains an x-coordinate and y-coordinate.
Fill in the blanks to write ordered pairs for the points that are labeled.
ORIGIN
Ordered Pairs for Quadrant II
Ordered Pairs for Quadrant I
Points on an axis do not lie in any QUADRANT
Try plotting (3, -2). Start at the origin. Count 3 units to the right, then 2 units down. Now try (-1, 5).
Ordered Pairs for Quadrant III:
Ordered Pairs for Quadrant IV:
The plural of "axis" is axes
X-COORDINATE
(7, -6)
Y-COORDINATE
ordered pair
PLOT A POINT
Right and UP are POSITIVE movements,
and Left & Down are NEGATIVE movements.
Remember
Which quadrants will have positive x- coordinates? y-coordinates? Place a + or – in each circle to show what the ordered pairs will look like in each quadrant. (Like this):

Properties of Addition & Multiplication
These number properties, or rules, allow us to: REWRITE an expression differently... (but EQUIVALENT)
"EQUALS"
CAN BE REWRITTEN as
"commute" means... to TRAVEL or MOVE ~back and forth~
or in this case change ORDER
COMMUTATIVE property
1
Commutative Property of Addition
a+b=b+a
Commutative Property of Multiplication
ab=ba
EXAMPLE
7 + 6 = 6 + 7
f · 144 = 144 f
travel
back + forth
move
change order
Name:

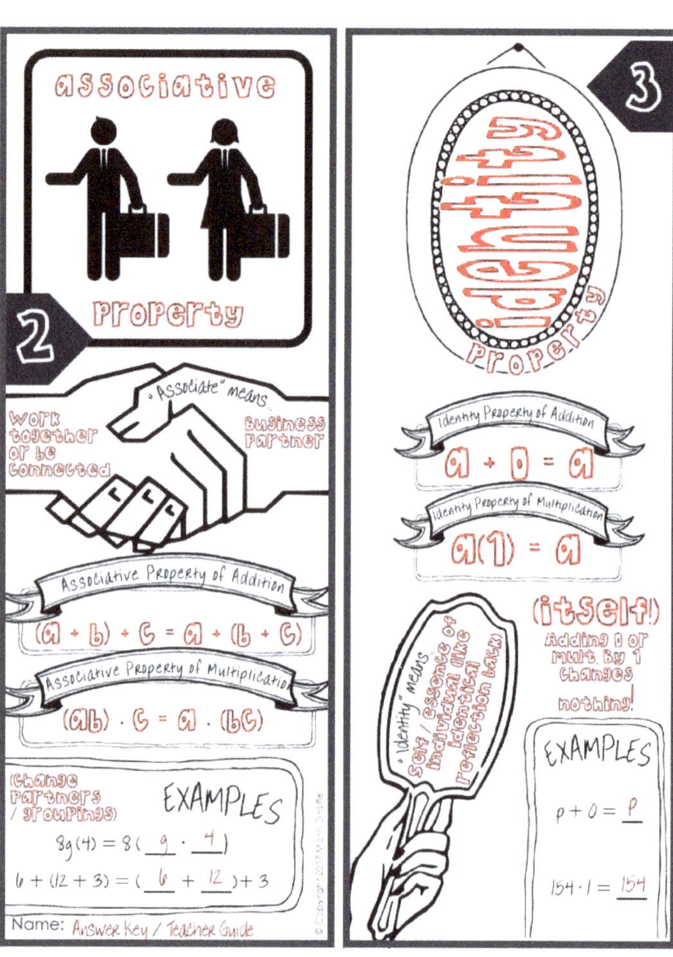
associative
property
2
"Associate" means
work together or be connected
business partner
Associative Property of Addition
(a + b) + c = a + (b + c)
Associative Property of Multiplication
(ab) · c = a · (bc)
change partners / groupings
EXAMPLES
8g(4) = 8(g · 4)
6 + (12 + 3) = (6 + 12) + 3
Name: Answer Key / Teacher Guide
3
identity
property
Identity Property of Addition
a + 0 = a
Identity Property of Multiplication
a(1) = a
"identity" means
self / essence of individual (like identical reflection back)
(itself!)
Adding 0 or mult. by 1 changes nothing!
EXAMPLES
p + 0 = p
154 · 1 = 154

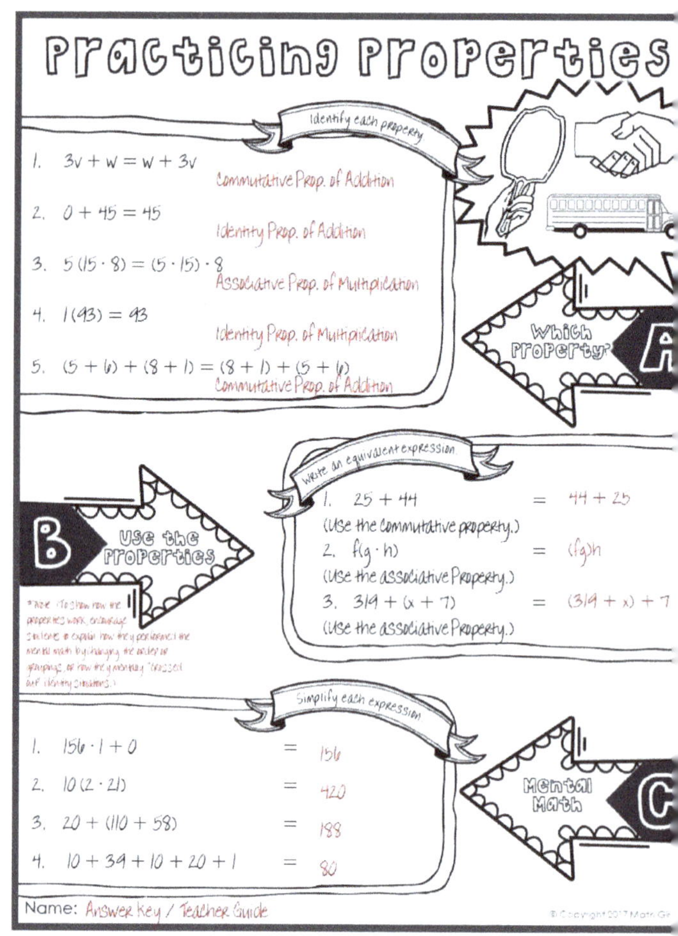
Practicing Properties
Identify each property
1. 3v + w = w + 3v
Commutative Prop. of Addition
2. 0 + 45 = 45
Identity Prop. of Addition
3. 5(15 · 8) = (5 · 15) · 8
Associative Prop. of Multiplication
4. 1(93) = 93
Identity Prop. of Multiplication
5. (5 + 6) + (8 + 1) = (8 + 1) + (5 + 6)
Commutative Prop. of Addition
Which property?
A
B
Use the properties
Write an equivalent expression
1. 25 + 44 = 44 + 25
(Use the commutative property.)
2. f(g · h) = (fg)h
(Use the associative Property.)
3. 319 + (x + 7) = (319 + x) + 7
(Use the associative Property.)
Simplify each expression
1. 156 · 1 + 0 = 156
2. 10(2 · 21) = 420
3. 20 + (110 + 58) = 188
4. 10 + 39 + 10 + 20 + 1 = 80
Mental Math
C
Name: Answer Key / Teacher Guide

ASSOCIATIVE
PROPERTY
2
"Associate" means
PARTNER
WORK -or- JOIN together
or in this case changing GROUPING
Associative Property of Addition
(a+b)+c=a+(b+c)
Associative Property of Multiplication
(ab)·c=a·(bc)
change groupings
EXAMPLES
8g(4) = 8(g · 4)
6 + (12 + 3) = (6 + 12) + 3
Name:
3
IDENTITY
property
Identity Property of Addition
a+0=a
Identity Property of Multiplication
a(1)=a
"identity" means
SELF
reflection-itself
in this case, it's identity is UNCHANGED
EXAMPLES
p + 0 = p
154 · 1 = 154

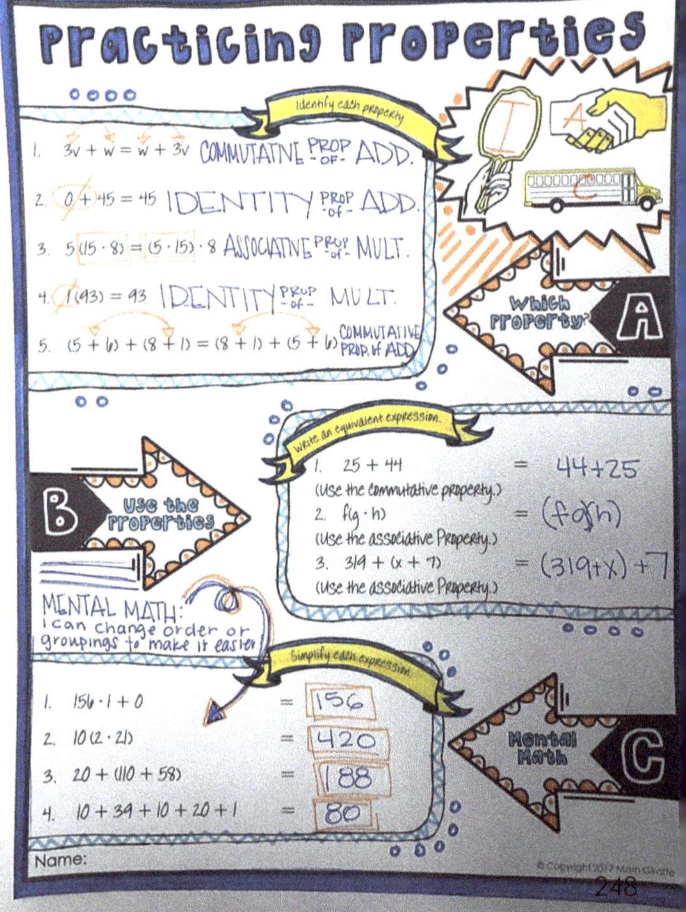
Practicing Properties
Identify each property
1. 3v + w = w + 3v COMMUTATIVE PROP OF ADD.
2. 0 + 45 = 45 IDENTITY PROP OF ADD.
3. 5(15 · 8) = (5 · 15) · 8 ASSOCIATIVE PROP OF MULT.
4. 1(93) = 93 IDENTITY PROP OF MULT.
5. (5 + 6) + (8 + 1) = (8 + 1) + (5 + 6) COMMUTATIVE PROP. of ADD
Which property?
A
B
Use the properties
Write an equivalent expression
1. 25 + 44 = 44+25
(Use the commutative property.)
2. f(g · h) = (fg)h
(Use the associative Property.)
3. 319 + (x + 7) = (319+x)+7
(Use the associative Property.)
MENTAL MATH:
I can change order or groupings to make it easier
Simplify each expression
1. 156 · 1 + 0 = 156
2. 10(2 · 21) = 420
3. 20 + (110 + 58) = 188
4. 10 + 39 + 10 + 20 + 1 = 80
Mental Math
C
Name:

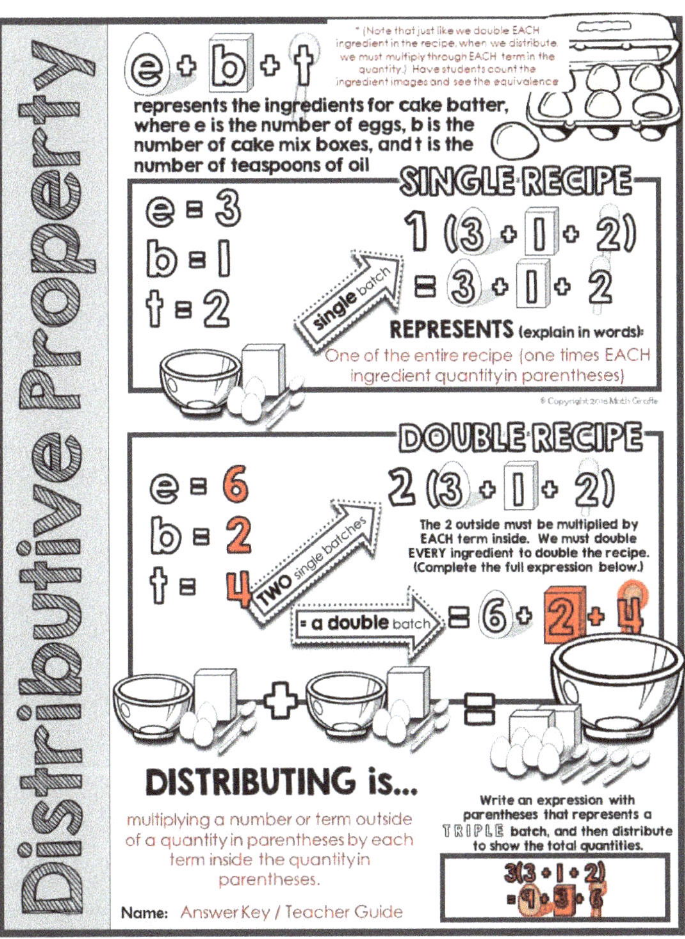
Distributive Property
e + b + t
* (Note that just like we double EACH ingredient in the recipe, when we distribute we must multiply through EACH term in the quantity.) Have students count the ingredient images and see the equivalence.
represents the ingredients for cake batter, where e is the number of eggs, b is the number of cake mix boxes, and t is the number of teaspoons of oil
SINGLE RECIPE
e = 3
b = 1
t = 2
single batch
1 (3 + 1 + 2)
= 3 + 1 + 2
REPRESENTS (explain in words):
One of the entire recipe (one times EACH ingredient quantity in parentheses)
© Copyright 2016 Math Giraffe
DOUBLE RECIPE
e = 6
b = 2
t = 4
TWO single batches
2 (3 + 1 + 2)
The 2 outside must be multiplied by EACH term inside. We must double EVERY ingredient to double the recipe. (Complete the full expression below.)
= a double batch
= 6 + 2 + 4
DISTRIBUTING is...
multiplying a number or term outside of a quantity in parentheses by each term inside the quantity in parentheses.
Write an expression with parentheses that represents a TRIPLE batch, and then distribute to show the total quantities.
3(3 + 1 + 2)
= 9 + 3 + 6
Name: Answer Key / Teacher Guide

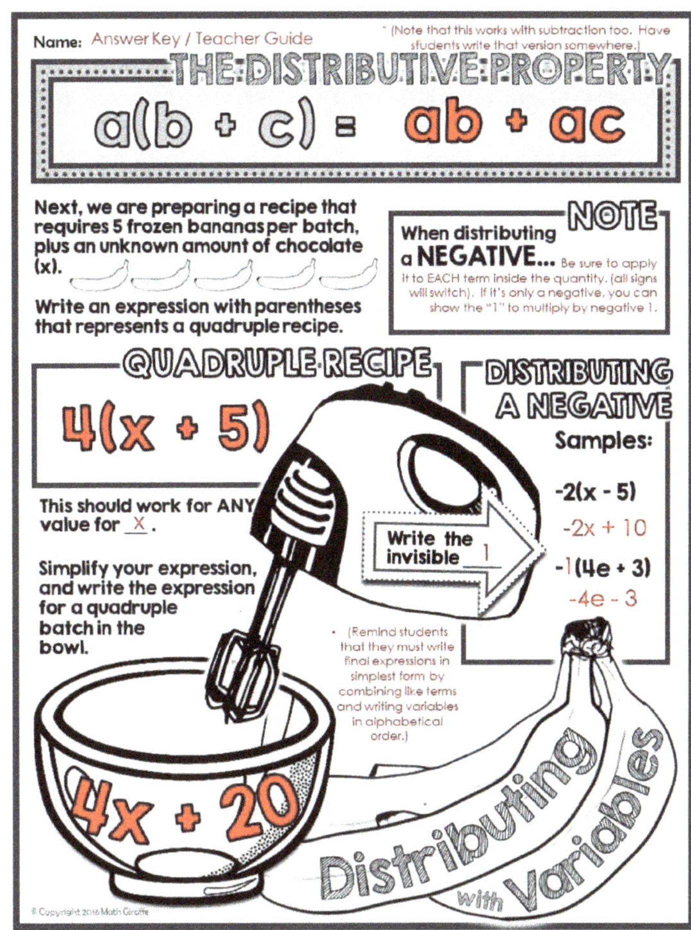
Name: Answer Key / Teacher Guide
* (Note that this works with subtraction too. Have students write that version somewhere.)
THE DISTRIBUTIVE PROPERTY
a(b + c) = ab + ac
Next, we are preparing a recipe that requires 5 frozen bananas per batch, plus an unknown amount of chocolate (x).
Write an expression with parentheses that represents a quadruple recipe.
NOTE
When distributing a NEGATIVE... Be sure to apply it to EACH term inside the quantity. (all signs will switch). If it's only a negative, you can show the "1" to multiply by negative 1.
QUADRUPLE RECIPE
4(x + 5)
This should work for ANY value for x.
Simplify your expression, and write the expression for a quadruple batch in the bowl.
Write the invisible 1
DISTRIBUTING A NEGATIVE
Samples:
-2(x - 5)
-2x + 10
-1(4e + 3)
-4e - 3
* (Remind students that they must write final expressions in simplest form by combining like terms and writing variables in alphabetical order.)
4x + 20
Distributing with Variables
© Copyright 2016 Math Giraffe

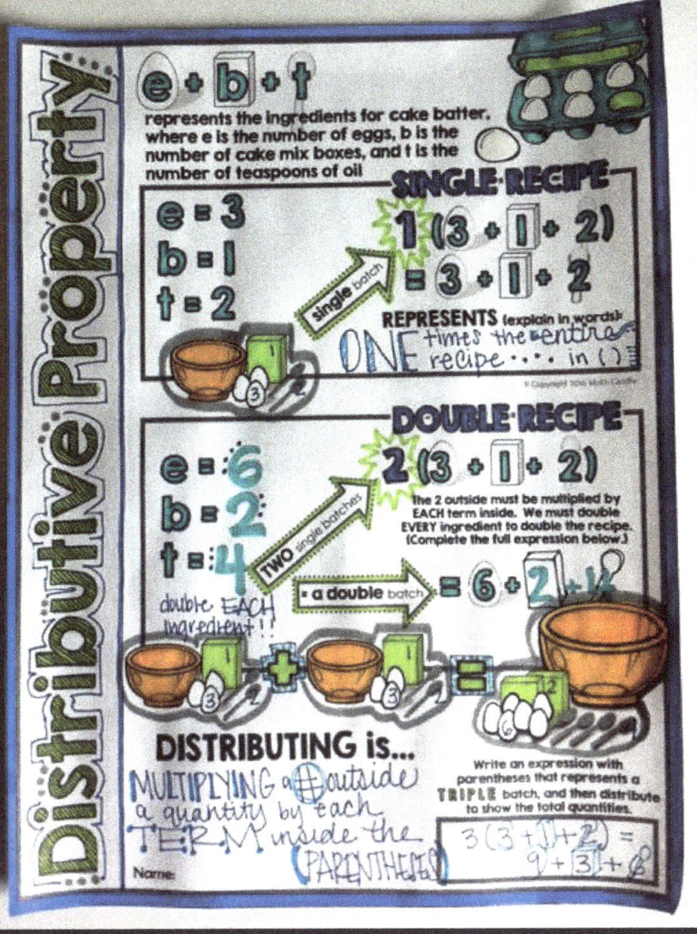
Distributive Property
e + b + t
represents the ingredients for cake batter, where e is the number of eggs, b is the number of cake mix boxes, and t is the number of teaspoons of oil
SINGLE RECIPE
e = 3
b = 1
t = 2
single batch
1 (3 + 1 + 2)
= 3 + 1 + 2
REPRESENTS (explain in words):
ONE times the entire recipe ... in ()
DOUBLE RECIPE
e = 6
b = 2
t = 4
TWO single batches
2 (3 + 1 + 2)
The 2 outside must be multiplied by EACH term inside. We must double EVERY ingredient to double the recipe. (Complete the full expression below.)
double EACH ingredient!!
= a double batch
= 6 + 2 + 4
DISTRIBUTING is...
MULTIPLYING a # outside a quantity by each TERM inside the PARENTHESIS
Write an expression with parentheses that represents a TRIPLE batch, and then distribute to show the total quantities.
3 (3 + 1 + 2) =
9 + 3 + 6
Name:

Name:
a(b − c) = ab − ac also
THE DISTRIBUTIVE PROPERTY
a(b + c) = ab + ac
Next, we are preparing a recipe that requires 5 frozen bananas per batch, plus an unknown amount of chocolate (x).
Write an expression with parentheses that represents a quadruple recipe.
*works w. subtract too
NOTE
When distributing a NEGATIVE... be sure to APPLY it to EACH TERM (all signs will switch)
QUADRUPLE RECIPE
4(x+5)
"FOUR BATCHES OF X AMT OF CHOCOLATE PLUS FIVE BANANAS"
This should work for ANY value for x.
Simplify your expression, and write the expression for a quadruple batch in the bowl.
Write the invisible ONE
be sure to MULTIPLY the outside term by EACH & EVERY term inside the ()
DISTRIBUTING A NEGATIVE
Samples:
-2(x - 5)
-2x + 10
-1(4e + 3)
-4e - 3
QUADRUPLE RECIPE
4X+20
SIMPLIFY
COMBINE LIKE TERMS
Distributing with Variables

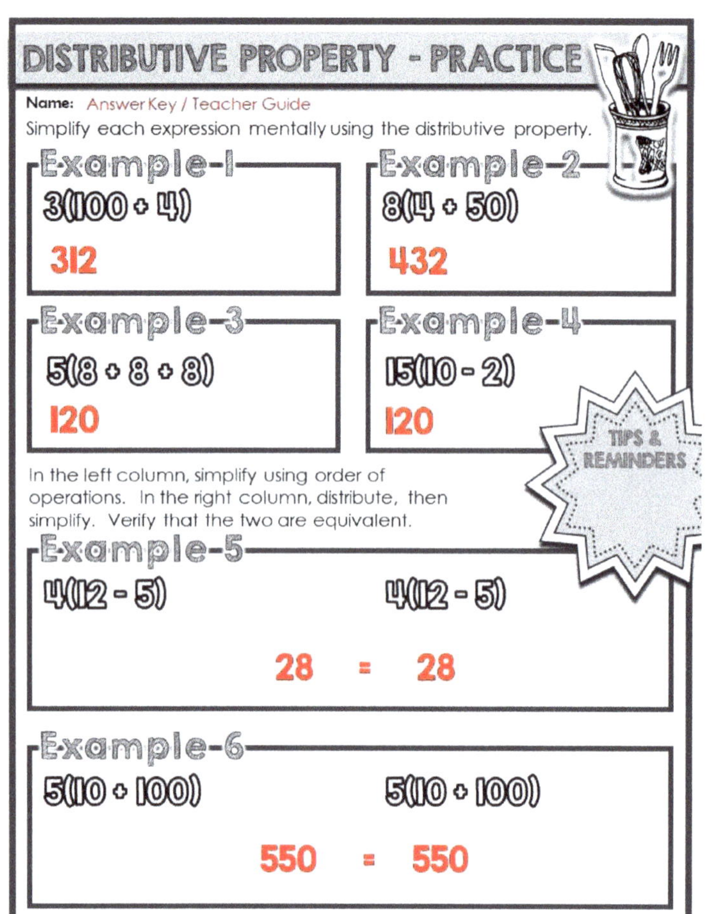

DISTRIBUTIVE PROPERTY - PRACTICE

Name: Answer Key / Teacher Guide

Simplify each expression mentally using the distributive property.

Example-1
$3(100 + 4)$
312

Example-2
$8(4 + 50)$
432

Example-3
$5(8 + 8 + 8)$
120

Example-4
$15(10 - 2)$
120

TIPS & REMINDERS

In the left column, simplify using order of operations. In the right column, distribute, then simplify. Verify that the two are equivalent.

Example-5
$4(12 - 5)$ | $4(12 - 5)$
28 = 28

Example-6
$5(10 + 100)$ | $5(10 + 100)$
550 = 550

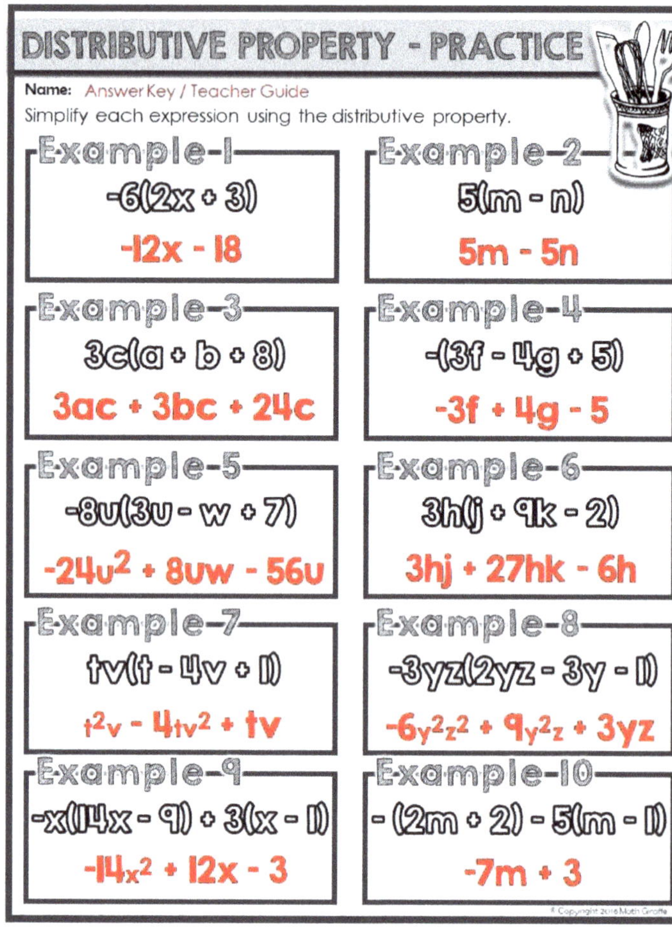

DISTRIBUTIVE PROPERTY - PRACTICE

Name: Answer Key / Teacher Guide

Simplify each expression using the distributive property.

Example-1
$-6(2x + 3)$
$-12x - 18$

Example-2
$5(m - n)$
$5m - 5n$

Example-3
$3c(a + b + 8)$
$3ac + 3bc + 24c$

Example-4
$-(3f - 4g + 5)$
$-3f + 4g - 5$

Example-5
$-8u(3u - w + 7)$
$-24u^2 + 8uw - 56u$

Example-6
$3h(j + 9k - 2)$
$3hj + 27hk - 6h$

Example-7
$tv(t - 4v + 1)$
$t^2v - 4tv^2 + tv$

Example-8
$-3yz(2yz - 3y - 1)$
$-6y^2z^2 + 9y^2z + 3yz$

Example-9
$-x(14x - 9) + 3(x - 1)$
$-14x^2 + 12x - 3$

Example-10
$-(2m + 2) - 5(m - 1)$
$-7m + 3$

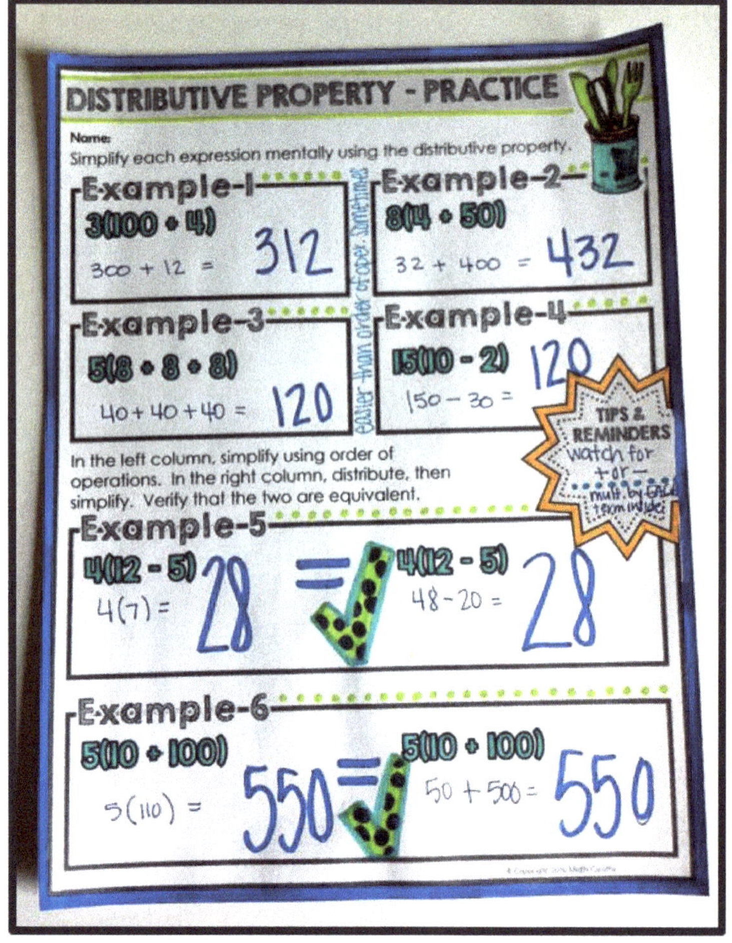

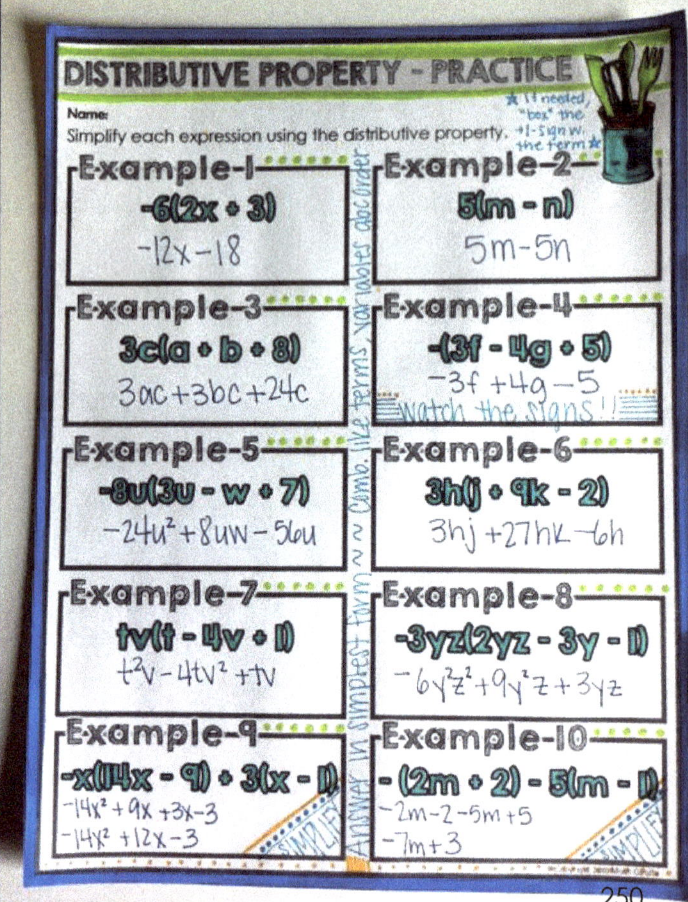

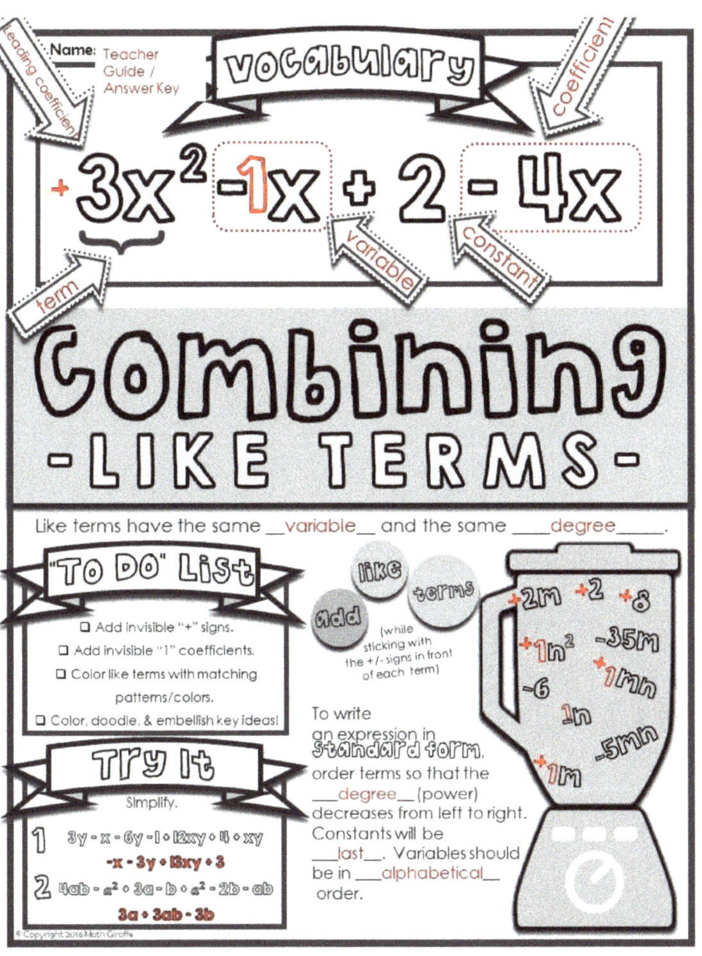
Name: Teacher Guide / Answer Key
vocabulary
leading coefficient
coefficient
$+3x^2 - 1x + 2 - 4x$
term
variable
constant
Combining
-LIKE TERMS-
Like terms have the same __variable__ and the same ___degree____.
"To Do" List
Add invisible "+" signs.
Add invisible "1" coefficients.
Color like terms with matching patterns/colors.
Color, doodle, & embellish key ideas!
like
terms
add
(while sticking with the +/- signs in front of each term)
To write an expression in Standard form, order terms so that the __degree__(power) decreases from left to right. Constants will be __last__. Variables should be in __alphabetical__ order.
Try It
Simplify.
1 $3y - x - 6y - 1 + 12xy + 4 + xy$
$-x - 3y + 13xy + 3$
2 $4ab - a^2 + 3a - b + a^2 - 2b - ab$
$3a + 3ab - 3b$
+2m +2 +8 +1n² -35m +1mn -6 1n -5mn +1m

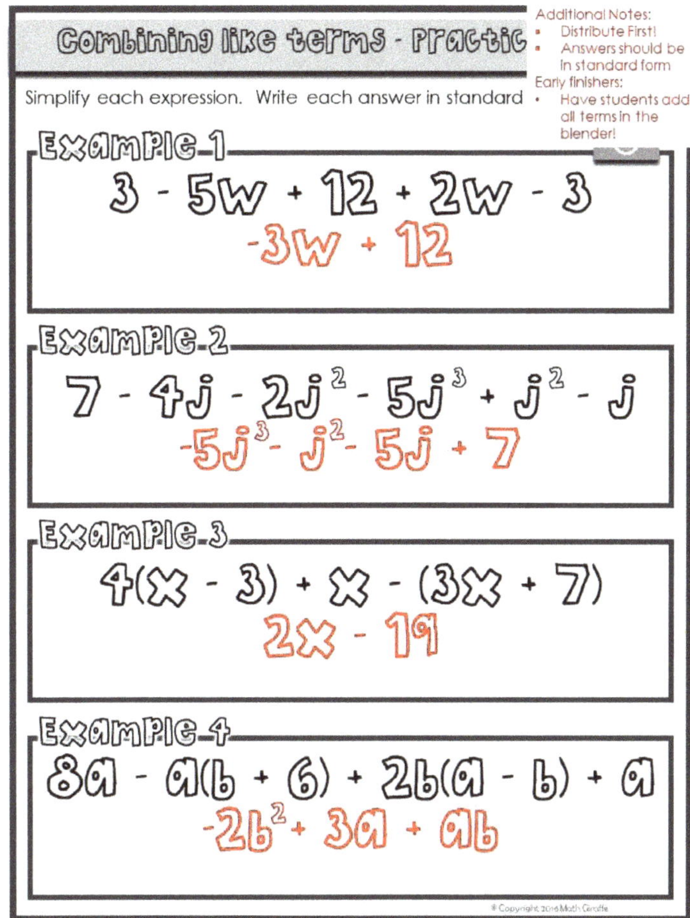
Combining like terms - Practic
Additional Notes:
Distribute First!
Answers should be in standard form
Early finishers:
Have students add all terms in the blender!
Simplify each expression. Write each answer in standard
Example 1
$3 - 5w + 12 + 2w - 3$
$-3w + 12$
Example 2
$7 - 4j - 2j^2 - 5j^3 + j^2 - j$
$-5j^3 - j^2 - 5j + 7$
Example 3
$4(x - 3) + x - (3x + 7)$
$2x - 19$
Example 4
$8a - a(b + 6) + 2b(a - b) + a$
$-2b^2 + 3a + ab$

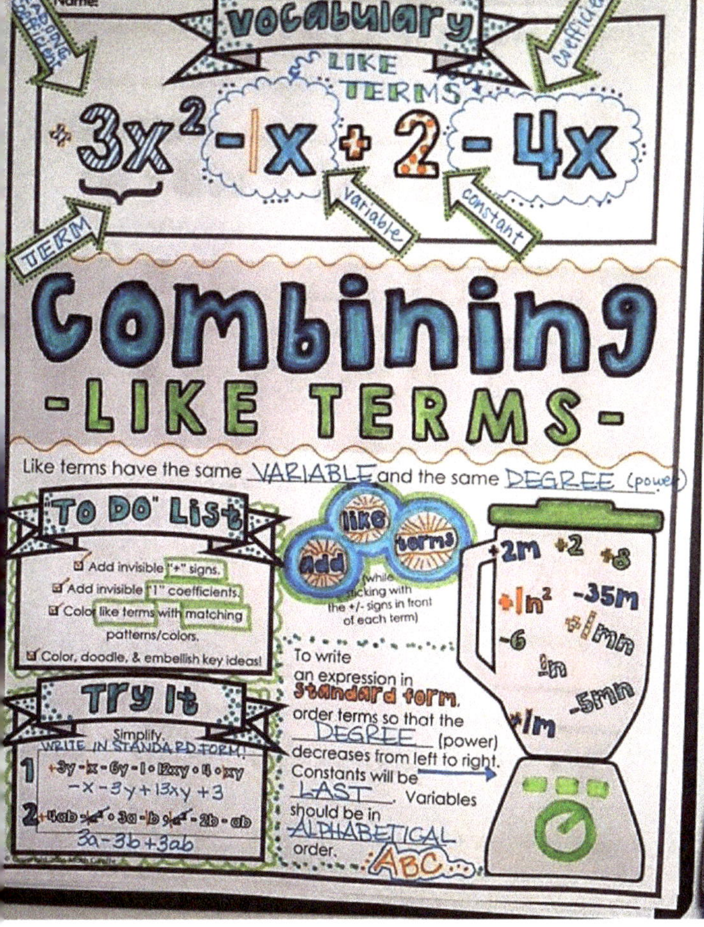
Name:
vocabulary
Leading coefficient
coefficient
LIKE TERMS
$+3x^2 - 1x + 2 - 4x$
TERM
variable
constant
Combining
-LIKE TERMS-
Like terms have the same VARIABLE and the same DEGREE (power)
"To Do" List
Add invisible "+" signs.
Add invisible "1" coefficients.
Color like terms with matching patterns/colors.
Color, doodle, & embellish key ideas!
like
terms
add
(while sticking with the +/- signs in front of each term)
To write an expression in Standard form, order terms so that the DEGREE (power) decreases from left to right. Constants will be LAST. Variables should be in ALPHABETICAL order.
ABC
Try It
Simplify.
WRITE IN STANDARD FORM!
$-x - 3y + 13xy + 3$
$3a - 3b + 3ab$

Combining like terms - Practice
Simplify each expression. Write each answer in standard form.
Example 1
$+3 - 5w + 12 + 2w - 3$
Add + signs and invisible 1s.
$-3w - 12$
Example 2
$7 - 4j - 2j^2 - 5j^3 + j^2 - j$
Standard Form - Degree decreases L→R
$-5j^3 - j^2 - 5j + 7$
Example 3
$4(x - 3) + x - (3x +$
$+4x - 12 + x - 3x - 7$
DISTRIBUTE 1st
2x
Example 4
$8a - a(b + 6) + 2b(a - b)$
$8a - ab + 6a + 2ab - 2b^2$
Like terms must have Same VARIABLE and DEGREE

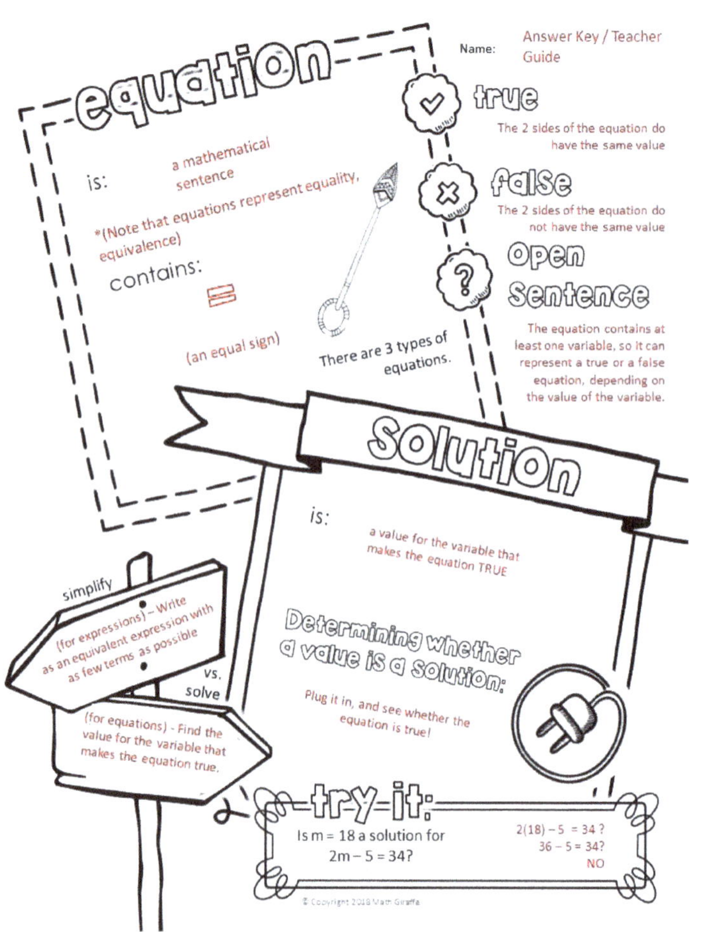

equation
Name: Answer Key / Teacher Guide
is: a mathematical sentence
*(Note that equations represent equality, equivalence)
contains: = (an equal sign)
There are 3 types of equations.
true
The 2 sides of the equation do have the same value
false
The 2 sides of the equation do not have the same value
Open Sentence
The equation contains at least one variable, so it can represent a true or a false equation, depending on the value of the variable.
Solution
is: a value for the variable that makes the equation TRUE
Determining whether a value is a solution:
Plug it in, and see whether the equation is true!
simplify
(for expressions) - Write as an equivalent expression with as few terms as possible
vs.
solve
(for equations) - Find the value for the variable that makes the equation true.
try it:
Is m = 18 a solution for 2m – 5 = 34?
2(18) – 5 = 34 ?
36 – 5 = 34?
NO
© Copyright 2018 Math Giraffe

Name: Answer Key / Teacher Guide
Substitution:
replacing the variable with the value you want to test
7k = k + 24
Is 4 a solution?
7(4) = 4 + 24?
28 = 28?
yes
Classify each as true, false, or an open sentence.
any value for the variable that makes the equation true
Solution
3m – 2 = 14
15 = 7(2) + 3
72 ÷ 3 = 8(3)
solutions for equations
How many solutions?
w + 1 = 6
one solution (w = 5)
3b = 3b
infinitely many solutions! Any value makes it true.
A team has 9 players on the field plus a group on the bench. If their roster shows 23 players total, how many are on the bench? Write an equation. Can there be 15 on the bench?
9 + b = 23
9 + 15 = 23?
NO
© Copyright 2018 Math Giraffe

REPRESENTS equality equivalence
equation
Name:
is: a MATHEMATICAL Sentence
that contains: EQUAL SIGN
ex: 4=1+3 x=3 m=4 a+b=c
There are 3 types of equations.
true
The 2 sides of the equation DO have the same value.
false
The 2 sides of the equation DO NOT have = values
Open Sentence
The equation contains at least one VARIABLE, so it can represent a true or a false equation, depending on the value of the variable.
Solution
is: a value for the VARIABLE that makes the equation TRUE
Determining whether a value is a solution:
Replace (Plug in) & see if the equation is TRUE!
PLUG IT IN
simplify
(for expressions) Write as an equivalent expression with as few terms as possible
vs.
solve
(for equations) Find the value for the variable(s) that makes the equation true.
try it:
Is m = 18 a solution for 2m – 5 = 34?
2(18)–5=34?
36–5=34?
NO
does NOT make the equation true
© Copyright 2018 Math Giraffe

Name:
Substitution:
REPLACING the variable with the value you want to test
7k = k + 24
Is 4 a solution?
7(4)=4+24 ?
28=28 ?
YES
SWAP
PLUG IN & check
Classify each as true, false, or an open sentence.
any value for the VARIABLE that makes the EQUATION TRUE
Solution
VARIABLE
3m – 2 = 14 O
15 = 7(2) + 3 F
72 ÷ 3 = 8(3) T
solutions for equations
An equation can have one solution, no solution, or many.
How many solutions?
w + 1 = 6
ONE solution (w=5)
3b = 3b
INFINITELY MANY SOLUTIONS
any value makes it true!
A team has 9 players on the field plus a group on the bench. If their roster shows 23 players total, how many are on the bench? Write an equation. Can there be 15 on the bench?
Test b=15
9+b=23
9+15=23?
NO

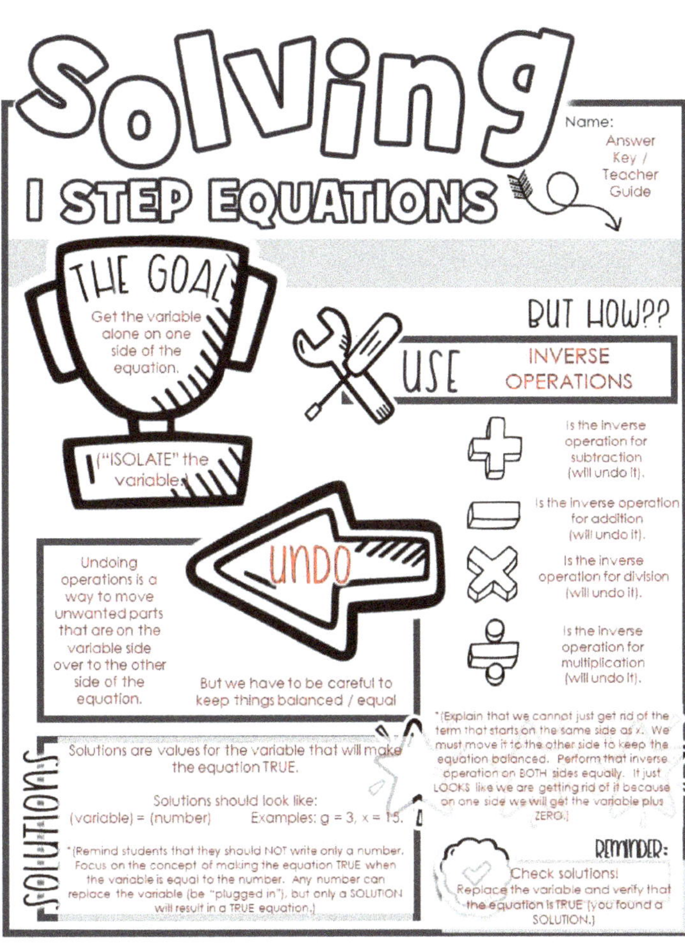

Solving
1 STEP EQUATIONS
Name: Answer Key / Teacher Guide
THE GOAL
Get the variable alone on one side of the equation.
("ISOLATE" the variable)
BUT HOW??
USE INVERSE OPERATIONS
is the inverse operation for subtraction (will undo it).
is the inverse operation for addition (will undo it).
is the inverse operation for division (will undo it).
is the inverse operation for multiplication (will undo it).
UNDO
Undoing operations is a way to move unwanted parts that are on the variable side over to the other side of the equation.
But we have to be careful to keep things balanced / equal
*(Explain that we cannot just get rid of the term that starts on the same side as x. We must move it to the other side to keep the equation balanced. Perform that inverse operation on BOTH sides equally. It just LOOKS like we are getting rid of it because on one side we will get the variable plus ZERO.)
SOLUTIONS
Solutions are values for the variable that will make the equation TRUE.
Solutions should look like:
(variable) = (number)
Examples: g = 3, x = 15.
*(Remind students that they should NOT write only a number. Focus on the concept of making the equation TRUE when the variable is equal to the number. Any number can replace the variable (be "plugged in"), but only a SOLUTION will result in a TRUE equation.)
REMINDER:
Check solutions!
Replace the variable and verify that the equation is TRUE (you found a SOLUTION.)

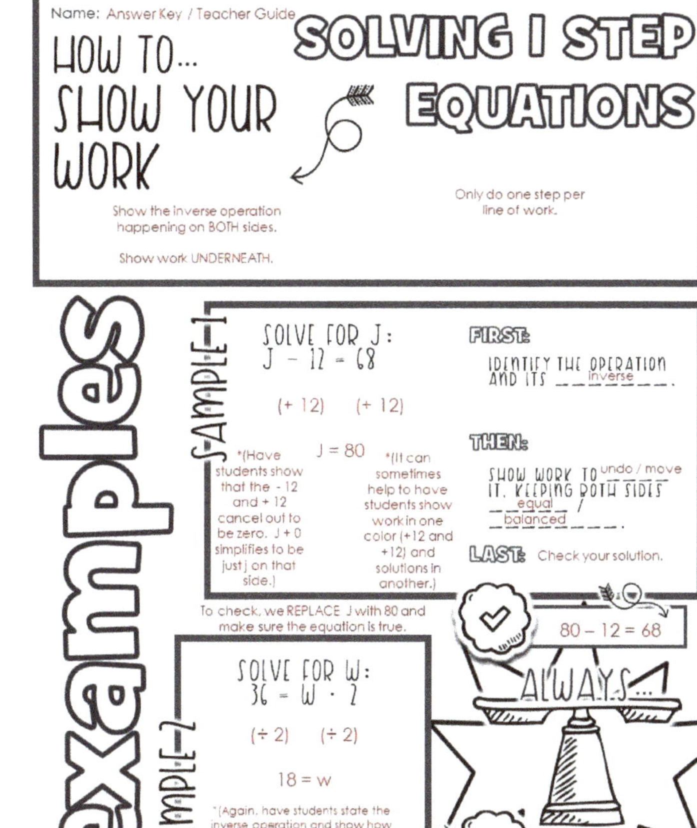

Name: Answer Key / Teacher Guide
HOW TO... SHOW YOUR WORK
SOLVING 1 STEP EQUATIONS
Show the inverse operation happening on BOTH sides.
Show work UNDERNEATH.
Only do one step per line of work.
examples
SAMPLE-1
SOLVE FOR J:
J – 12 = 68
(+ 12) (+ 12)
J = 80
*(Have students show that the - 12 and + 12 cancel out to be zero. J + 0 simplifies to be just J on that side.)
*(It can sometimes help to have students show work in one color (+12 and +12) and solutions in another.)
FIRST:
IDENTIFY THE OPERATION AND ITS inverse.
THEN:
SHOW WORK TO undo / move IT, KEEPING BOTH SIDES equal / balanced.
LAST:
Check your solution.
To check, we REPLACE J with 80 and make sure the equation is true.
80 – 12 = 68
SAMPLE-2
SOLVE FOR W:
36 = W · 2
(÷ 2) (÷ 2)
18 = w
*(Again, have students state the inverse operation and show how multiplying by 2 and dividing by 2 undo one another to leave us with w · 1, which simplifies to just w.)
ALWAYS...
36 = 18 (2)

Solving
1 STEP EQUATIONS
Name:
MOVE everything AWAY from the VARIABLE to the other side of the EQUATION
THE GOAL
Get the VARIABLE alone on one SIDE
ISOLATE the variable
BUT HOW??
USE INVERSE OPERATIONS
is the OPPOSITE of – will UNDO
is the OPPOSITE of + will UNDO
is the OPPOSITE of ÷ will UNDO
is the OPPOSITE of × will UNDO
INVERSE operations UNDO one another to get rid of a term near the VARIABLE.
UNDO
the OPPOSITE of adding two is taking it back away (subtract 2)
THINK about it!
SOLUTIONS
ARE values for the VARIABLE that make the EQUATION TRUE!
LOOK LIKE variable = #
ex: x=2, y=15
BUT
REMINDER:
CHECK work by testing solution - Plug back in!

Name:
HOW TO... SHOW YOUR WORK
SOLVING 1 STEP EQUATIONS
Show the INVERSE OPERATION on BOTH sides
with ONE Step per LINE
examples
SAMPLE-1
SOLVE FOR J:
J – 12 = 68
+12 +12
J = 80
the opposite of subtraction is addition
How to check: If J = 80, then we REPLACE J with 80 and it should be TRUE
FIRST:
IDENTIFY THE OPERATION AND ITS INVERSE.
(in this case + so we use –)
THEN:
SHOW WORK TO UNDO IT, KEEPING BOTH SIDES EQUAL / BALANCED.
LAST:
CHECK solution!
80–12=68
SAMPLE-2
SOLVE FOR W:
36 = W · 2
÷2 ÷2
18 = w
The opposite of multiplication is division.
ALWAYS...
KEEP it equal BALANCE
SHOW OPERATION on both sides
36 = 18·2

Practice

solving 1 step equations

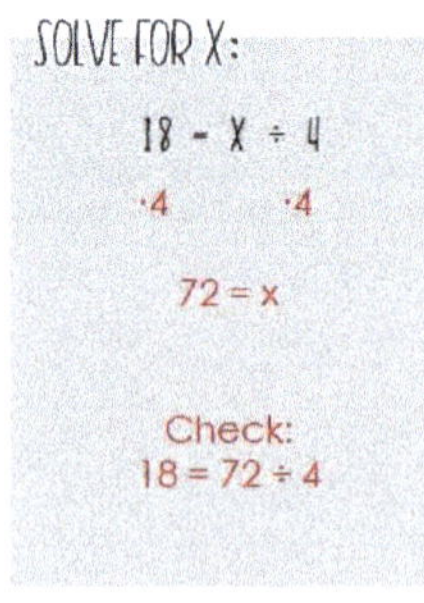

REMINDERS FOR WORK & ANSWERS

Only do/show one thing per line.

Show inverse operations on BOTH SIDES.

Check all solutions.

Name: Answer Key / Teacher Guide

a value for the variable that makes the equation true

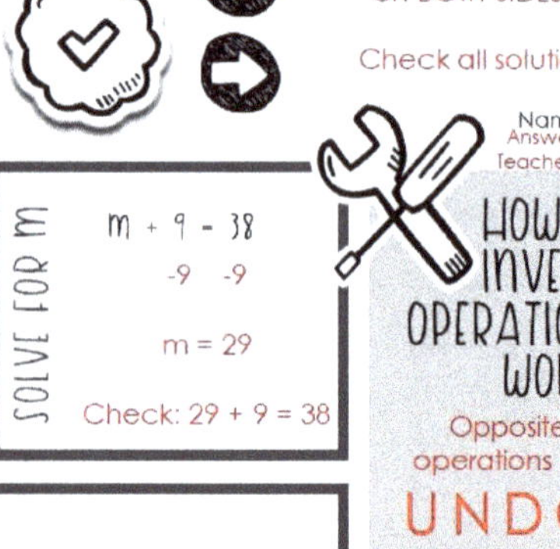

Check: 29 + 9 = 38

CHECK: 29+9=38✓
HOW DO INVERSE OPERATIONS WORK?
Opposite operations will
UNDO
one another.
(+ & -)
(x & ÷)
They will result in the variable times 1 or the variable plus 0. (which simplifies to the variable itself, alone.)

WHEN SOLVING, WHAT IS THE GOAL?

Isolate the variable.

(Get it alone on one side of the equation.)

We do this by moving the other terms using inverse operations.

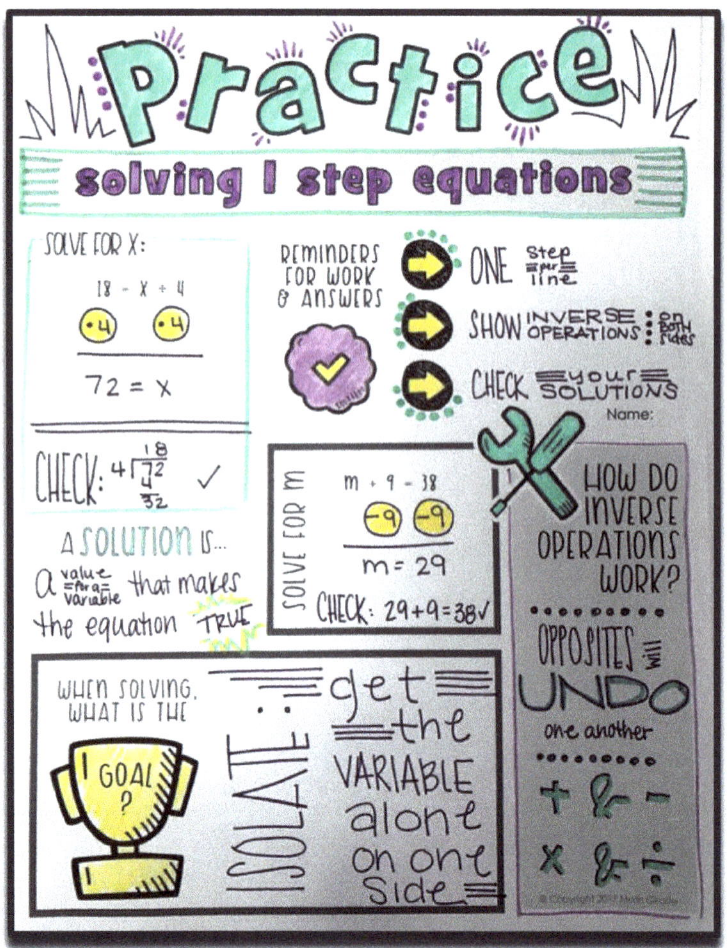

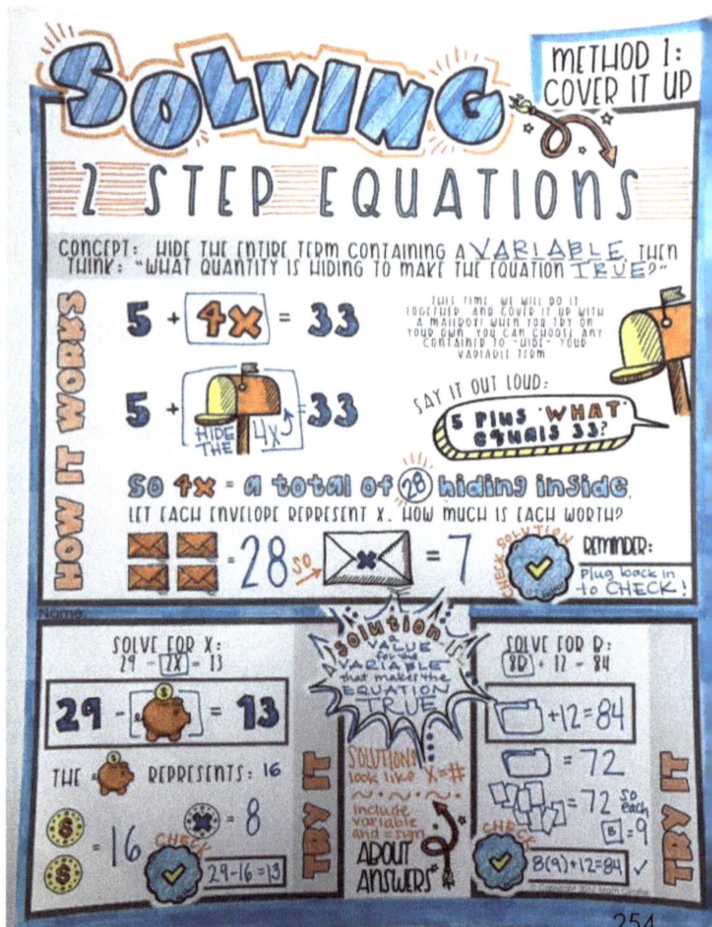

SOLVING
METHOD 1: COVER IT UP
*(This method is for helping students understand the concept behind a two-step equation and what's really happening when we solve it. It's great for introducing the idea, and helps to lead into WHY we do inverse operations. Use this page first to build understanding, then explain that this method is not as helpful with large numbers or decimals, etc. when mental math gets tricky. This leads into method #2 and inverse operations.)
2 STEP EQUATIONS
CONCEPT: HIDE THE ENTIRE TERM CONTAINING A VARIABLE, THEN THINK: "WHAT QUANTITY IS HIDING TO MAKE THE EQUATION TRUE?"
HOW IT WORKS
5 + 4x = 33
5 + HIDE THE 4x = 33
THIS TIME, WE WILL DO IT TOGETHER, AND COVER IT UP WITH A MAILBOX! WHEN YOU TRY ON YOUR OWN, YOU CAN CHOOSE ANY CONTAINER TO "HIDE" YOUR VARIABLE TERM.
SAY IT OUT LOUD:
5 plus "what" equals 33?
So 4x = a total of 28 hiding inside.
LET EACH ENVELOPE REPRESENT X. HOW MUCH IS EACH WORTH?
= 28 so = 7
REMINDER:
Plug back in to CHECK!
SOLVE FOR X:
29 - 2X = 13
29 - [piggy bank] = 13
29 - = 13
THE REPRESENTS: 16
= 16
= 8
29-16=13
TRY IT
solution: a value for a variable that makes the EQUATION TRUE
SOLUTIONS look like X=#
include variable and = sign
ABOUT ANSWERS
SOLVE FOR B:
8B + 12 = 84
+12=84
CHECK: 8(9)+12=84
= 72
= 72 so each = 9
8(9)+12=84 ✓
TRY IT

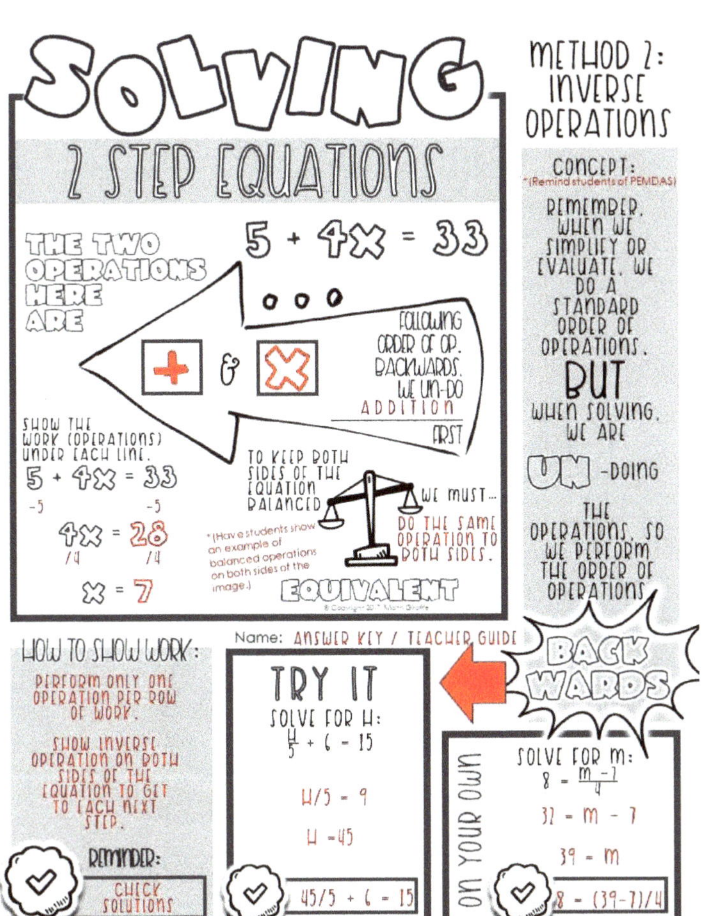
SOLVING
2 STEP EQUATIONS
METHOD 2: INVERSE OPERATIONS
THE TWO OPERATIONS HERE ARE
5 + 4x = 33
FOLLOWING ORDER OF OP. BACKWARDS, WE UN-DO ADDITION FIRST
SHOW THE WORK (OPERATIONS) UNDER EACH LINE.
5 + 4x = 33
-5 -5
4x = 28
/4 /4
x = 7
TO KEEP BOTH SIDES OF THE EQUATION BALANCED
WE MUST... DO THE SAME OPERATION TO BOTH SIDES.
*(Have students show an example of balanced operations on both sides of the image.)
EQUIVALENT
CONCEPT: *(Remind students of PEMDAS)
REMEMBER, WHEN WE SIMPLIFY OR EVALUATE, WE DO A STANDARD ORDER OF OPERATIONS.
BUT WHEN SOLVING, WE ARE UN-DOING THE OPERATIONS, SO WE PERFORM THE ORDER OF OPERATIONS
BACKWARDS
HOW TO SHOW WORK:
PERFORM ONLY ONE OPERATION PER ROW OF WORK.
SHOW INVERSE OPERATION ON BOTH SIDES OF THE EQUATION TO GET TO EACH NEXT STEP.
REMINDER: CHECK SOLUTIONS
Name: ANSWER KEY / TEACHER GUIDE
TRY IT
SOLVE FOR H: H/5 + 6 = 15
H/5 = 9
H = 45
45/5 + 6 = 15
ON YOUR OWN
SOLVE FOR m: 8 = (m - 7)/4
32 = m - 7
39 = m
8 = (39-7)/4

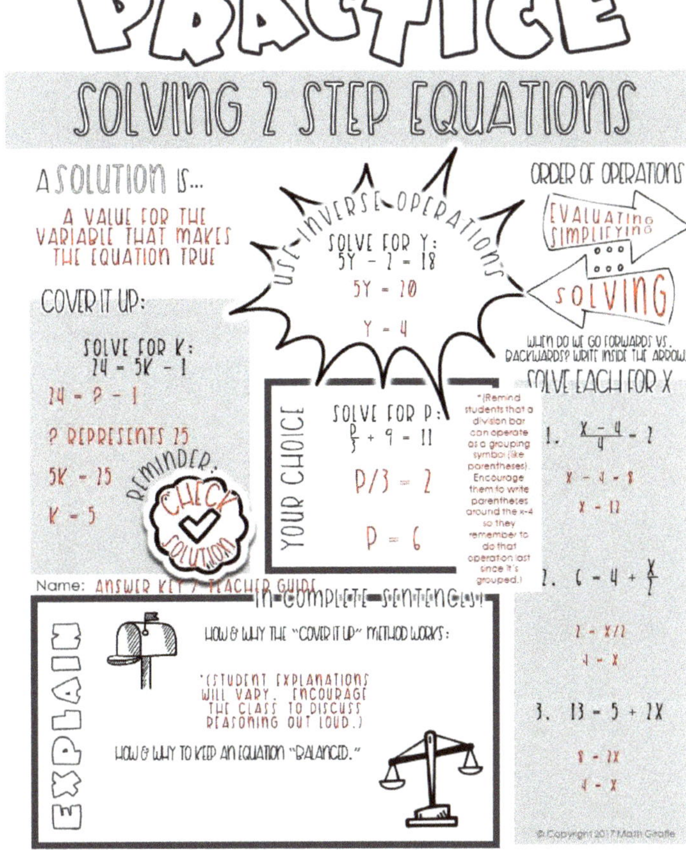
PRACTICE
SOLVING 2 STEP EQUATIONS
A SOLUTION IS...
A VALUE FOR THE VARIABLE THAT MAKES THE EQUATION TRUE
COVER IT UP:
SOLVE FOR K: 24 = 5K - 1
24 = ? - 1
? REPRESENTS 25
5K = 25
K = 5
USE INVERSE OPERATIONS
SOLVE FOR Y: 5Y - 2 = 18
5Y = 20
Y = 4
ORDER OF OPERATIONS
EVALUATING SIMPLIFYING
SOLVING
WHEN DO WE GO FORWARDS VS. BACKWARDS? WRITE INSIDE THE ARROWS.
SOLVE EACH FOR X
1. (X - 4)/4 = 2
X - 4 = 8
X = 12
2. 6 = 4 + X/2
2 = X/2
4 = X
3. 13 = 5 + 2X
8 = 2X
4 = X
YOUR CHOICE
SOLVE FOR P: P/3 + 9 = 11
P/3 = 2
P = 6
*(Remind students that a division bar can operate as a grouping symbol (like parentheses). Encourage them to write parentheses around the x-4 so they remember to do that operation last since it's grouped.)
REMINDER: CHECK SOLUTION
Name: ANSWER KEY / TEACHER GUIDE
IN COMPLETE SENTENCES!
EXPLAIN
HOW & WHY THE "COVER IT UP" METHOD WORKS:
*(STUDENT EXPLANATIONS WILL VARY. ENCOURAGE THE CLASS TO DISCUSS REASONING OUT LOUD.)
HOW & WHY TO KEEP AN EQUATION "BALANCED."
© Copyright 2017 Math Giraffe

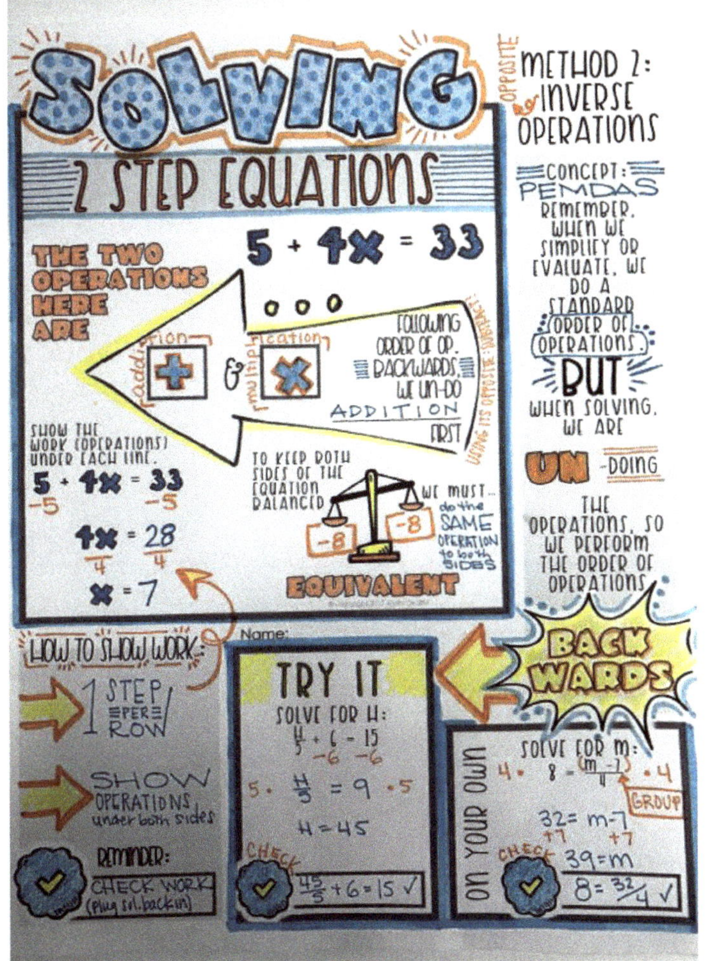
SOLVING
2 STEP EQUATIONS
METHOD 2: OPPOSITE or INVERSE OPERATIONS
CONCEPT: PEMDAS
THE TWO OPERATIONS HERE ARE
addition multiplication
USING ITS OPPOSITE SUBTRACT
-8 -8
WE MUST... do the SAME OPERATION to both SIDES
HOW TO SHOW WORK:
1 STEP PER ROW
SHOW OPERATIONS under both sides
REMINDER: CHECK WORK (Plug it back in)
5 · H/5 = 9 · 5
H = 45
45/5 + 6 = 15 ✓
4 · 8 = (m - 1)/4 · 4
GROUP
32 = m - 7
+7 +7
39 = m
CHECK
8 = 32/4 ✓

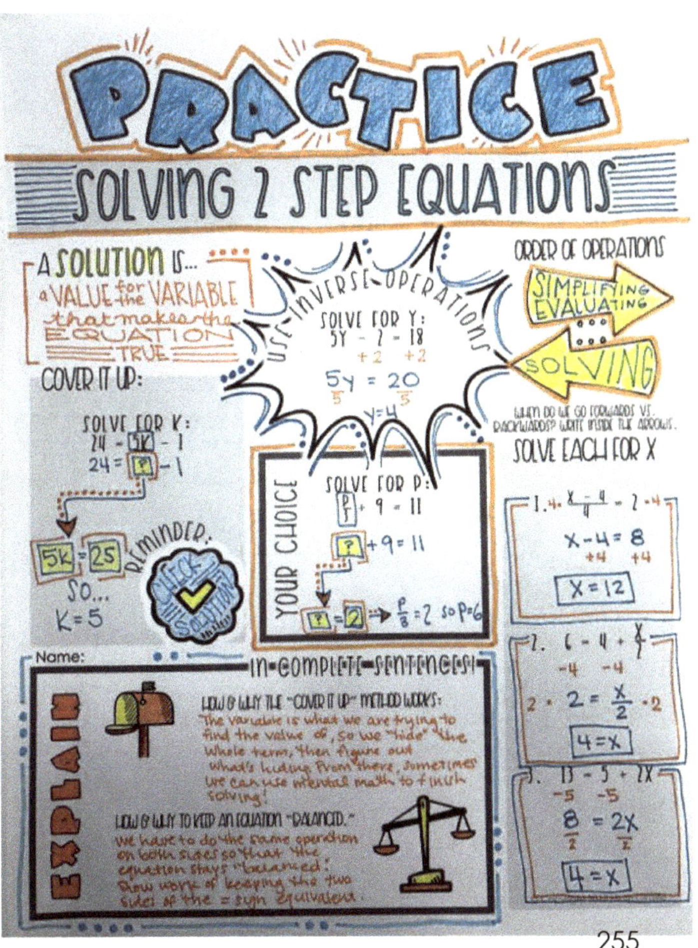
PRACTICE
SOLVING 2 STEP EQUATIONS
A SOLUTION IS...
a VALUE for the VARIABLE that makes the EQUATION TRUE
+2 +2
5y = 20
y = 4
SIMPLIFYING EVALUATING
SOLVING
SO... K = 5
P/3 = 2 so P = 6
X - 4 = 8
+4 +4
X = 12
-4 -4
2 · 2 = X/2 · 2
4 = X
-5 -5
8 = 2X
4 = X
The variable is what we are trying to find the value of, so we "hide" the whole term, then figure out what's hiding from there. Sometimes we can use mental math to finish solving.
We have to do the same operation on both sides so that the equation stays "balanced." Show work of keeping the two sides of the = sign equivalent.

inequalities
"is less than"
"is greater than"
"is less than or equal to"
"is greater than or equal to"
"is not equal to"
*(Note that "inequality" means "NOT equal.")
Name: Teacher Guide / Answer Key
The carnival has more than 30 rides and attractions!
inequality:
a mathematical sentence containing one of the above symbols
r > 30 where "r" represents: # of rides & attractions
There may be 31, 32, 33... (multiple solutions!)
solution:
Any value for the variable that makes the inequality true.
example
Write an inequality to represent: Only children under 5 years old may enter the bounce house.
a < 5
*(Help students think it through. 4 year old kids would be ok (a solution), even 4 and a half, but not 5 years old or older.)
try it:
Write an inequality to represent: The steepness of a wooden roller coaster hill cannot exceed 85 degrees.
s ≤ 85
*(Help students think it through. "Cannot exceed" means it can be less or it can be exactly that value, but not higher. So we include the exact value of 85 by using the ≤ symbol instead of just the <.)

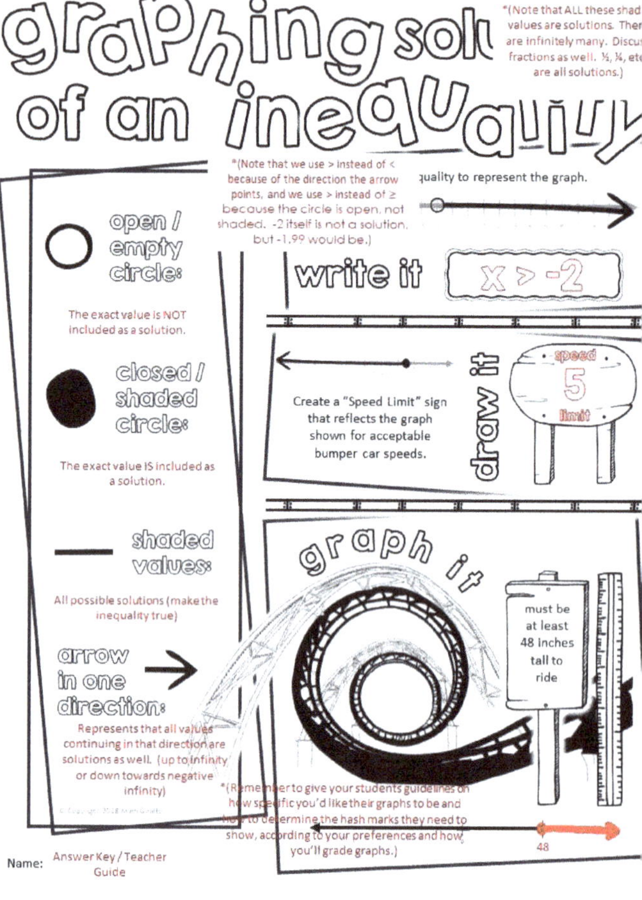
graphing solutions of an inequality
*(Note that ALL these shaded values are solutions. There are infinitely many. Discuss fractions as well.)
*(Note that we use > instead of < because of the direction the arrow points, and we use > instead of ≥ because the circle is open, not shaded. -2 itself is not a solution, but -1.99 would be.)
open / empty circle:
The exact value is NOT included as a solution.
closed / shaded circle:
The exact value IS included as a solution.
shaded values:
All possible solutions (make the inequality true)
arrow in one direction:
Represents that all values continuing in that direction are solutions as well. (up to infinity or down towards negative infinity)
write it
x > -2
Create a "Speed Limit" sign that reflects the graph shown for acceptable bumper car speeds.
draw it
speed 5 limit
graph it
must be at least 48 inches tall to ride
*(Remember to give your students guidelines on how specific you'd like their graphs to be and how to determine the hash marks they need to show, according to your preferences and how you'll grade graphs.)
48
Name: Answer Key / Teacher Guide

Solving one step inequalities by adding or subtracting
The process / steps of solving an inequality has some similarities to solving an equation.
We need to find the SOLUTIONS (all values for the variable that make the inequality TRUE).
inverse operations
+ will undo -
- will undo +
x will undo ÷
÷ will undo x
goal
Get the variable alone on one side of the inequality.
FINISH
Check solutions! (by plugging in)
Name: Answer Key / Teacher Guide
*(Remind students to show operations – one step per line.)
example
Solve for w and graph the solutions.
show work
w + 3 ≤ 1
(-3) (-3)
w ≤ -2
check work
Test multiple values (some that should be true and some that should be false)
try it
Solve for x and graph the solution.
-2 > x - 7
(+7) (+7)
5 > x
(or x < 5)
*(Specifically show students how to test & check solutions. Try plugging in -8, then -2, then 0 together.)
*(Discuss graphing when the variable is on the right. (5 > x instead of x < 5).

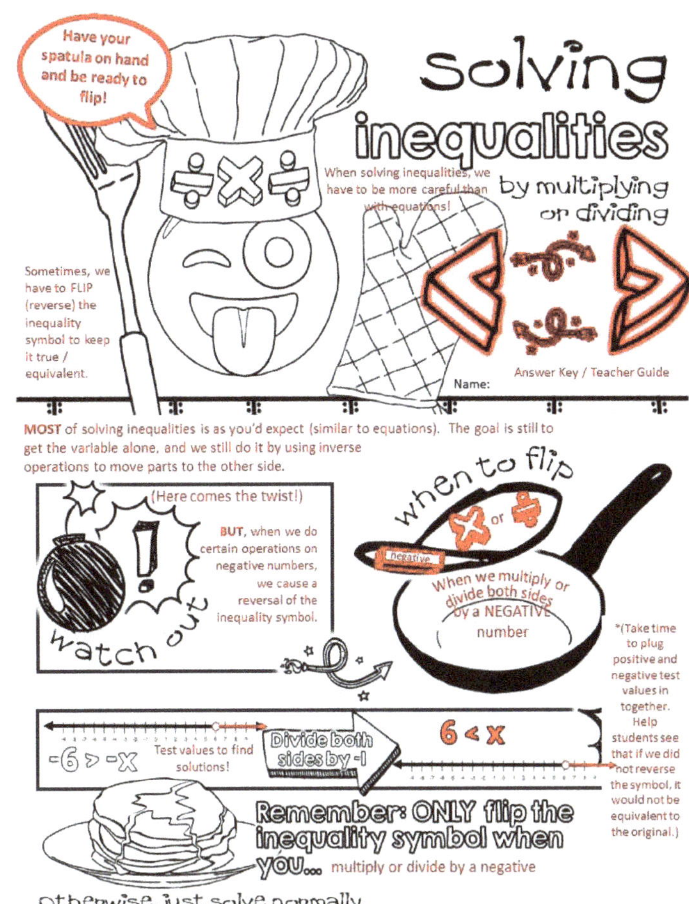

Have your spatula on hand and be ready to flip!
Solving inequalities by multiplying or dividing
When solving inequalities, we have to be more careful than with equations!
Sometimes, we have to FLIP (reverse) the inequality symbol to keep it true / equivalent.
Name: Answer Key / Teacher Guide
MOST of solving inequalities is as you'd expect (similar to equations). The goal is still to get the variable alone, and we still do it by using inverse operations to move parts to the other side.
(Here comes the twist!)
BUT, when we do certain operations on negative numbers, we cause a reversal of the inequality symbol.
watch out
when to flip
negative
When we multiply or divide both sides by a NEGATIVE number
*(Take time to plug positive and negative test values in together. Help students see that if we did not reverse the symbol, it would not be equivalent to the original.)
-6 > -x
Test values to find solutions!
Divide both sides by -1
6 < x
Remember: ONLY flip the inequality symbol when you... multiply or divide by a negative
otherwise, just solve normally.

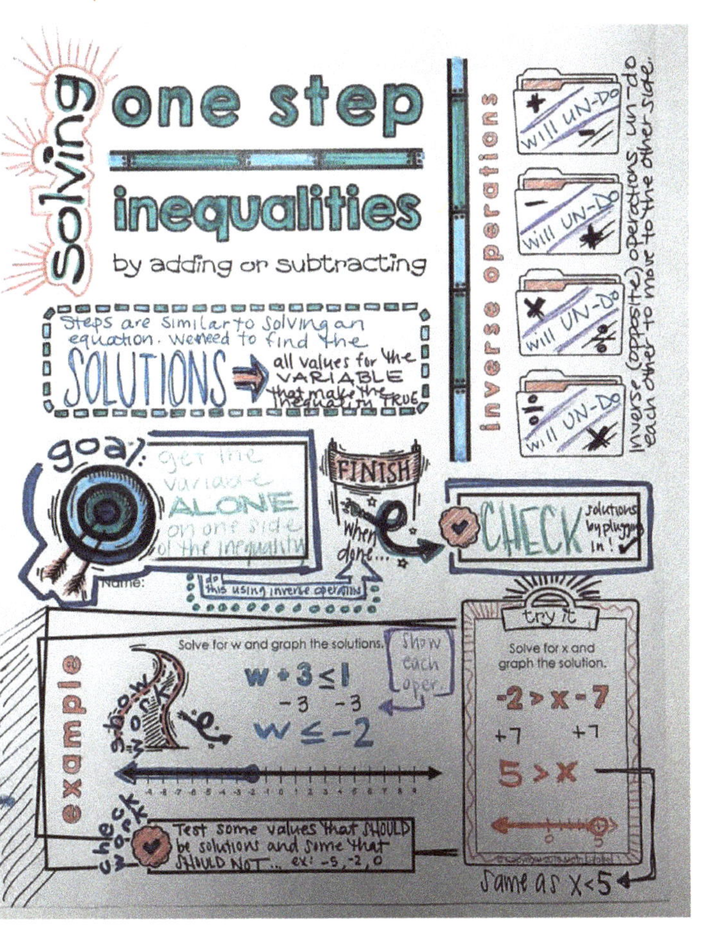

Solving one step inequalities by adding or subtracting
Steps are similar to solving an equation. We need to find the SOLUTIONS ⇒ all values for the VARIABLE that make the inequality TRUE
inverse operations
+ will UN-DO −
− will UN-DO +
× will UN-DO ÷
÷ will UN-DO ×
Inverse (opposite) operations un-do each other to move to the other side.
goal: get the variable ALONE on one side of the inequality
FINISH
when done...
CHECK solutions by plugging in!
do this using inverse operations
Name:
example
Solve for w and graph the solutions.
show each oper.
w + 3 ≤ 1
−3 −3
w ≤ −2
check work
Test some values that SHOULD be solutions and some that SHOULD NOT... ex: -5, -2, 0
try it
Solve for x and graph the solution.
-2 > x - 7
+7 +7
5 > x
Same as x < 5

Have a spatula on hand & be ready to FLIP the symbol!
by a NEGATIVE
Solving inequalities by multiplying or dividing
We have to be more careful!
Sometimes we have to REVERSE (FLIP) the inequality symbol to keep it true / equivalent.
REVERSE
FLIP!
Name:
MOST of solving inequalities is as you would expect (similar to solving an equation. The GOAL is still to get the variable alone, and we still do it by using inverse operations to move pieces to the other side...
(Here comes the twist!)
... BUT... when we do certain operations on NEGATIVE #s, we cause a reversal of the inequal. symbol
watch out
when to flip
by a NEGATIVE
ONLY!
WHEN WE MULTIPLY OR DIVIDE BOTH SIDES BY A NEGATIVE #
Test: 0: NOT SOL. -7: NOT SOL.
7: SOLUTION 8: SOLUTION
GRAPH:
-6 > -x
NOT SOL
Divide both sides by -1
6 < x
why?
Remember: ONLY flip the inequality symbol when you... multiply or divide by a NEGATIVE while solving
otherwise, just solve normally.

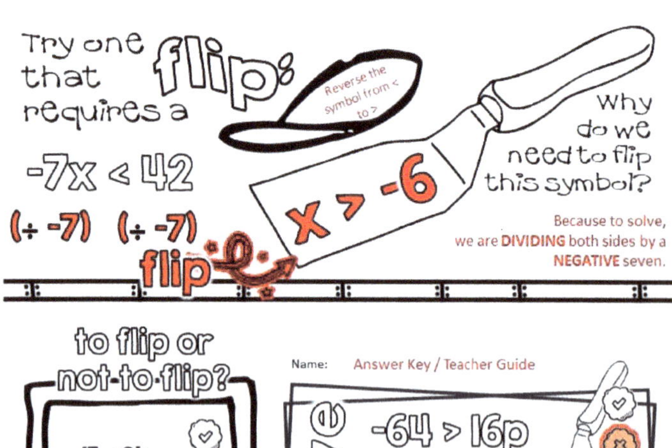
Try one that requires a flip:
Reverse the symbol from < to >
-7x < 42
(÷ -7) (÷ -7)
flip
x > -6
Why do we need to flip this symbol?
Because to solve, we are DIVIDING both sides by a NEGATIVE seven.

to flip or not-to-flip?
15 < 2b
-4 + m > 2
g/-8 ≥ 2
p(3) ≤ -2
-12 < v/4
Name: Answer Key / Teacher Guide
solve
-64 > 16p
(÷ 16) (÷ 16)
-4 > p
flip?
NO. Even though we see a negative, we are dividing by a positive.
graph
solve
m/-3 ≤ 1
(·-3) (·-3)
m ≥ -3
flip?
YES. We are multiplying both sides by a negative.
graph

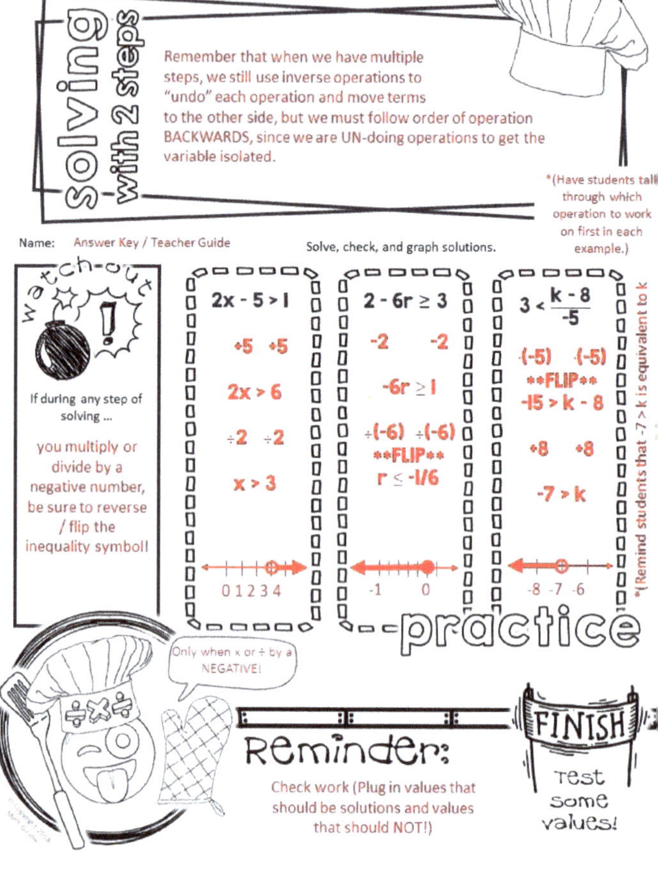
Solving with 2 steps
Remember that when we have multiple steps, we still use inverse operations to "undo" each operation and move terms to the other side, but we must follow order of operation BACKWARDS, since we are UN-doing operations to get the variable isolated.
*(Have students talk through which operation to work on first in each example.)
Name: Answer Key / Teacher Guide
Solve, check, and graph solutions.
watch-out
If during any step of solving ... you multiply or divide by a negative number, be sure to reverse / flip the inequality symbol!
2x - 5 > 1
+5 +5
2x > 6
÷2 ÷2
x > 3
0 1 2 3 4
2 - 6r ≥ 3
-2 -2
-6r ≥ 1
÷(-6) ÷(-6)
FLIP
r ≤ -1/6
-1 0
3 < (k - 8)/-5
(·-5) (·-5)
FLIP
-15 > k - 8
+8 +8
-7 > k
-8 -7 -6
*(Remind students that -7 > k is equivalent to k < -7)
practice
Only when x or ÷ by a NEGATIVE!
Reminder:
Check work (Plug in values that should be solutions and values that should NOT!)
FINISH
Test some values!

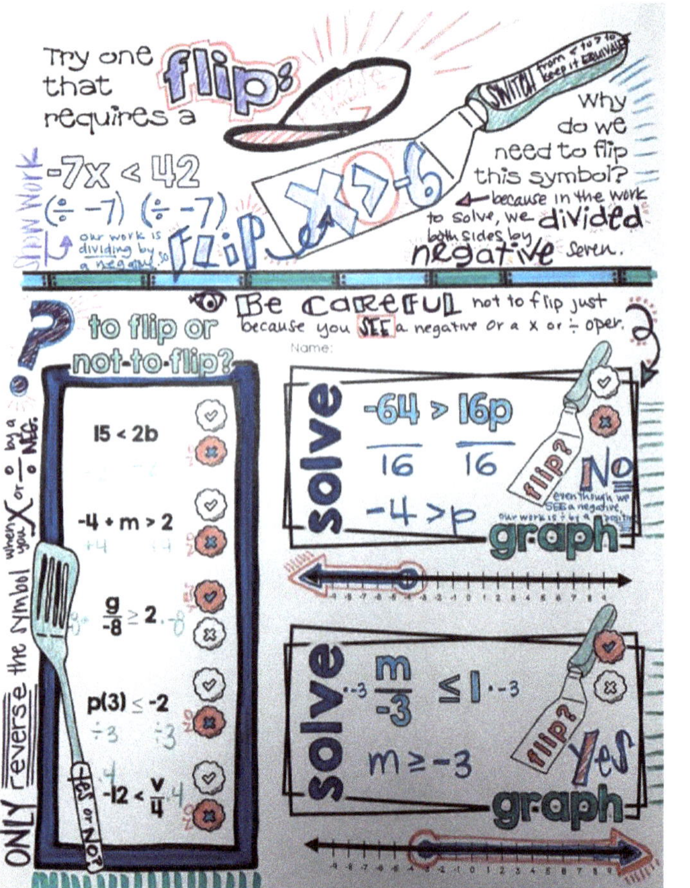
Try one that requires a flip:
-7x < 42
(÷ -7) (÷ -7)
FLIP
Why do we need to flip this symbol?
because in the work to solve, we divided both sides by negative seven.
Be careful not to flip just because you SEE a negative or a x or ÷ oper.
to flip or not-to-flip?
15 < 2b
-4 + m > 2
g/-8 ≥ 2
p(3) ≤ -2
-12 < v/4
ONLY reverse the symbol when you X or ÷ by a NEG.
solve
-64 > 16p
16 16
-4 > p
NO
graph
solve
m/-3 ≤ 1
m ≥ -3
Yes
graph

< > ≤ ≥ ≠
INEQUALITIES:
Solving with 2 steps
INVERSE OPERATIONS
UN-DOING
REMEMBER
Name:
Solve, check, and graph solutions.
watch-out
If during any step of solving ...
You mult. or divide by a Negative #, be sure to flip / reverse the inequal. symbol
2x - 5 > 1
+5 +5
2x > 6
÷2 ÷2
x > 3
2 - 6r ≥ 3
-2 -2
-6r ≥ 1
FLIP
r ≤ -1/6
3 < (k - 8)/-5
FLIP
-15 > k-8
+8 +8
-7 > k
is equivalent to k < -7
practice
ONLY when x or ÷ by a NEGATIVE!
FLIP
Reminder:
CHECK SOLUTIONS
Plug in values that SHOULD be solutions and some that should NOT
after you FINISH
Test some values!

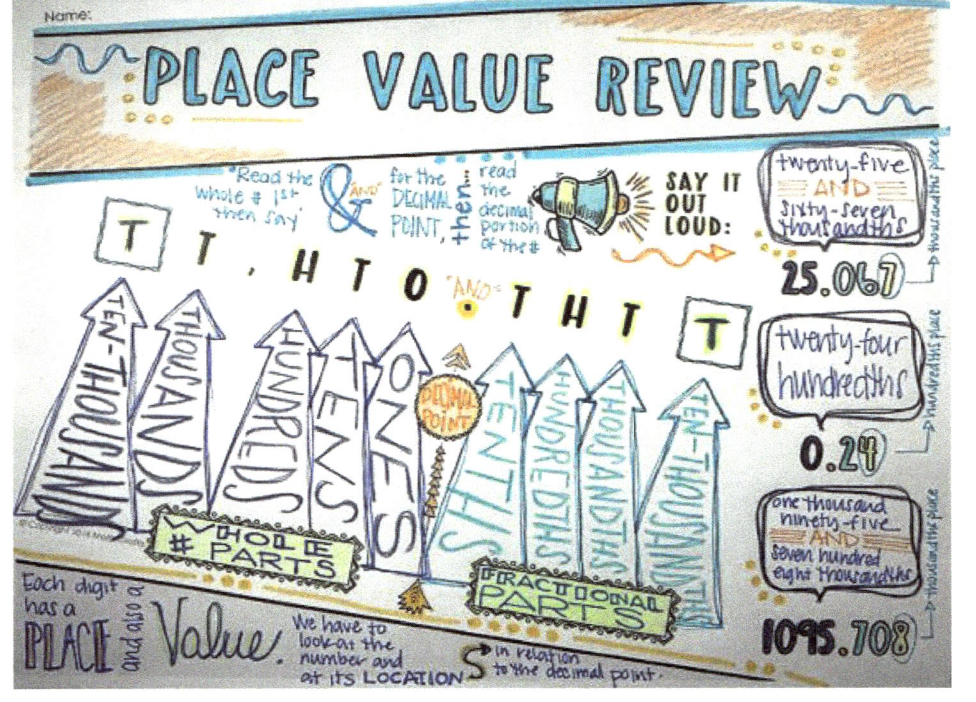
Name:
PLACE VALUE REVIEW
SAY IT OUT LOUD:
twenty-five AND sixty-seven thousandths
25.067
twenty-four hundredths
0.24
one thousand ninety-five AND seven hundred eight thousandths
1095.708
T T , H T O . T H T T
TEN-THOUSANDS
THOUSANDS
HUNDREDS
TENS
ONES
TENTHS
HUNDREDTHS
THOUSANDTHS
TEN-THOUSANDTHS
WHOLE # PARTS
FRACTIONAL PARTS
Each digit has a PLACE and a Value.

Name:
Answer Key / Teacher Guide

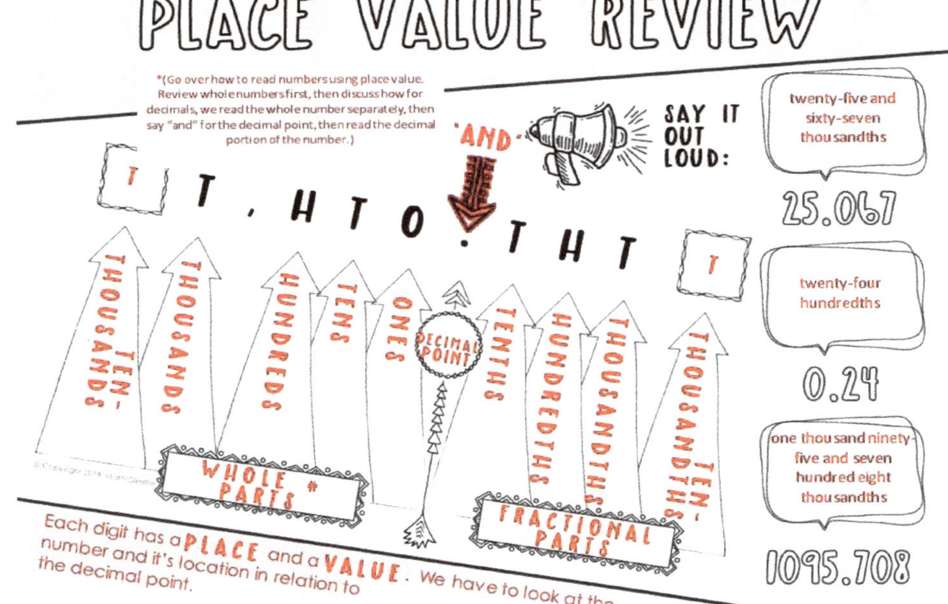
PLACE VALUE REVIEW
*(Go over how to read numbers using place value. Review whole numbers first, then discuss how for decimals, we read the whole number separately, then say "and" for the decimal point, then read the decimal portion of the number.)
'AND'
SAY IT OUT LOUD:
twenty-five and sixty-seven thousandths
25.067
twenty-four hundredths
0.24
one thousand ninety-five and seven hundred eight thousandths
1095.708
T T , H T O . T H T T
TEN-THOUSANDS
THOUSANDS
HUNDREDS
TENS
ONES
DECIMAL POINT
TENTHS
HUNDREDTHS
THOUSANDTHS
TEN-THOUSANDTHS
WHOLE # PARTS
FRACTIONAL PARTS
Each digit has a PLACE and a VALUE. We have to look at the number and it's location in relation to the decimal point.

MODELING DECIMALS
& COMPARING & ORDERING
0.68
Color the grid to represent 0.68. Each small square represents: ONE HUNDREDTH
We can use visual grid models to help see the values.
TIP: Add the missing zero digits
1.206
1.260
1.026
1.200
1.020
Order from least to greatest:
1.206
1.26
1.026
1.2
1.02
Using the models or our understanding of place value, we can determine which numbers are greatest or least.
Look at each place from left to right when comparing each digit.
For models, numbers with LESS of the grid colored are LOWER.
Order from least to greatest:
0.62, 0.66, 0.26
GREATEST
LEAST
0.26, 0.62, 0.66
least
1.02
1.026
1.2
1.206
1.26
greatest
1
2
3
4
5
DECIMALS
Name:

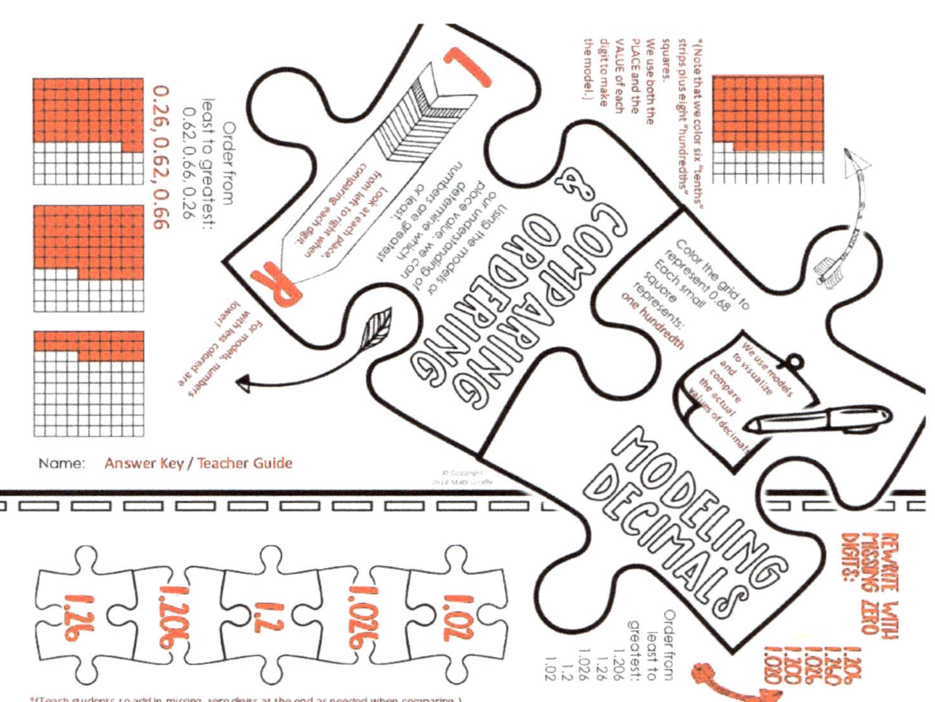
MODELING DECIMALS
& COMPARING & ORDERING
Color the grid to represent 0.68. Each small square represents: one hundredth
We use models to visualize and compare the actual values of decimals.
*(Note that we color six "tenths" strips plus eight "hundredths" squares. We use both the PLACE and the VALUE of each digit to make the model.)
REWRITE WITH MISSING ZERO DIGITS:
1.206
1.260
1.026
1.200
1.020
Order from least to greatest:
1.206
1.26
1.026
1.2
1.02
Using the models or our understanding of place value, we can determine which numbers are greatest or least.
Look at each place, from left to right when comparing each digit.
L
R
For models, numbers with less colored are lower!
Order from least to greatest:
0.62, 0.66, 0.26
0.26, 0.62, 0.66
1.02
1.026
1.2
1.206
1.26
*(Teach students to add in missing zero digits at the end as needed when comparing.)
Name:
Answer Key / Teacher Guide

ROUNDING with DECIMALS
Name: Answer Key / Teacher Guide
VALUE IS CLOSER TO THE LOWER NUMBER
VALUE IS CLOSER TO THE HIGHER NUMBER
0 1 2 3 4 5 6 7 8 9
KEEP the lower previous digit
INCREASE the previous digit by 1
2.942
7.8068
Look
UNDERLINE at the place you're asked to round to.
Use the next digit TO THE RIGHT to decide... (CIRCLE IT)
Try it
Round to the nearest hundredth. 2.942 (use the thousandth digit (2) to decide) 2.94
Round to the nearest tenth. 2.942 (use the hundredths digit (4) to decide) 2.9
Round to the nearest whole (one). 7.8068 (use the tenths digit (8) to decide) 8
Round to the nearest thousandth. 7.8068 (use the ten thousandths digit (8) to decide) 7.807
Which is it closer to?

Name:
Answer Key / Teacher Guide
I've got a diamond worth half a million dollars!
I just found a diamond that's worth $499,999.97!! What do I tell my friends it's worth?
I make about $200 a month.
*(Explain that in some cases, we want to give a less precise number. It can help to round numbers to approximate. Ask students to come up with more examples.)
(student example here)
Sketch it
Give your own real-world example and show how to round it to give a reasonable approximation.
6.4
why?
I ran 6.383 miles. I'll round to the nearest tenth.
6.383
*(Explain that to make our rounding decision, we use the place AFTER the digit we are rounding to. If that number (circled) is 5 or greater, we increase the underlined digit, but if it's 4 or less, we keep the underlined digit.)
I burned 127.0146 calories during my workout. I'll round to the nearest hundredth.
127.0146
Key idea
WHICH (LESS PRECISE) VALUE IS IT CLOSER TO?
The 5 and greater / 4 and less rule is to determine if its closer to the higher or lower number.

ROUNDING with DECIMALS
Name:
KEEP the lower previous digit
INCREASE the previous digit by 1.
2.942
7.8068
Look
underline the place you're asked to round to.
Use the next digit to the right to decide. (circle it!)
Try it
Which is it closer to?

Name:
I've got a diamond worth 1/2 a MILLION dollars
I just found a diamond that's worth $499,999.97!! What do I tell my friends it's worth?
I make about $200 per month
In some cases, we want to give a less precise number: ROUND to approximate
Sketch it
why?
I ran 6.383 miles. I'll round to the nearest tenth.
6.383
tenth
I burned 127.0146 calories during my workout. I'll round to the nearest hundredth.
127.0146
hundredth
ROUNDING DECIMALS
Key idea
Which (Less Precise) value is it CLOSER to?

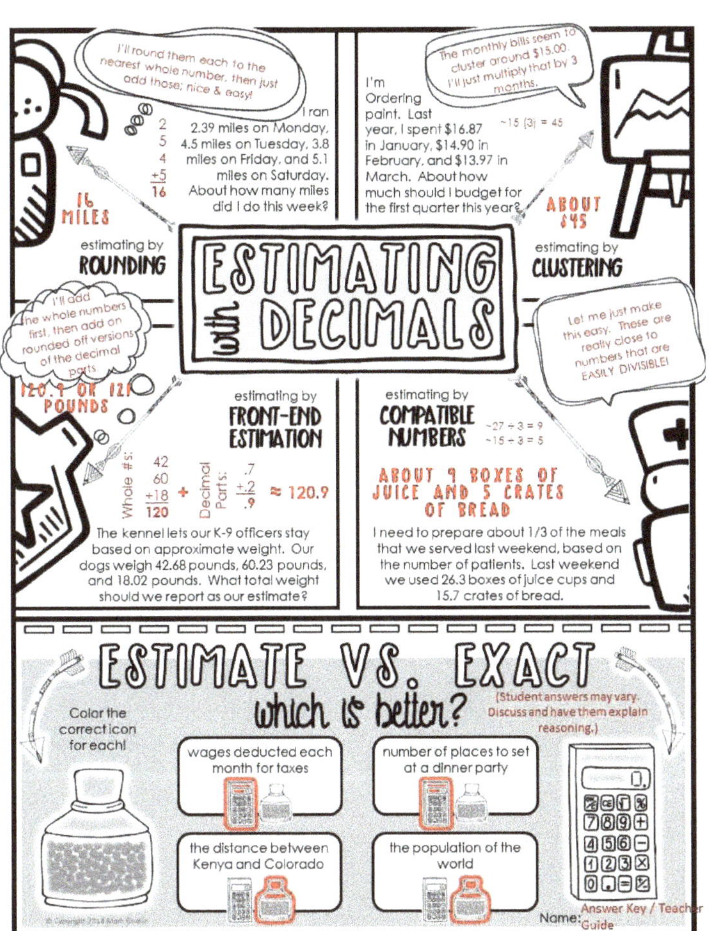
I'll round them each to the nearest whole number, then just add those; nice & easy!
I ran 2.39 miles on Monday, 4.5 miles on Tuesday, 3.8 miles on Friday, and 5.1 miles on Saturday. About how many miles did I do this week?
2
5
4
+5
16
16 MILES
estimating by ROUNDING
The monthly bills seem to cluster around $15.00. I'll just multiply that by 3 months.
I'm Ordering paint. Last year, I spent $16.87 in January, $14.90 in February, and $13.97 in March. About how much should I budget for the first quarter this year?
~15 (3) = 45
ABOUT $45
estimating by CLUSTERING
ESTIMATING with DECIMALS
I'll add the whole numbers first, then add on rounded off versions of the decimal parts.
120.9 OR 121 POUNDS
Let me just make this easy. These are really close to numbers that are EASILY DIVISIBLE!
estimating by FRONT-END ESTIMATION
Whole #s: 42 60 +18 120
Decimal Parts: .7 +.2 .9
≈ 120.9
The kennel lets our K-9 officers stay based on approximate weight. Our dogs weigh 42.68 pounds, 60.23 pounds, and 18.02 pounds. What total weight should we report as our estimate?
estimating by COMPATIBLE NUMBERS
~27 ÷ 3 = 9
~15 ÷ 3 = 5
ABOUT 9 BOXES OF JUICE AND 5 CRATES OF BREAD
I need to prepare about 1/3 of the meals that we served last weekend, based on the number of patients. Last weekend we used 26.3 boxes of juice cups and 15.7 crates of bread.
ESTIMATE VS. EXACT
which is better?
(Student answers may vary. Discuss and have them explain reasoning.)
Color the correct icon for each!
wages deducted each month for taxes
number of places to set at a dinner party
the distance between Kenya and Colorado
the population of the world
Name: Answer Key / Teacher Guide

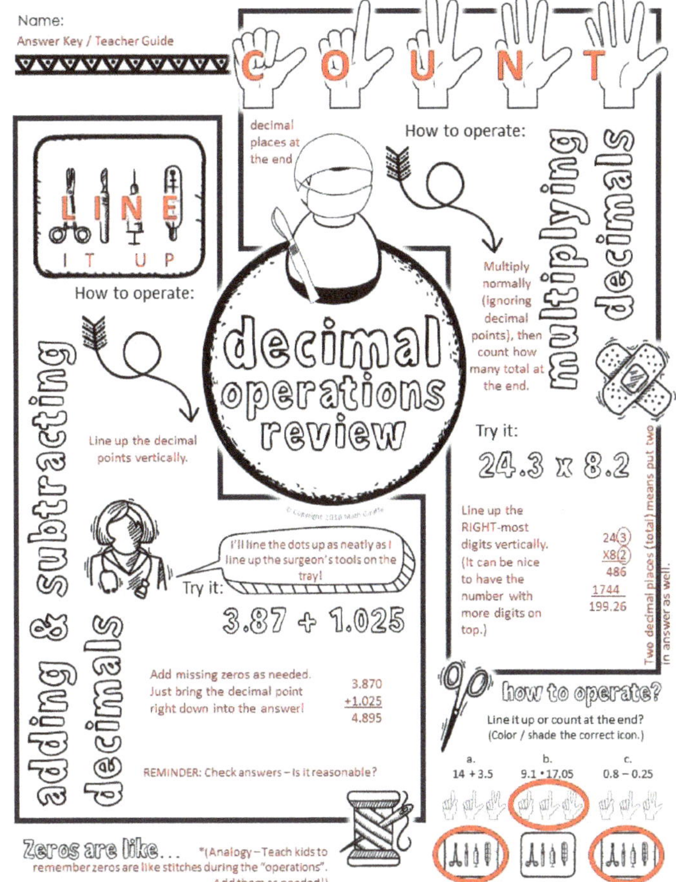
Name:
Answer Key / Teacher Guide
COUNT
decimal places at the end
How to operate:
LINE IT UP
How to operate:
Line up the decimal points vertically.
decimal operations review
Multiply normally (ignoring decimal points), then count how many total at the end.
multiplying decimals
Try it:
24.3 x 8.2
Line up the RIGHT-most digits vertically. (It can be nice to have the number with more digits on top.)
24(3)
X8(2)
486
1744
199.26
Two decimal places (total) means put two in answer as well.
I'll line the dots up as neatly as I line up the surgeon's tools on the tray!
Try it:
3.87 + 1.025
Add missing zeros as needed. Just bring the decimal point right down into the answer!
3.870
+1.025
4.895
REMINDER: Check answers – Is it reasonable?
adding & subtracting decimals
how to operate?
Line it up or count at the end? (Color / shade the correct icon.)
a. 14 + 3.5
b. 9.1 • 17.05
c. 0.8 – 0.25
Zeros are like...
*(Analogy – Teach kids to remember zeros are like stitches during the "operations". Add them as needed!)

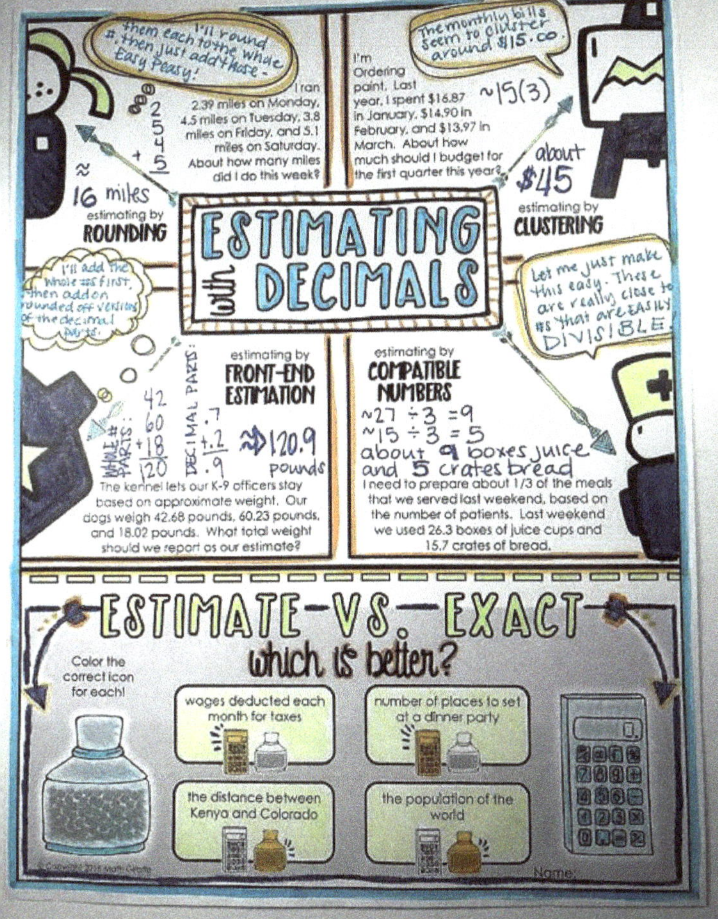
I'll round them each to the whole #, then just add those - Easy Peasy!
I ran 2.39 miles on Monday, 4.5 miles on Tuesday, 3.8 miles on Friday, and 5.1 miles on Saturday. About how many miles did I do this week?
≈ 16 miles
estimating by ROUNDING
The monthly bills seem to cluster around $15.00
I'm Ordering paint. Last year, I spent $16.87 in January, $14.90 in February, and $13.97 in March. About how much should I budget for the first quarter this year?
~15(3)
about $45
estimating by CLUSTERING
ESTIMATING with DECIMALS
I'll add the whole #s first, then add on rounded off versions of the decimal parts.
Let me just make this easy. These are really close to #s that are EASILY DIVISIBLE.
estimating by FRONT-END ESTIMATION
WHOLE PARTS: 42 60 +18 120
DECIMAL PARTS: .7 +.2 .9
≈120.9 pounds
The kennel lets our K-9 officers stay based on approximate weight. Our dogs weigh 42.68 pounds, 60.23 pounds, and 18.02 pounds. What total weight should we report as our estimate?
estimating by COMPATIBLE NUMBERS
~27 ÷ 3 = 9
~15 ÷ 3 = 5
about 9 boxes juice and 5 crates bread
I need to prepare about 1/3 of the meals that we served last weekend, based on the number of patients. Last weekend we used 26.3 boxes of juice cups and 15.7 crates of bread.
ESTIMATE VS. EXACT
which is better?
Color the correct icon for each!
wages deducted each month for taxes
number of places to set at a dinner party
the distance between Kenya and Colorado
the population of the world
Name:

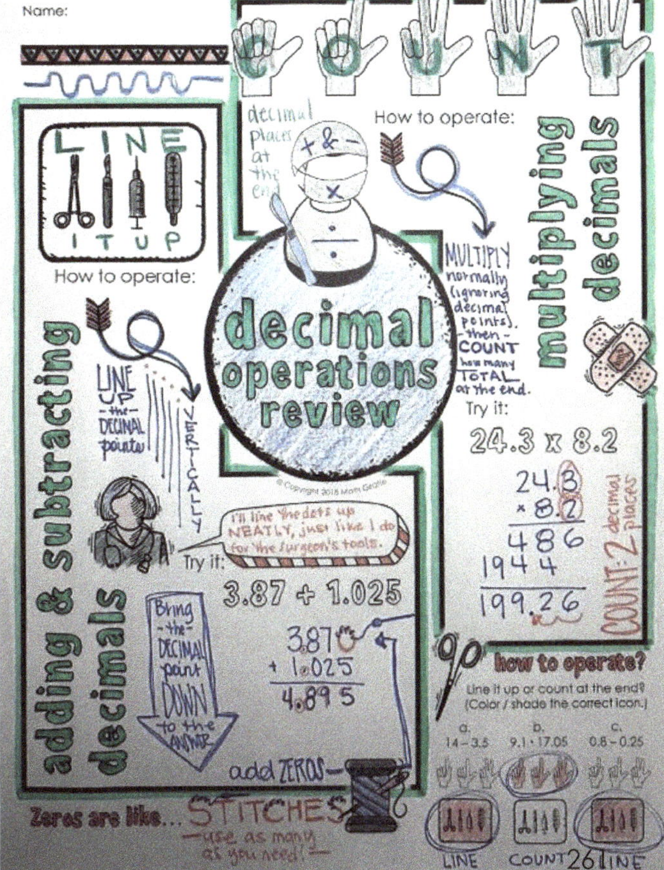
Name:
COUNT
decimal places at the end
How to operate:
LINE IT UP
How to operate:
LINE UP the DECIMAL points VERTICALLY
decimal operations review
MULTIPLY normally (ignoring decimal points), then COUNT how many TOTAL at the end.
multiplying decimals
Try it:
24.3 x 8.2
24.3
×8.2
486
1944
199.26
COUNT: 2 decimal places
I'll line the dots up NEATLY, just like I do for the surgeon's tools.
Try it:
3.87 + 1.025
Bring the DECIMAL point DOWN to the ANSWER
3.870
+ 1.025
4.895
add ZEROS
adding & subtracting decimals
how to operate?
Line it up or count at the end? (Color / shade the correct icon.)
a. 14 – 3.5
b. 9.1 • 17.05
c. 0.8 – 0.25
LINE
COUNT
LINE
Zeros are like... STITCHES – use as many as you need! –

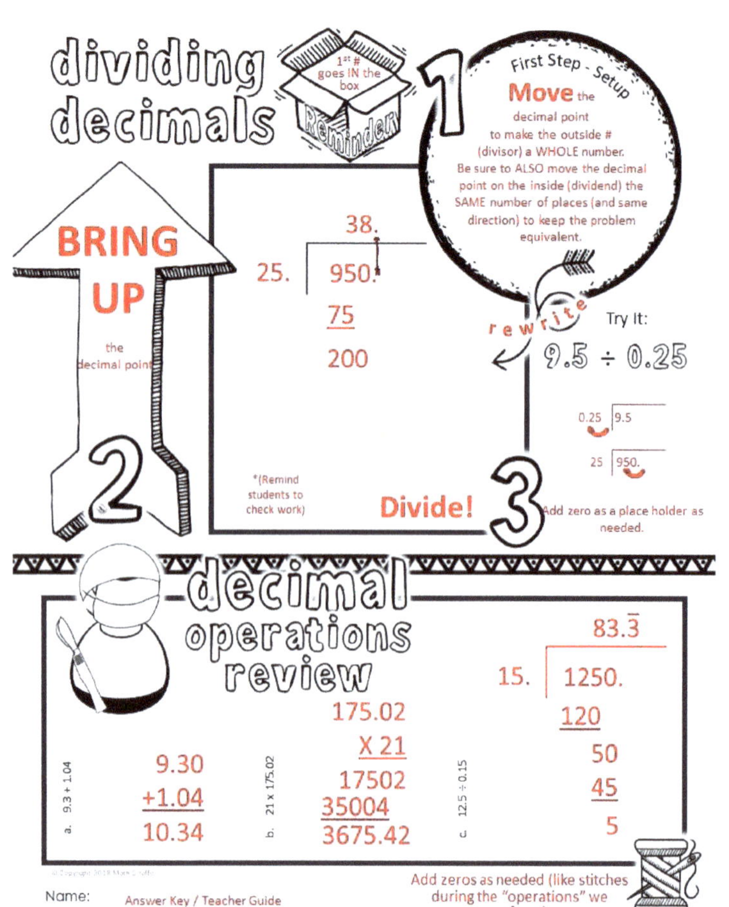
dividing decimals
1st # goes IN the box
Reminder
1
First Step - Setup
Move the decimal point to make the outside # (divisor) a WHOLE number. Be sure to ALSO move the decimal point on the inside (dividend) the SAME number of places (and same direction) to keep the problem equivalent.
BRING UP the decimal point
2
38.
25. 950.
75
200
rewrite
Try It:
9.5 ÷ 0.25
*(Remind students to check work)
Divide!
3
Add zero as a place holder as needed.
decimal operations review
a. 9.3 + 1.04
9.30
+1.04
10.34
b. 21 x 175.02
175.02
X 21
17502
35004
3675.42
c. 12.5 ÷ 0.15
83.3
15. 1250.
120
50
45
5
Name: Answer Key / Teacher Guide
Add zeros as needed (like stitches during the "operations" we perform).

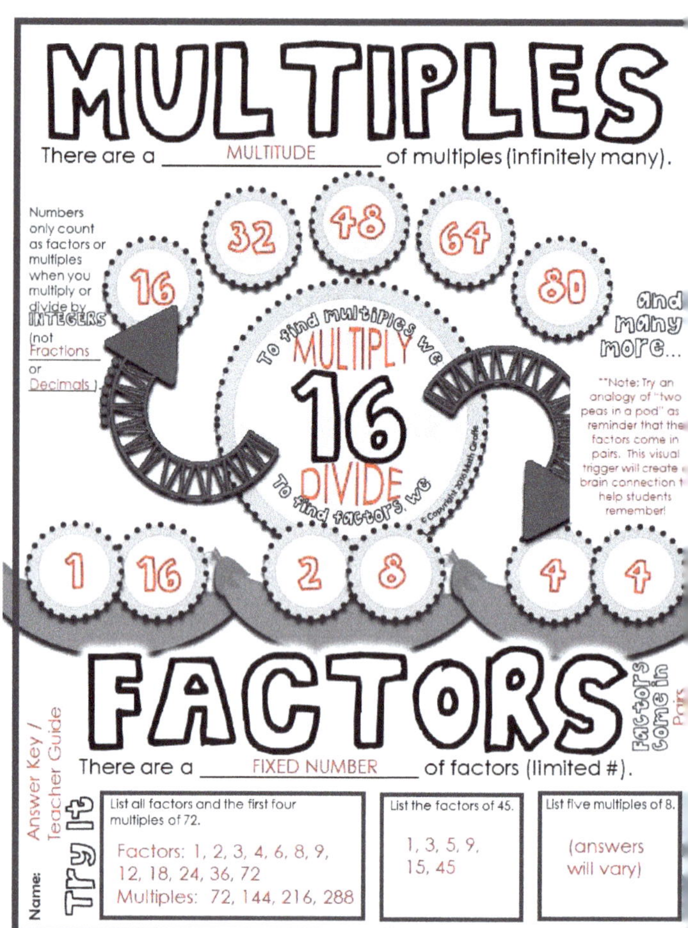
MULTIPLES
There are a MULTITUDE of multiples (infinitely many).
Numbers only count as factors or multiples when you multiply or divide by INTEGERS (not Fractions or Decimals).
16 32 48 64 80
and many more...
To find multiples we MULTIPLY
16
To find factors we DIVIDE
**Note: Try an analogy of "two peas in a pod" as reminder that the factors come in pairs.
1 16 2 8 4 4
FACTORS
Factors come in Pairs
There are a FIXED NUMBER of factors (limited #).
Answer Key / Teacher Guide
Try It
List all factors and the first four multiples of 72.
Factors: 1, 2, 3, 4, 6, 8, 9, 12, 18, 24, 36, 72
Multiples: 72, 144, 216, 288
List the factors of 45.
1, 3, 5, 9, 15, 45
List five multiples of 8.
(answers will vary)
Name:

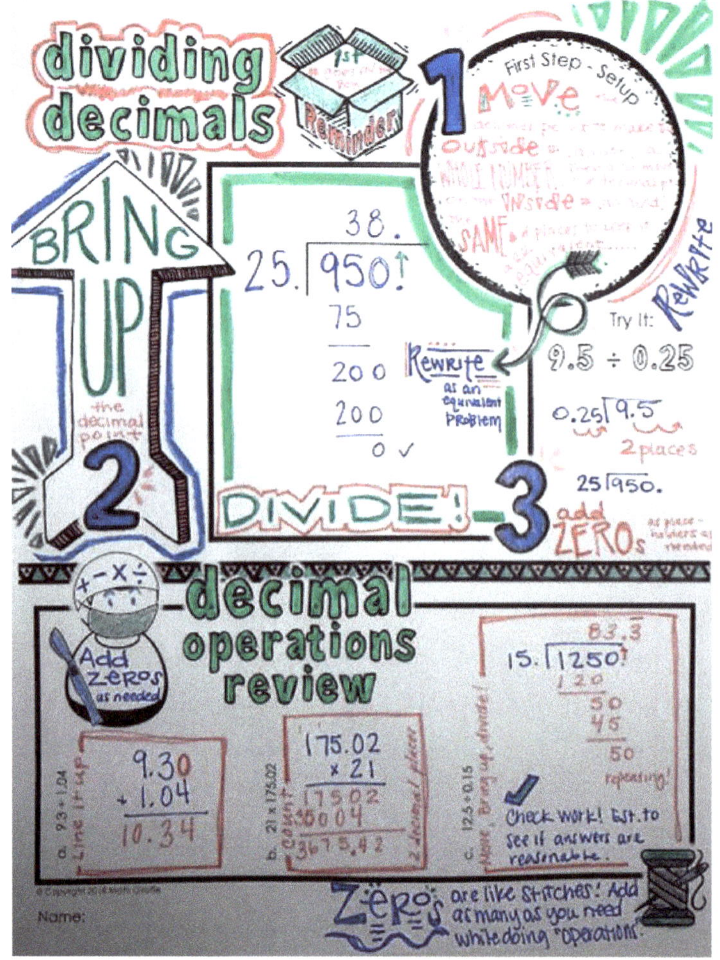
dividing decimals
Reminder
1
First Step - Setup
Move
BRING UP the decimal point
2
38.
25. 950.
75
200
200
0
Rewrite as an equivalent problem
Try It:
9.5 ÷ 0.25
0.25 9.5
2 places
25 950.
DIVIDE!
3
add ZEROs
decimal operations review
Add zeros as needed
a. 9.3 + 1.04
9.30
+ 1.04
10.34
b. 21 x 175.02
175.02
x 21
17502
35004
3675.42
2 decimal places
c. 12.5 ÷ 0.15
83.3
15. 1250
120
50
45
50
repeating!
Check work! Est. to see if answers are reasonable.
Zeros are like stitches! Add as many as you need while doing "operations."
Name:

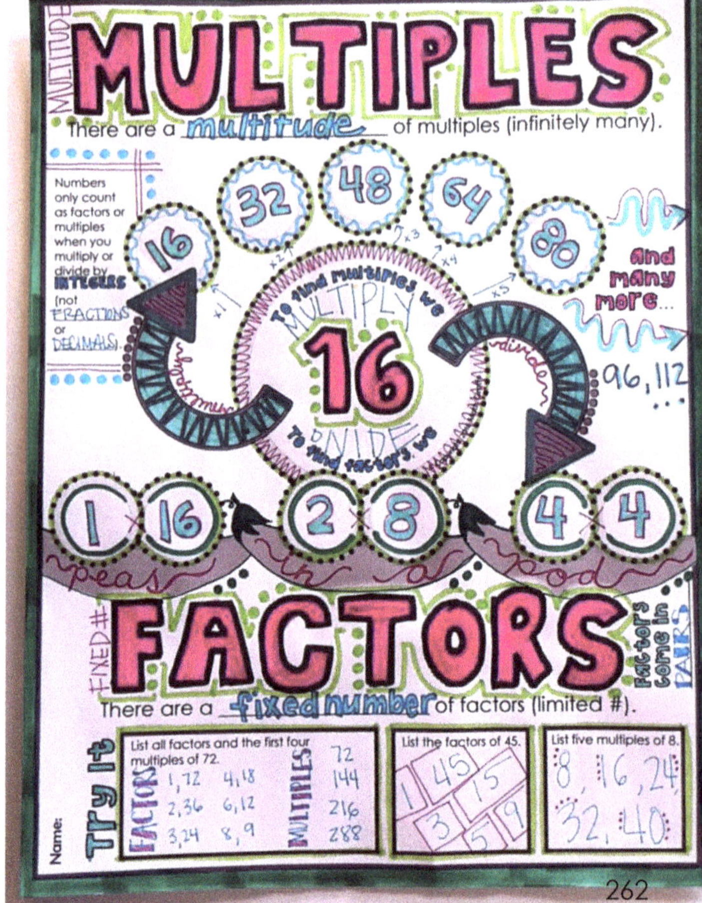
MULTITUDE
MULTIPLES
There are a multitude of multiples (infinitely many).
Numbers only count as factors or multiples when you multiply or divide by INTEGERS (not FRACTIONS or DECIMALS).
16 32 48 64 80
and many more...
96, 112
To find multiples we MULTIPLY
16
To find factors we DIVIDE
1 16 2 8 4 4
peas in a pod
FIXED #
FACTORS
Factors come in PAIRS
There are a fixed number of factors (limited #).
Try It
List all factors and the first four multiples of 72.
FACTORS 1,72 2,36 3,24 4,18 6,12 8,9
MULTIPLES 72 144 216 288
List the factors of 45.
1 45 3 15 5 9
List five multiples of 8.
8, 16, 24, 32, 40
Name:

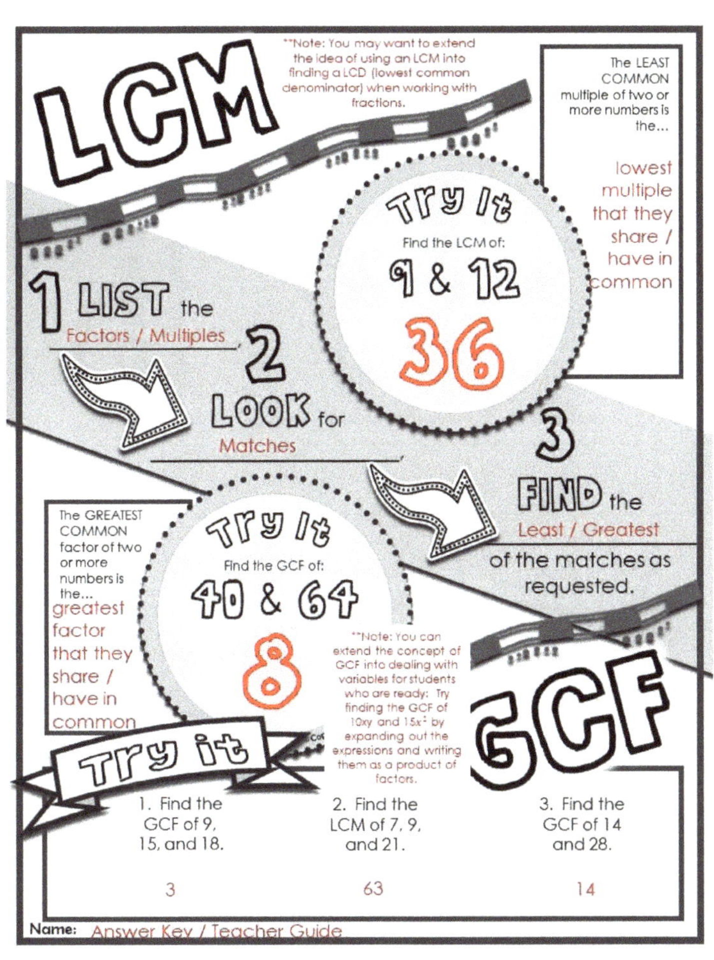
LCM
**Note: You may want to extend the idea of using an LCM into finding a LCD (lowest common denominator) when working with fractions.
The LEAST COMMON multiple of two or more numbers is the...
lowest multiple that they share / have in common
Try It
Find the LCM of:
9 & 12
36
1 LIST the Factors / Multiples
2 LOOK for Matches
3 FIND the Least / Greatest of the matches as requested.
The GREATEST COMMON factor of two or more numbers is the...
greatest factor that they share / have in common
Try It
Find the GCF of:
40 & 64
8
**Note: You can extend the concept of GCF into dealing with variables for students who are ready: Try finding the GCF of 10xy and 15x² by expanding out the expressions and writing them as a product of factors.
GCF
Try it
1. Find the GCF of 9, 15, and 18.
2. Find the LCM of 7, 9, and 21.
3. Find the GCF of 14 and 28.
3
63
14
Name: Answer Key / Teacher Guide

examples
(student examples may vary) – 4, 6, 8, 9, 10, 12
*(You can compare a composite number to an element to continue the analogy if you wish – It can be broken into whole parts (atoms) just like a composite number can be broken into whole parts (integer factors).)
COMPOSITE numbers
are integers greater than 1 that have
more than 2 positive factors
TRY IT
Color all prime numbers blue & composite numbers red.
examples:
(student examples may vary) – 2, 3, 5, 7, 11, 13, 17
PRIME numbers
are integers greater than 1 that have exactly
TWO positive factors (1 and the # itself)
are like atoms because...
If you split an atom, you get an explosion, and millions of tiny parts, no whole atoms! If you split a prime, you can only get fractional parts
they can't be split into whole parts (cannot be broken down further).
the number 1 is NEITHER prime nor composite
Name: Answer Key / Teacher Guide

LCM
LEAST COMMON MULTIPLE
The LEAST COMMON multiple of two or more numbers is the...
SMALLEST multiple that they SHARE
these come in handy when working w/ fractions
Try It
Find the LCM of:
9 & 12
36
1 LIST the mult/fact.
2 LOOK for matches
3 FIND the least/great. of the matches as requested.
The GREATEST COMMON factor of two or more numbers is the...
LARGEST factor that they SHARE
Try It
Find the GCF of:
40 & 64
1. Find the GCF of 9, 15, and 18.
9: 1, 3, 9
15: 1, 3, 5, 15
18: 1, 2, 3, 6, 9, 18
3
14
Name:

examples
4, 6, 8, 9, 10, 12
6 ← composite
2 · 3 ← prime
factor factor
can't be broken down further
TRY IT
Color all prime numbers blue & composite numbers red.
examples:
2, 3, 5, 7, 11, 13, 17
COMPOSITE numbers are integers >1 that have MORE than 2 positive factors
PRIME numbers
are integers >1 that have EXACTLY
2 positive factors
1 & the # itself...
are like atoms because...
If you split an atom, you get an explosion + millions of tiny parts - no whole pieces! If you split a prime, you get only FRACTIONAL parts!
they can't be split into smaller whole parts
the number 1 is neither PRIME nor composite
Name:

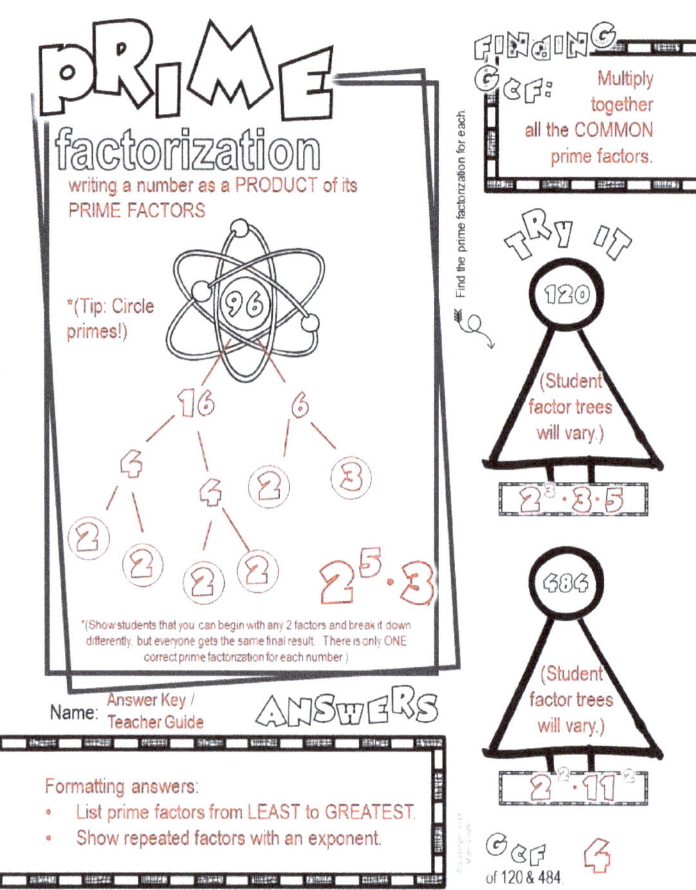

PRIME factorization
writing a number as a PRODUCT of its PRIME FACTORS
*(Tip: Circle primes!)
96
16
6
4
4
2
3
2
2
2
2
$2^5 \cdot 3$
*(Show students that you can begin with any 2 factors and break it down differently, but everyone gets the same final result. There is only ONE correct prime factorization for each number.)
Name: Answer Key / Teacher Guide
ANSWERS
Formatting answers:
• List prime factors from LEAST to GREATEST.
• Show repeated factors with an exponent.
FINDING GCF: Multiply together all the COMMON prime factors.
Find the prime factorization for each.
TRY IT
120
(Student factor trees will vary.)
$2^3 \cdot 3 \cdot 5$
486
(Student factor trees will vary.)
$2^2 \cdot 11^2$
GCF of 120 & 484
4

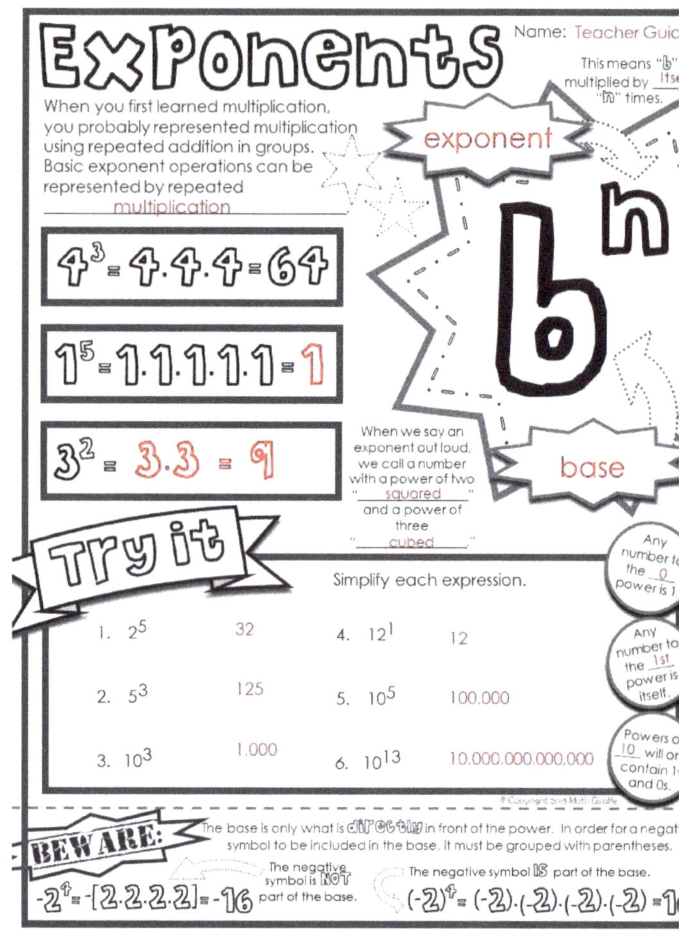

Exponents
Name: Teacher Guid
When you first learned multiplication, you probably represented multiplication using repeated addition in groups. Basic exponent operations can be represented by repeated multiplication
This means "b" multiplied by itse "n" times.
exponent
base
b^n
$4^3 = 4 \cdot 4 \cdot 4 = 64$
$1^5 = 1 \cdot 1 \cdot 1 \cdot 1 \cdot 1 = 1$
$3^2 = 3 \cdot 3 = 9$
When we say an exponent out loud, we call a number with a power of two "squared" and a power of three "cubed"
Try it
Simplify each expression.
1. 2^5 32
2. 5^3 125
3. 10^3 1,000
4. 12^1 12
5. 10^5 100,000
6. 10^{13} 10,000,000,000,000
Any number to the 0 power is 1
Any number to the 1st power is itself.
Powers of 10 will on contain 1s and 0s.
BEWARE: The base is only what is directly in front of the power. In order for a negat symbol to be included in the base, it must be grouped with parentheses.
The negative symbol is NOT part of the base.
$-2^4 = -[2 \cdot 2 \cdot 2 \cdot 2] = -16$
The negative symbol IS part of the base.
$(-2)^4 = (-2) \cdot (-2) \cdot (-2) \cdot (-2) = 1$

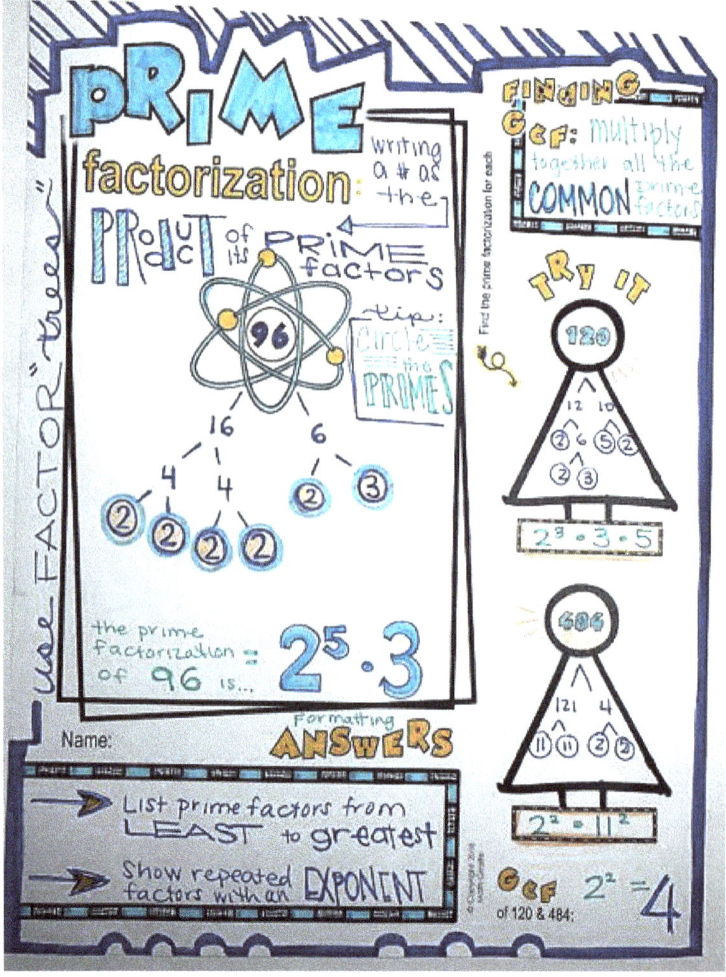

PRIME factorization
writing a # as the PRODUCT of its PRIME factors
tip: circle the PRIMES
use "FACTOR" trees
96
16
6
4
4
2
3
2
2
2
2
the prime factorization of 96 is... $2^5 \cdot 3$
Name:
Formatting ANSWERS
List prime factors from LEAST to greatest
Show repeated factors with an EXPONENT
FINDING GCF: multiply together all the COMMON prime factors
Find the prime factorization for each.
TRY IT
120
12 10
$2^3 \cdot 3 \cdot 5$
484
121 4
11 11 2 2
$2^2 \cdot 11^2$
GCF of 120 & 484: $2^2 = 4$

Exponents
Name:
This means "b" multiplied by ITSELF "n" times.
When you first learned multiplication, you probably represented multiplication using repeated addition in groups. Basic exponent operations can be represented by repeated multiplication
EXPONENT
BASE
b^n
$4^3 = 4 \cdot 4 \cdot 4 = 64$
3 times
$1^5 = 1 \cdot 1 \cdot 1 \cdot 1 \cdot 1 = 1$
5 times
$3^2 = 3 \cdot 3 = 9$
2 times
When we say an exponent out loud, we call a number with a power of two "squared" and a power of three "cubed"
Try it
Simplify each expression.
1. 2^5 2·2·2·2·2 = 32
2. 5^3 5·5·5 = 125
3. 10^3 10·10·10 = 1000
4. 12^1 12 = 12
5. 10^5 10·10·10·10·10 = 100,000
6. 10^{13} = 10,000,000,000,000
Any number to the 0 power is 1
Any number to the 1st power is itself.
Powers of 10 will only contain 1s and 0s.
BEWARE: The base is only what is directly in front of the power. In order for a negative symbol to be included in the base, it must be grouped with parentheses.
The negative symbol is NOT part of the base.
$-2^4 = -[2 \cdot 2 \cdot 2 \cdot 2] = -16$
The negative symbol IS part of the base.
$(-2)^4 = (-2) \cdot (-2) \cdot (-2) \cdot (-2) = 16$

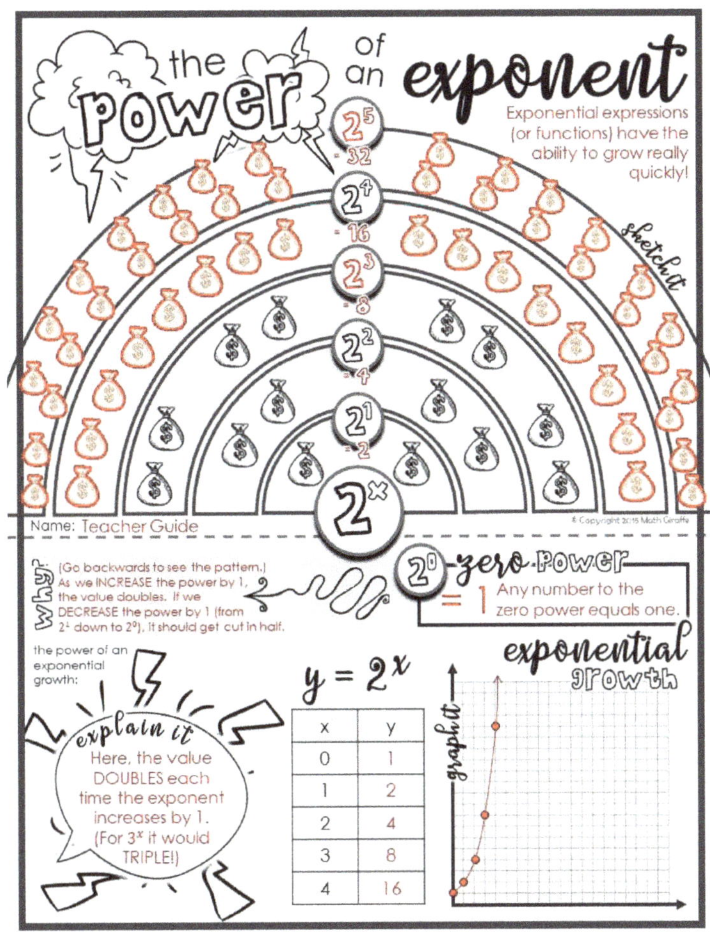

the power of an exponent
Exponential expressions (or functions) have the ability to grow really quickly!
2^5 = 32
2^4 = 16
2^3 = 8
2^2 = 4
2^1 = 2
2^x
sketch it
Name: Teacher Guide
© Copyright 2015 Math Giraffe
why? (Go backwards to see the pattern.) As we INCREASE the power by 1, the value doubles. If we DECREASE the power by 1 (from 2^1 down to 2^0), it should get cut in half.
2^0 zero power
= 1 Any number to the zero power equals one.
the power of an exponential growth:
explain it
Here, the value DOUBLES each time the exponent increases by 1. (For 3^x it would TRIPLE!)
$y = 2^x$
x y
0 1
1 2
2 4
3 8
4 16
exponential growth
graph it

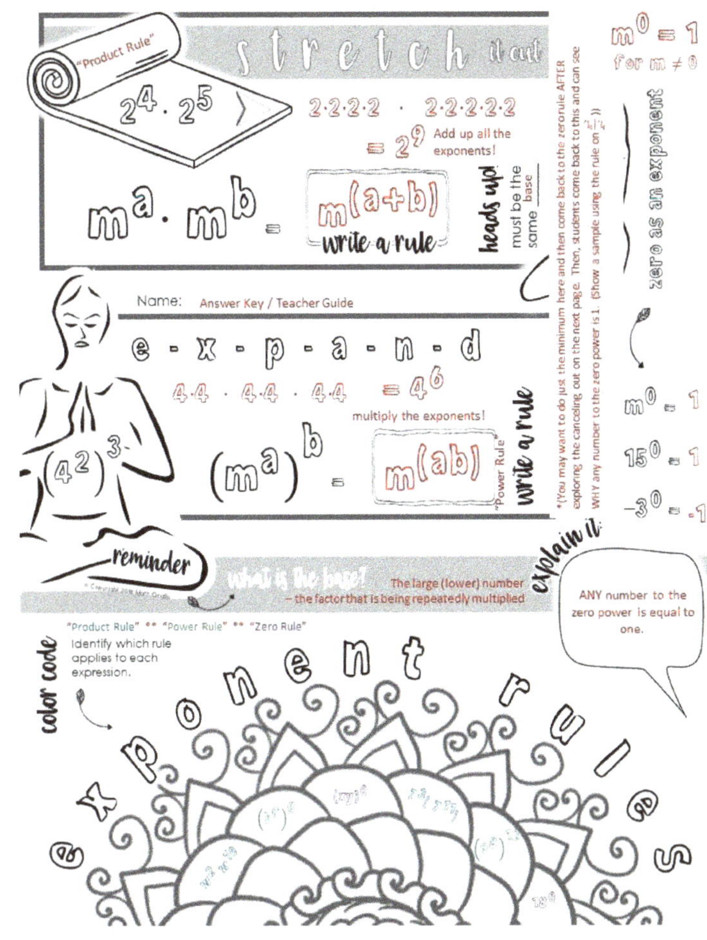

"Product Rule"
stretch it out
$2^4 \cdot 2^5$
$2\cdot2\cdot2\cdot2 \cdot 2\cdot2\cdot2\cdot2\cdot2$
$= 2^9$ Add up all the exponents!
$m^a \cdot m^b = m^{(a+b)}$
write a rule
heads up! must be the same base
Name: Answer Key / Teacher Guide
e-x-p-a-n-d
$4\cdot4 \cdot 4\cdot4 \cdot 4\cdot4 = 4^6$
$(4^2)^3$
multiply the exponents!
$(m^a)^b = m^{(ab)}$
"Power Rule"
write a rule
reminder
what is the base? The large (lower) number – the factor that is being repeatedly multiplied
explain it
ANY number to the zero power is equal to one.
"Product Rule" ** "Power Rule" ** "Zero Rule"
color code
Identify which rule applies to each expression.
exponent rules
$m^0 = 1$ for $m \neq 0$
zero as an exponent
*(You may want to do just the minimum here and then come back to the zero rule AFTER exploring the canceling out on the next page. Then, students come back to this and can see WHY any number to the zero power is 1. (Show a sample using the rule on
$m^0 = 1$
$15^0 = 1$
$-3^0 = -1$

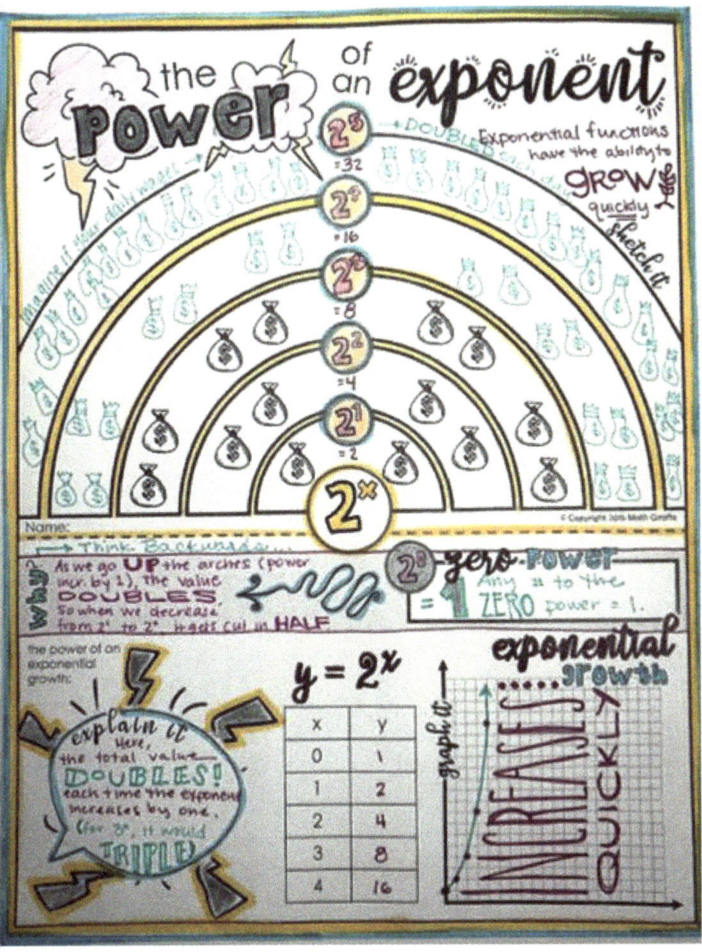

the power of an exponent
Exponential functions have the ability to GROW quickly
DOUBLES
2^5 = 32
2^4 = 16
2^3 = 8
2^2 = 4
2^1 = 2
2^x
sketch it
Name:
Think Backwards
why? As we go UP the arches (power incr. by 1), the value DOUBLES. So when we decrease from 2^1 to 2^0, it gets cut in HALF
2^0 zero power
= 1 Any # to the ZERO power = 1.
the power of an exponential growth:
explain it
Here, the total value DOUBLES! each time the exponent increases by one. (For 3^x, it would TRIPLE)
$y = 2^x$
x y
0 1
1 2
2 4
3 8
4 16
exponential growth
graph it
INCREASES QUICKLY

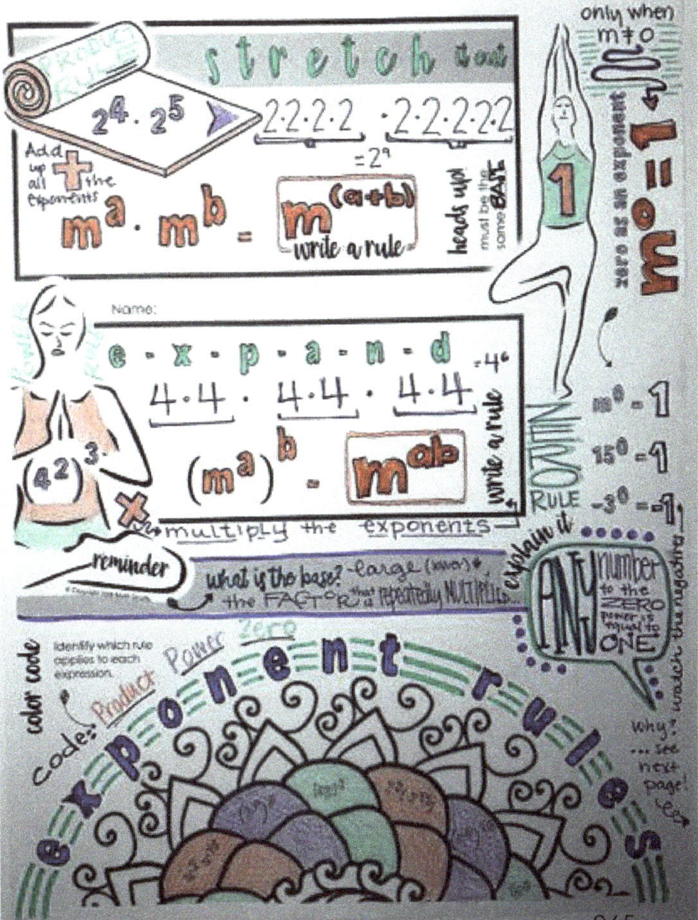

only when m≠0
stretch it out
PRODUCT RULE
$2^4 \cdot 2^5$
$2\cdot2\cdot2\cdot2 \cdot 2\cdot2\cdot2\cdot2\cdot2$
$= 2^9$
Add up all the exponents
$m^a \cdot m^b = m^{(a+b)}$
write a rule
heads up! must be the same BASE
zero as an exponent
$m^0 = 1$
Name:
e-x-p-a-n-d
$4\cdot4 \cdot 4\cdot4 \cdot 4\cdot4 = 4^6$
$(4^2)^3$
$(m^a)^b = m^{ab}$
write a rule
POWER RULE
multiply the exponents
$m^0 = 1$
$15^0 = 1$
$-3^0 = -1$
reminder
what is the base? large (lower) the FACTOR that's repeatedly MULTIPLIED
explain it
ANY number to the ZERO power is equal to ONE
watch the negatives
color code
Identify which rule applies to each expression.
code: Product Power Zero
exponent rules
why? ... see next page!

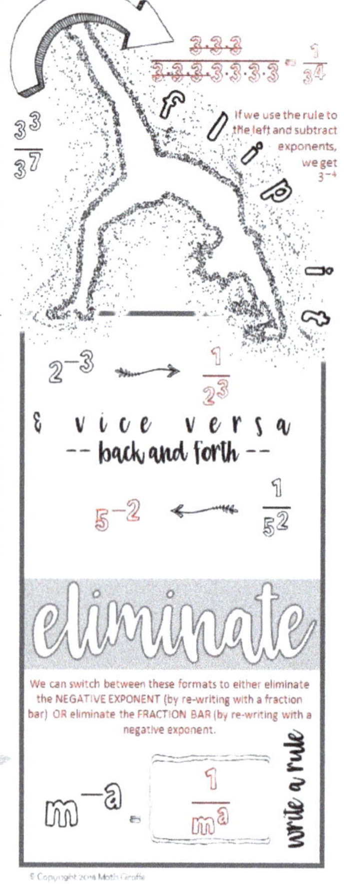

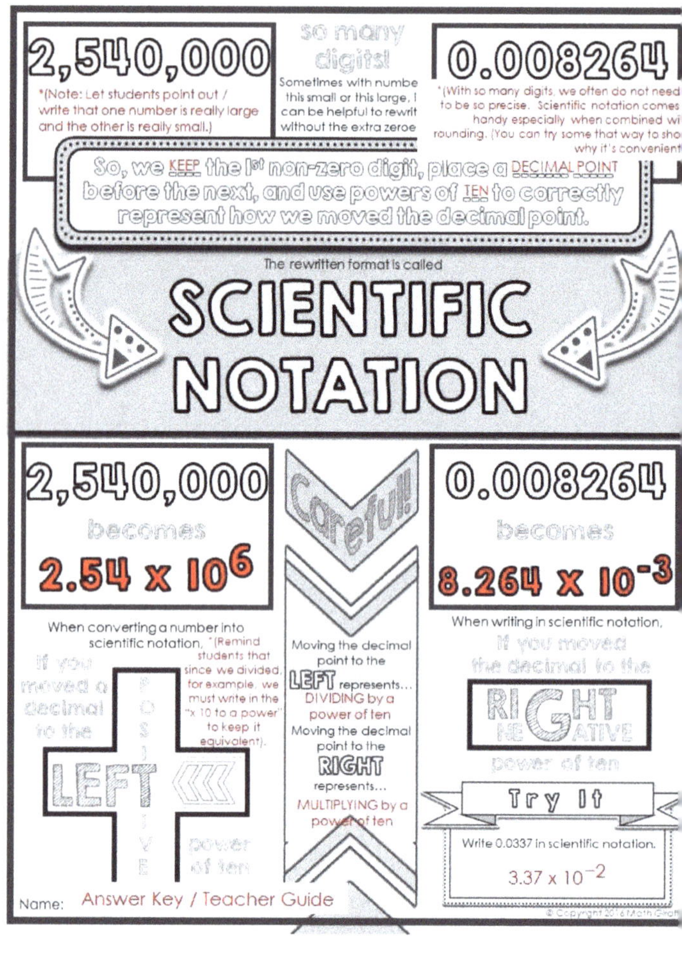

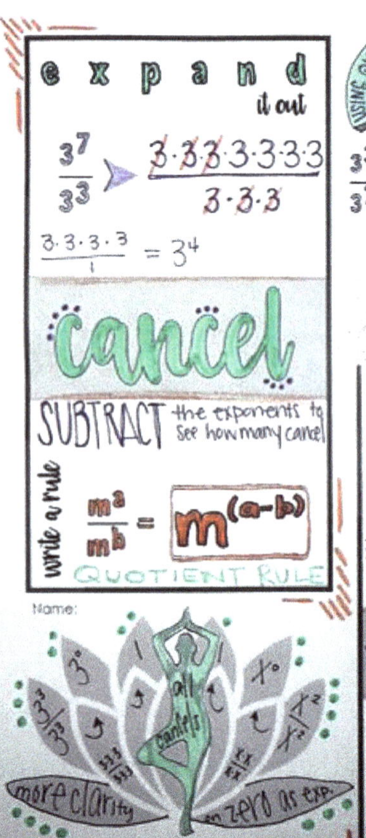

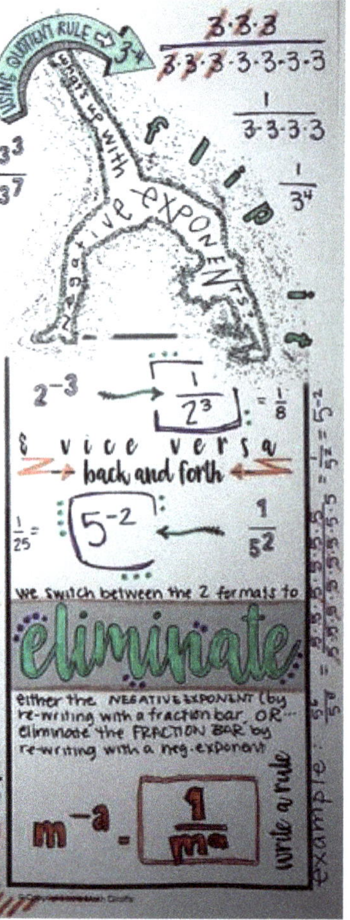

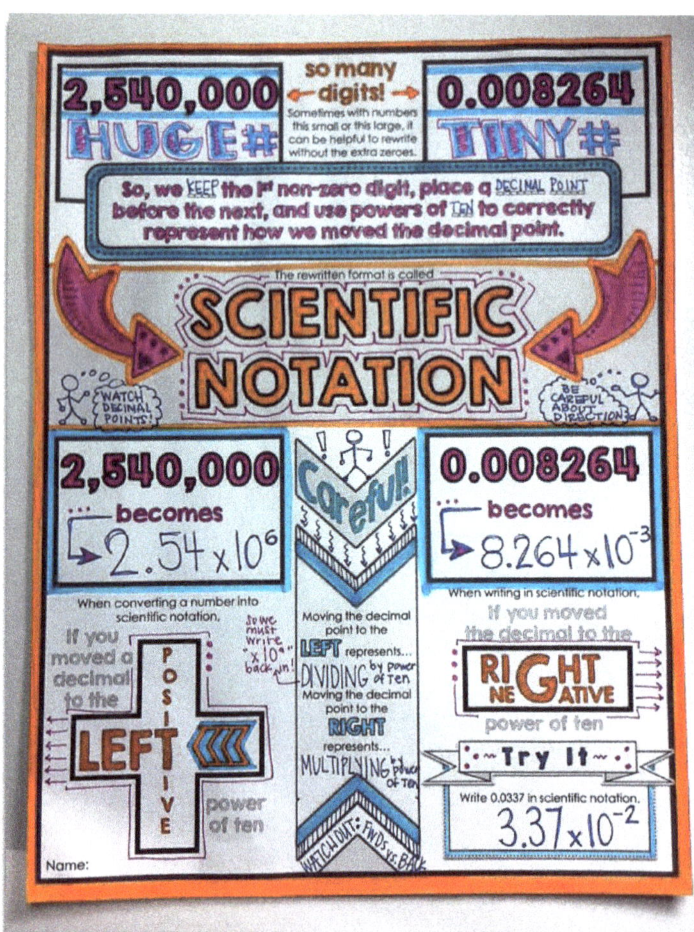

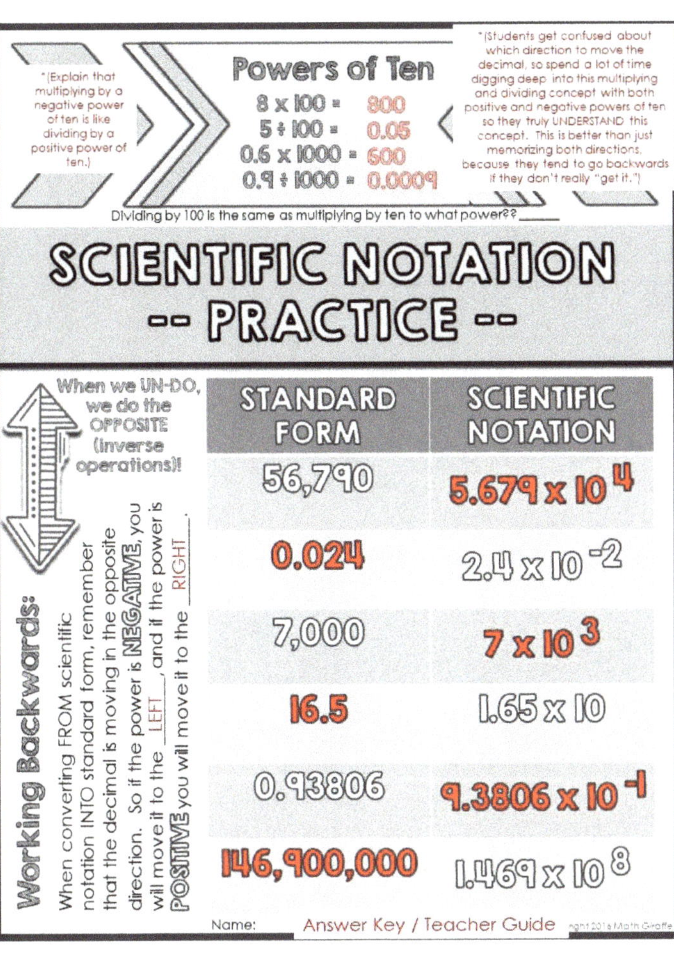
Powers of Ten
8 x 100 = 800
5 ÷ 100 = 0.05
0.6 x 1000 = 600
0.9 ÷ 1000 = 0.0009
Dividing by 100 is the same as multiplying by ten to what power??
SCIENTIFIC NOTATION
-- PRACTICE --
When we UN-DO, we do the OPPOSITE (inverse operations)!
STANDARD FORM
SCIENTIFIC NOTATION
56,790
5.679 x 10 4
0.024
2.4 x 10 -2
7,000
7 x 10 3
16.5
1.65 x 10
0.93806
9.3806 x 10 -1
146,900,000
1.469 x 10 8
Working Backwards:
When converting FROM scientific notation INTO standard form, remember that the decimal is moving in the opposite direction. So if the power is NEGATIVE, you will move it to the LEFT, and if the power is POSITIVE you will move it to the RIGHT.
Name:
Answer Key / Teacher Guide

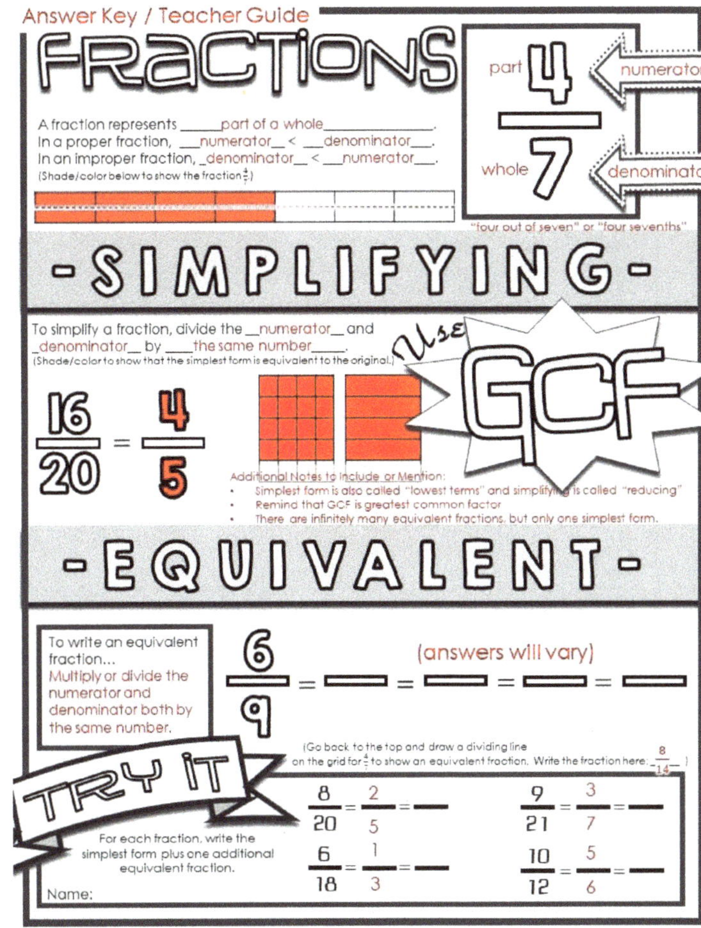
Answer Key / Teacher Guide
FRACTIONS
A fraction represents part of a whole.
In a proper fraction, numerator < denominator.
In an improper fraction, denominator < numerator.
part 4
whole 7
numerator
denominator
"four out of seven" or "four sevenths"
- SIMPLIFYING -
To simplify a fraction, divide the numerator and denominator by the same number.
16/20 = 4/5
Use GCF
Additional Notes to Include or Mention:
Simplest form is also called "lowest terms" and simplifying is called "reducing"
Remind that GCF is greatest common factor
There are infinitely many equivalent fractions, but only one simplest form.
- EQUIVALENT -
To write an equivalent fraction... Multiply or divide the numerator and denominator both by the same number.
6/9
(answers will vary)
TRY IT
For each fraction, write the simplest form plus one additional equivalent fraction.
8/20 = 2/5
9/21 = 3/7
6/18 = 1/3
10/12 = 5/6
Name:

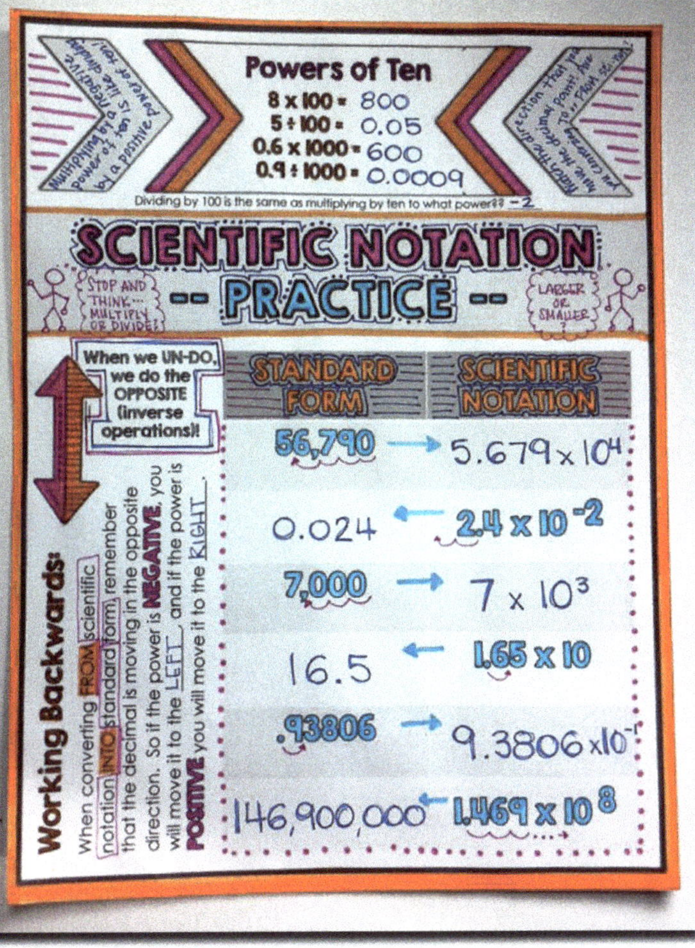
Powers of Ten
8 x 100 = 800
5 ÷ 100 = 0.05
0.6 x 1000 = 600
0.9 ÷ 1000 = 0.0009
Dividing by 100 is the same as multiplying by ten to what power?? -2
SCIENTIFIC NOTATION
-- PRACTICE --
STOP AND THINK... MULTIPLY OR DIVIDE?
LARGER OR SMALLER?
When we UN-DO, we do the OPPOSITE (inverse operations)!
STANDARD FORM
SCIENTIFIC NOTATION
56,790
5.679 x 10⁴
0.024
2.4 x 10 -2
7,000
7 x 10³
16.5
1.65 x 10
.93806
9.3806 x 10⁻¹
146,900,000
1.469 x 10 8
Working Backwards:
When converting FROM scientific notation INTO standard form, remember that the decimal is moving in the opposite direction. So if the power is NEGATIVE, you will move it to the LEFT, and if the power is POSITIVE you will move it to the RIGHT.

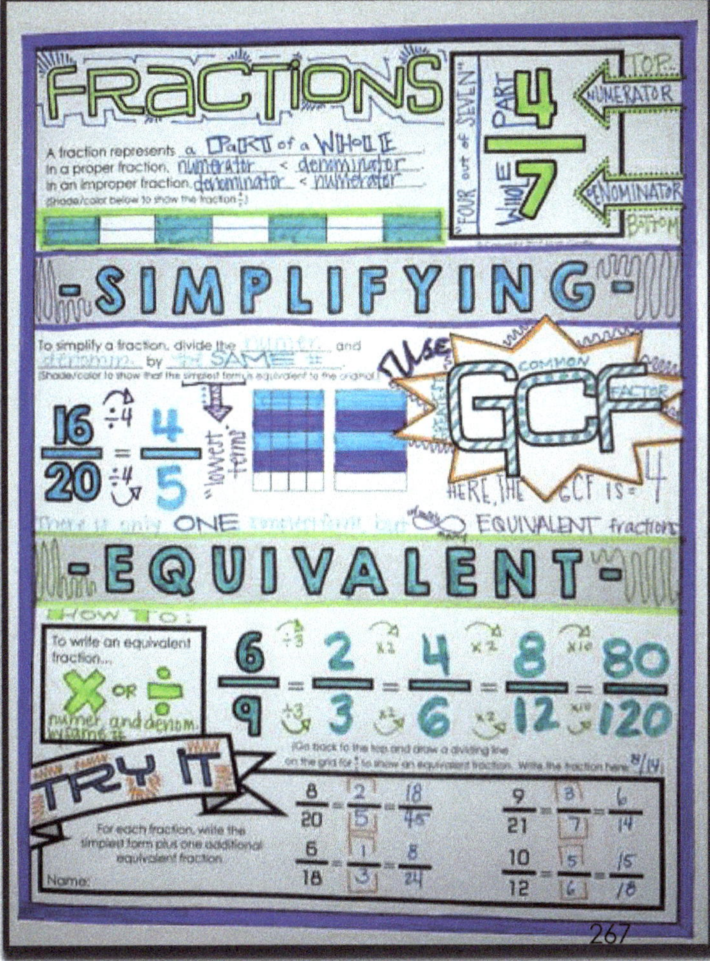
FRACTIONS
A fraction represents a PART of a WHOLE
In a proper fraction, numerator < denominator
In an improper fraction, denominator < numerator
TOP
NUMERATOR
DENOMINATOR
BOTTOM
PART 4
WHOLE 7
FOUR out of SEVEN
- SIMPLIFYING -
To simplify a fraction, divide the numer. and denom. by the SAME #
16/20 = 4/5
"lowest terms"
Use GCF
COMMON
FACTOR
HERE, THE GCF IS = 4
There is only ONE simplest form but many EQUIVALENT fractions
- EQUIVALENT -
HOW TO:
To write an equivalent fraction...
numer. and denom.
6/9 = 2/3 = 4/6 = 8/12 = 80/120
TRY IT
For each fraction, write the simplest form plus one additional equivalent fraction.
8/20 = 2/5 = 18/45
9/21 = 3/7 = 6/14
6/18 = 1/3 = 8/24
10/12 = 5/6 = 15/18
Name:

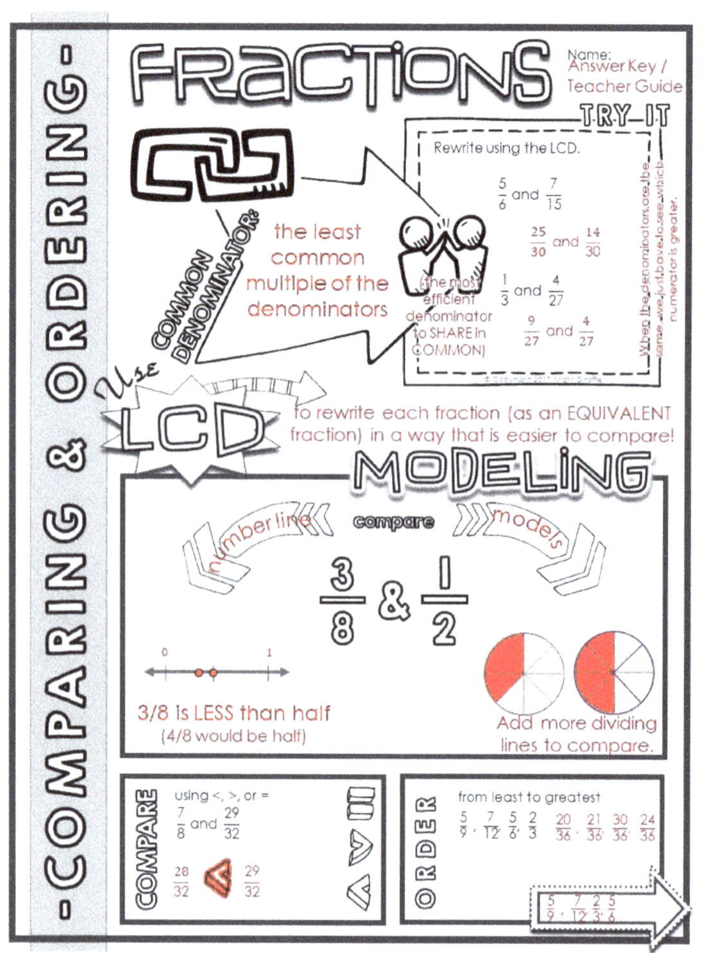
-COMPARING & ORDERING-
FRACTIONS
Name: Answer Key / Teacher Guide
TRY IT
Rewrite using the LCD.
5/6 and 7/15
25/30 and 14/30
1/3 and 4/27
9/27 and 4/27
When the denominators are the same, we just have to see which numerator is greater.
COMMON DENOMINATOR:
the least common multiple of the denominators
(the most efficient denominator to SHARE in COMMON)
Use
LCD
to rewrite each fraction (as an EQUIVALENT fraction) in a way that is easier to compare!
MODELING
number line
compare
models
3/8 & 1/2
0
1
3/8 is LESS than half
(4/8 would be half)
Add more dividing lines to compare.
COMPARE
using <, >, or =
7/8 and 29/32
28/32 < 29/32
ORDER
from least to greatest
5/9, 7/12, 5/6, 2/3
20/36, 21/36, 30/36, 24/36
5/9, 7/12, 2/3, 5/6

Name:
Answer Key / Teacher Guide
2 1/3
MULTIPL
the stripe
ADD
WHY? MODEL IT
What are they good for?
Mixed numbers are nicer for visualizing or understanding a quantity.
MIXED NUMBERS
IMPROPER FRACTIONS
*(Remind students that the numerator / top # goes "in the box" when dividing! They can sketch a memory trigger for this if they need to.)
7/3
DIVIDE.
2 R1
3 | 7
6
1
remainder
denominator
becomes the fractional part
It's TOP-HEAVY!! So IMPROPER!
What are they good for?
Performing certain operations is easier in improper fraction form. They can also be clearer in equations or formulas.

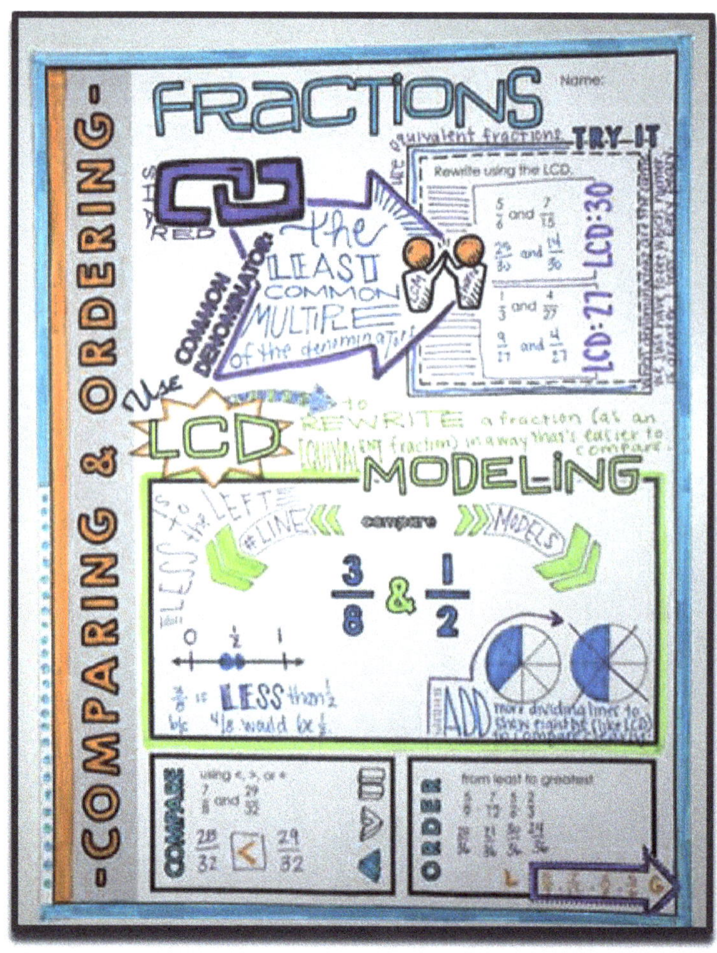
-COMPARING & ORDERING-
FRACTIONS
Name:
TRY IT
Rewrite using the LCD.
LCD:30
LCD:27
COMMON DENOMINATOR:
the LEAST COMMON MULTIPLE of the denominators
Use
LCD
to REWRITE a fraction (as an EQUIVALENT fraction) in a way that's easier to compare.
MODELING
compare
MODELS
3/8 & 1/2
3/8 is LESS than 1/2
b/c 4/8 would be 1/2
COMPARE
using <, >, or =
28/32 < 29/32
ORDER
from least to greatest

MULTIPLY
the striped
ADD
CONVERT TO IMPROPER FRACTION
WHY? MODEL IT
What are they good for?
easier to
VISUALIZE & UNDERSTAND
MIXED NUMBERS
IMPROPER FRACTIONS
DIVIDE.
CONVERT TO MIXED NUMBER
2 R1
= 2 1/3
remainder
denominator
becomes the fractional part (left over)
shh...
IT'S TOP-Heavy
So (improper)
What are they good for?
→ easier to do OPERATIONS
→ clearer in equations

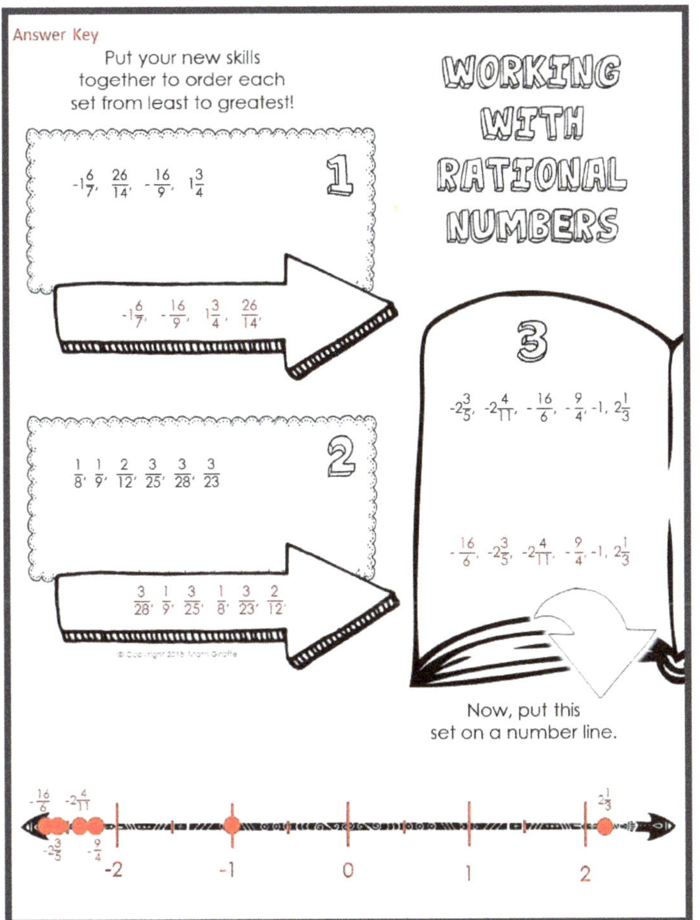
Answer Key
Put your new skills together to order each set from least to greatest!
WORKING WITH RATIONAL NUMBERS
1
-1 6/7, 26/14, -16/9, 1 3/4
-1 6/7, -16/9, 1 3/4, 26/14
2
1/8, 1/9, 2/12, 3/25, 3/28, 3/23
3/28, 1/9, 3/25, 1/8, 3/23, 2/12
3
-2 3/5, -2 4/11, -16/6, -9/4, -1, 2 1/3
-16/6, -2 3/5, -2 4/11, -9/4, -1, 2 1/3
Now, put this set on a number line.
-2
-1
0
1
2

converting a fraction to a decimal

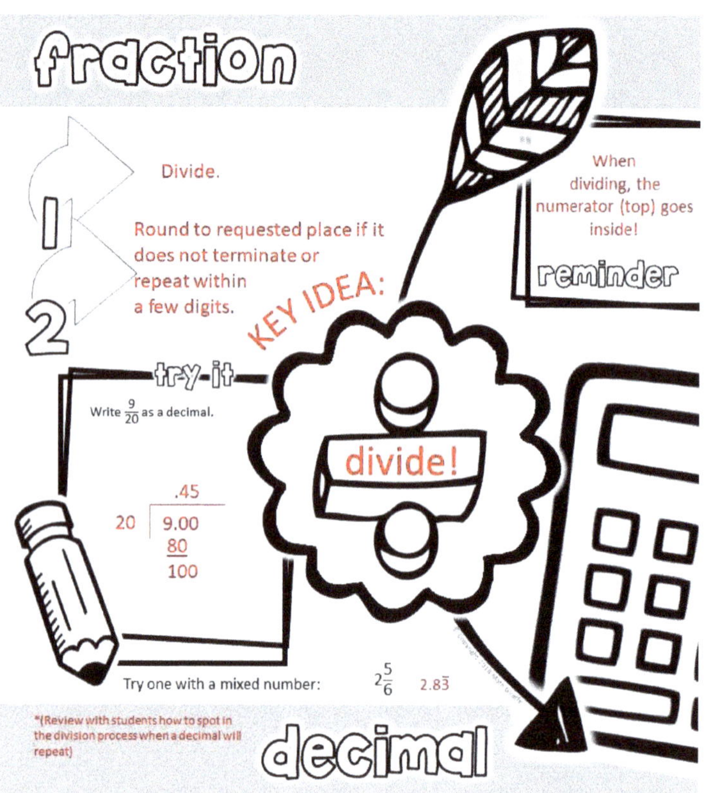

fraction
1
Divide.
2
Round to requested place if it does not terminate or repeat within a few digits.
When dividing, the numerator (top) goes inside!
reminder
KEY IDEA:
divide!
try-it
Write $\frac{9}{20}$ as a decimal.
.45
20 | 9.00
80
100
Try one with a mixed number: $2\frac{5}{6}$ 2.83
*(Review with students how to spot in the division process when a decimal will repeat)
decimal

Name: Answer Key / Teacher Guide

*(Start at the bottom on this page, then the two sheets can be laid side-by-side for a full study guide of the cycle and forth fraction to deci

converting a decimal to a fraction

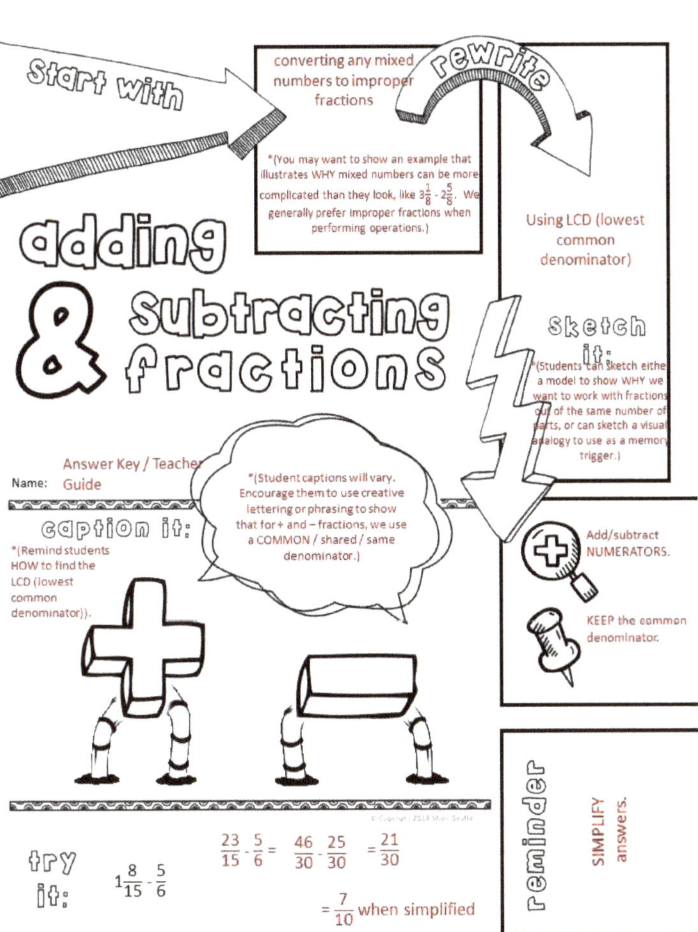
Start with
converting any mixed numbers to improper fractions
rewrite
*(You may want to show an example that illustrates WHY mixed numbers can be more complicated than they look, like $3\frac{1}{8} - 2\frac{5}{8}$. We generally prefer improper fractions when performing operations.)
Using LCD (lowest common denominator)
adding & subtracting fractions
Sketch it:
*(Students can sketch either a model to show WHY we want to work with fractions out of the same number of parts, or can sketch a visual analogy to use as a memory trigger.)
Name: Answer Key / Teacher Guide
*(Student captions will vary. Encourage them to use creative lettering or phrasing to show that for + and – fractions, we use a COMMON / shared / same denominator.)
caption it:
*(Remind students HOW to find the LCD (lowest common denominator)).
Add/subtract NUMERATORS.
KEEP the common denominator.
try it: $1\frac{8}{15} - \frac{5}{6}$
$\frac{23}{15} - \frac{5}{6} = \frac{46}{30} - \frac{25}{30} = \frac{21}{30}$
$= \frac{7}{10}$ when simplified
reminder
SIMPLIFY answers.

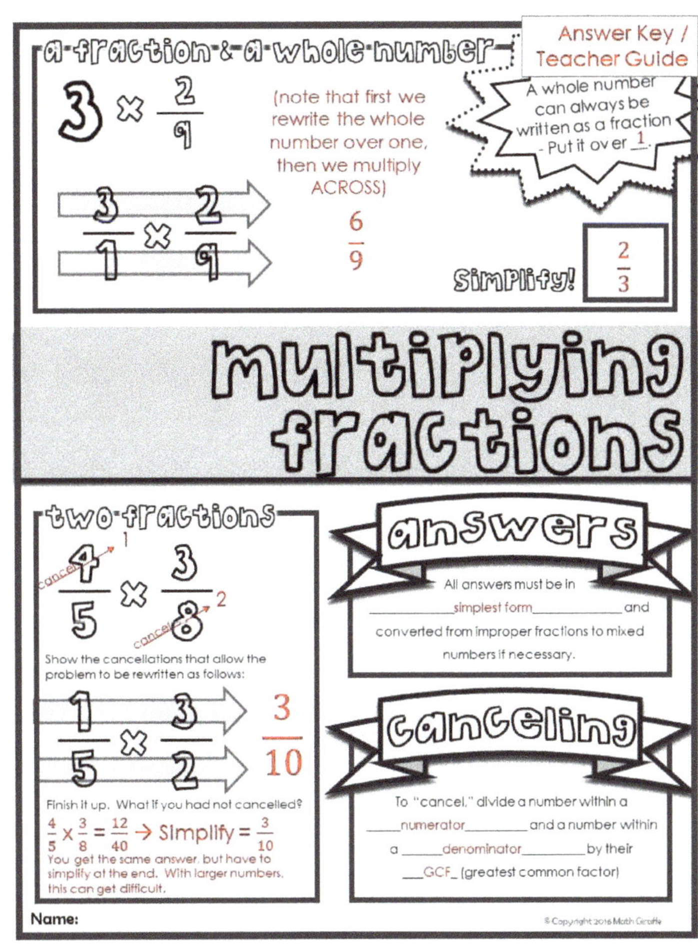
Answer Key / Teacher Guide
a fraction & a whole number
$3 \times \frac{2}{9}$
(note that first we rewrite the whole number over one, then we multiply ACROSS)
A whole number can always be written as a fraction - Put it over 1.
$\frac{3}{1} \times \frac{2}{9}$
$\frac{6}{9}$
Simplify! $\frac{2}{3}$
multiplying fractions
two fractions
$\frac{4}{5} \times \frac{3}{8}$
Show the cancellations that allow the problem to be rewritten as follows:
$\frac{1}{5} \times \frac{3}{2}$
$\frac{3}{10}$
Finish it up. What if you had not cancelled?
$\frac{4}{5} \times \frac{3}{8} = \frac{12}{40} \rightarrow$ Simplify $= \frac{3}{10}$
You get the same answer, but have to simplify at the end. With larger numbers, this can get difficult.
answers
All answers must be in simplest form and converted from improper fractions to mixed numbers if necessary.
canceling
To "cancel," divide a number within a numerator and a number within a denominator by their GCF (greatest common factor)
Name:

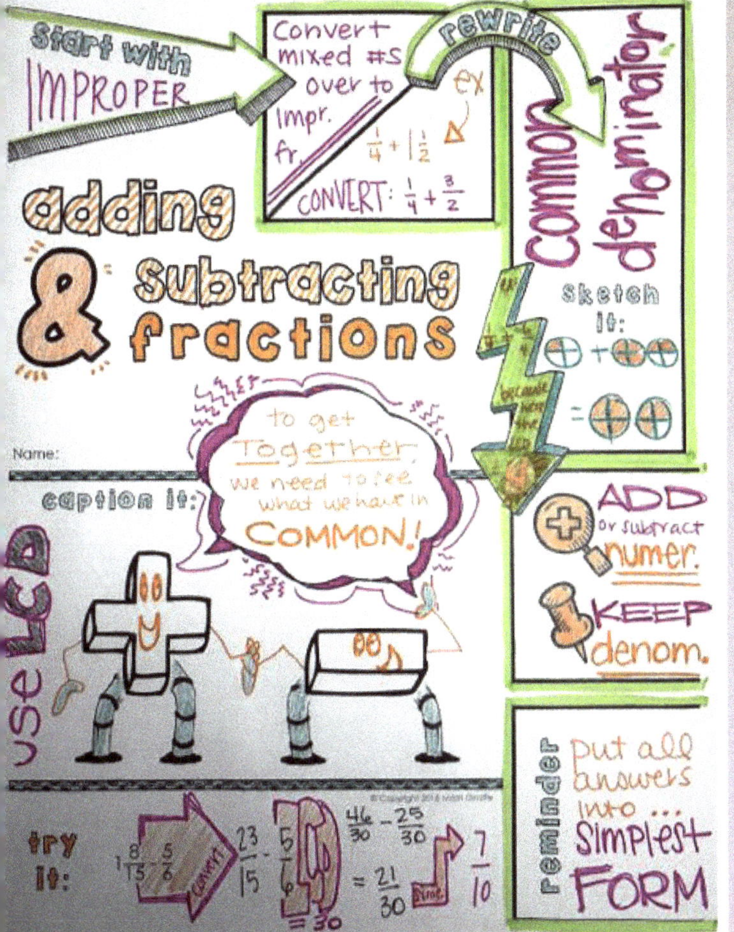
Start with IMPROPER
Convert mixed #s over to Impr. fr.
rewrite
CONVERT: $\frac{1}{4} + \frac{3}{2}$
common denominator
adding & subtracting fractions
Sketch it:
to get Together we need to see what we have in COMMON!
Name:
caption it:
use LCD
ADD or subtract numer.
KEEP denom.
reminder
put all answers into ... simplest FORM
try it:

a fraction & a whole number
1 REWRITE
2 multiply numer. multiply denom.
A whole number can always be written as a fraction - Put it over 1.
multiply ACROSS
3 Simplify! 2/3
multiplying fractions
two fractions
answers
All answers must be in SIMPLEST FORM and converted from improper fractions to mixed numbers if necessary. REDUCED MIXED
canceling
To "cancel," divide a number within a NUMERATOR and a number within a DENOMINATOR by their GCF (greatest common factor)
HAVE TO SIMPLIFY
SIMPLIFY!
Name:

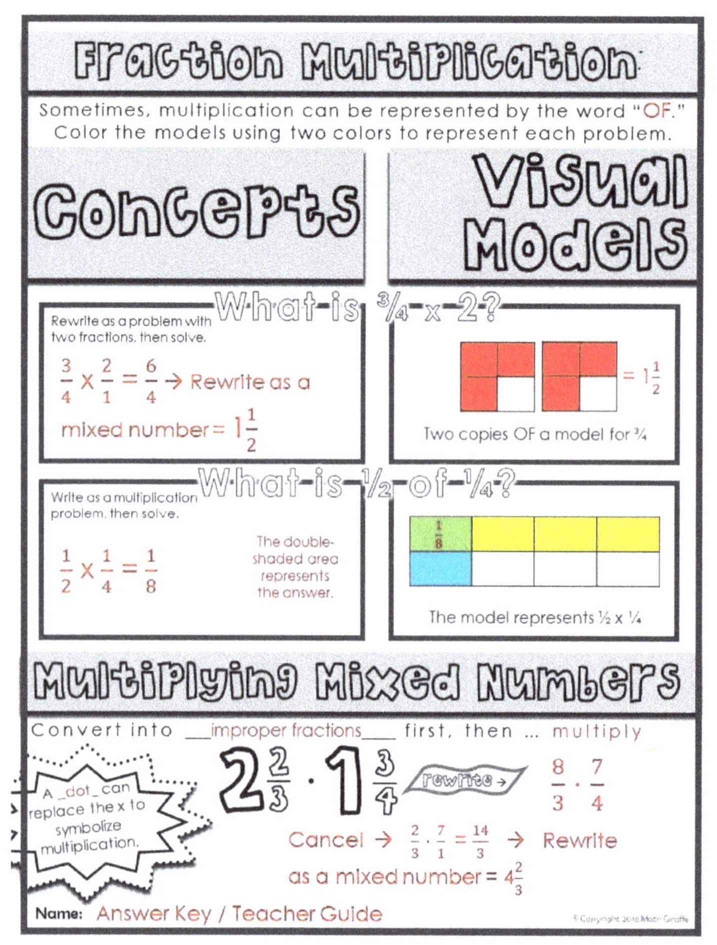
Fraction Multiplication
Sometimes, multiplication can be represented by the word "OF."
Color the models using two colors to represent each problem.
Concepts
Visual Models
What is ¾ x 2?
Rewrite as a problem with two fractions, then solve.
$\frac{3}{4} \times \frac{2}{1} = \frac{6}{4}$ → Rewrite as a mixed number = $1\frac{1}{2}$
= $1\frac{1}{2}$
Two copies OF a model for ¾
What is ½ of ¼?
Write as a multiplication problem, then solve.
$\frac{1}{2} \times \frac{1}{4} = \frac{1}{8}$
The double-shaded area represents the answer.
$\frac{1}{8}$
The model represents ½ x ¼
Multiplying Mixed Numbers
Convert into __improper fractions___ first, then ... multiply
A _dot_ can replace the x to symbolize multiplication.
$2\frac{2}{3} \cdot 1\frac{3}{4}$ rewrite → $\frac{8}{3} \cdot \frac{7}{4}$
Cancel → $\frac{2}{3} \cdot \frac{7}{1} = \frac{14}{3}$ → Rewrite as a mixed number = $4\frac{2}{3}$
Name: Answer Key / Teacher Guide
© Copyright 2016 Math Giraffe

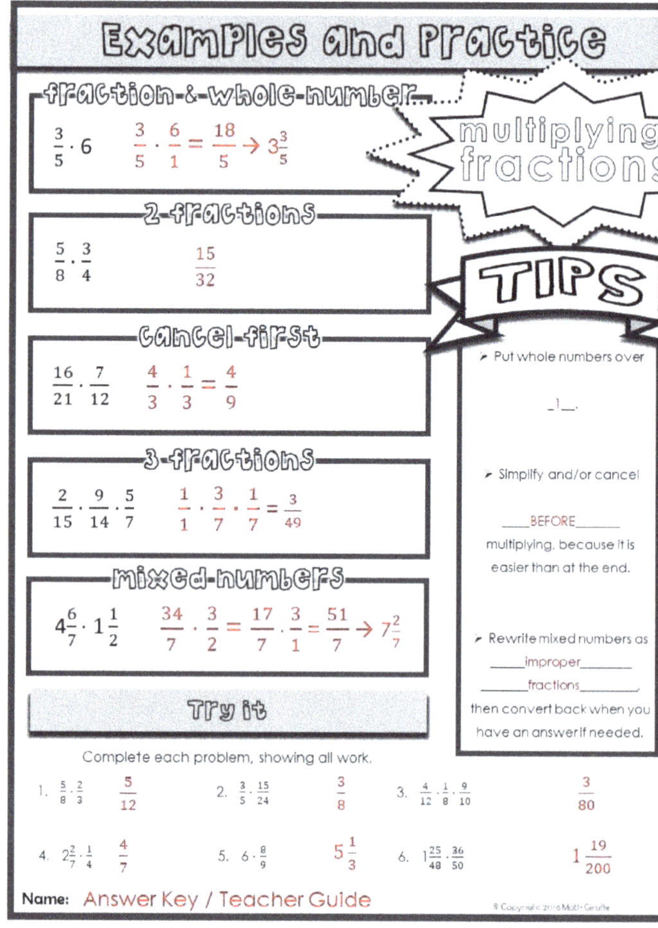
Examples and Practice
fraction & whole number
$\frac{3}{5} \cdot 6$ $\frac{3}{5} \cdot \frac{6}{1} = \frac{18}{5} \rightarrow 3\frac{3}{5}$
multiplying fractions
2 fractions
$\frac{5}{8} \cdot \frac{3}{4}$ $\frac{15}{32}$
TIPS
cancel first
$\frac{16}{21} \cdot \frac{7}{12}$ $\frac{4}{3} \cdot \frac{1}{3} = \frac{4}{9}$
3 fractions
$\frac{2}{15} \cdot \frac{9}{14} \cdot \frac{5}{7}$ $\frac{1}{1} \cdot \frac{3}{7} \cdot \frac{1}{7} = \frac{3}{49}$
mixed numbers
$4\frac{6}{7} \cdot 1\frac{1}{2}$ $\frac{34}{7} \cdot \frac{3}{2} = \frac{17}{7} \cdot \frac{3}{1} = \frac{51}{7} \rightarrow 7\frac{2}{7}$
Put whole numbers over _1_.
Simplify and/or cancel ___BEFORE___ multiplying, because it is easier than at the end.
Rewrite mixed numbers as ___improper___ ___fractions___, then convert back when you have an answer if needed.
Try it
Complete each problem, showing all work.
1. $\frac{5}{8} \cdot \frac{2}{3}$ $\frac{5}{12}$
2. $\frac{3}{5} \cdot \frac{15}{24}$ $\frac{3}{8}$
3. $\frac{4}{12} \cdot \frac{1}{8} \cdot \frac{9}{10}$ $\frac{3}{80}$
4. $2\frac{2}{7} \cdot \frac{1}{4}$ $\frac{4}{7}$
5. $6 \cdot \frac{8}{9}$ $5\frac{1}{3}$
6. $1\frac{25}{48} \cdot \frac{36}{50}$ $1\frac{19}{200}$
Name: Answer Key / Teacher Guide
© Copyright 2016 Math Giraffe

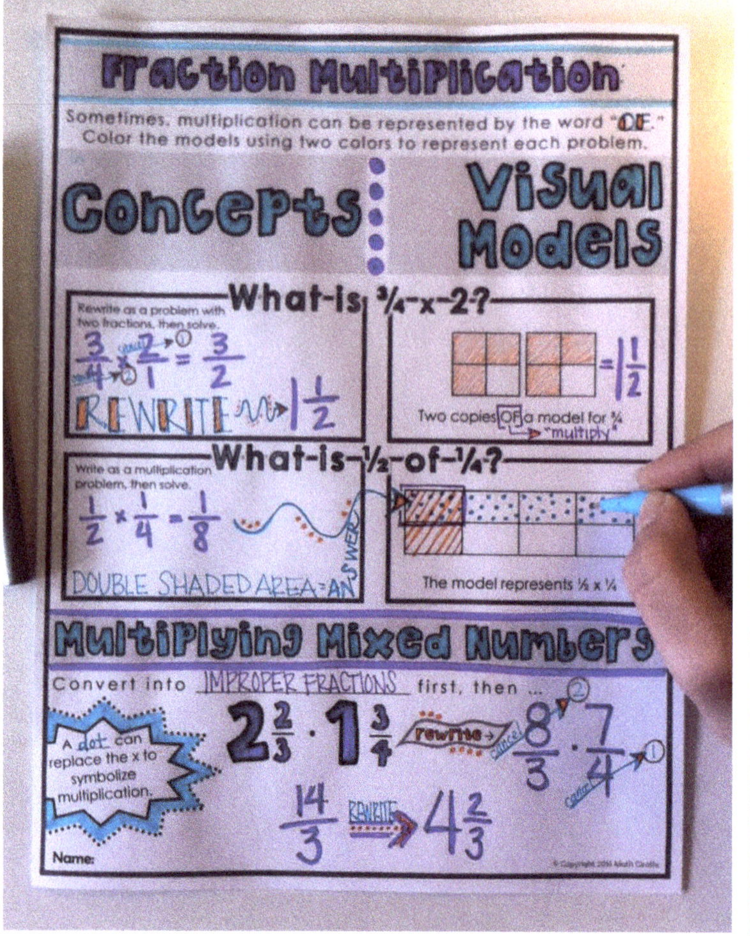
Fraction Multiplication
Sometimes, multiplication can be represented by the word "OF."
Color the models using two colors to represent each problem.
Concepts
Visual Models
What is ¾ x 2?
REWRITE $1\frac{1}{2}$
Two copies OF a model for ¾
"multiply"
What is ½ of ¼?
DOUBLE SHADED AREA = ANSWER
The model represents ½ x ¼
Multiplying Mixed Numbers
Convert into IMPROPER FRACTIONS first, then ...
A dot can replace the x to symbolize multiplication.
rewrite
cancel
REWRITE
$\frac{14}{3}$ → $4\frac{2}{3}$
Name:

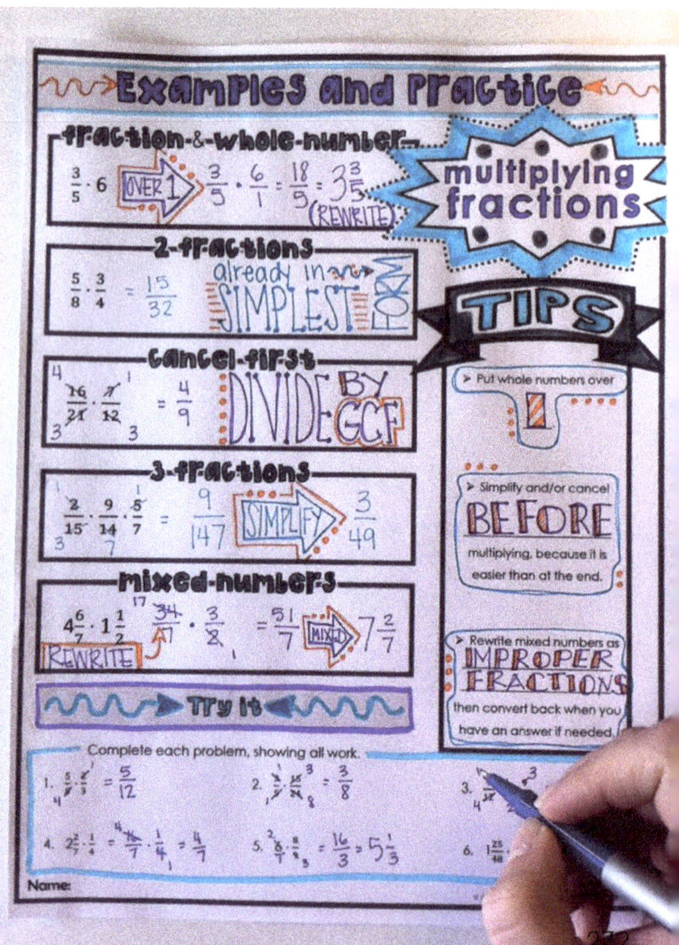
Examples and Practice
fraction & whole number
OVER 1
REWRITE
multiplying fractions
2 fractions
already in SIMPLEST FORM
TIPS
cancel first
DIVIDE BY GCF
3 fractions
SIMPLIFY
mixed numbers
REWRITE
MIXED
Put whole numbers over 1
Simplify and/or cancel BEFORE multiplying, because it is easier than at the end.
Rewrite mixed numbers as IMPROPER FRACTIONS then convert back when you have an answer if needed.
Try it
Complete each problem, showing all work.
Name:

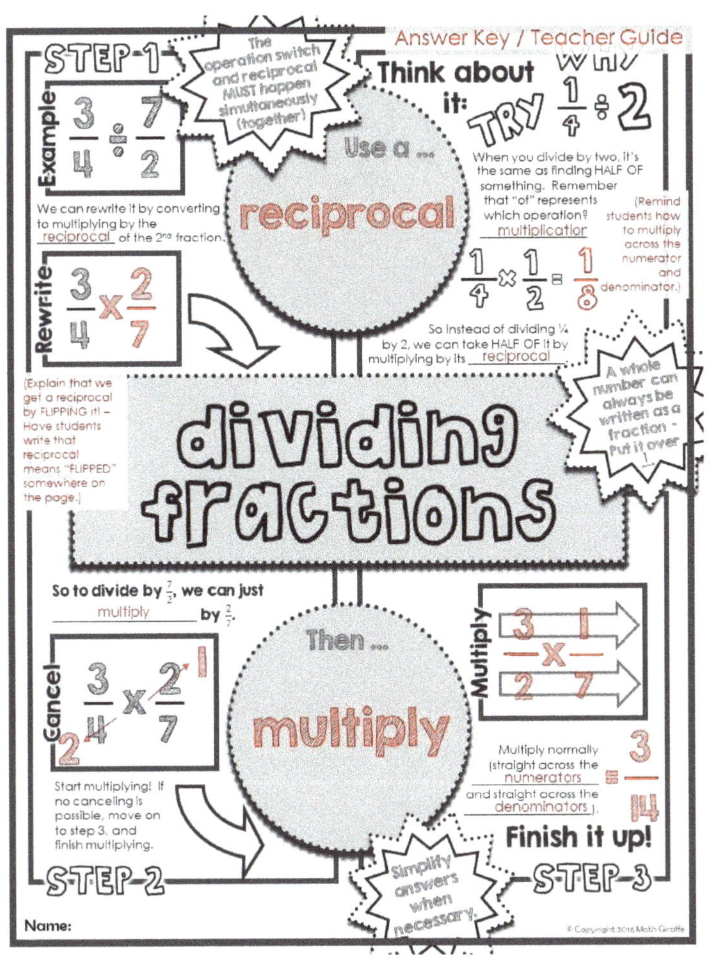
Answer Key / Teacher Guide
STEP-1
The operation switch and reciprocal MUST happen simultaneously (together)
Example
3/4 ÷ 7/2
We can rewrite it by converting to multiplying by the reciprocal of the 2nd fraction.
Rewrite
3/4 x 2/7
Think about it:
WHY
TRY 1/4 ÷ 2
Use a ...
reciprocal
When you divide by two, it's the same as finding HALF OF something. Remember that "of" represents which operation? multiplication
(Remind students how to multiply across the numerator and denominator.)
1/4 x 1/2 = 1/8
So instead of dividing ¼ by 2, we can take HALF OF it by multiplying by its reciprocal
(Explain that we get a reciprocal by FLIPPING it! – Have students write that reciprocal means "FLIPPED" somewhere on the page.)
A whole number can always be written as a fraction - Put it over 1
dividing fractions
So to divide by 7/2, we can just multiply by 2/7.
Cancel
Then ...
multiply
Multiply
Start multiplying! If no canceling is possible, move on to step 3, and finish multiplying.
Multiply normally (straight across the numerators and straight across the denominators).
= 3/14
Finish it up!
Simplify answers when necessary.
STEP-2
STEP-3
Name:
© Copyright 2016 Math Giraffe

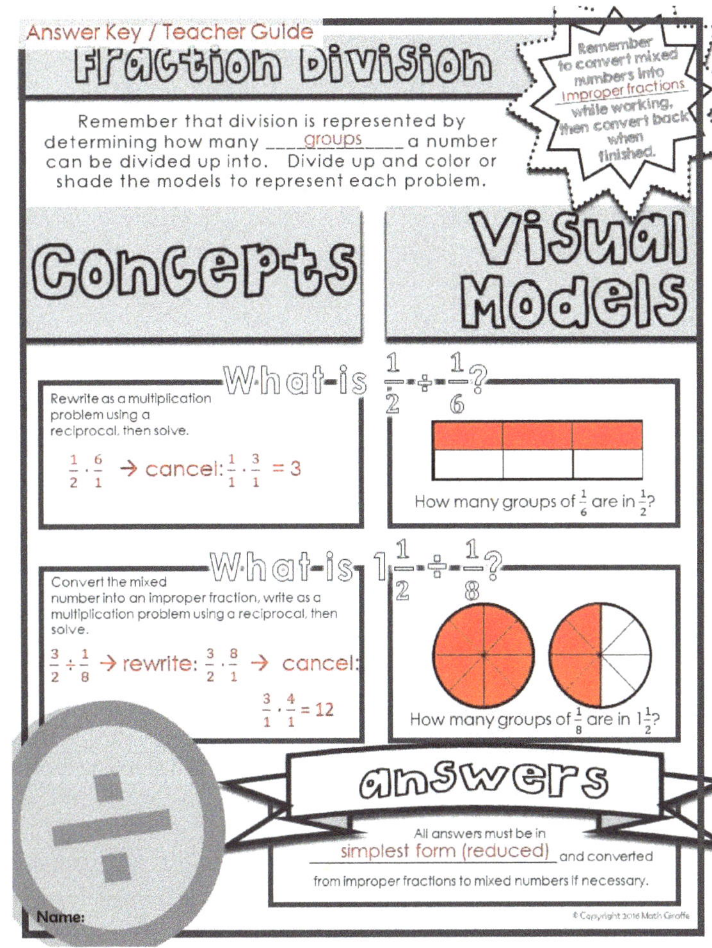
Answer Key / Teacher Guide
Fraction Division
Remember to convert mixed numbers into improper fractions while working, then convert back when finished.
Remember that division is represented by determining how many groups a number can be divided up into. Divide up and color or shade the models to represent each problem.
Concepts
Visual Models
What is 1/2 ÷ 1/6?
Rewrite as a multiplication problem using a reciprocal, then solve.
1/2 · 6/1 → cancel: 1/1 · 3/1 = 3
How many groups of 1/6 are in 1/2?
What is 1 1/2 ÷ 1/8?
Convert the mixed number into an improper fraction, write as a multiplication problem using a reciprocal, then solve.
3/2 ÷ 1/8 → rewrite: 3/2 · 8/1 → cancel: 3/1 · 4/1 = 12
How many groups of 1/8 are in 1 1/2?
answers
All answers must be in simplest form (reduced) and converted from improper fractions to mixed numbers if necessary.
Name:
© Copyright 2016 Math Giraffe

STEP-1
Example
Rewrite
Think about it:
WHY
Use a ...
RECIPROCAL
FLIP
dividing fractions
MULTIPLY
Then ...
MULTIPLY ACROSS
Cancel
Multiply
Finish it up!
STEP-2
STEP-3
Name:

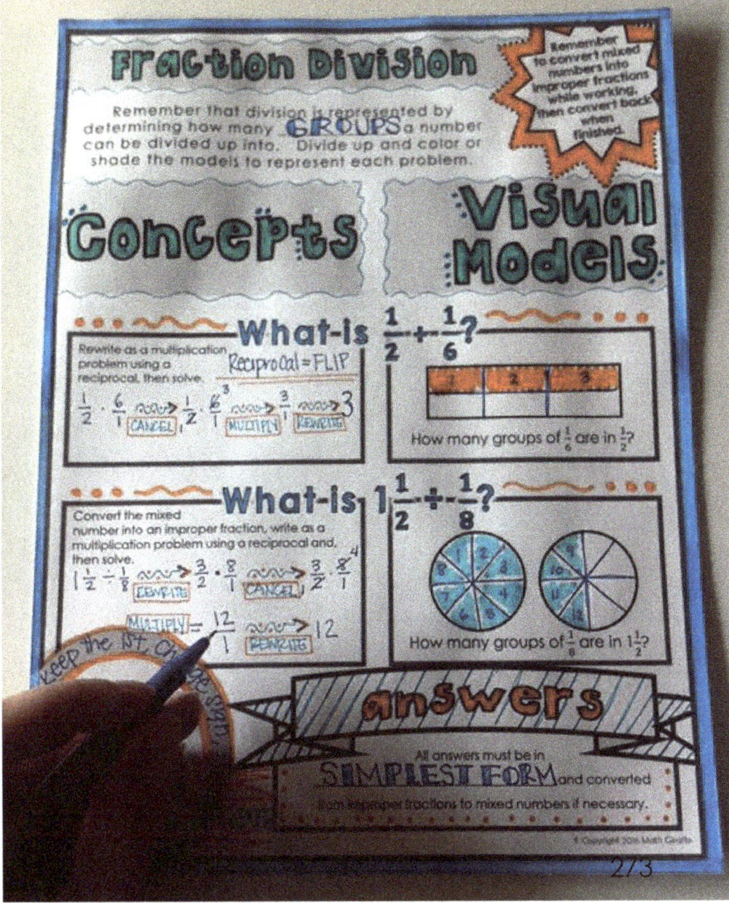
Fraction Division
Concepts
Visual Models
What is
Reciprocal = FLIP
CANCEL
MULTIPLY
REWRITE
CONVERT
answers
SIMPLEST FORM

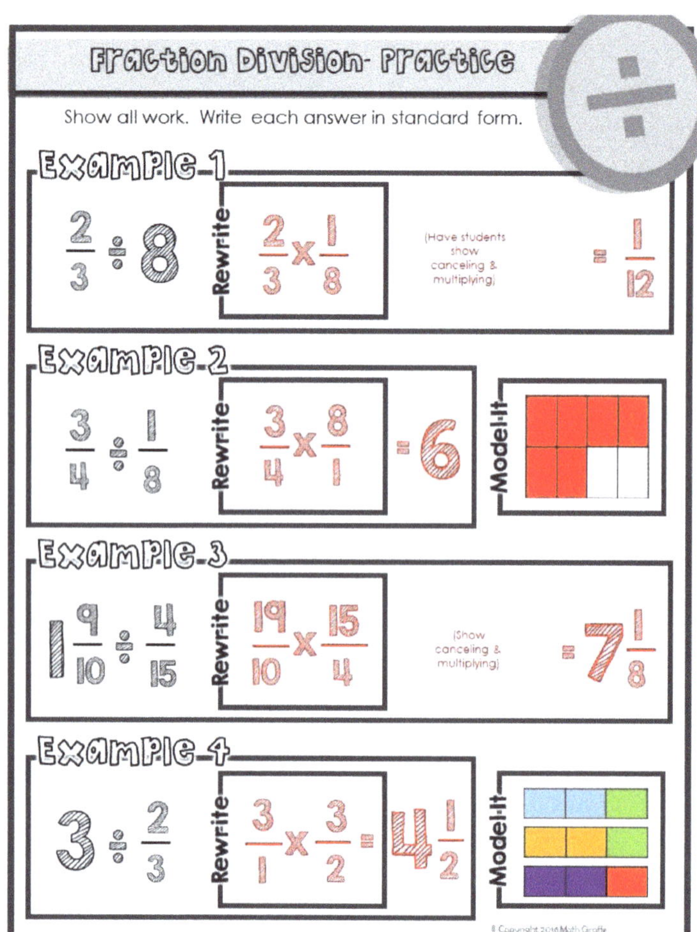
Fraction Division- Practice
Show all work. Write each answer in standard form.
Example 1
$\frac{2}{3} \div 8$
Rewrite
$\frac{2}{3} \times \frac{1}{8}$
(Have students show canceling & multiplying)
$= \frac{1}{12}$
Example 2
$\frac{3}{4} \div \frac{1}{8}$
Rewrite
$\frac{3}{4} \times \frac{8}{1}$
$= 6$
Model-It
Example 3
$1\frac{9}{10} \div \frac{4}{15}$
Rewrite
$\frac{19}{10} \times \frac{15}{4}$
(Show canceling & multiplying)
$= 7\frac{1}{8}$
Example 4
$3 \div \frac{2}{3}$
Rewrite
$\frac{3}{1} \times \frac{3}{2} = 4\frac{1}{2}$
Model-It
© Copyright 2016 Math Giraffe

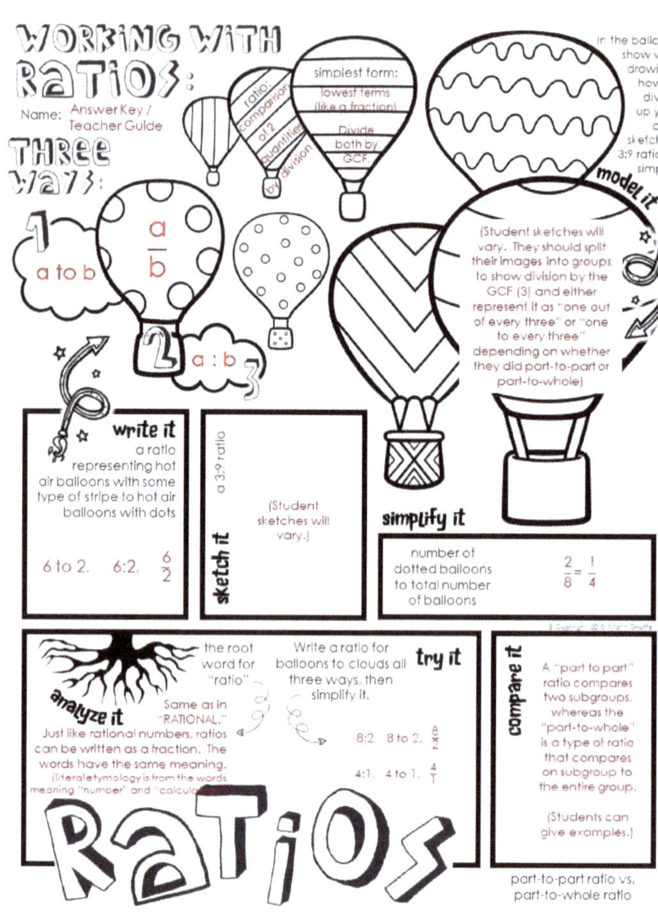
Working with Ratios:
Name: Answer Key / Teacher Guide
Three Ways:
1 a to b
$\frac{a}{b}$
2 a : b
3
ratio: comparison of 2 quantities by division
simplest form: lowest terms (like a fraction) Divide both by GCF
model it
(Student sketches will vary. They should split their images into groups to show division by the GCF (3) and either represent it as "one out of every three" or "one to every three" depending on whether they did part-to-part or part-to-whole)
write it
a ratio representing hot air balloons with some type of stripe to hot air balloons with dots
6 to 2, 6:2, $\frac{6}{2}$
sketch it
a 3:9 ratio
(Student sketches will vary.)
simplify it
number of dotted balloons to total number of balloons
$\frac{2}{8} = \frac{1}{4}$
analyze it
the root word for "ratio"
Same as in "RATIONAL."
Just like rational numbers, ratios can be written as a fraction. The words have the same meaning.
try it
Write a ratio for balloons to clouds all three ways, then simplify it.
8:2, 8 to 2, $\frac{8}{2}$
4:1, 4 to 1, $\frac{4}{1}$
compare it
A "part to part" ratio compares two subgroups, whereas the "part-to-whole" is a type of ratio that compares on subgroup to the entire group.
(Students can give examples.)
part-to-part ratio vs. part-to-whole ratio
Ratios

Fraction Division- Practice
Show all work. Write each answer in standard form.
Example 1
$\frac{2}{3} \div 8$
Rewrite
MULTIPLY ACROSS
$\frac{1}{12}$
Example 2
$\frac{3}{4} \div \frac{1}{8}$
Rewrite
$= 6$
Model-It
Example 3
$1\frac{9}{10} \div \frac{4}{15}$
Rewrite
$= \frac{57}{8}$
REWRITE
CONVERT
$7\frac{1}{8}$
Example 4
$3 \div \frac{2}{3}$
Rewrite
$\frac{3}{1} \cdot \frac{3}{2} = \frac{9}{2}$
REWRITE
$4\frac{1}{2}$
Model-It
© Copyright 2016 Math Giraffe

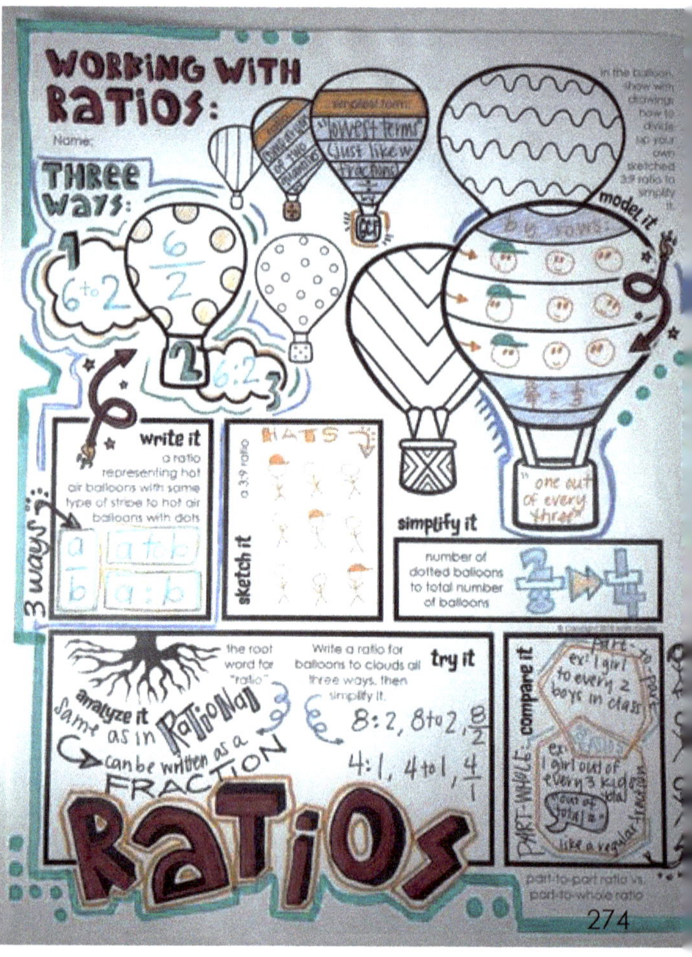
Working with Ratios:
Name:
Three Ways:
6 to 2
$\frac{6}{2}$
6:2
simplest form:
lowest terms (just like a fraction)
model it
one out of every three
write it
a ratio representing hot air balloons with some type of stripe to hot air balloons with dots
3 ways
sketch it
a 3:9 ratio
HATS
simplify it
number of dotted balloons to total number of balloons
analyze it
the root word for "ratio"
Same as in Rational
can be written as a FRACTION
try it
Write a ratio for balloons to clouds all three ways, then simplify it.
8:2, 8 to 2, $\frac{8}{2}$
4:1, 4 to 1, $\frac{4}{1}$
compare it
part-to-part ratio vs. part-to-whole ratio
Ratios

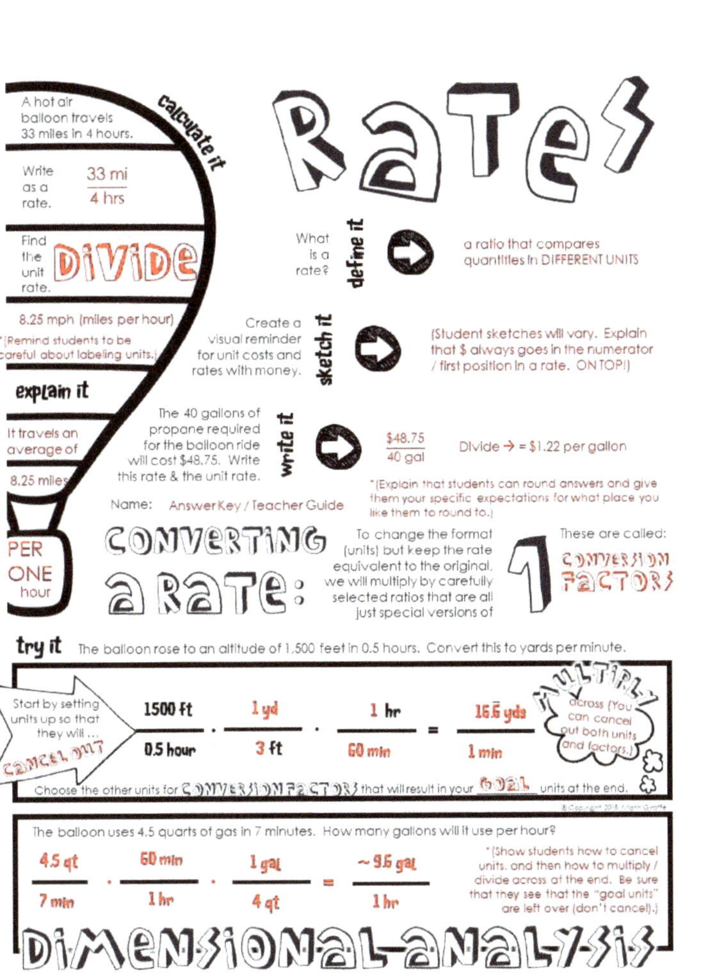
Rates
A hot air balloon travels 33 miles in 4 hours.
calculate it
Write as a rate.
33 mi / 4 hrs
Find the unit rate.
DIVIDE
8.25 mph (miles per hour)
*(Remind students to be careful about labeling units.)
explain it
It travels an average of 8.25 miles PER ONE hour
What is a rate?
define it
a ratio that compares quantities in DIFFERENT UNITS
Create a visual reminder for unit costs and rates with money.
sketch it
(Student sketches will vary. Explain that $ always goes in the numerator / first position in a rate. ON TOP!)
The 40 gallons of propane required for the balloon ride will cost $48.75. Write this rate & the unit rate.
write it
$48.75 / 40 gal
Divide → = $1.22 per gallon
*(Explain that students can round answers and give them your specific expectations for what place you like them to round to.)
Name: Answer Key / Teacher Guide
CONVERTING a RATE:
To change the format (units) but keep the rate equivalent to the original, we will multiply by carefully selected ratios that are all just special versions of
1
These are called: CONVERSION FACTORS
try it The balloon rose to an altitude of 1,500 feet in 0.5 hours. Convert this to yards per minute.
Start by setting units up so that they will ... CANCEL OUT
1500 ft / 0.5 hour · 1 yd / 3 ft · 1 hr / 60 min = 16.6 yds / 1 min
MULTIPLY across (You can cancel out both units and factors.)
Choose the other units for CONVERSION FACTORS that will result in your GOAL units at the end.
© Copyright 2016 Math Giraffe
The balloon uses 4.5 quarts of gas in 7 minutes. How many gallons will it use per hour?
4.5 qt / 7 min · 60 min / 1 hr · 1 gal / 4 qt = ~9.6 gal / 1 hr
*(Show students how to cancel units, and then how to multiply / divide across at the end. Be sure that they see that the "goal units" are left over (don't cancel).)
DIMENSIONAL ANALYSIS

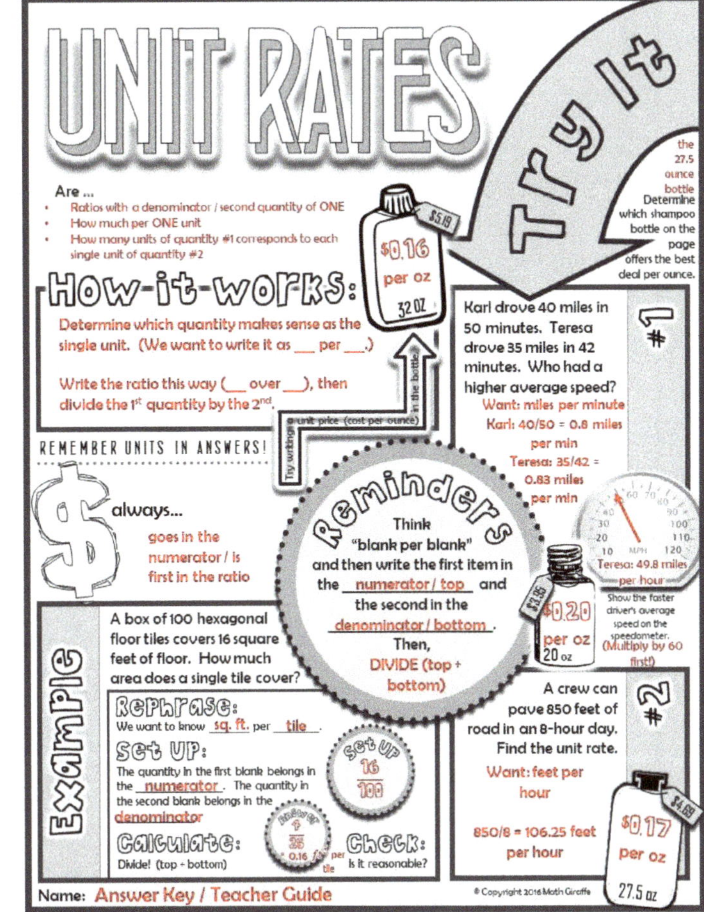
UNIT RATES
Try it
Are ...
Ratios with a denominator / second quantity of ONE
How much per ONE unit
How many units of quantity #1 corresponds to each single unit of quantity #2
$5.19
$0.16 per oz
32 OZ
How-it-works:
Determine which quantity makes sense as the single unit. (We want to write it as ___ per ___.)
Write the ratio this way (___ over ___), then divide the 1st quantity by the 2nd.
Try writing the unit price (cost per ounce) in the bottle.
the 27.5 ounce bottle
Determine which shampoo bottle on the page offers the best deal per ounce.
#1
Karl drove 40 miles in 50 minutes. Teresa drove 35 miles in 42 minutes. Who had a higher average speed?
Want: miles per minute
Karl: 40/50 = 0.8 miles per min
Teresa: 35/42 = 0.83 miles per min
Teresa: 49.8 miles per hour
Show the faster driver's average speed on the speedometer. (Multiply by 60 first!)
REMEMBER UNITS IN ANSWERS!
$ always... goes in the numerator / is first in the ratio
Reminders
Think "blank per blank" and then write the first item in the numerator / top and the second in the denominator / bottom. Then, DIVIDE (top ÷ bottom)
$3.95
$0.20 per oz
20 oz
Example
A box of 100 hexagonal floor tiles covers 16 square feet of floor. How much area does a single tile cover?
Rephrase: We want to know sq. ft. per tile.
Set Up: The quantity in the first blank belongs in the numerator. The quantity in the second blank belongs in the denominator.
Set up 16/100
Calculate: Divide! (top ÷ bottom)
Answer 0.16 ft² per tile
Check: Is it reasonable?
#2
A crew can pave 850 feet of road in an 8-hour day. Find the unit rate.
Want: feet per hour
850/8 = 106.25 feet per hour
$4.69
$0.17 per oz
27.5 oz
Name: Answer Key / Teacher Guide
© Copyright 2016 Math Giraffe

Watch UNITS closely!
Rates
33 mi / 4 hrs
DIVIDE
8.25 mph
RATIO that compares quantities in DIFFERENT UNITS
MONEY always on the TOP BUNKS
$1.22 (ROUND) $1.22 per gallon
CONVERSION FACTORS
GOAL
1500 ft / 0.5 hour · 1 yds / 3 ft · 1 hr / 60 min = 16.6 yds / 1 min
MULTIPLY ACROSS
CANCEL OUT
4.5 qts / 7 min · 1 gal / 4 qts · 60 min / 1 hr. ≈ 9.6 gal / 1 hr.
goal units
DIMENSIONAL ANALYSIS

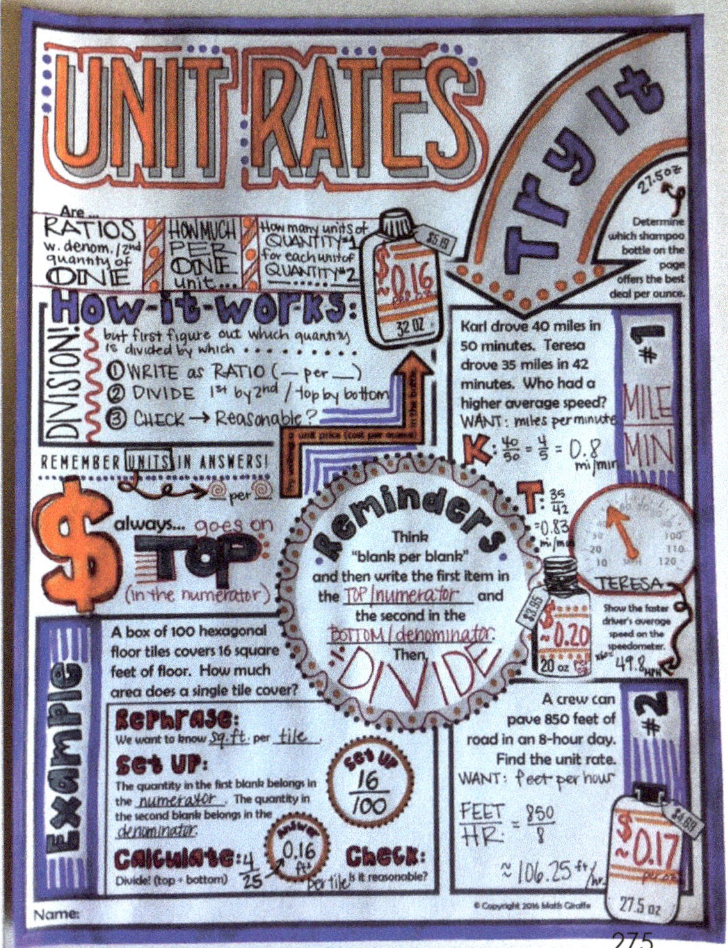
UNIT RATES
Try it
RATIOS w. denom. / 2nd quantity of ONE
HOW MUCH PER ONE unit
How many units of QUANTITY #1 for each unit of QUANTITY #2
How-it-works:
DIVISION!
but first figure out which quantity is divided by which
① WRITE as RATIO (— per —)
② DIVIDE 1st by 2nd / top by bottom
③ CHECK → Reasonable?
REMEMBER UNITS IN ANSWERS!
$ always... goes on TOP (in the numerator)
Reminders
the TOP / numerator
BOTTOM / denominator
DIVIDE
WANT: miles per minute
K: 40/50 = 4/5 = 0.8 mi/min
T: 35/42 = 0.83 mi/min
TERESA 49.8
MILE / MIN
Rephrase: We want to know sq. ft. per tile.
Set Up: the numerator, the denominator
16/100
0.16 ft² per tile
WANT: feet per hour
FEET/HR = 850/8
≈ 106.25 ft/hr
$0.17
© Copyright 2016 Math Giraffe

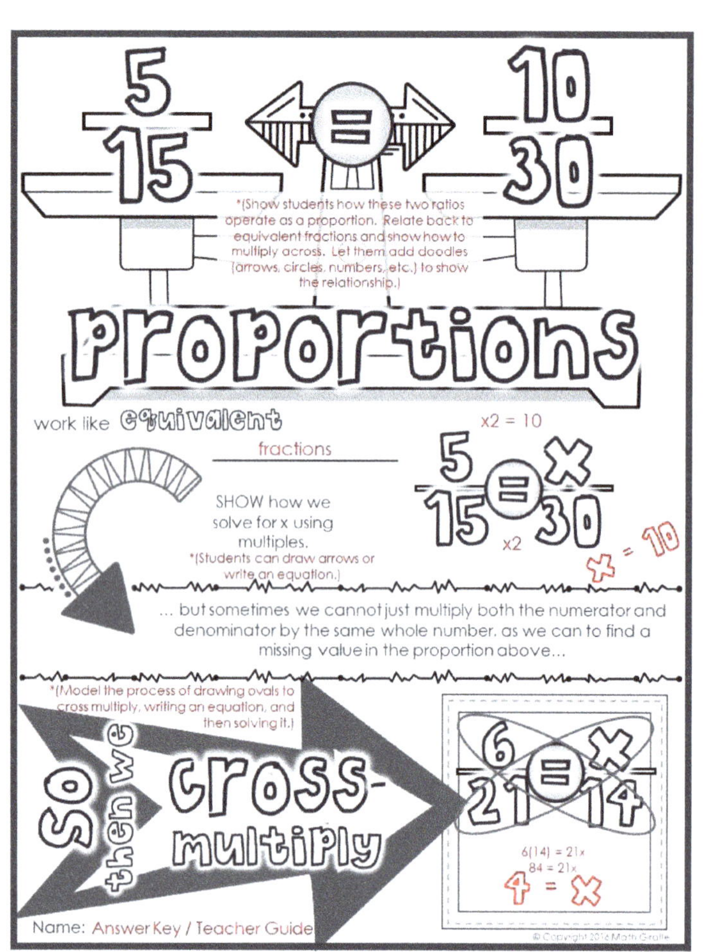
5/15 = 10/30
*(Show students how these two ratios operate as a proportion. Relate back to equivalent fractions and show how to multiply across. Let them add doodles (arrows, circles, numbers, etc.) to show the relationship.)
proportions
work like equivalent fractions
x2 = 10
SHOW how we solve for x using multiples.
*(Students can draw arrows or write an equation.)
x = 10
... but sometimes we cannot just multiply both the numerator and denominator by the same whole number, as we can to find a missing value in the proportion above...
*(Model the process of drawing ovals to cross multiply, writing an equation, and then solving it.)
So then we cross-multiply
6(14) = 21x
84 = 21x
4 = x
Name: Answer Key / Teacher Guide

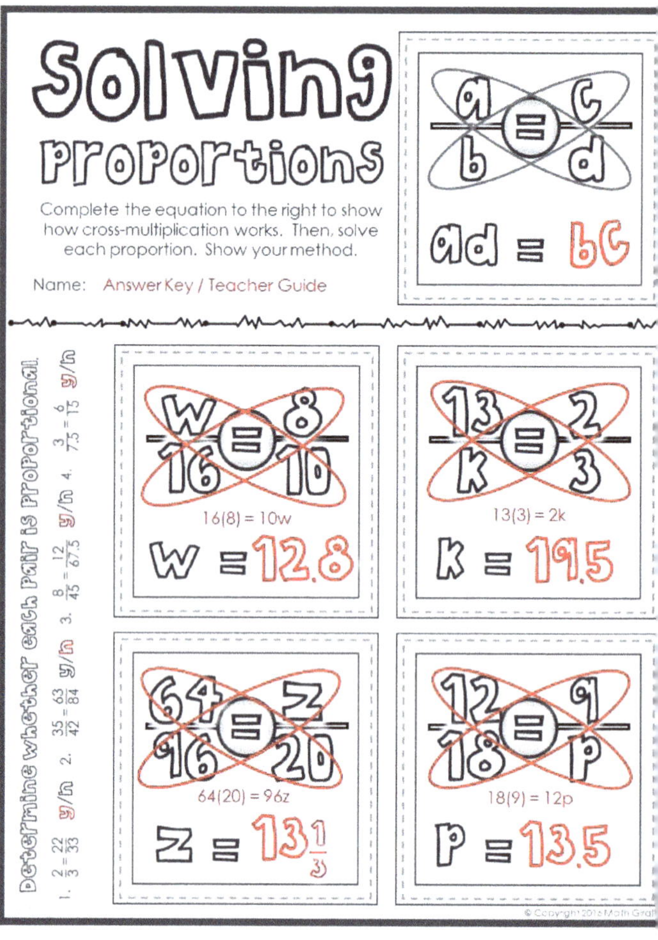
Solving proportions
Complete the equation to the right to show how cross-multiplication works. Then, solve each proportion. Show your method.
Name: Answer Key / Teacher Guide
ad = bc
Determine whether each pair is proportional.
1. 2/3 = 22/33 y/n 2. 35/42 = 63/84 y/n 3. 8/45 = 12/67.5 y/n 4. 3/7.5 = 6/15 y/n
16(8) = 10w
w = 12.8
13(3) = 2k
k = 19.5
64(20) = 96z
z = 13 1/3
18(9) = 12p
p = 13.5

proportions
work like equivalent FRACTIONS
5(2)=10
x=10
SHOW how we solve for x using multiples.
MULTIPLY numerator & denom. by same number
... but sometimes we cannot just multiply both the numerator and denominator by the same whole number, as we can to find a missing value in the proportion above...
1 WRITE EQUATION with diagonals
2 SOLVE
So then we cross-multiply
6•14 = 21•x
84 = 21x
4=x
Name:

Solving proportions
Complete the equation to the right to show how cross-multiplication works. Then, solve each proportion. Show your method.
Name:
ad = bc
16(8)=10w 128=10w
w = 12.8
3(13)=2k 39=2k
k = 19.5
CROSS MULTIPLY!
64(20)=96z
z = 13 1/3
9(18)=12p
p = 13.5

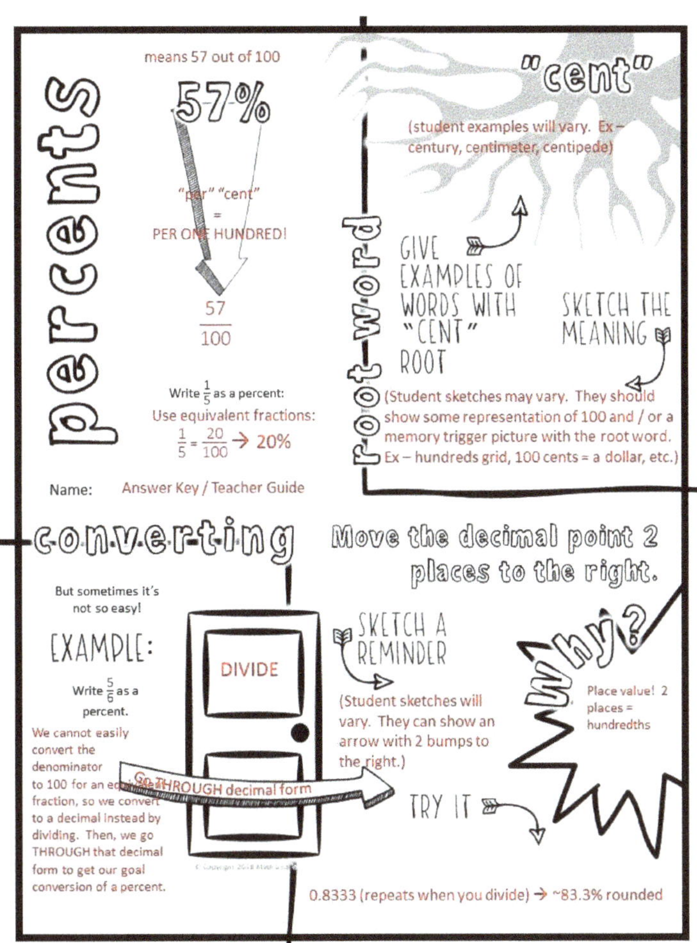

percents
means 57 out of 100
57%
"per" "cent"
=
PER ONE HUNDRED!
57/100
Write 1/5 as a percent:
Use equivalent fractions:
1/5 = 20/100 → 20%
Name: Answer Key / Teacher Guide
root-word
"cent"
(student examples will vary. Ex – century, centimeter, centipede)
GIVE EXAMPLES OF WORDS WITH "CENT" ROOT
SKETCH THE MEANING
(Student sketches may vary. They should show some representation of 100 and / or a memory trigger picture with the root word. Ex – hundreds grid, 100 cents = a dollar, etc.)
converting
Move the decimal point 2 places to the right.
But sometimes it's not so easy!
EXAMPLE:
Write 5/6 as a percent.
We cannot easily convert the denominator to 100 for an equivalent fraction, so we convert to a decimal instead by dividing. Then, we go THROUGH that decimal form to get our goal conversion of a percent.
DIVIDE
Go THROUGH decimal form
SKETCH A REMINDER
(Student sketches will vary. They can show an arrow with 2 bumps to the right.)
Why?
Place value! 2 places = hundredths
TRY IT
0.8333 (repeats when you divide) → ~83.3% rounded

divide
decimals
Say it out loud
fractions
Move decimal point 2 places to the left
Move decimal point 2 places to the right
*(Encourage students to add visuals and sketches that will help them remember. They can even add examples too.)
converting
BETWEEN FRACTIONS, DECIMALS, AND PERCENTS
percents
Name: Answer Key / Teacher Guide

"out of 100"
percents
57%
means
"PER"
100
57/100
Write 1/5 as a percent:
Use EQUIVALENT FRACTIONS
1/5 = 20/100 = 20%
root-word
centipede
"cent"
century
centigrade
cent
centimeter
GIVE EXAMPLES OF WORDS WITH "CENT" ROOT
SKETCH THE MEANING
100
Name:
converting
Move the decimal point 2 places to the right.
But sometimes it's not so easy!
EXAMPLE:
Write 5/6 as a percent.
DIVIDE
go THROUGH decimal form
FOR DECIMAL FORM
5/6 = ?/100
(not so easy to write an equivalent fraction)
SKETCH A REMINDER
Why?
PLACE VALUE
TRY IT
.83
6⟌5.0
48
20
18
Repeats
0.83 ~ 83.3%

0.6
decimals
Say it OUT LOUD
fractions
3/5
MOVE the decimal point
2 places to the Left
2 places to the Right
They Work together as different FORMS of the same NUMBER...
converting
BETWEEN FRACTIONS, DECIMALS, AND PERCENTS
percents
%
60%
Name:

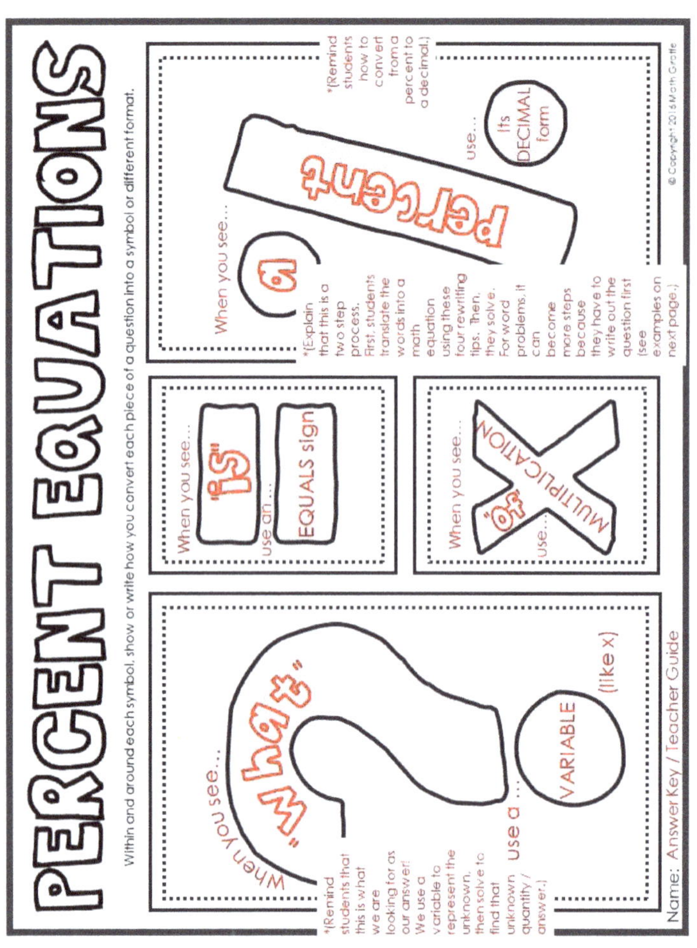
PERCENT EQUATIONS
Within and around each symbol, show or write how you convert each piece of a question into a symbol or different format.
When you see... "what"
use a ... VARIABLE (like x)
*(Remind students that this is what we are looking for as our answer! We use a variable to represent the unknown, then solve to find that unknown quantity / answer.)
When you see... "is"
use an ... EQUALS sign
When you see... "of"
use... MULTIPLICATION
When you see... a percent
use... its DECIMAL form
*(Remind students how to convert from a percent to a decimal.)
*(Explain that this is a two step process. First, students translate the words into a math equation using these four rewriting tips. Then, they solve. For word problems, it can become more steps because they have to write out the question first (see examples on next page.)
Name: Answer Key / Teacher Guide
© Copyright 2016 Math Giraffe

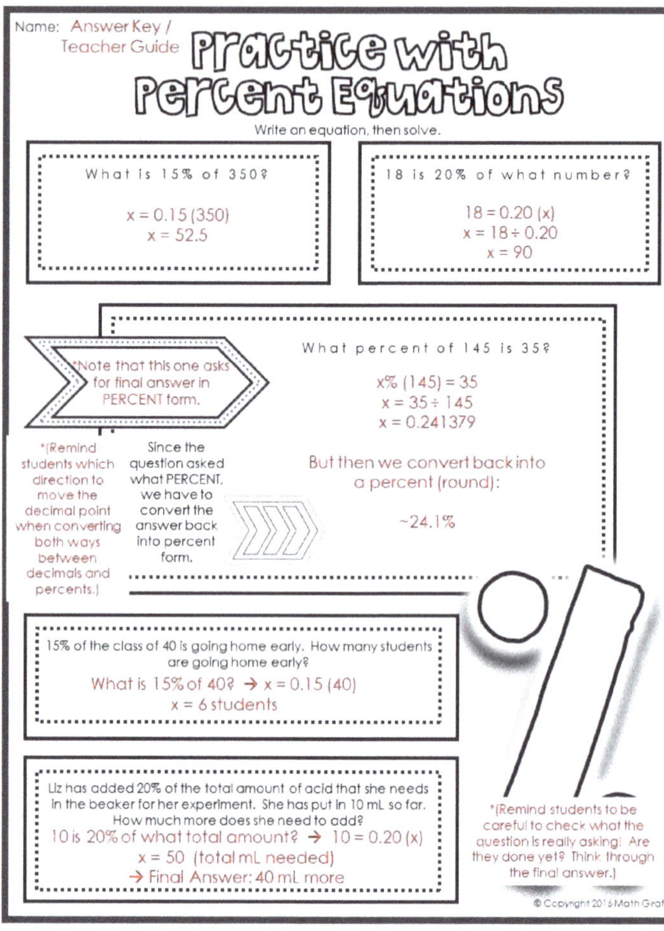
Name: Answer Key / Teacher Guide
Practice with Percent Equations
Write an equation, then solve.
What is 15% of 350?
x = 0.15 (350)
x = 52.5
18 is 20% of what number?
18 = 0.20 (x)
x = 18 ÷ 0.20
x = 90
*Note that this one asks for final answer in PERCENT form.
What percent of 145 is 35?
x% (145) = 35
x = 35 ÷ 145
x = 0.241379
*(Remind students which direction to move the decimal point when converting both ways between decimals and percents.)
Since the question asked what PERCENT, we have to convert the answer back into percent form.
But then we convert back into a percent (round):
~24.1%
15% of the class of 40 is going home early. How many students are going home early?
What is 15% of 40? → x = 0.15 (40)
x = 6 students
Liz has added 20% of the total amount of acid that she needs in the beaker for her experiment. She has put in 10 mL so far. How much more does she need to add?
10 is 20% of what total amount? → 10 = 0.20 (x)
x = 50 (total mL needed)
→ Final Answer: 40 mL more
*(Remind students to be careful to check what the question is really asking! Are they done yet? Think through the final answer.)
© Copyright 2016 Math Giraffe

PERCENT EQUATIONS
Within and around each symbol, show or write how you convert each piece of a question into a symbol or different format.
TRANSLATION KEY
When you see... "what"
That is the # we are looking for as an ANSWER
use a ... VARIABLE
x y z
When you see... "is"
use an ... EQUALS SIGN
When you see... "of"
use... MULTIPLICATION (times)
multiply them!
1 Rewrite words as equation.
2 Solve the equation!
When you see... a percent
use... its DECIMAL form
CONVERT
(move decimal point 2 places to the left.)
Name:
© Copyright 2016 Math Giraffe

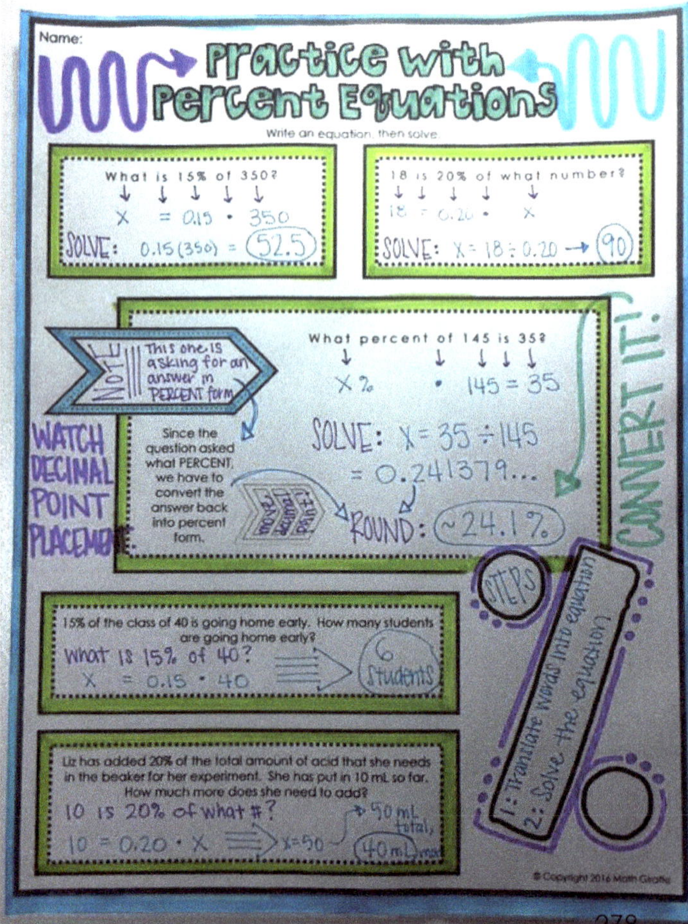
Name:
Practice with Percent Equations
Write an equation, then solve.
What is 15% of 350?
x = 0.15 · 350
SOLVE: 0.15(350) = 52.5
18 is 20% of what number?
18 = 0.20 · x
SOLVE: x = 18 ÷ 0.20 → 90
NOTE This one is asking for an answer in PERCENT form
What percent of 145 is 35?
x% · 145 = 35
SOLVE: x = 35 ÷ 145
= 0.241379...
ROUND: ~24.1%
WATCH DECIMAL POINT PLACEMENT
Since the question asked what PERCENT, we have to convert the answer back into percent form.
move decimal point
CONVERT IT!
STEPS
1: Translate words into equation
2: Solve the equation
15% of the class of 40 is going home early. How many students are going home early?
What is 15% of 40?
x = 0.15 · 40
6 students
Liz has added 20% of the total amount of acid that she needs in the beaker for her experiment. She has put in 10 mL so far. How much more does she need to add?
10 is 20% of what #?
10 = 0.20 · x
x = 50
50 mL total
40 mL more
© Copyright 2016 Math Giraffe

percent change
=
amount of change
original amount
*(Remind students that the fraction bar means DIVISION.)
Name:
Answer Key / Teacher Guide
A credit card bill that was $346 went up to $380 the next month after not being paid. By what percent did it increase?
34/346 = 0.098 → ~9.8%
*(Remind students how to convert from decimal to percent by moving the decimal point.)
define it
the difference (increase) between the cost to the seller and the price they actually charge the consumer. (so they can get a profit!)
Find the cost for a consumer if a coffee shop marks up a $0.40 cup of coffee by 260%.
$1.44
*(Warn students to be careful to check whether they are being asked for the change or for the final total. In this case, they must find the markup ($1.04) and add it to the cost to produce ($0.40))
markup
A grocery store purchases crates of oranges for their produce section for $3.00 per crate of a dozen. They sell the oranges individually for $0.49 each. Find the percent of markup.
Store buys for: $0.25 each & sells them for: $0.49
Markup: 96%
A store offers a 15% discount to its own employees.
discount
How much is the employee discount on a book with a price of $18.50?
0.15($18.50) = ~$2.78 discount
What would the employee pay for the book?
$18.50 – 2.78 = $15.72
What would an employee pay for a computer with a price of $489.95?
0.15($489.95) = $73.49 discount
Employee pays: $416.46 total
SALE

"point P"
P
a location in space (no size)
Notation: named with just a capital letter (P)
point
"line AB"
A
B
an infinite number of points extending in opposite directions (no end)
Notation: named with a lowercase letter or two points with a line above (n or $\overleftrightarrow{AB}$)
line
"plane $\mathcal{M}$"
$\mathcal{M}$
a flat surface containing an infinite number of lines (no thickness)
Notation: named with a script capital letter ($\mathcal{M}$)
plane
define it
"CO"
"TOGETHER"/"WITH"
collinear:
lying on the same line
coplanar:
lying on the same plane
identify it
*Answers will vary. Have students find and label a few points, lines, and planes in the figure and practice naming them both aloud and with written notation.)
sketch it
B and C are collinear.
$\overline{AC}$ is non coplanar with $\overline{MN}$.
(Student diagrams will vary. It can take a lot of practice to draw planes. It sometimes helps to allow a few students to try at the board before having everyone try final sketches.)
points, lines, & planes
the building blocks of geometry
Name: Answer Key / Teacher Guide
© Copyright 2018 Math Giraffe

percent change
=
amount of CHANGE
ORIGINAL AMOUNT
discount
markup

point
line
plane
define it
"CO"
TOGETHER/WITH
collinear:
coplanar:
identify it
sketch it
B and C are collinear.
points, lines, & planes
the building blocks of geometry
Name:

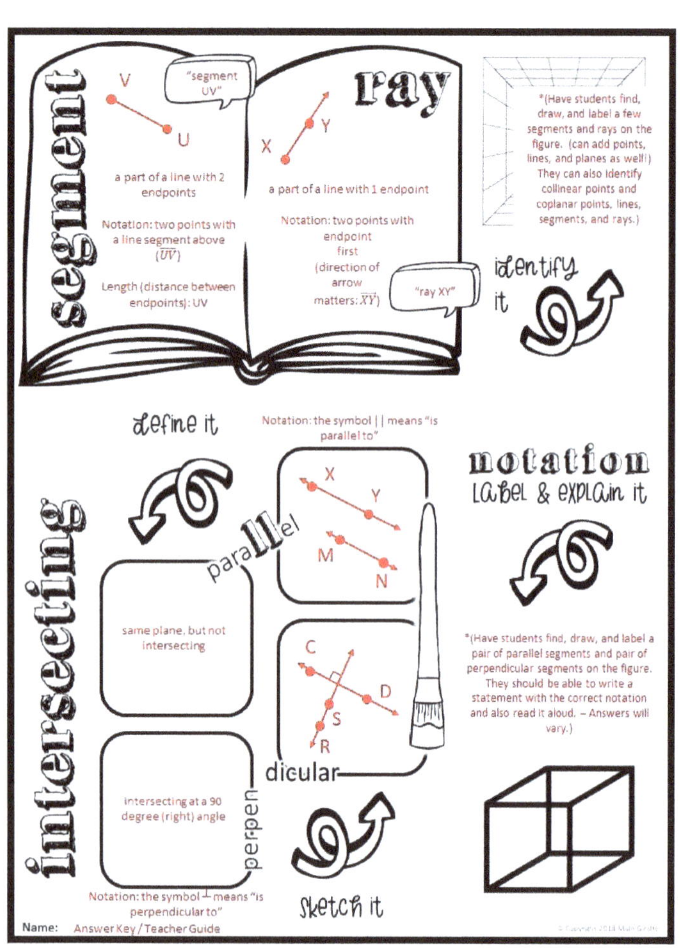
segment
"segment UV"
a part of a line with 2 endpoints
Notation: two points with a line segment above ($\overline{UV}$)
Length (distance between endpoints): UV
ray
a part of a line with 1 endpoint
Notation: two points with endpoint first (direction of arrow matters: $\overrightarrow{XY}$)
"ray XY"
*(Have students find, draw, and label a few segments and rays on the figure. (can add points, lines, and planes as well!) They can also identify collinear points and coplanar points, lines, segments, and rays.)
identify it
define it
Notation: the symbol || means "is parallel to"
notation
label & explain it
parallel
same plane, but not intersecting
perpendicular
intersecting at a 90 degree (right) angle
Notation: the symbol ⊥ means "is perpendicular to"
intersecting
*(Have students find, draw, and label a pair of parallel segments and pair of perpendicular segments on the figure. They should be able to write a statement with the correct notation and also read it aloud. – Answers will vary.)
sketch it
Name: Answer Key / Teacher Guide
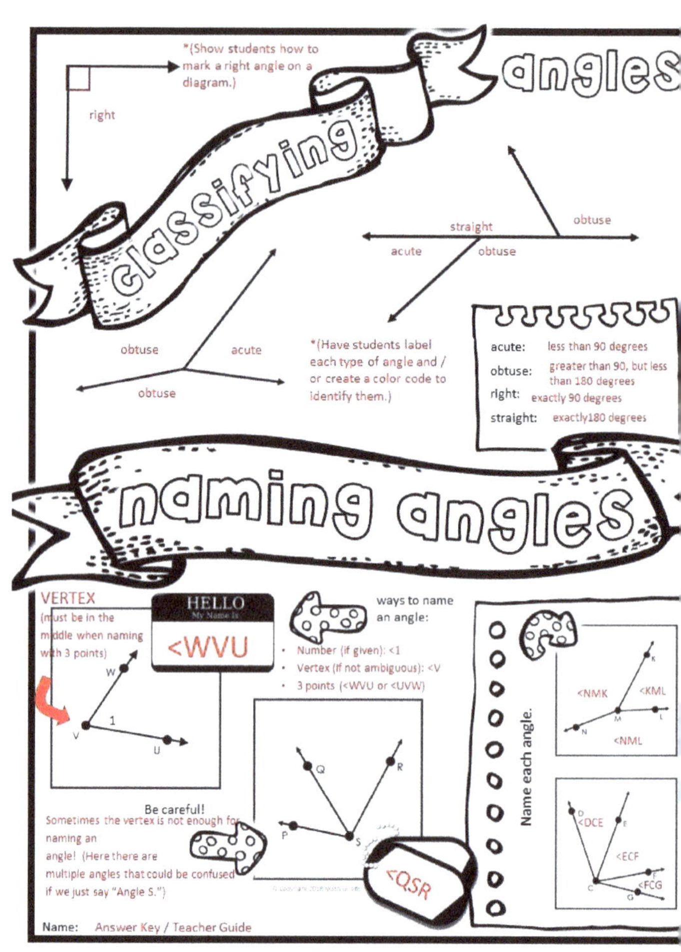
*(Show students how to mark a right angle on a diagram.)
right
classifying
angles
straight
obtuse
acute
obtuse
obtuse
acute
obtuse
*(Have students label each type of angle and / or create a color code to identify them.)
acute: less than 90 degrees
obtuse: greater than 90, but less than 180 degrees
right: exactly 90 degrees
straight: exactly180 degrees
naming angles
VERTEX
(must be in the middle when naming with 3 points)
HELLO
My Name Is
<WVU
ways to name an angle:
Number (if given): <1
Vertex (if not ambiguous): <V
3 points (<WVU or <UVW)
Be careful!
Sometimes the vertex is not enough for naming an angle! (Here there are multiple angles that could be confused if we just say "Angle S.")
<QSR
Name each angle.
<NMK
<KML
<NML
<DCE
<ECF
<FCG
Name: Answer Key / Teacher Guide

segment
ray
2 ENDPOINTS
ONE
NOTATION:
2 points w. line above $\overline{UV}$
(length/dist: UV)
NOTATION: 2 pts.
w ray above $\overrightarrow{XY}$
(direction of arrow matters!)
ray XY
identify it
MN and NP are coplanar.
A and C are collinear.
define it
notation
label & explain it
parallel
same plane but do NOT intersect
DJ || FG
HG ⊥ GF
INTERSECT at a RIGHT angle 90°
perpendicular
intersecting
sketch it
Name:
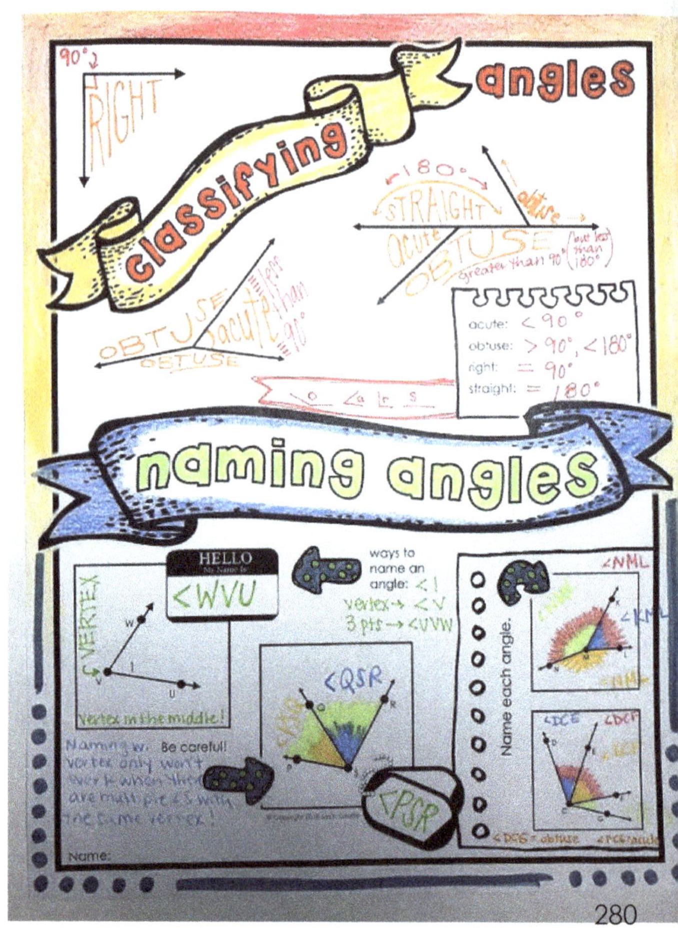
90°
RIGHT
classifying
angles
180°
STRAIGHT
acute
OBTUSE
greater than 90°
OBTUSE
acute
acute: < 90°
obtuse: > 90°, < 180°
right: = 90°
straight: = 180°
naming angles
VERTEX
HELLO
My Name Is
<WVU
ways to name an angle: <1
vertex → <V
3 pts → <UVW
Vertex in the middle!
Be careful!
<QSR
Name each angle.
<NML
<KML
Name:

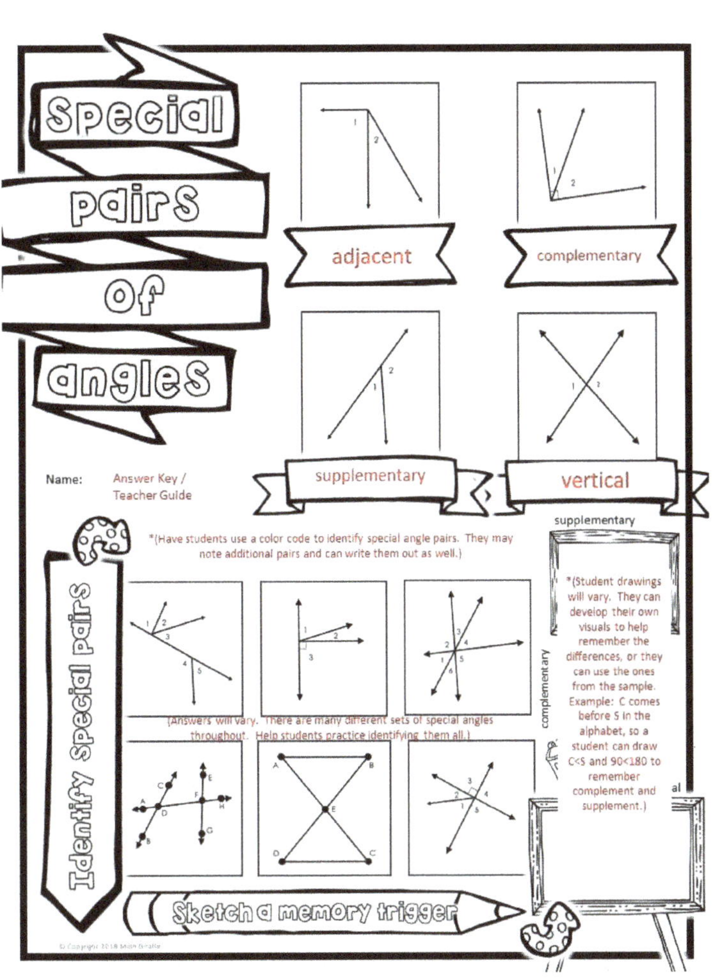
Special pairs of angles
adjacent
complementary
supplementary
vertical
Name: Answer Key / Teacher Guide
*(Have students use a color code to identify special angle pairs. They may note additional pairs and can write them out as well.)
Identify Special pairs
*(Answers will vary. There are many different sets of special angles throughout. Help students practice identifying them all.)
supplementary
*(Student drawings will vary. They can develop their own visuals to help remember the differences, or they can use the ones from the sample. Example: C comes before S in the alphabet, so a student can draw C<S and 90<180 to remember complement and supplement.)
complementary
Sketch a memory trigger

is congruent to
bisect
to divide into two equal parts
*(Student sketches will vary. Have them label and identify parts.)
sketch it:
an angle bisector
≅
congruence
"Congruent" means that two figures are the same SIZE and SHAPE.
ex: segments that are the same LENGTH or angles with the same # of DEGREES (measures)
*(Explain to students that vertical angles are always congruent. They can test this with a hands-on manipulative, then mark the angles on the diagram.)
example
marking congruent parts
angles
arc marks
hash marks
segments
writing congruence statements
segments
AB ≅ CD
<1 ≅ <2
angles
Name: Answer Key / Teacher Guide
*(Show students how to use double marks to differentiate between sets of congruent segments.)
PR and TS bisect one another. Mark congruent parts and write as many congruence statements as possible.
*(Students should have a few congruence statements about segments and a few about angles. Point out the reasoning, using properties like vertical angles, the definition of bisector, etc.)

Special pairs of angles
adjacent
COMPLEMENTARY
SUPPLEMENTARY
vertical
Name:
... Color Code
supplementary
180°
complementary
90°
Identify Special pairs
vertical
KISSING FISH
Sketch a memory trigger

is CoNgRueNT to...
bisect
to DIVIDE into 2 equal parts
sketch it:
an angle bisector
BX bisects <ABC.
CONGRUENT
≅
figures that are the same Size & SHAPE
congruence
marking congruent parts
angles
ARC marks
HASH marks
segments
writing congruency statements
segments
AB ≅ CD
<1 ≅ <2
angles
vertical <s are ≅
example
2 pairs of ≅ angles
Name:
PR and TS bisect one another. Mark congruent parts and write as many congruency statements as possible.
"midpoint"
<BCD ≅ <SQR
<SQR ≅ <PQT
<PQT ≅ <BCD
<TQR ≅ <SQR
SQ ≅ QT, PQ ≅ QR

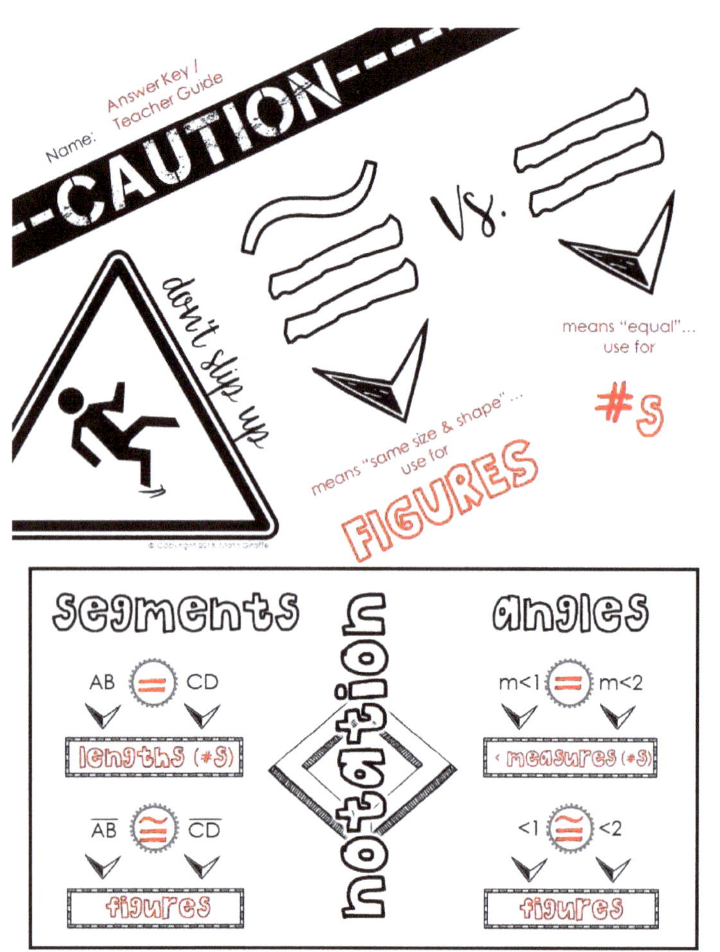
Name: Answer Key / Teacher Guide
CAUTION
don't slip up
vs.
means "equal"... use for
#s
means "same size & shape"... use for
FIGURES
segments
notation
angles
AB = CD
lengths (#s)
AB ≅ CD
figures
m<1 = m<2
measures (#s)
<1 ≅ <2
figures

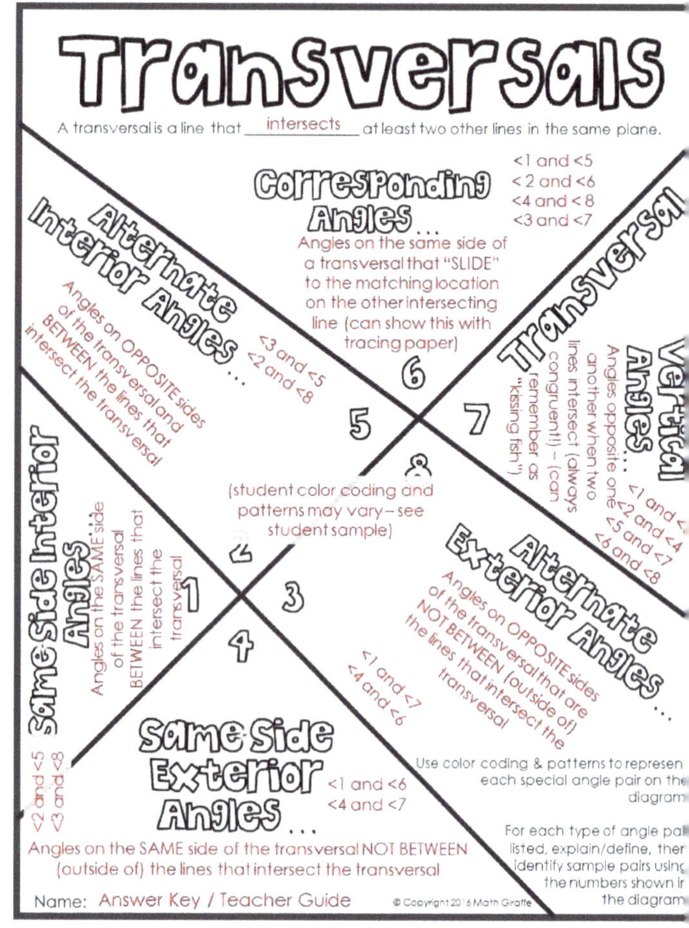
Transversals
A transversal is a line that intersects at least two other lines in the same plane.
Corresponding Angles...
<1 and <5
<2 and <6
<4 and <8
<3 and <7
Angles on the same side of a transversal that "SLIDE" to the matching location on the other intersecting line (can show this with tracing paper)
Alternate Interior Angles...
Angles on OPPOSITE sides of the transversal and BETWEEN the lines that intersect the transversal
<3 and <5
<2 and <8
Transversal
Vertical Angles...
Angles opposite one another when two lines intersect (always congruent!) - (can remember as "kissing fish")
<1 and <3
<2 and <4
<5 and <7
<6 and <8
(student color coding and patterns may vary - see student sample)
Alternate Exterior Angles...
Angles on OPPOSITE sides of the transversal that are NOT BETWEEN (outside of) the lines that intersect the transversal
<1 and <7
<4 and <6
Same Side Interior Angles
Angles on the SAME side of the transversal BETWEEN the lines that intersect the transversal
<2 and <5
<3 and <8
Same Side Exterior Angles...
<1 and <6
<4 and <7
Angles on the SAME side of the transversal NOT BETWEEN (outside of) the lines that intersect the transversal
Use color coding & patterns to represent each special angle pair on the diagram
For each type of angle pair listed, explain/define, then identify sample pairs using the numbers shown in the diagram
Name: Answer Key / Teacher Guide
1 2 3 4 5 6 7 8

Be careful not to confuse notation for congruence statements & equivalence statements
Name:
CAUTION
"is congruent to"
"is equal to"
vs.
equivalent
don't slip up
same size shape
use for figures
use for #s
segments
notation
angles
AB = CD
LENGTH (#)
AB ≅ CD
FIGURES
m<1 = m<2
<measures (#)
<1 ≅ <2
figures

Transversals
A transversal is a line that INTERSECTS at least two other lines in the same plane.
Corresponding Angles
<3 and <7
"SLIDE" ALONG transversal
Alternate Interior Angles
<3 and <5
OPPOSITE sides of transversal BETWEEN lines intersecting transversal
Transversal
Vertical Angles
<5 and <7
KISSING FISH
Alternate Exterior Angles
<4 and <6
OPPOSITE sides of the transversal OUTSIDE (not between) lines intersecting transversal
Same Side Interior Angles
<2 and <5
Same Side Exterior Angles
<1 and <6
Use color coding & patterns to represent each special angle pair on the diagram.
For each type of angle pair listed, explain/define, then identify sample pairs using the numbers shown in the diagram.
Name:
1 2 3 4 5 6 7 8

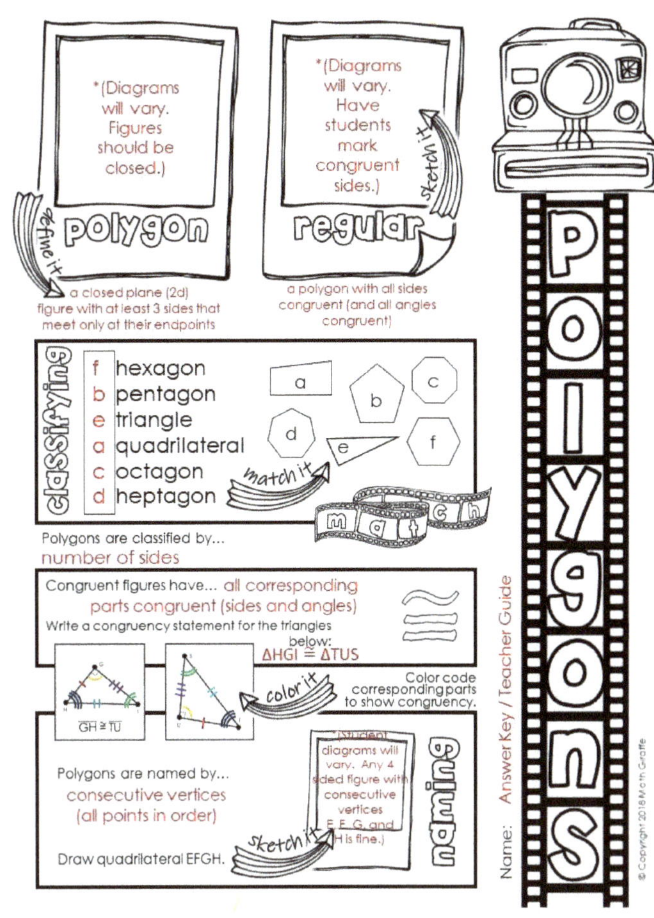
Polygons
polygon
define it
*(Diagrams will vary. Figures should be closed.)
a closed plane (2d) figure with at least 3 sides that meet only at their endpoints
regular
sketch it
*(Diagrams will vary. Have students mark congruent sides.)
a polygon with all sides congruent (and all angles congruent)
classifying
f hexagon
b pentagon
e triangle
a quadrilateral
c octagon
d heptagon
match it
match
Polygons are classified by...
number of sides
Congruent figures have... all corresponding parts congruent (sides and angles)
Write a congruency statement for the triangles below:
ΔHGI ≅ ΔTUS
color it
Color code corresponding parts to show congruency.
GH ≅ TU
naming
Polygons are named by...
consecutive vertices (all points in order)
Draw quadrilateral EFGH.
sketch it
(student diagrams will vary. Any 4 sided figure with consecutive vertices E, F, G, and H is fine.)
Name: Answer Key / Teacher Guide
© Copyright 2018 Math Giraffe

Parallel
lines intersected by a transversal
Name:
© Copyright 2018 Math Giraffe
= corresponding <s
= vertical <s
properties / special angle pairs used:
All of the obtuse <s end up ≅
All of the acute <s end up ≅
TRANSVERSAL
m<1 + m<5 = 180°
= alternate interior <s
= vertical <s
= corresponding <s
= alternate exterior <s
ONLY when lines are parallel!
color code to show congruent angles
What if m<1 = 110°? → then all obtuse (orange <s) = 110° and all acute (purple <s) = 70°
Write a BIG list of congruency statements. Then convert each into an equivalence statement using "measure of" notation for each angle.
<2 ≅ <5
<5 ≅ <7
<6 ≅ <8
<4 ≅ <7
<1 ≅ <3
<4 ≅ <5
is congruent to
m<2 = m<5
m<5 = m<7
m<6 = m<8
m<4 = m<7
m<1 = m<3
m<4 = m<5
is equal to

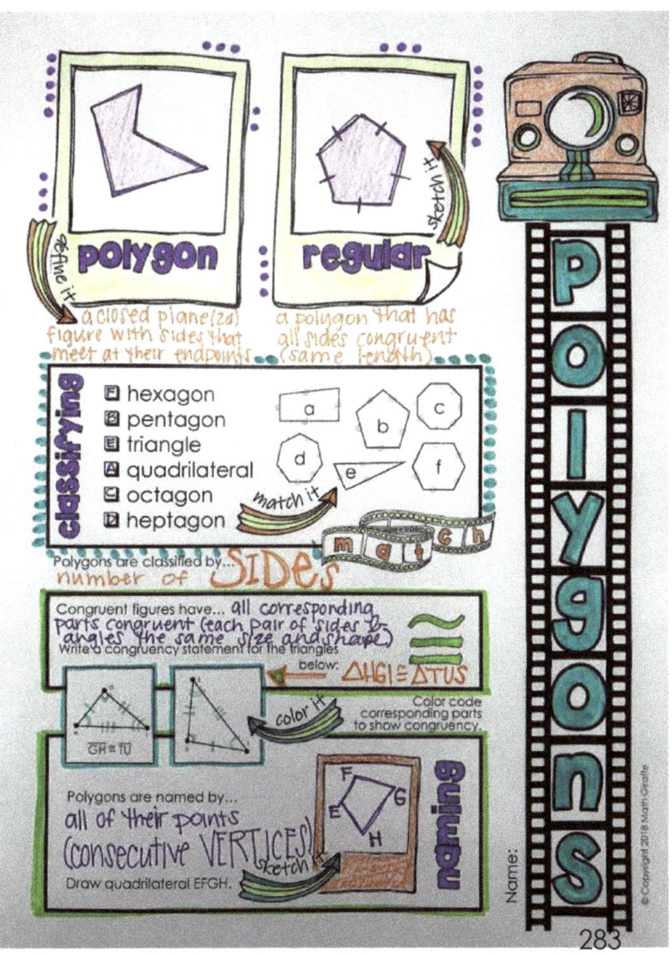
Polygons
polygon
define it
a closed plane (2d) figure with sides that meet at their endpoints
regular
sketch it
a polygon that has all sides congruent (same length)
classifying
f hexagon
b pentagon
e triangle
a quadrilateral
c octagon
d heptagon
match it
match
Polygons are classified by...
number of SIDES
Congruent figures have... all corresponding parts congruent (each pair of sides & angles the same size and shape)
Write a congruency statement for the triangles below:
ΔHGI ≅ ΔTUS
color it
Color code corresponding parts to show congruency.
GH ≅ TU
naming
Polygons are named by...
all of their points (consecutive VERTICES!)
Draw quadrilateral EFGH.
sketch it
F G E H
Name:
© Copyright 2018 Math Giraffe

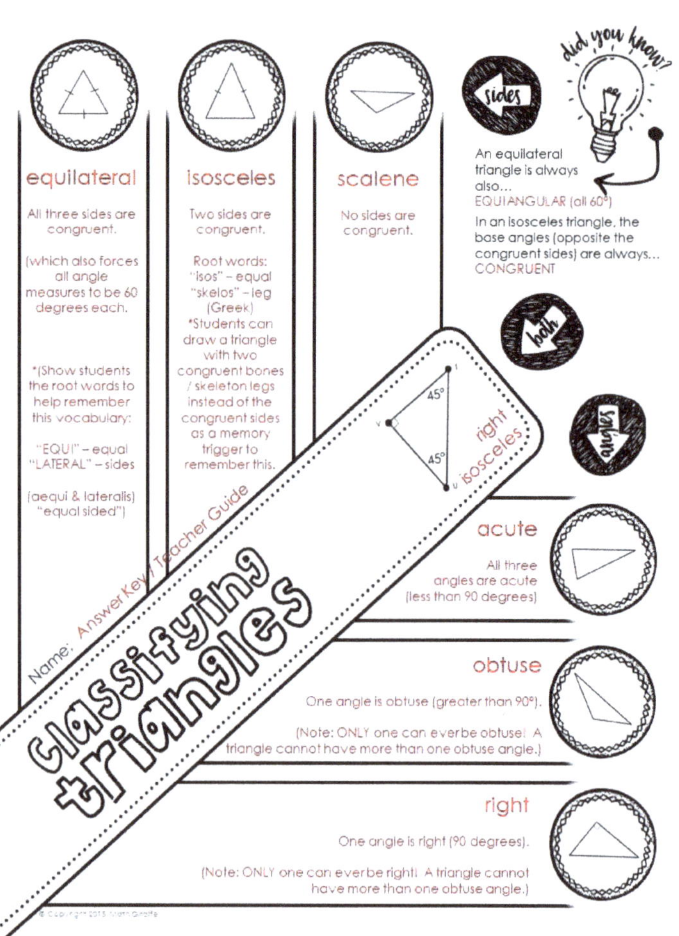
Classifying triangles
Name: Answer Key / Teacher Guide
equilateral
All three sides are congruent.
(which also forces all angle measures to be 60 degrees each.
*(Show students the root words to help remember this vocabulary:
"EQUI" – equal
"LATERAL" – sides
(aequi & lateralis) "equal sided")
isosceles
Two sides are congruent.
Root words: "isos" – equal "skelos" – leg (Greek)
*Students can draw a triangle with two congruent bones / skeleton legs instead of the congruent sides as a memory trigger to remember this.
scalene
No sides are congruent.
did you know?
sides
An equilateral triangle is always also... EQUIANGULAR (all 60°)
In an isosceles triangle, the base angles (opposite the congruent sides) are always... CONGRUENT
both
angles
right isosceles
acute
All three angles are acute (less than 90 degrees)
obtuse
One angle is obtuse (greater than 90°).
(Note: ONLY one can everbe obtuse! A triangle cannot have more than one obtuse angle.)
right
One angle is right (90 degrees).
(Note: ONLY one can everbe right! A triangle cannot have more than one obtuse angle.)

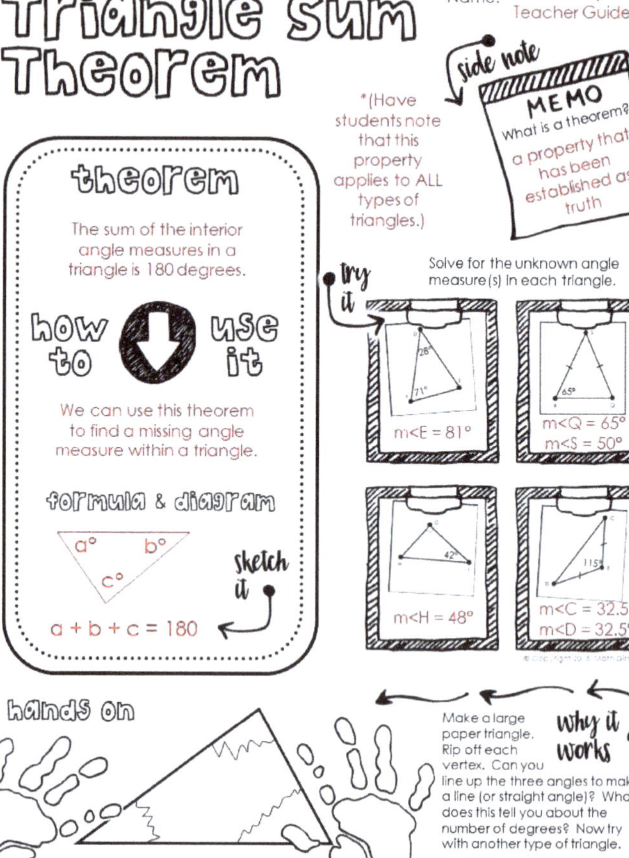
Triangle Sum Theorem
Name: Answer Key / Teacher Guide
side note
MEMO
What is a theorem?
a property that has been established as truth
*(Have students note that this property applies to ALL types of triangles.)
theorem
The sum of the interior angle measures in a triangle is 180 degrees.
how to use it
We can use this theorem to find a missing angle measure within a triangle.
formula & diagram
a° b° c°
a + b + c = 180
sketch it
try it
Solve for the unknown angle measure(s) in each triangle.
m<E = 81°
m<Q = 65°
m<S = 50°
m<H = 48°
m<C = 32.5°
m<D = 32.5°
hands on
why it works
Make a large paper triangle. Rip off each vertex. Can you line up the three angles to make a line (or straight angle)? What does this tell you about the number of degrees? Now try with another type of triangle.

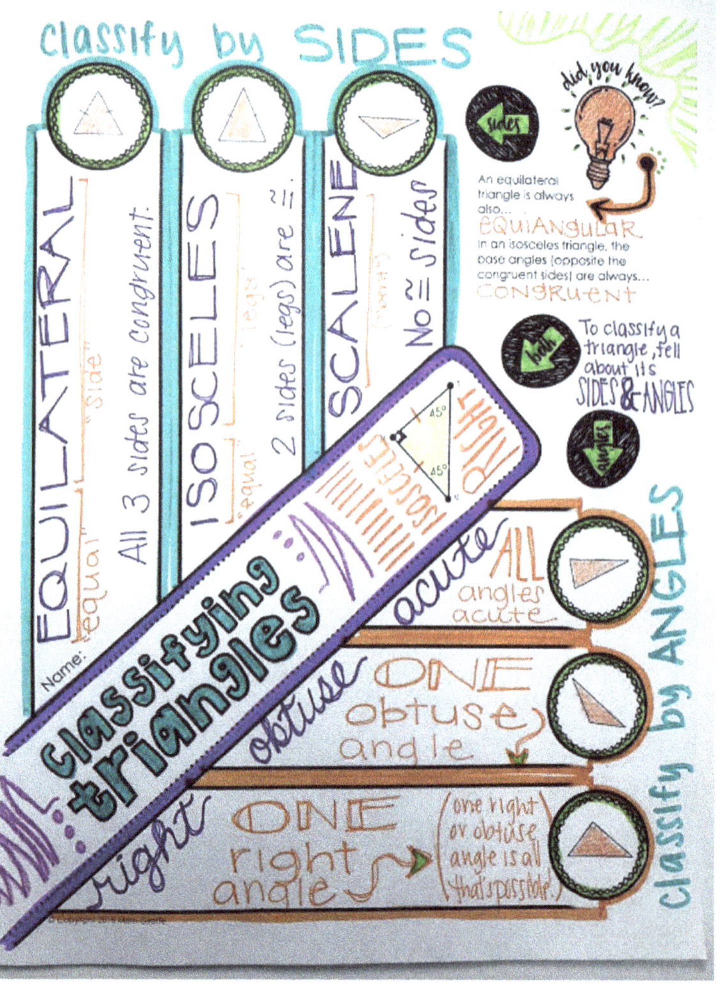
classify by SIDES
EQUILATERAL
"equal" "side"
All 3 sides are congruent.
ISOSCELES
"equal" "legs"
2 sides (legs) are ≅
SCALENE
No ≅ sides
did you know?
An equilateral triangle is always also... EQUIANGULAR
In an isosceles triangle, the base angles (opposite the congruent sides) are always... CONGRUENT
To classify a triangle, tell about its SIDES & ANGLES
Classifying triangles
Name:
right isosceles
acute
ALL angles acute
obtuse
ONE obtuse angle
right
ONE right angle
(one right or obtuse angle is all that's possible.)
classify by ANGLES

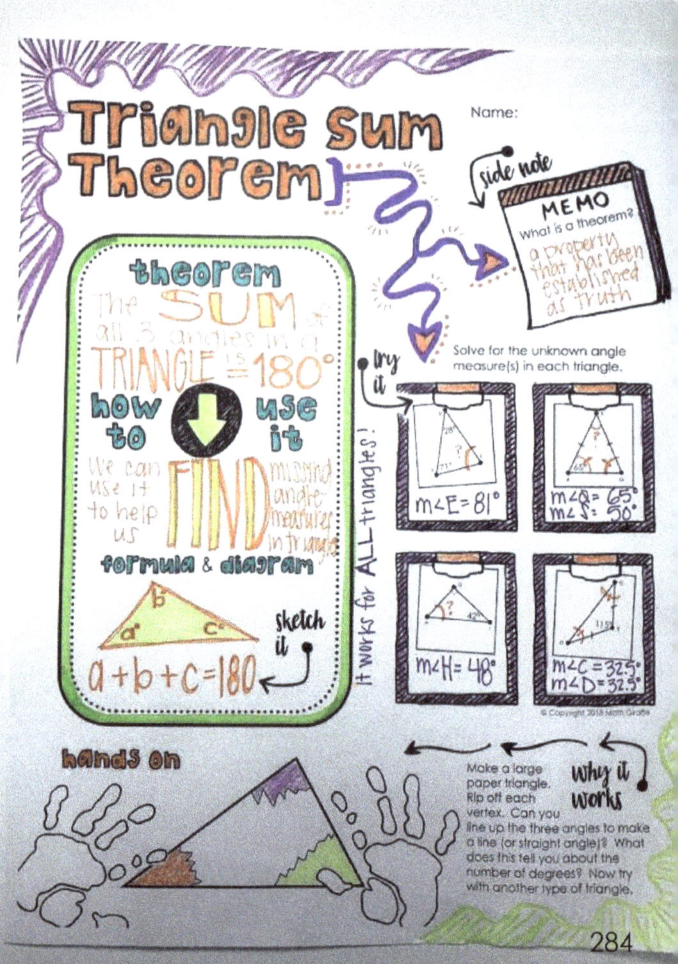
Triangle Sum Theorem
Name:
side note
MEMO
What is a theorem?
a property that has been established as truth
theorem
The SUM of all 3 angles in a TRIANGLE is 180°
how to use it
We can use it to help us FIND missing angle measures in triangles
formula & diagram
a + b + c = 180
sketch it
It works for ALL triangles!
try it
Solve for the unknown angle measure(s) in each triangle.
m<E = 81°
m<Q = 65°
m<S = 50°
m<H = 48°
m<C = 32.5°
m<D = 32.5°
hands on
why it works
Make a large paper triangle. Rip off each vertex. Can you line up the three angles to make a line (or straight angle)? What does this tell you about the number of degrees? Now try with another type of triangle.

impossible triangles
example
a triangle with all 3 angles obtuse
*(Encourage students to try drawing this!!)
define it
Sometimes a triangle cannot possibly be drawn with a certain set of criteria, because one of the inherent properties of triangles is violated.
Name: Answer Key / Teacher Guide
Complete the chart by sketching triangles. Represent impossible triangles instead in any boxes that cannot exist.
sketch it
isosceles
equilateral
scalene
obtuse
acute
right
impossible
impossible
Can you think of any other impossible triangle classifications?
list it
*(Student answers will vary. Example: a right, obtuse triangle; a scalene equiangular triangle, etc.)

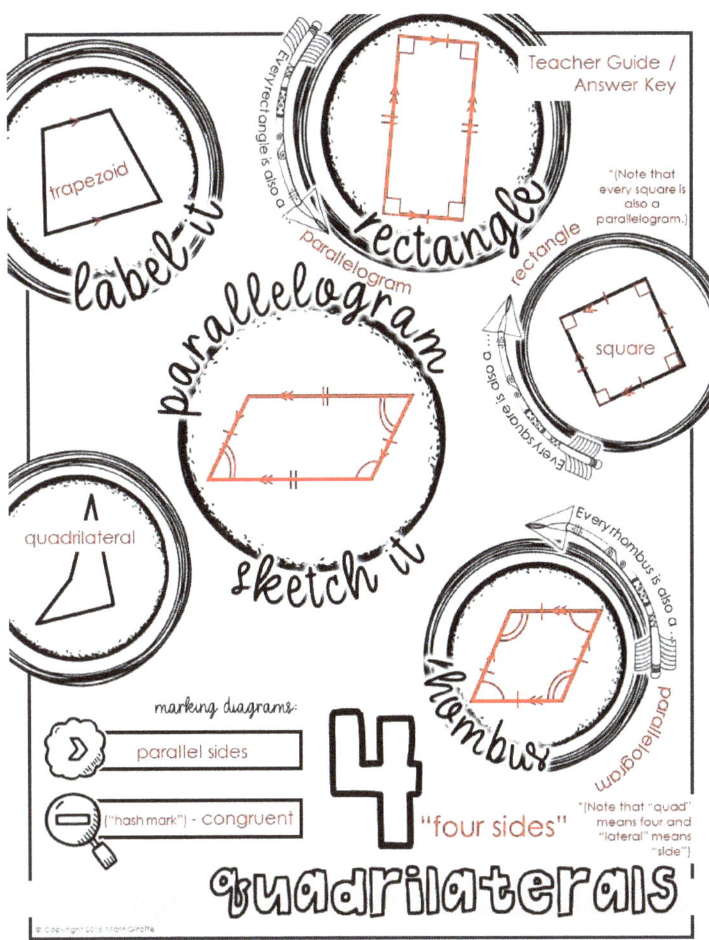
Teacher Guide / Answer Key
trapezoid
label it
Every rectangle is also a parallelogram
rectangle
*(Note that every square is also a parallelogram.)
Every square is also a rectangle
square
parallelogram
quadrilateral
sketch it
Every rhombus is also a parallelogram
rhombus
marking diagrams:
parallel sides
("hash mark") - congruent
4
"four sides"
*(Note that "quad" means four and "lateral" means "side")
quadrilaterals

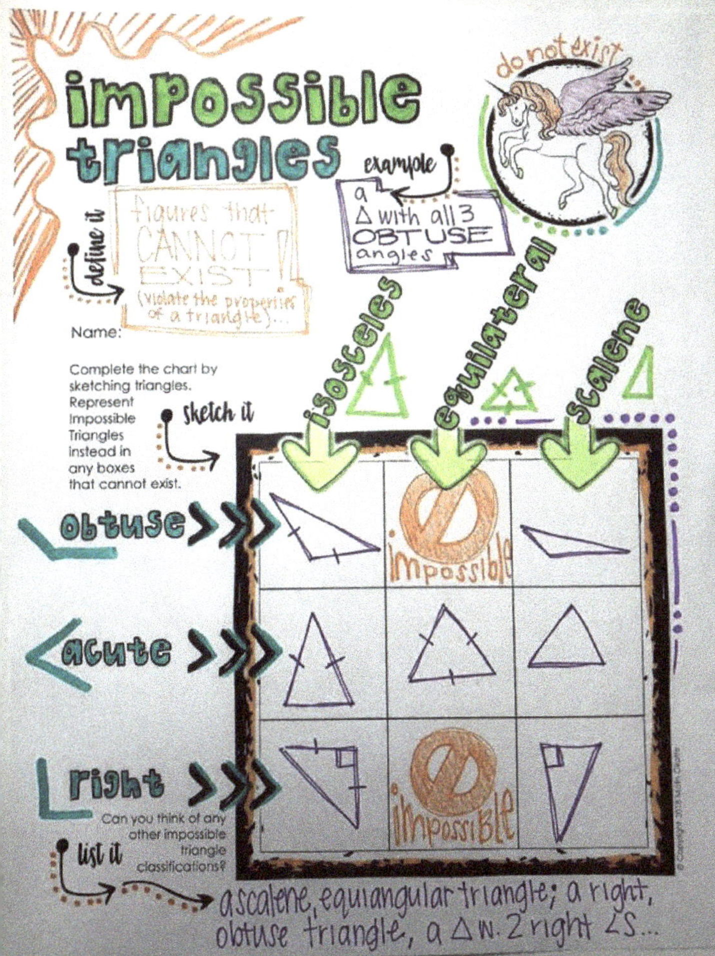
impossible triangles
do not exist
example
a Δ with all 3 OBTUSE angles
define it
figures that CANNOT EXIST! (violate the properties of a triangle)...
Name:
Complete the chart by sketching triangles. Represent impossible Triangles instead in any boxes that cannot exist.
sketch it
isosceles
equilateral
scalene
obtuse
acute
right
impossible
impossible
Can you think of any other impossible triangle classifications?
list it
a scalene, equiangular triangle; a right, obtuse triangle, a Δ w. 2 right ∠s...

TRAPEZOID
one pair ∥ sides
label it
Every rectangle is also a
all right ∠s
opposite sides are ≅ and ∥
rectangle
So, of course every square is also a PARALLELOGRAM
PARALLELOGRAM
RECTANGLE
all right angles
square
all sides ≅
2 pairs ∥ sides
Every square is also a
parallelogram
opposite ∠s are ≅
2 pairs ∥ sides
2 pairs ≅ sides
opposite sides are ∥ and ≅
QUADRILATERAL
sketch it
Every rhombus is also a
PARALLELOGRAM
2 pairs ∥ sides
all sides ≅
opposite ∠s ≅
rhombus
marking diagrams:
PARALLEL SIDES
"hash mark"
CONGRUENT parts
(or ∠s - "arc mark")
FOUR SIDES
"SIDE"
"FOUR"
quadrilaterals
4-sided figures

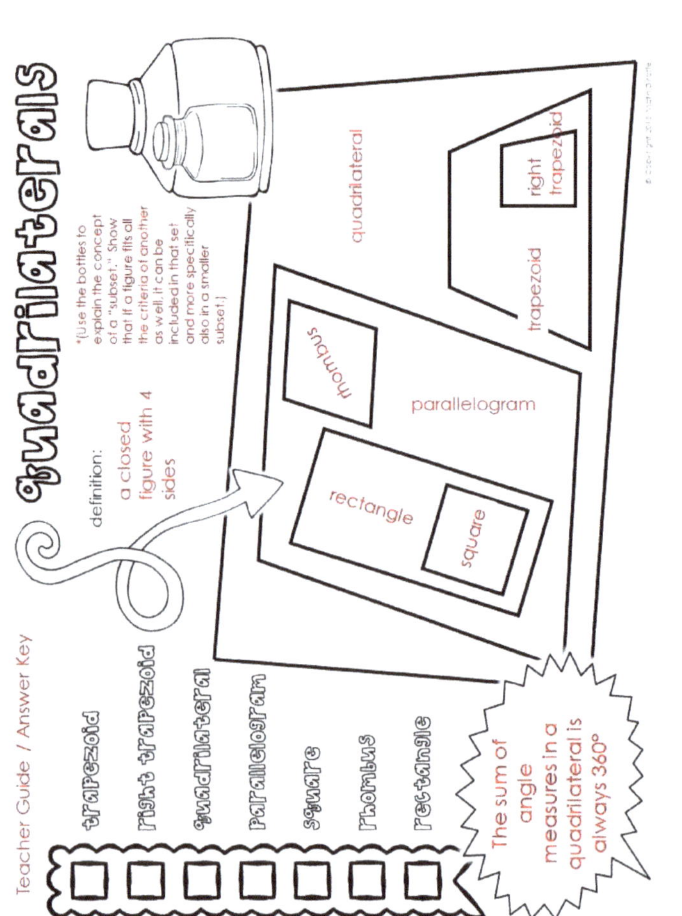
quadrilaterals
definition:
a closed figure with 4 sides
*(use the bottles to explain the concept of a "subset." Show that if a figure fits all the criteria of another as well, it can be included in that set and more specifically also in a smaller subset.)
Teacher Guide / Answer Key
trapezoid
right trapezoid
quadrilateral
parallelogram
square
rhombus
rectangle
quadrilateral
trapezoid
right trapezoid
rhombus
parallelogram
rectangle
square
The sum of angle measures in a quadrilateral is always 360°

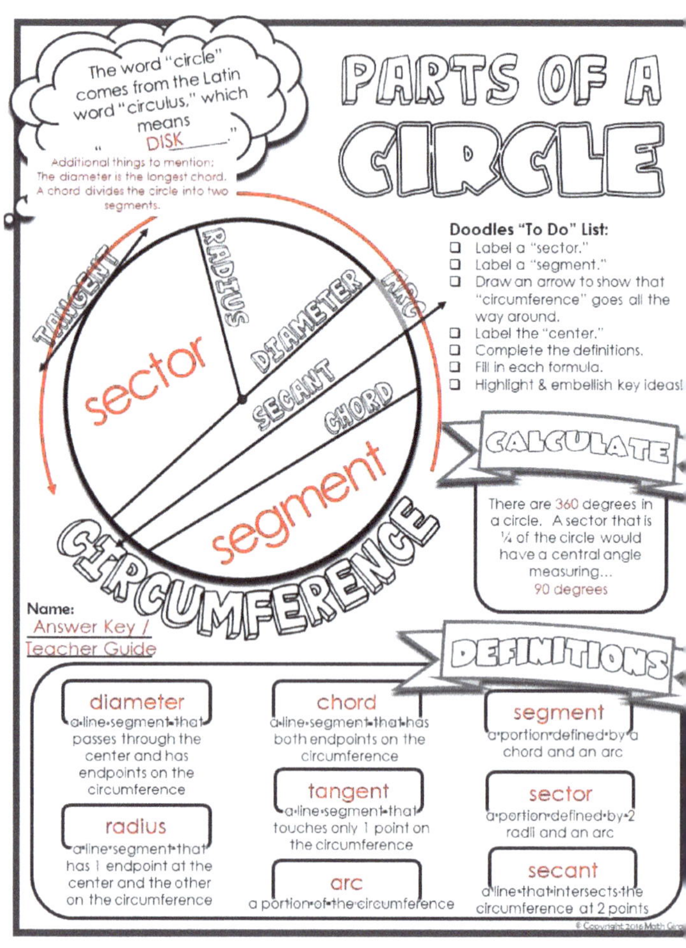
The word "circle" comes from the Latin word "circulus," which means "DISK."
Additional things to mention: The diameter is the longest chord. A chord divides the circle into two segments.
PARTS OF A CIRCLE
Doodles "To Do" List:
Label a "sector."
Label a "segment."
Draw an arrow to show that "circumference" goes all the way around.
Label the "center."
Complete the definitions.
Fill in each formula.
Highlight & embellish key ideas!
TANGENT
RADIUS
DIAMETER
ARC
sector
SECANT
CHORD
segment
CIRCUMFERENCE
CALCULATE
There are 360 degrees in a circle. A sector that is ¼ of the circle would have a central angle measuring... 90 degrees
Name: Answer Key / Teacher Guide
DEFINITIONS
diameter
a line segment that passes through the center and has endpoints on the circumference
chord
a line segment that has both endpoints on the circumference
segment
a portion defined by a chord and an arc
radius
a line segment that has 1 endpoint at the center and the other on the circumference
tangent
a line segment that touches only 1 point on the circumference
sector
a portion defined by 2 radii and an arc
arc
a portion of the circumference
secant
a line that intersects the circumference at 2 points

quadrilaterals
definition: a closed FIGURE with 4 SIDES
find & label:
isosceles trapezoid
right trapezoid
quadrilateral
parallelogram
square
rhombus
rectangle
The SUM of all 4 angle measures in a quadrilateral is always 360°

The word "circle" comes from the Latin word "circulus," which means "DISC."
PARTS OF A CIRCLE
DEFINITIONS
diameter
chord
segment
radius
tangent
sector
arc
secant

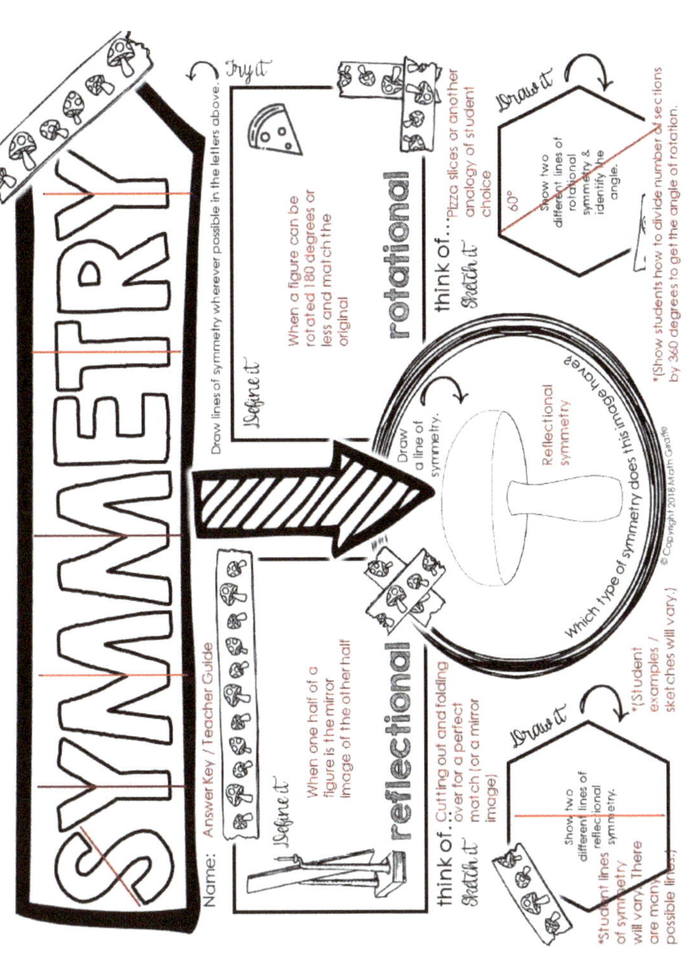
SYMMETRY
Name: Answer Key / Teacher Guide
Draw lines of symmetry wherever possible in the letters above.
Try it
Define it
When one half of a figure is the mirror image of the other half
reflectional
think of... Cutting out and folding over for a perfect match (or a mirror image)
Sketch it
Define it
When a figure can be rotated 180 degrees or less and match the original
rotational
think of... Pizza slices or another analogy of student choice
Sketch it
Draw a line of symmetry.
Reflectional symmetry
Which type of symmetry does this image have?
Draw it
Show two different lines of reflectional symmetry.
*Student lines of symmetry will vary. There are many possible lines.)
*(Student examples / sketches will vary.)
Draw it
Show two different lines of rotational symmetry & identify the angle.
60°
*(Show students how to divide number of sections by 360 degrees to get the angle of rotation.
© Copyright 2018 Math Giraffe
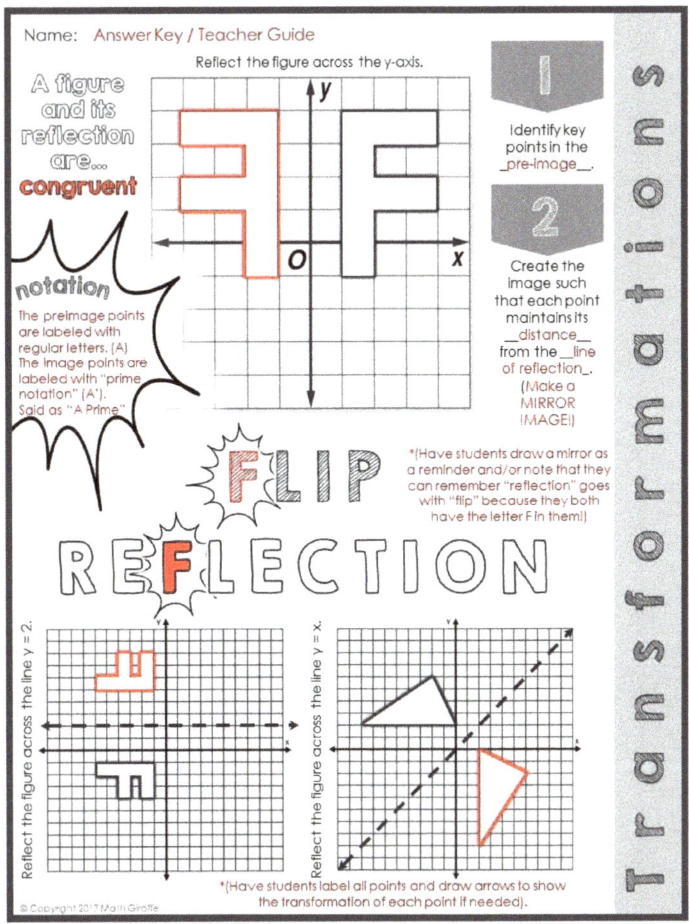
Name: Answer Key / Teacher Guide
Reflect the figure across the y-axis.
y
O
x
A figure and its reflection are... congruent
1
Identify key points in the _pre-image_.
2
Create the image such that each point maintains its _distance_ from the _line of reflection_. (Make a MIRROR IMAGE!)
notation
The preimage points are labeled with regular letters. (A)
The image points are labeled with "prime notation" (A').
Said as "A Prime"
FLIP
REFLECTION
*(Have students draw a mirror as a reminder and/or note that they can remember "reflection" goes with "flip" because they both have the letter F in them!)
Transformations
Reflect the figure across the line y = 2.
Reflect the figure across the line y = x.
*(Have students label all points and draw arrows to show the transformation of each point if needed).
© Copyright 2017 Math Giraffe

SYMMETRY
Name:
Draw lines of symmetry wherever possible in the letters above.
Try it
Define it
REFLECTIONAL SYMMETRY:
when ½ of a figure is the MIRROR image of the other ½.
reflectional
think of... FOLDING
Sketch it
Define it
ROTATIONAL SYMMETRY:
when a figure can be ROTATED 180° or less and match the original figure perfectly
rotational
think of... PIZZA SLICES
Sketch it
Draw a line of symmetry.
REFLECTIONAL
Which type of symmetry does this image have?
TIP: Divide # of sections by 360°
Draw it
Show two different lines of reflectional symmetry.
Draw it
Show two different lines of rotational symmetry & identify the angle.
60°

Reflect the figure across the y-axis.
A figure and its reflection are... CONGRUENT
1
Identify key points in the PREIMAGE.
2
Create the image such that each point maintains its DISTANCE from the LINE OF REFLECTION
notation
PRE-IMAGE: A "POINT A"
IMAGE: A' "A PRIME"
IMAGE (after reflect.)
PREIMAGE (before reflection)
FLIP
REFLECTION
Transformations
Reflect the figure across the line y = 2.
LINE OF REFLECTION
Reflect the figure across the line y = x.
y=x
LINE OF REFLECTION
© Copyright 2017 Math Giraffe

Rotate the figure 90° clockwise around the origin.
An image and its pre-image before a rotation are... congruent.
AROUND
ROTATION
1
Identify key points in the __preimage__.
2
Create the image such that each point is rotated the correct number of degrees in the correct direction.
Rotate the figure 180° around point N.
Rotate the figure 90° anticlockwise around the point (1, -2).
Transformations
Name: Answer Key / Teacher Guide
© Copyright 2017 Math Giraffe

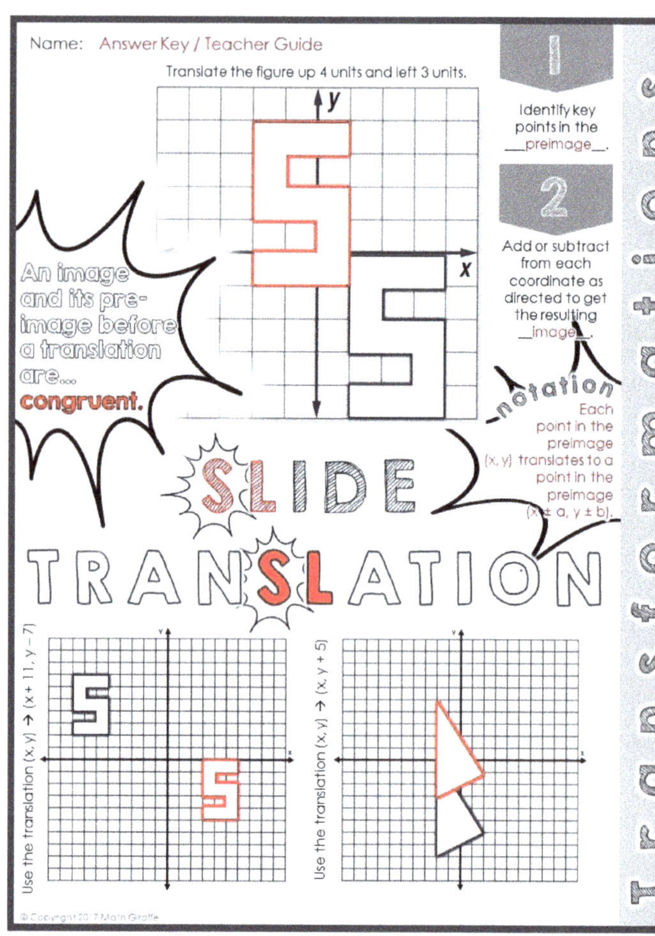

Name: Answer Key / Teacher Guide
Translate the figure up 4 units and left 3 units.
1
Identify key points in the __preimage__.
2
Add or subtract from each coordinate as directed to get the resulting __image__.
An image and its pre-image before a translation are... congruent.
notation
Each point in the preimage (x, y) translates to a point in the preimage (x ± a, y ± b).
SLIDE
TRANSLATION
Use the translation (x, y) → (x + 11, y - 7)
Use the translation (x, y) → (x, y + 5)
Transformations
© Copyright 2017 Math Giraffe

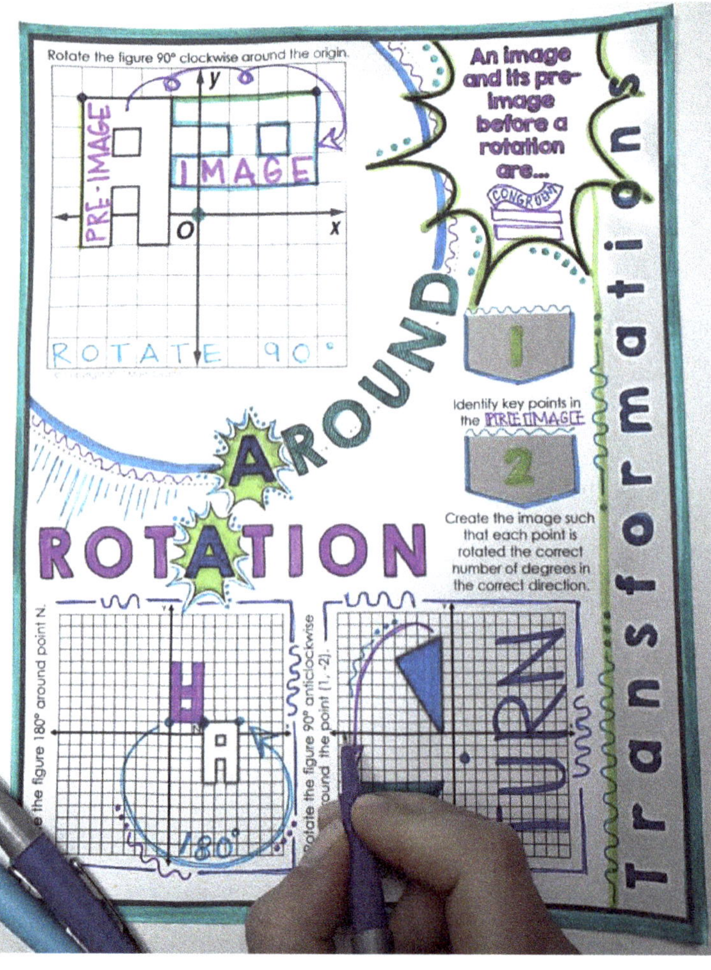

Rotate the figure 90° clockwise around the origin.
PRE-IMAGE
IMAGE
ROTATE 90°
An image and its pre-image before a rotation are... CONGRUENT
AROUND
ROTATION
1
Identify key points in the PRE-IMAGE
2
Create the image such that each point is rotated the correct number of degrees in the correct direction.
180°
TURN
Transformations

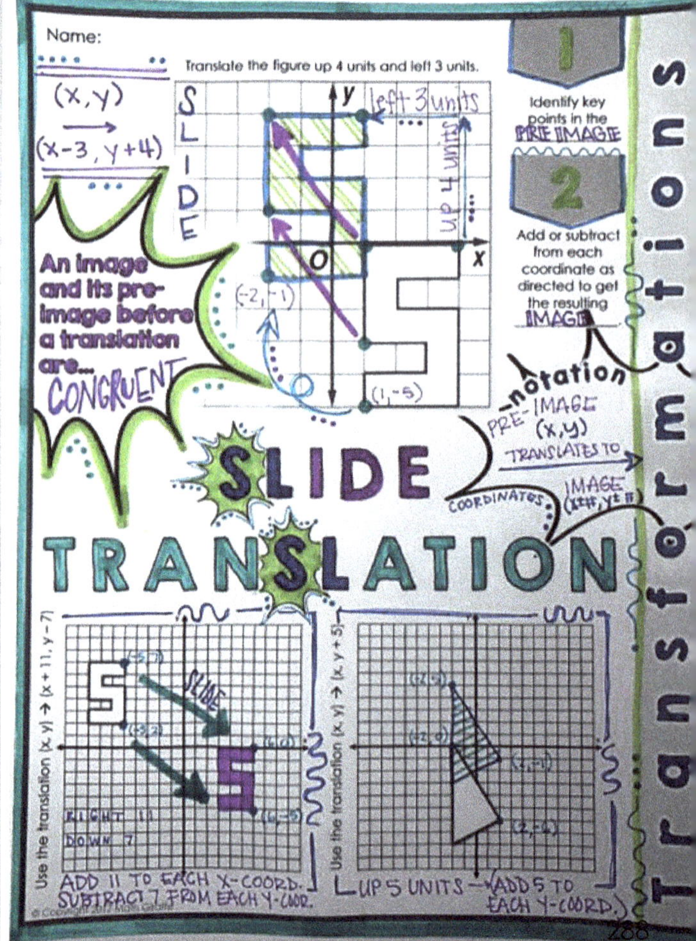

Name:
Translate the figure up 4 units and left 3 units.
(x, y)
(x-3, y+4)
SLIDE
left 3 units
up 4 units
(-2, -1)
(1, -5)
An image and its pre-image before a translation are... CONGRUENT
1
Identify key points in the PRE-IMAGE
2
Add or subtract from each coordinate as directed to get the resulting IMAGE
notation
PRE-IMAGE (x,y) TRANSLATES TO IMAGE COORDINATES
SLIDE
TRANSLATION
RIGHT 11
DOWN 7
ADD 11 TO EACH X-COORD. SUBTRACT 7 FROM EACH Y-COOR.
UP 5 UNITS — ADD 5 TO EACH Y-COORD.
Transformations

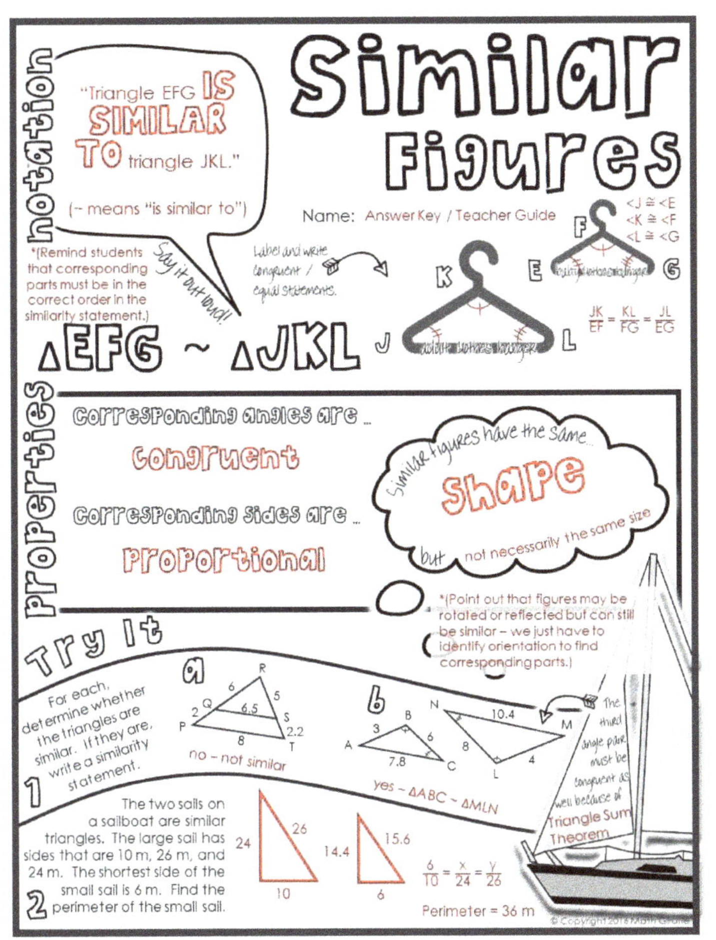

Similar Figures
notation
"Triangle EFG IS SIMILAR TO triangle JKL."
(~ means "is similar to")
*(Remind students that corresponding parts must be in the correct order in the similarity statement.)
Say it out loud!
Name: Answer Key / Teacher Guide
Label and write congruent / equal statements.
∠J ≅ ∠E
∠K ≅ ∠F
∠L ≅ ∠G
ΔEFG ~ ΔJKL
JK/EF = KL/FG = JL/EG
properties
Corresponding angles are ... congruent
Corresponding sides are ... proportional
Similar figures have the same... SHAPE but not necessarily the same size
*(Point out that figures may be rotated or reflected but can still be similar – we just have to identify orientation to find corresponding parts.)
Try It
1 For each, determine whether the triangles are similar. If they are, write a similarity statement.
a no – not similar
b yes ~ ΔABC ~ ΔMLN
The third angle pair must be congruent as well because of Triangle Sum Theorem
2 The two sails on a sailboat are similar triangles. The large sail has sides that are 10 m, 26 m, and 24 m. The shortest side of the small sail is 6 m. Find the perimeter of the small sail.
6/10 = x/24 = y/26
Perimeter = 36 m
© Copyright 2018 Math Giraffe

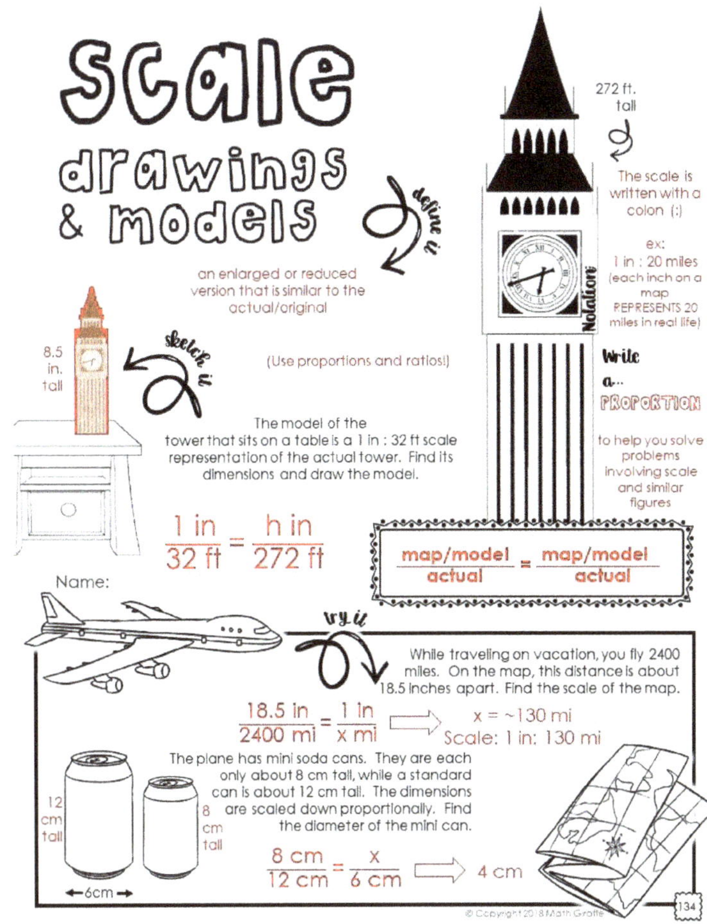

Scale drawings & models
define it
an enlarged or reduced version that is similar to the actual/original
sketch it
(Use proportions and ratios!)
8.5 in. tall
272 ft. tall
The scale is written with a colon (:)
ex: 1 in : 20 miles (each inch on a map REPRESENTS 20 miles in real life)
Notation
Write a... PROPORTION to help you solve problems involving scale and similar figures
The model of the tower that sits on a table is a 1 in : 32 ft scale representation of the actual tower. Find its dimensions and draw the model.
1 in / 32 ft = h in / 272 ft
map/model / actual = map/model / actual
Name:
try it
While traveling on vacation, you fly 2400 miles. On the map, this distance is about 18.5 inches apart. Find the scale of the map.
18.5 in / 2400 mi = 1 in / x mi
x = ~130 mi
Scale: 1 in: 130 mi
The plane has mini soda cans. They are each only about 8 cm tall, while a standard can is about 12 cm tall. The dimensions are scaled down proportionally. Find the diameter of the mini can.
12 cm tall
8 cm tall
6cm
8 cm / 12 cm = x / 6 cm
4 cm
134
© Copyright 2018 Math Giraffe

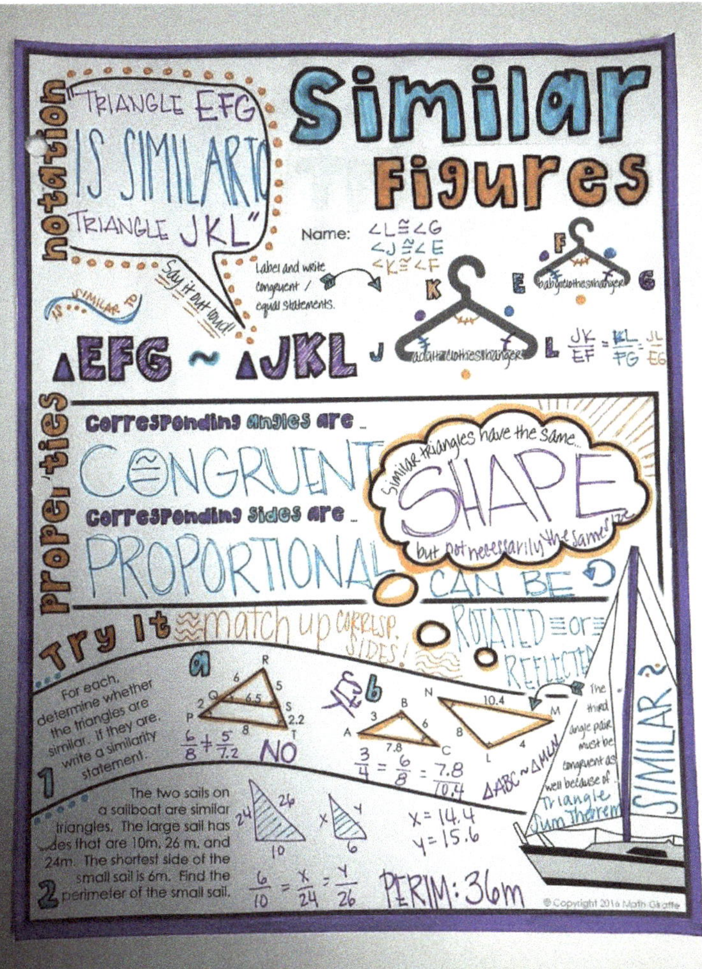

Similar Figures
notation
TRIANGLE EFG IS SIMILAR TO TRIANGLE JKL"
IS SIMILAR TO
Say it out loud!
Name:
∠L≅∠G
∠J≅∠E
∠K≅∠F
ΔEFG ~ ΔJKL
properties
Corresponding angles are ... CONGRUENT
Corresponding sides are ... PROPORTIONAL
Similar triangles have the same... SHAPE but not necessarily the same size
CAN BE ROTATED or REFLECTED
Try It match up corresp. sides!
NO
3/4 = 6/8 = 7.8/10.4
ΔABC ~ ΔMLN
SIMILAR
x = 14.4
y = 15.6
6/10 = x/24 = y/26
PERIM: 36m
© Copyright 2018 Math Giraffe

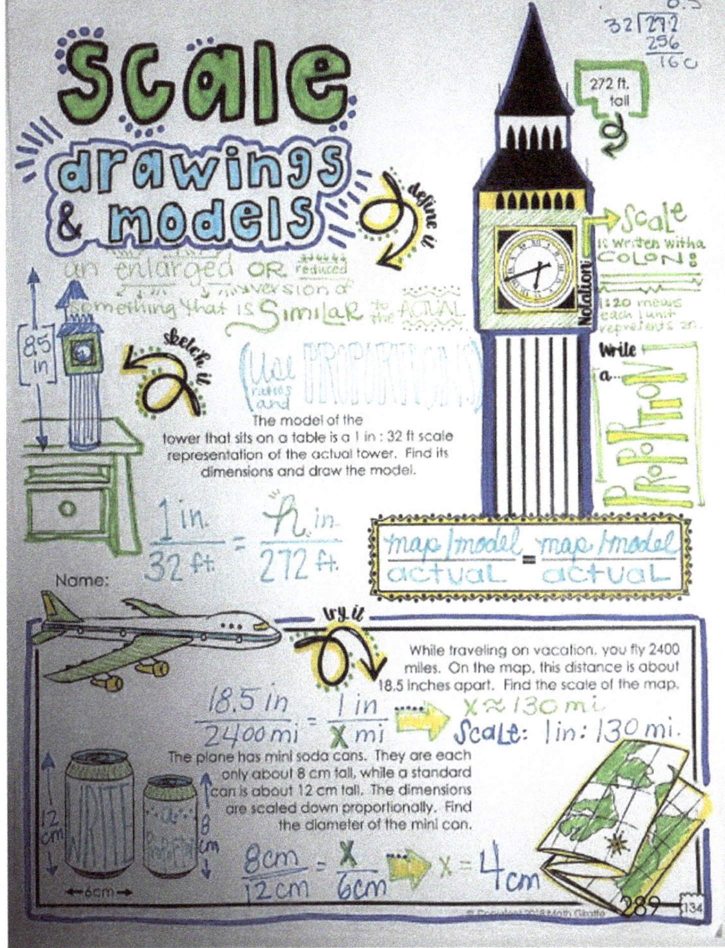

Scale drawings & models
an enlarged OR reduced version of something that is similar to the actual
Use ratios and PROPORTIONS
8.5 in
272 ft. tall
Scale is written with a COLON:
Notation
Write a... PROPORTION
1 in. / 32 ft. = h in. / 272 ft.
map/model / actual = map/model / actual
18.5 in / 2400 mi = 1 in / x mi
x ≈ 130 mi
Scale: 1in : 130 mi.
8cm / 12cm = x / 6cm
x = 4cm
289
134

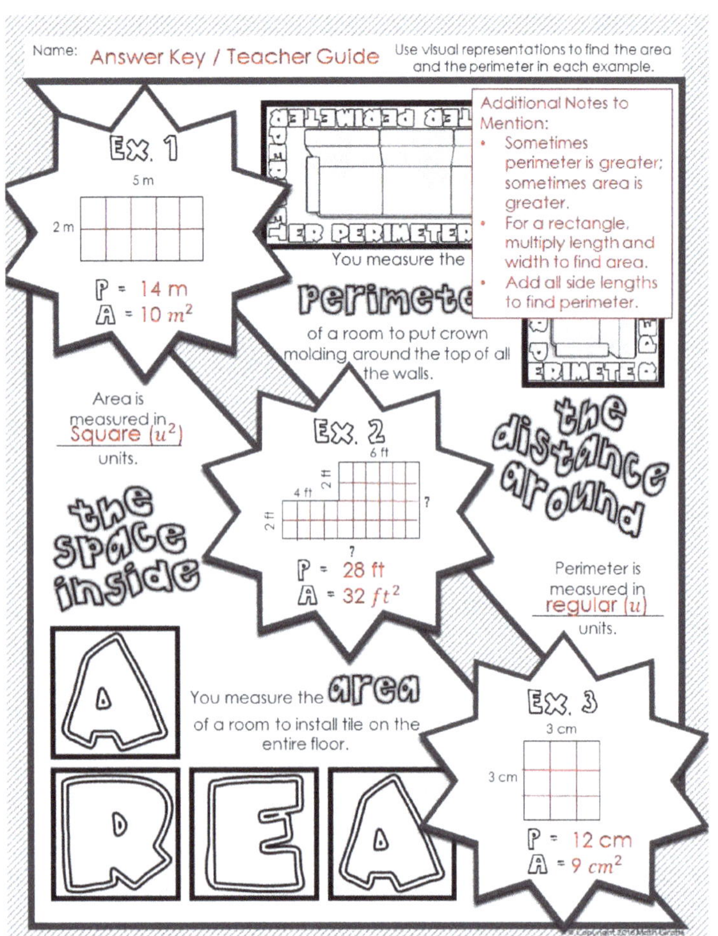

Name: Answer Key / Teacher Guide
Use visual representations to find the area and the perimeter in each example.
Additional Notes to Mention:
Sometimes perimeter is greater; sometimes area is greater.
For a rectangle, multiply length and width to find area.
Add all side lengths to find perimeter.
Ex. 1
5 m
2 m
P = 14 m
A = 10 m^2
You measure the perimeter of a room to put crown molding around the top of all the walls.
Area is measured in Square (u^2) units.
Ex. 2
6 ft
4 ft
2 ft
2 ft
?
?
P = 28 ft
A = 32 ft^2
the distance around
the space inside
Perimeter is measured in regular (u) units.
You measure the area of a room to install tile on the entire floor.
A
AREA
Ex. 3
3 cm
3 cm
P = 12 cm
A = 9 cm^2

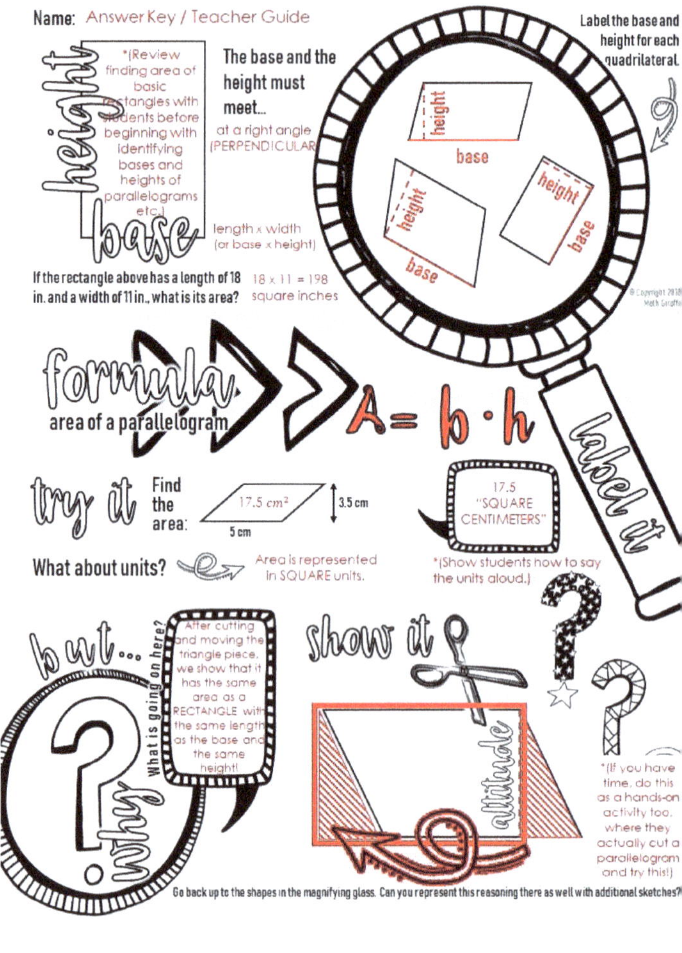

Name: Answer Key / Teacher Guide
Label the base and height for each quadrilateral.
height
base
*(Review finding area of basic rectangles with students before beginning with identifying bases and heights of parallelograms etc.)
The base and the height must meet...
at a right angle (PERPENDICULAR)
length x width (or base x height)
If the rectangle above has a length of 18 in. and a width of 11 in., what is its area?
18 x 11 = 198 square inches
formula
area of a parallelogram
A = b · h
label it
try it
Find the area:
17.5 cm^2
3.5 cm
5 cm
17.5 "SQUARE CENTIMETERS"
What about units?
Area is represented in SQUARE units.
*(Show students how to say the units aloud.)
but... why
What is going on here?
After cutting and moving the triangle piece, we show that it has the same area as a RECTANGLE with the same length as the base and the same height!
show it
altitude
*(If you have time, do this as a hands-on activity too, where they actually cut a parallelogram and try this!)
Go back up to the shapes in the magnifying glass. Can you represent this reasoning there as well with additional sketches?

Name:
Ex. 1
5 m
2 m
P = 14m
A = 10m^2
perimeter
of a room to put crown molding around the top of all the walls.
Area is measured in square units.
Ex. 2
6 ft
P = 28ft
A = 32ft^2
the distance around
the space inside
Perimeter is measured in regular units.
*for a rectangle, multiply length x width
You measure the area of a room to install tile on the entire floor.
Ex. 3
3 cm
3 cm
P = 12 cm
A = 9cm^2

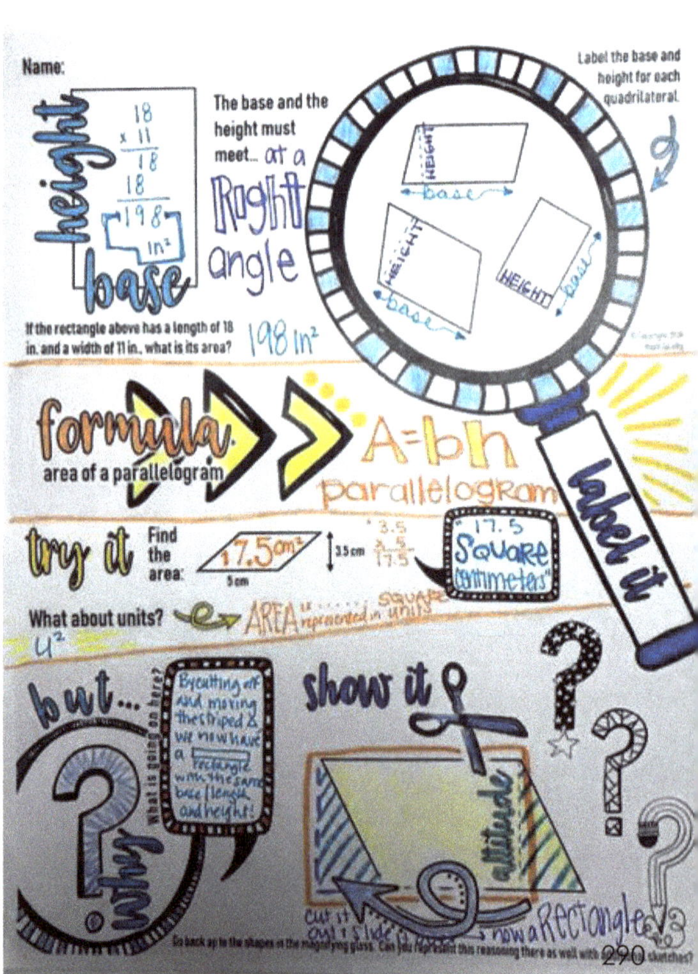

Name:
Label the base and height for each quadrilateral.
height
base
18 x 11
198 in^2
The base and the height must meet... at a Right angle
If the rectangle above has a length of 18 in. and a width of 11 in., what is its area?
198 in^2
formula
area of a parallelogram
A=bh
parallelogram
label it
try it
Find the area:
17.5cm^2
3.5 cm
5 cm
17.5 Square centimeters
What about units?
AREA is represented in SQUARE units
u^2
but... why
What is going on here?
By cutting off and moving the striped & we now have a rectangle with the same base length and height!
show it
altitude
cut it out + slide it
now a Rectangle!
Go back up to the shapes in the magnifying glass. Can you represent this reasoning there as well with additional sketches?

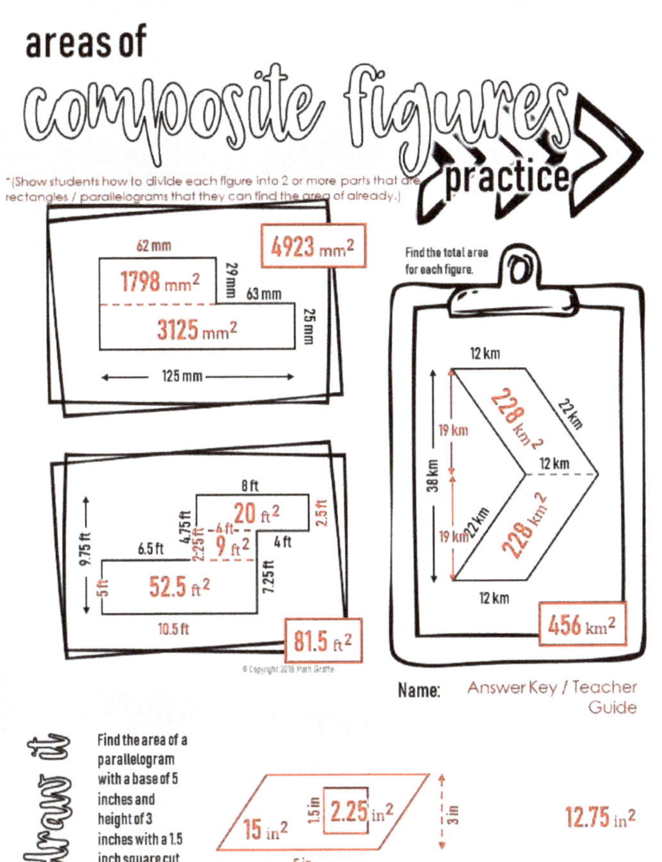

areas of
composite figures
practice
*(Show students how to divide each figure into 2 or more parts that are rectangles / parallelograms that they can find the area of already.)
62 mm
4923 mm²
1798 mm²
29 mm
63 mm
3125 mm²
25 mm
125 mm
Find the total area for each figure.
12 km
228 km²
22 km
19 km
38 km
12 km
19 km
22 km
228 km²
12 km
456 km²
8 ft
20 ft²
2.5 ft
9.75 ft
4.75 ft
4 ft
2.25 ft
9 ft²
4 ft
6.5 ft
5 ft
52.5 ft²
7.25 ft
10.5 ft
81.5 ft²
Name: Answer Key / Teacher Guide
draw it
Find the area of a parallelogram with a base of 5 inches and height of 3 inches with a 1.5 inch square cut out of it.
15 in²
1.5 in
2.25 in²
3 in
5 in
12.75 in²

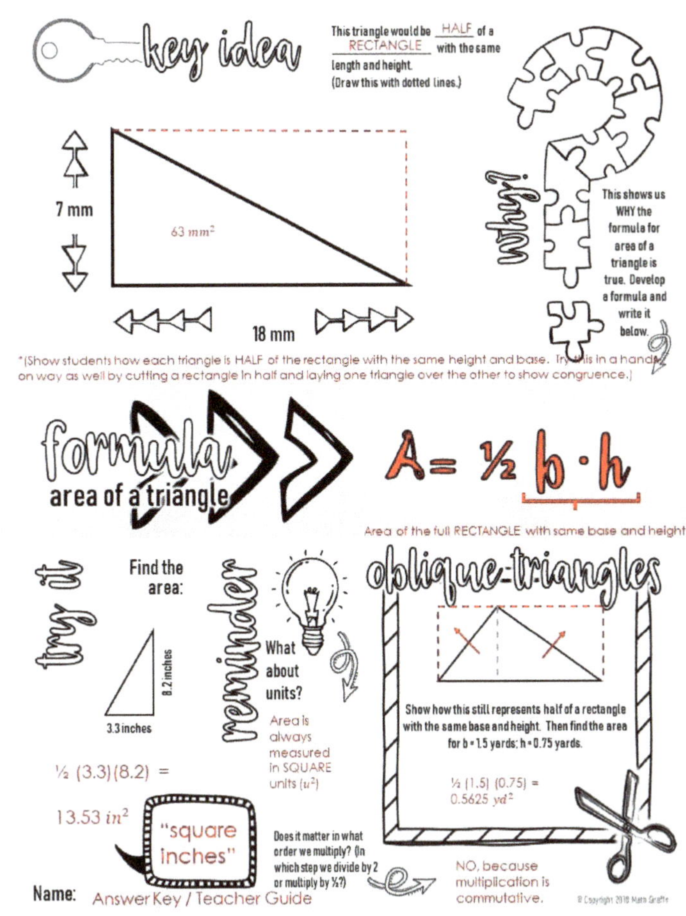

key idea
This triangle would be HALF of a RECTANGLE with the same length and height. (Draw this with dotted lines.)
7 mm
63 mm²
18 mm
why?
This shows us WHY the formula for area of a triangle is true. Develop a formula and write it below.
*(Show students how each triangle is HALF of the rectangle with the same height and base. Try this in a hands-on way as well by cutting a rectangle in half and laying one triangle over the other to show congruence.)
formula
area of a triangle
A = ½ b · h
Area of the full RECTANGLE with same base and height
try it
Find the area:
8.2 inches
3.3 inches
½ (3.3)(8.2) =
13.53 in²
reminder
What about units?
Area is always measured in SQUARE units (u²)
"square inches"
Does it matter in what order we multiply? (In which step we divide by 2 or multiply by ½?)
NO, because multiplication is commutative.
oblique triangles
Show how this still represents half of a rectangle with the same base and height. Then find the area for b = 1.5 yards; h = 0.75 yards.
½ (1.5) (0.75) = 0.5625 yd²
Name: Answer Key / Teacher Guide

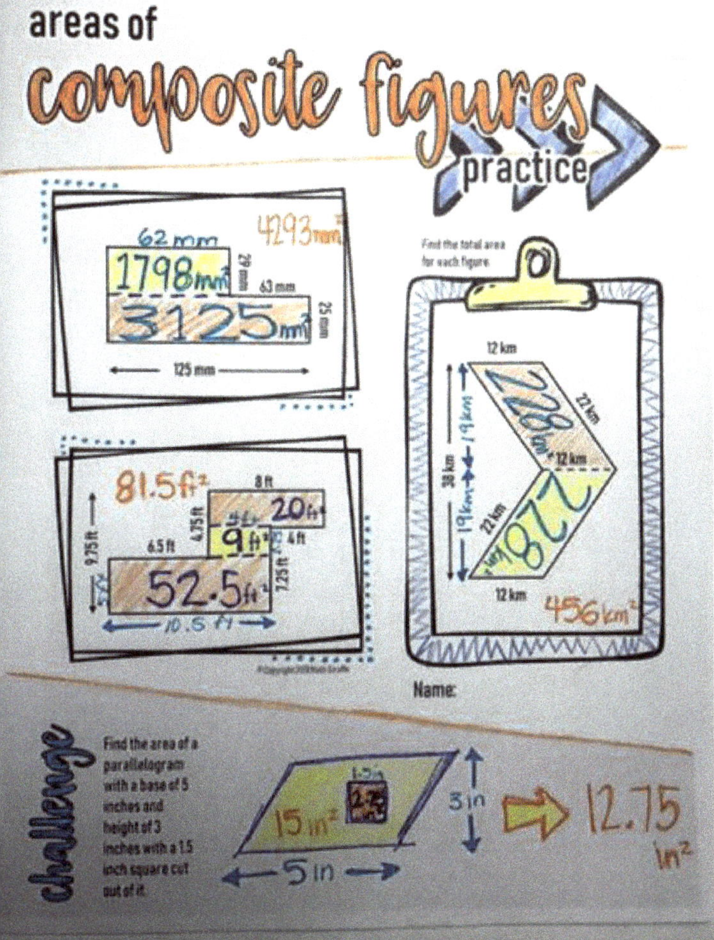

areas of
composite figures
practice
62 mm
4293 mm²
1798 mm²
29 mm
63 mm
3125 mm²
25 mm
125 mm
Find the total area for each figure.
12 km
22 km
38 km
19 km
12 km
228
228
12 km
456 km²
81.5 ft²
8 ft
20 ft²
9.75 ft
4.75 ft
9 ft²
4 ft
6.5 ft
52.5 ft²
7.25 ft
10.5 ft
Name:
challenge
Find the area of a parallelogram with a base of 5 inches and height of 3 inches with a 1.5 inch square cut out of it.
15 in²
3 in
5 in
12.75 in²

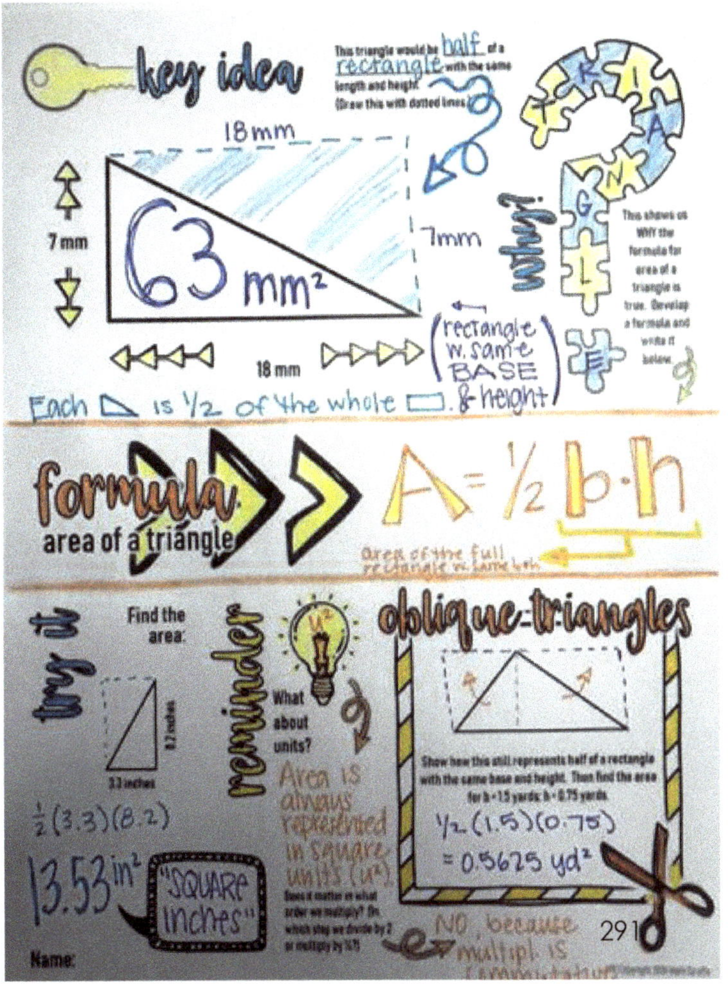

key idea
This triangle would be half of a rectangle with the same length and height. (Draw this with dotted lines.)
18 mm
7 mm
63 mm²
7 mm
18 mm
why?
This shows us WHY the formula for area of a triangle is true. Develop a formula and write it below.
rectangle w. same BASE & height
Each △ is ½ of the whole ▭.
formula
area of a triangle
A = ½ b · h
area of the full rectangle w. same b+h
try it
Find the area:
8.2 inches
3.3 inches
½ (3.3)(8.2)
13.53 in²
reminder
What about units?
Area is always represented in square units (u²)
"square inches"
Does it matter in what order we multiply? (In which step we divide by 2 or multiply by ½?)
NO, because multipl. is commutative
oblique triangles
Show how this still represents half of a rectangle with the same base and height. Then find the area for b = 1.5 yards; h = 0.75 yards.
½ (1.5)(0.75) = 0.5625 yd²
Name:

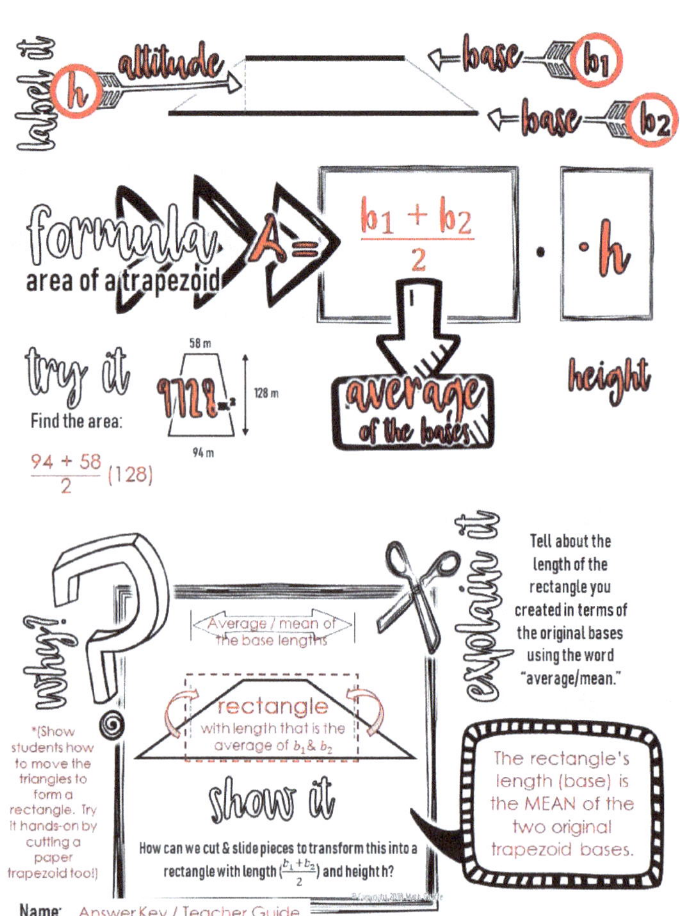
label it
h
altitude
base
b_1
base
b_2
formula
area of a trapezoid
A =
$\frac{b_1 + b_2}{2}$
· h
average of the bases
height
try it
Find the area:
58 m
128 m
94 m
9728 m²
$\frac{94+58}{2}(128)$
why?
explain it
Tell about the length of the rectangle you created in terms of the original bases using the word "average/mean."
Average / mean of the base lengths
rectangle
with length that is the average of b_1 & b_2
*(Show students how to move the triangles to form a rectangle. Try it hands-on by cutting a paper trapezoid too!)
show it
How can we cut & slide pieces to transform this into a rectangle with length $(\frac{b_1+b_2}{2})$ and height h?
The rectangle's length (base) is the MEAN of the two original trapezoid bases.
Name: Answer Key / Teacher Guide

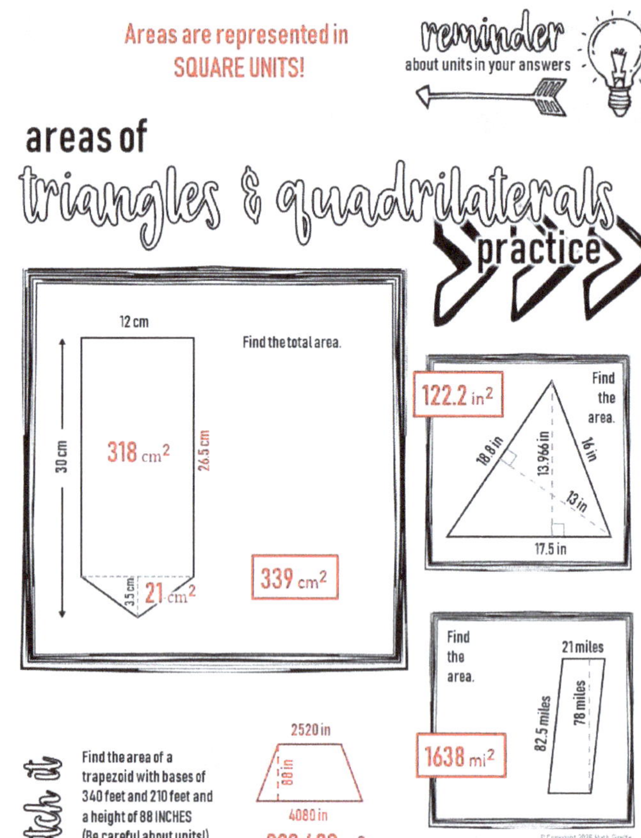
Areas are represented in SQUARE UNITS!
reminder
about units in your answers
areas of
triangles & quadrilaterals
practice
Find the total area.
12 cm
30 cm
318 cm²
26.5 cm
3.5 cm
21 cm²
339 cm²
Find the area.
122.2 in²
18.8 in
13.966 in
16 in
13 in
17.5 in
Find the area.
21 miles
82.5 miles
78 miles
1638 mi²
sketch it
Find the area of a trapezoid with bases of 340 feet and 210 feet and a height of 88 INCHES (Be careful about units!)
2520 in
88 in
4080 in
290,400 in²
or 2016.$\overline{6}$ ft²
Name: Answer Key / Teacher Guide

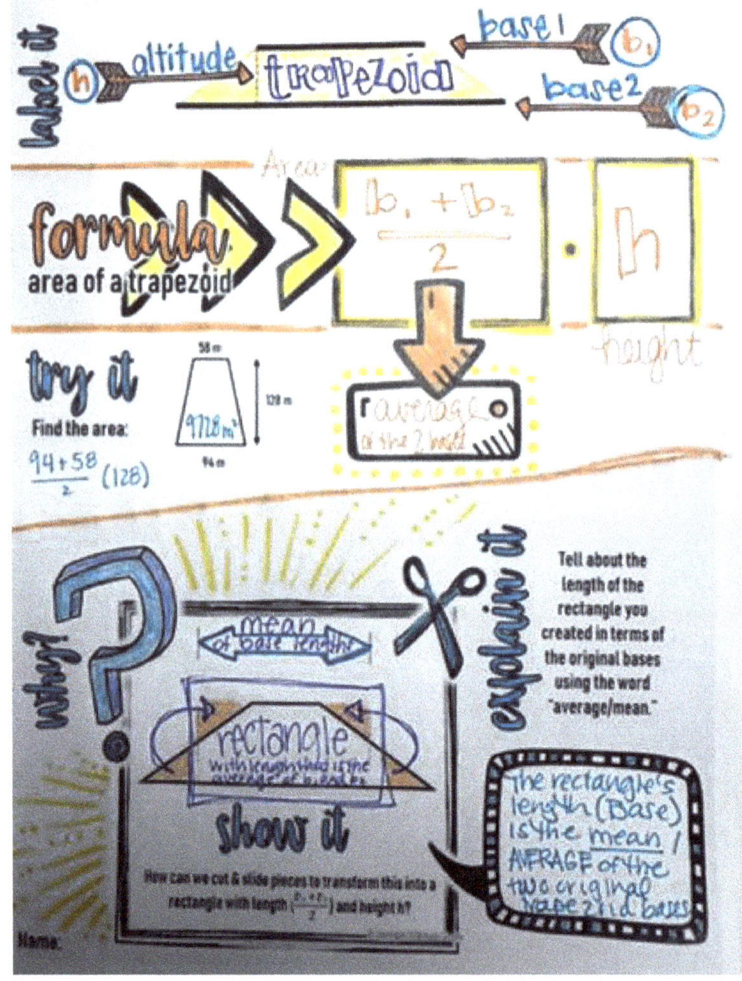
label it
h
altitude
trapezoid
base 1
b_1
base 2
b_2
Area:
formula
area of a trapezoid
$\frac{b_1 + b_2}{2}$
h
height
average of the 2 bases
try it
Find the area:
58 m
128 m
94 m
9728 m²
$\frac{94+58}{2}(128)$
why?
explain it
Tell about the length of the rectangle you created in terms of the original bases using the word "average/mean."
mean of base lengths
rectangle
show it
How can we cut & slide pieces to transform this into a rectangle with length $(\frac{b_1+b_2}{2})$ and height h?
The rectangle's length (Base) is the mean / AVERAGE of the two original trapezoid bases
Name:

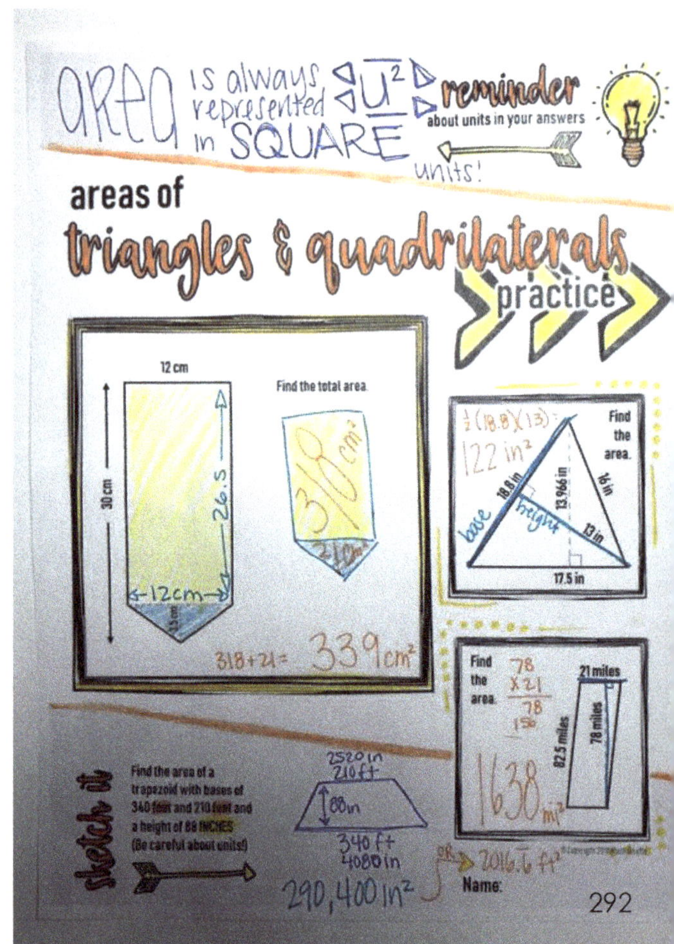
area is always represented in SQUARE units!
u²
reminder
about units in your answers
areas of
triangles & quadrilaterals
practice
12 cm
Find the total area.
30 cm
26.5
318 cm²
21 cm²
12cm
318+21= 339 cm²
½ (18.8)(13)
122 in²
Find the area.
18.8 in
13.966 in
16 in
13 in
17.5 in
base
height
Find the area.
78
x21
78
156
21 miles
82.5 miles
78 miles
1638 mi²
sketch it
Find the area of a trapezoid with bases of 340 feet and 210 feet and a height of 88 INCHES (Be careful about units!)
2520 in
210 ft
88 in
340 ft
4080 in
290,400 in²
or 2016.$\overline{6}$ ft²
Name:

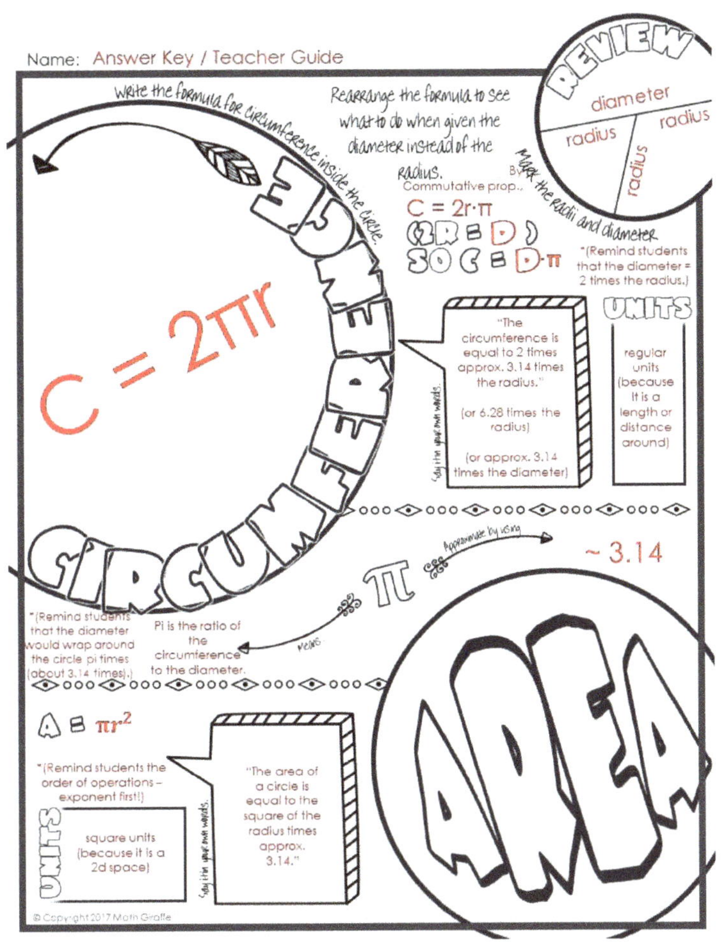

Name: Answer Key / Teacher Guide
REVIEW
diameter
radius
radius
radius
Mark the radii and diameter
Write the formula for circumference inside the circle.
C = 2πr
Rearrange the formula to see what to do when given the diameter instead of the radius.
By Commutative prop.,
C = 2r·π
(2R = D)
SO C = D·π
*(Remind students that the diameter = 2 times the radius.)
"The circumference is equal to 2 times approx. 3.14 times the radius."
(or 6.28 times the radius)
(or approx. 3.14 times the diameter)
Say it in your own words.
UNITS
regular units (because it is a length or distance around)
CIRCUMFERENCE
π
Approximate by using
~ 3.14
means
Pi is the ratio of the circumference to the diameter.
*(Remind students that the diameter would wrap around the circle pi times (about 3.14 times).)
A = πr²
*(Remind students the order of operations – exponent first!)
UNITS
square units (because it is a 2d space)
"The area of a circle is equal to the square of the radius times approx. 3.14."
AREA
© Copyright 2017 Math Giraffe

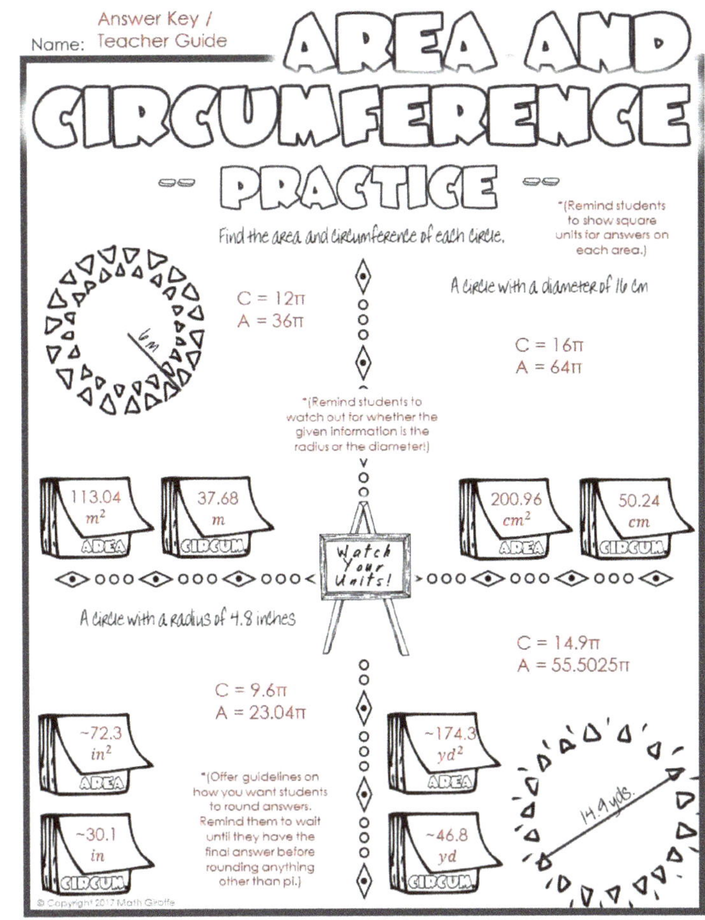

Name: Answer Key / Teacher Guide
AREA AND CIRCUMFERENCE
-- PRACTICE --
Find the area and circumference of each circle.
*(Remind students to show square units for answers on each area.)
6 m
C = 12π
A = 36π
A circle with a diameter of 16 cm
C = 16π
A = 64π
*(Remind students to watch out for whether the given information is the radius or the diameter!)
113.04 m² AREA
37.68 m CIRCUM
200.96 cm² AREA
50.24 cm CIRCUM
Watch Your Units!
A circle with a radius of 4.8 inches
C = 9.6π
A = 23.04π
C = 14.9π
A = 55.5025π
~72.3 in² AREA
~30.1 in CIRCUM
*(Offer guidelines on how you want students to round answers. Remind them to wait until they have the final answer before rounding anything other than pi.)
~174.3 yd² AREA
~46.8 yd CIRCUM
14.9 yds.
© Copyright 2017 Math Giraffe

Name:
REVIEW
2·r
diameter
radius
radius
radius
Write the formula for circumference inside the circle.
DISTANCE AROUND
C=2πr
CIRCUMFERENCE
Rearrange the formula to see what to do when given the diameter instead of the radius.
C = 2r · π
(2R = d)
SO C = d · π
Mark the radii and diameter
The CIRCUMFERENCE is equal to 2 times 3.14 times the... RADIUS
OR 6.28 times the RADIUS!
Say it in your own words.
UNITS
regular units because it's a LENGTH
π
Approximate by using
ROUNDED ~3.14
means
the RATIO of the CIRCUMFERENCE to the diameter
A = πr²
ORDER OF OP. Do r² FIRST, then MULTIPLY by π
UNITS
SQUARE UNITS (u²) because it's a 2-D AREA
The AREA of a circle is equal to the SQUARE of the RADIUS times 3.14
AREA
© Copyright 2017 Math Giraffe

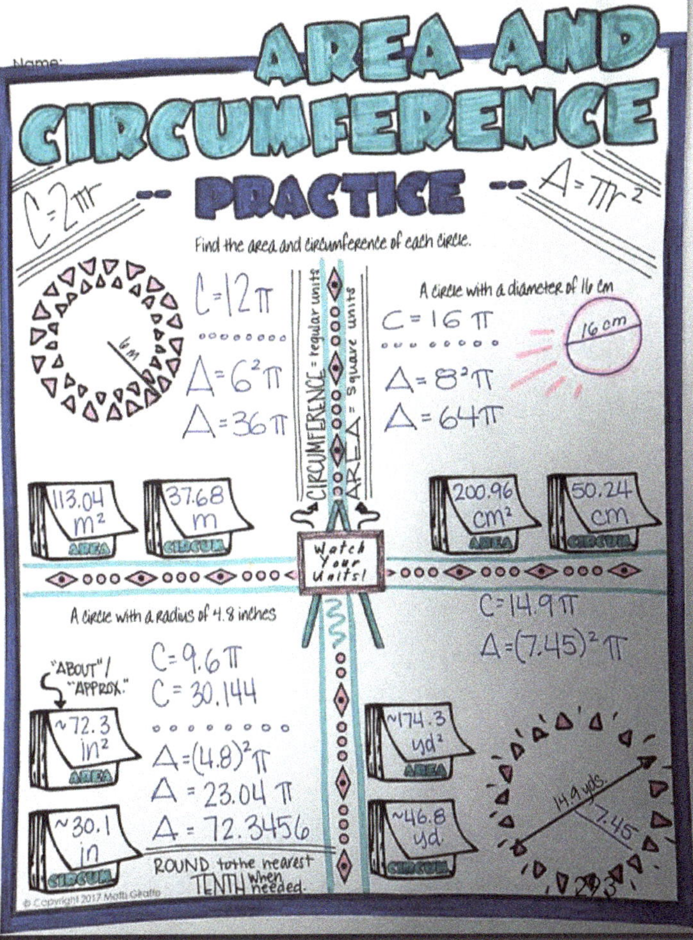

Name:
AREA AND CIRCUMFERENCE
-- PRACTICE --
C=2πr
A=πr²
Find the area and circumference of each circle.
6 m
C=12π
A=6²π
A=36π
CIRCUMFERENCE = regular units
AREA = square units
A circle with a diameter of 16 cm
16 cm
C=16π
A=8²π
A=64π
113.04 m² AREA
37.68 m CIRCUM
200.96 cm² AREA
50.24 cm CIRCUM
Watch Your Units!
A circle with a radius of 4.8 inches
"ABOUT"/ "APPROX."
C=9.6π
C=30.144
A=(4.8)²π
A=23.04π
A=72.3456
ROUND to the nearest TENTH when needed.
C=14.9π
A=(7.45)²π
~72.3 in² AREA
~30.1 in CIRCUM
~174.3 yd² AREA
~46.8 yd CIRCUM
14.9 yds.
7.45
© Copyright 2017 Math Giraffe

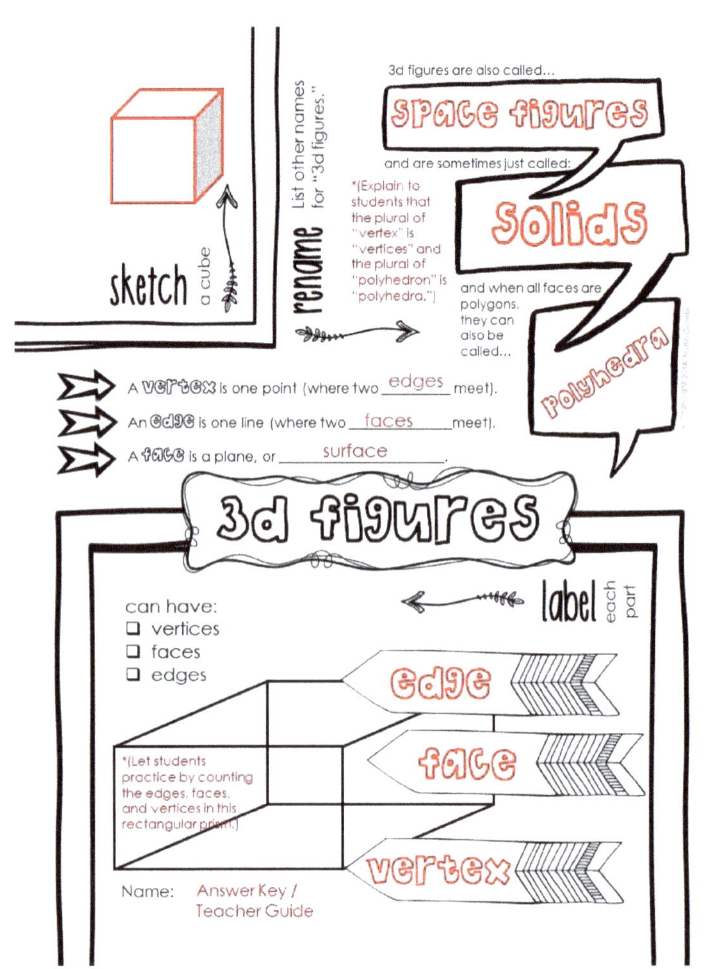
sketch a cube
rename
List other names for "3d figures."
3d figures are also called...
space figures
and are sometimes just called:
solids
and when all faces are polygons, they can also be called...
polyhedra
*(Explain to students that the plural of "vertex" is "vertices" and the plural of "polyhedron" is "polyhedra.")
A vertex is one point (where two edges meet).
An edge is one line (where two faces meet).
A face is a plane, or surface.
3d figures
label each part
can have:
vertices
faces
edges
edge
face
vertex
*(Let students practice by counting the edges, faces, and vertices in this rectangular prism.)
Name: Answer Key / Teacher Guide

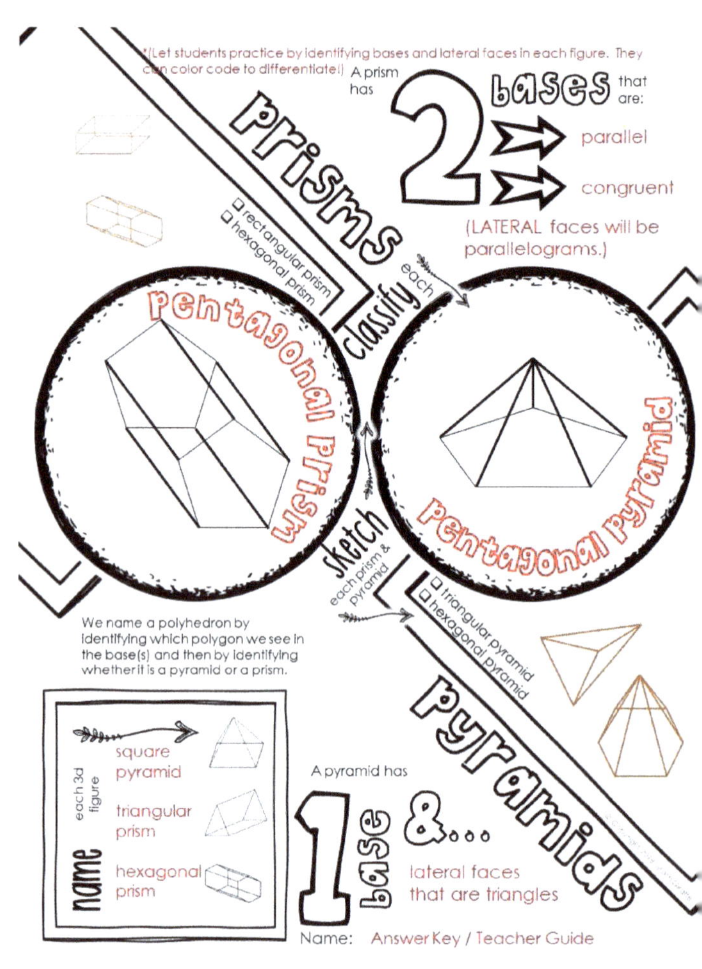
*(Let students practice by identifying bases and lateral faces in each figure. They can color code to differentiate!)
A prism has
2 bases that are:
parallel
congruent
(LATERAL faces will be parallelograms.)
prisms
rectangular prism
hexagonal prism
classify each
pentagonal prism
pentagonal pyramid
sketch each prism & pyramid
triangular pyramid
hexagonal pyramid
pyramids
We name a polyhedron by identifying which polygon we see in the base(s) and then by identifying whether it is a pyramid or a prism.
name each 3d figure
square pyramid
triangular prism
hexagonal prism
A pyramid has
1 base &...
lateral faces that are triangles
Name: Answer Key / Teacher Guide

3d figures are also called...
SPACE FIGURES
and are sometimes just called:
SOLIDS
and when all faces are polygons, they can also be called...
Polyhedra
sketch a cube
rename
List other names for "3d figures."
3d figures
label each part
can have:
vertices
faces
edges
edge
FACE
vertex
Name:

A prism has
2 bases that are:
Parallel
Congruent
prisms
rectangular prism
hexagonal prism
classify each
Pentagonal Prism
lateral faces
2 bases
Pentagonal pyramid
1 base
sketch each prism & pyramid
triangular pyramid
hexagonal pyramid
pyramids
We name a polyhedron by identifying which polygon we see in the base(s) and then by identifying whether it is a pyramid or a prism.
name each 3d figure
square pyramid
triangular prism
hexagonal prism
A pyramid has
1 base &...
lateral faces that are TRIANGLES
Name:

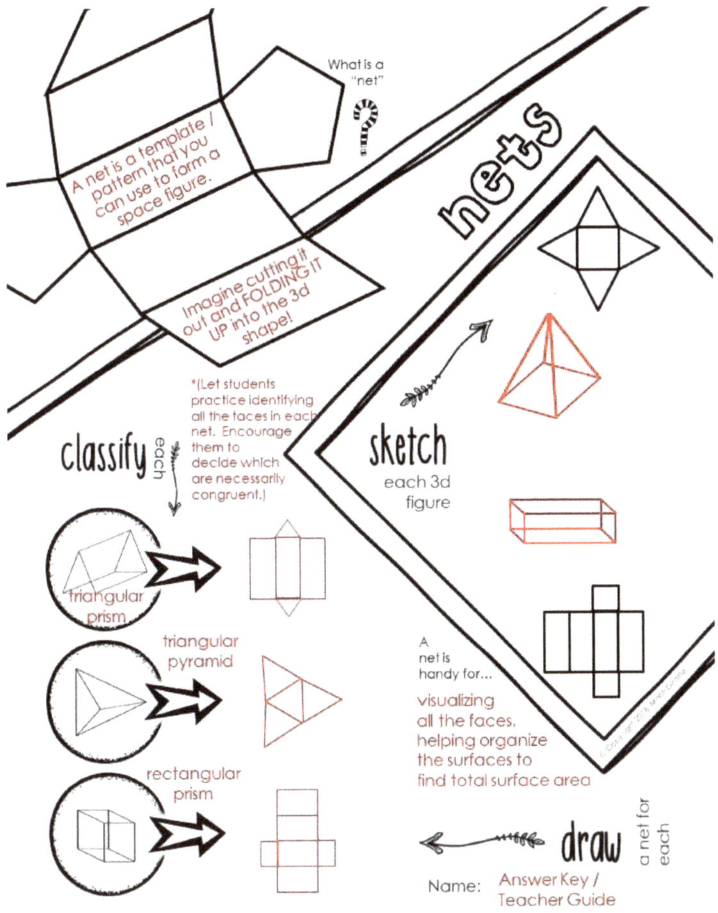

nets

What is a "net"?

A net is a template / pattern that you can use to form a space figure.

Imagine cutting it out and FOLDING IT UP into the 3d shape!

*(Let students practice identifying all the faces in each net. Encourage them to decide which are necessarily congruent.)

classify each

- triangular prism
- triangular pyramid
- rectangular prism

sketch each 3d figure

A net is handy for...

visualizing all the faces, helping organize the surfaces to find total surface area

draw a net for each

Name: Answer Key / Teacher Guide

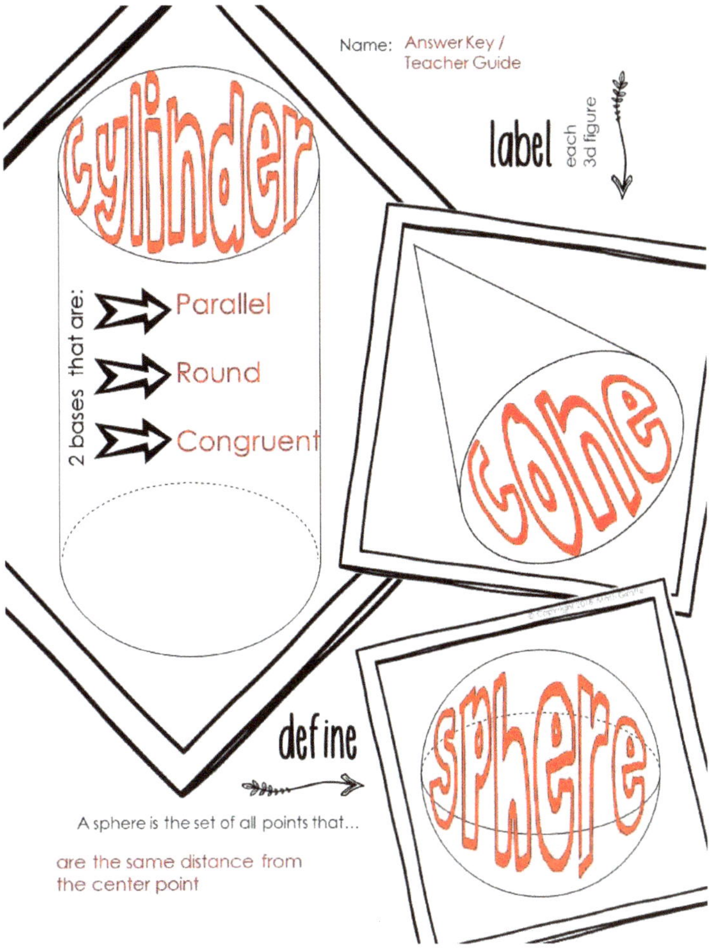

Name: Answer Key / Teacher Guide

label each 3d figure

Cylinder

2 bases that are:

- Parallel
- Round
- Congruent

Cone

Sphere

define

A sphere is the set of all points that...

are the same distance from the center point

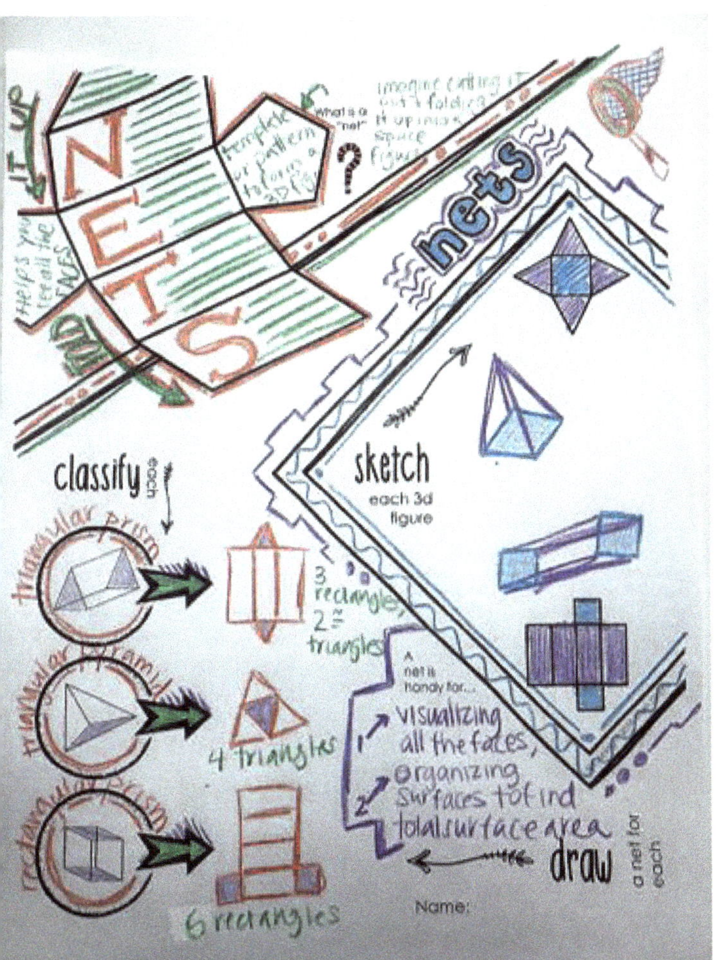

Name: Answer Key / Teacher Guide
You find the volume to pack or fill up the entire inside of a 3d shape.
How much can it hold?
VOLUME
dimensions
Length
Width
Height
To find how much a rectangular prism can hold, you need to know the dimensions.
3724 cm³
28 cm
7 cm
19 cm
Color & label LENGTH.
length
Color & label WIDTH.
width
Color & label HEIGHT.
height
Volume is measured in cubic units
u³
2 ways to calculate the volume
Find the area of the base then multiply by the height.
OR
Multiply: l x w x h (for rectangular prisms only)
Try it
Find the volume of each.
EX.1
1 cm
2 cm
3 cm
6 cm³
EX.2
5 in.
125 in³

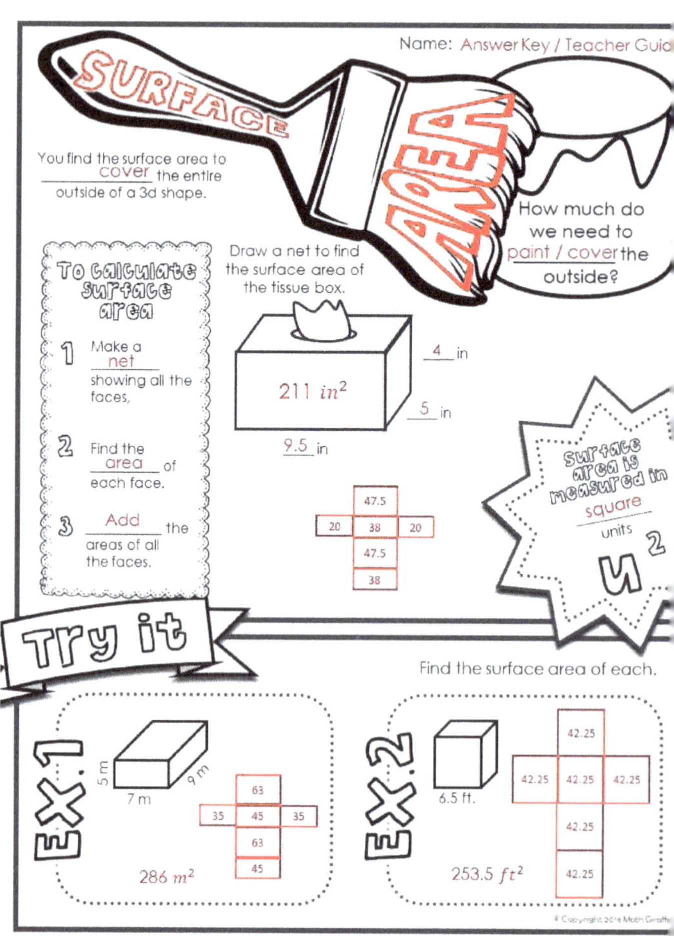

Name: Answer Key / Teacher Guid
SURFACE
AREA
You find the surface area to cover the entire outside of a 3d shape.
How much do we need to paint / cover the outside?
To calculate surface area
1 Make a net showing all the faces.
2 Find the area of each face.
3 Add the areas of all the faces.
Draw a net to find the surface area of the tissue box.
211 in²
4 in
5 in
9.5 in
47.5
20
38
20
47.5
38
Surface area is measured in square units
u²
Try it
Find the surface area of each.
EX.1
5 m
7 m
9 m
63
35
45
35
63
45
286 m²
EX.2
6.5 ft.
42.25
42.25
42.25
42.25
42.25
42.25
253.5 ft²

Name:
You find the volume to PACK or FILL UP the entire inside of a 3d shape.
How much can it hold?
VOLUME
dimensions
Length
Width
Height
To find how much a rectangular prism can hold, you need to know the DIMENSIONS.
3724 cm³
CEREAL
28 cm
7 cm
19 cm
Color & label LENGTH.
LEN.
Color & label WIDTH.
WID.
Color & label HEIGHT.
HEI.
Volume is measured in cubic units
u³
2 ways to calculate the volume
Find the area of the base then multiply by the height.
OR
Multiply: l x w x h (for rectangular prisms only)
Try it
Find the volume of each.
EX.1
1 cm
2 cm
3 cm
3(2)(1)
= 6cm³
EX.2
5 in.
5(5)(5)
= 125 in³

Name:
SURFACE
AREA
You find the surface area to paint / cover the entire outside of a 3d shape.
How much do we need to COVER the outside?
To calculate surface area
1 Make a NET showing all the faces.
2 Find the AREA of each face.
3 ADD the areas of all the faces.
Draw a net to find the surface area of the tissue box.
211 in²
4 in
5 in
9.5 in
"SQUARE INCHES"
Draw a NET
47.5
20
38
20
47.5
38
ADD AREA of each FACE
Surface area is measured in SQUARE units
u²
Try it
Find the surface area of each.
EX.1
5 m
7 m
9 m
63
45
63
45
35
35
286 m²
EX.2
6.5 ft.
42.25
x 6
253.5
253.5 ft²

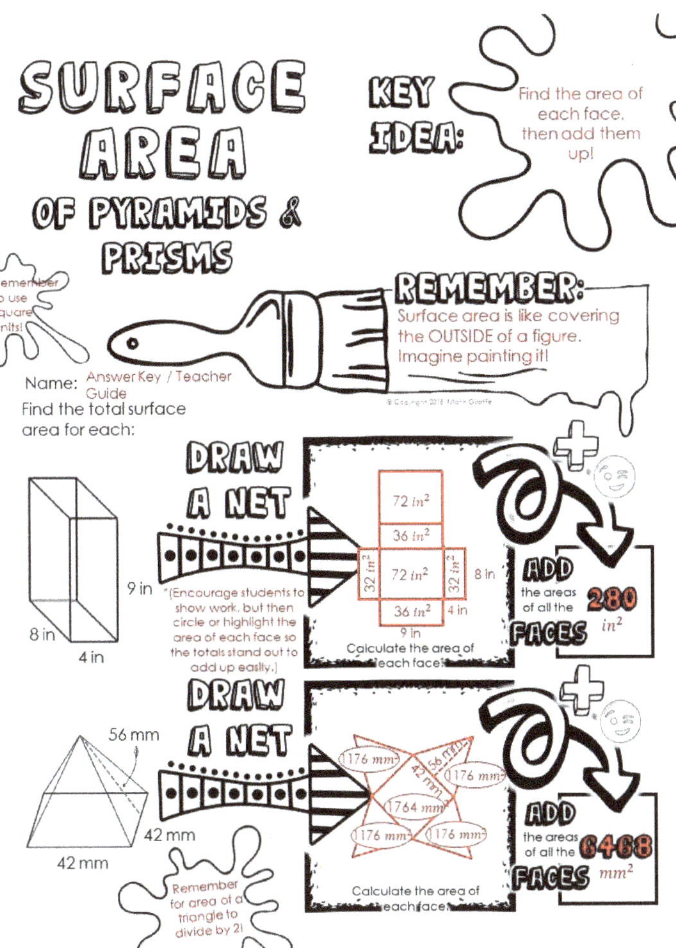

SURFACE AREA OF PYRAMIDS & PRISMS
KEY IDEA:
Find the area of each face, then add them up!
Remember to use square units!
REMEMBER:
Surface area is like covering the OUTSIDE of a figure. Imagine painting it!
Name: Answer Key / Teacher Guide
Find the total surface area for each:
DRAW A NET
9 in
8 in
4 in
*(Encourage students to show work, but then circle or highlight the area of each face so the totals stand out to add up easily.)
72 in^2
36 in^2
32 in^2
72 in^2
32 in^2
36 in^2
8 in
4 in
9 in
Calculate the area of each face
ADD the areas of all the FACES
280 in^2
56 mm
42 mm
42 mm
1176 mm^2
1176 mm^2
1764 mm^2
1176 mm^2
1176 mm^2
Remember for area of a triangle to divide by 2!
6468 mm^2

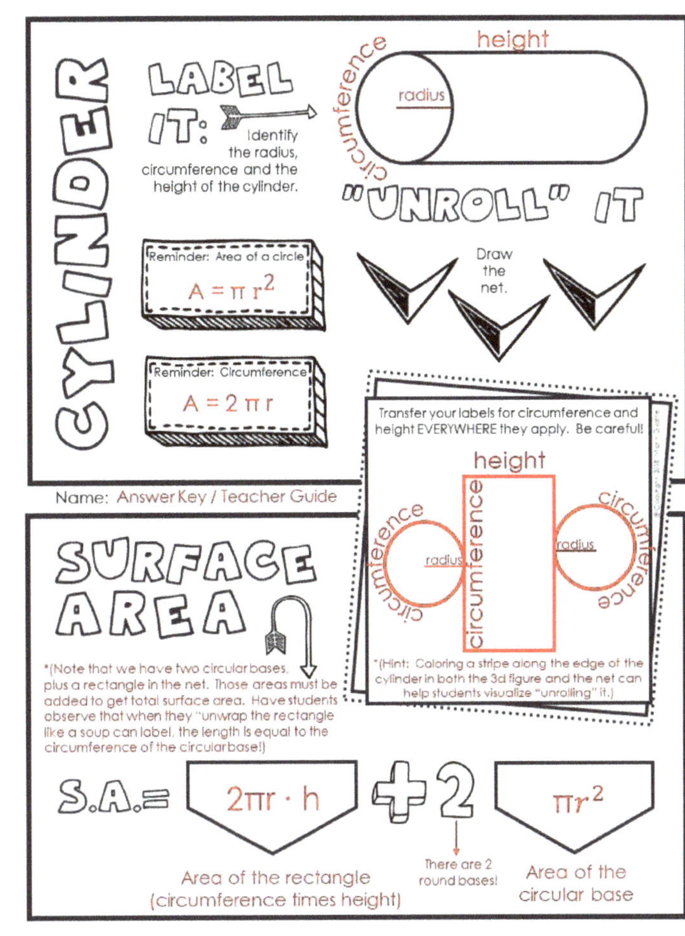

CYLINDER
LABEL IT:
Identify the radius, circumference and the height of the cylinder.
height
radius
circumference
"UNROLL" IT
Reminder: Area of a circle
$A = \pi r^2$
Reminder: Circumference
$A = 2\pi r$
Draw the net.
Transfer your labels for circumference and height EVERYWHERE they apply. Be careful!
height
circumference
radius
radius
circumference
circumference
Name: Answer Key / Teacher Guide
SURFACE AREA
*(Note that we have two circular bases, plus a rectangle in the net. Those areas must be added to get total surface area. Have students observe that when they "unwrap the rectangle like a soup can label, the length is equal to the circumference of the circular base!)
*(Hint: Coloring a stripe along the edge of the cylinder in both the 3d figure and the net can help students visualize "unrolling" it.)
S.A. = $2\pi r \cdot h$ + 2 πr^2
Area of the rectangle (circumference times height)
There are 2 round bases!
Area of the circular base

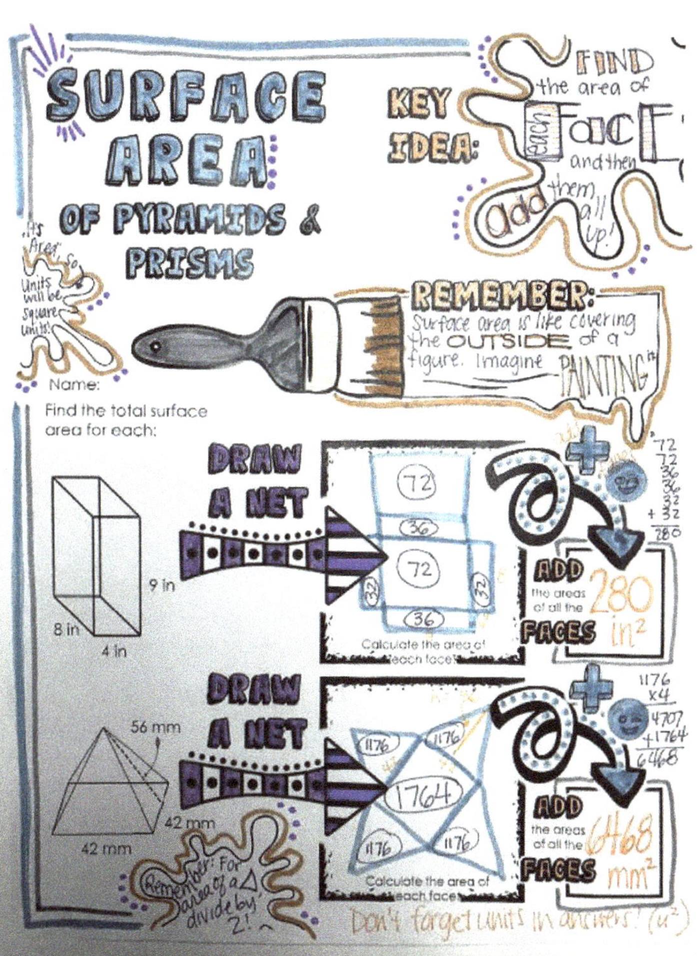

SURFACE AREA OF PYRAMIDS & PRISMS
KEY IDEA:
FIND the area of each FACE and then add them all up!
It's Area so units will be square units!
REMEMBER:
Surface area is like covering the OUTSIDE of a figure. Imagine PAINTING it!
Name:
Find the total surface area for each:
DRAW A NET
72
36
72
32
32
36
9 in
8 in
4 in
72
72
36
36
32
+ 32
280
Calculate the area of each face
ADD the areas of all the FACES
280 in^2
1176
x4
4707
+1764
6468
1176
1176
1764
1176
1176
56 mm
42 mm
42 mm
Remember: For area of a △ divide by 2!
6468 mm^2
Don't forget units in answers! (u^2)

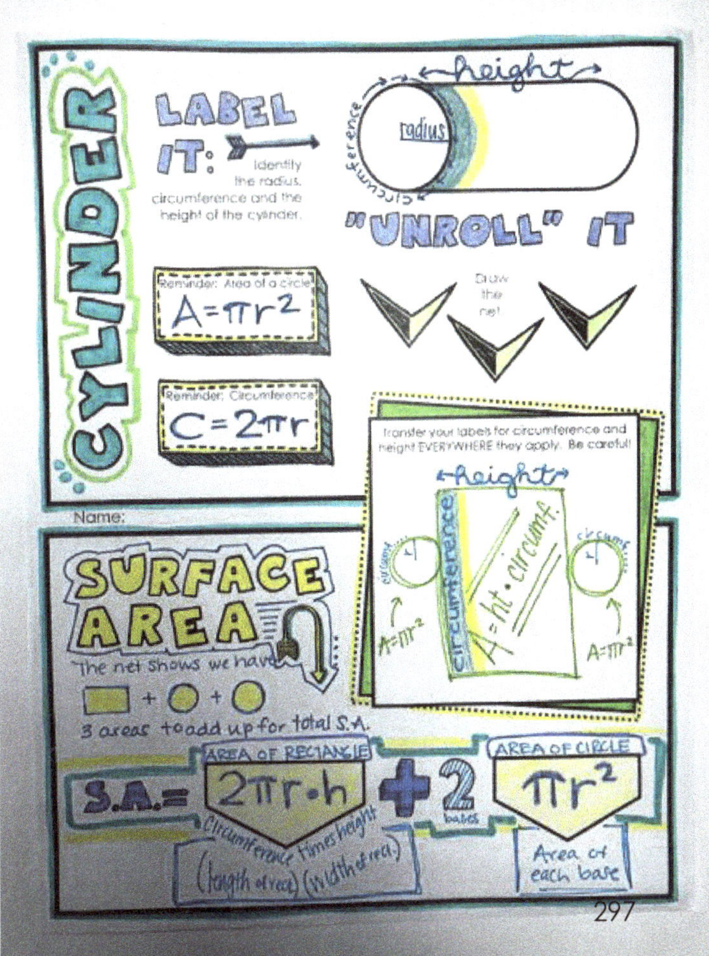

CYLINDER
LABEL IT:
Identify the radius, circumference and the height of the cylinder.
height
radius
circumference
"UNROLL" IT
Reminder: Area of a circle
$A = \pi r^2$
Reminder: Circumference
$C = 2\pi r$
Draw the net!
Transfer your labels for circumference and height EVERYWHERE they apply. Be careful!
height
circumference
A = ht · circumf.
$A = \pi r^2$
$A = \pi r^2$
Name:
SURFACE AREA
The net shows we have
3 areas to add up for total S.A.
AREA OF RECTANGLE
AREA OF CIRCLE
S.A. = $2\pi r \cdot h$ + 2 πr^2
bases
Circumference times height
(length of rect.) (width of rect.)
Area of each base

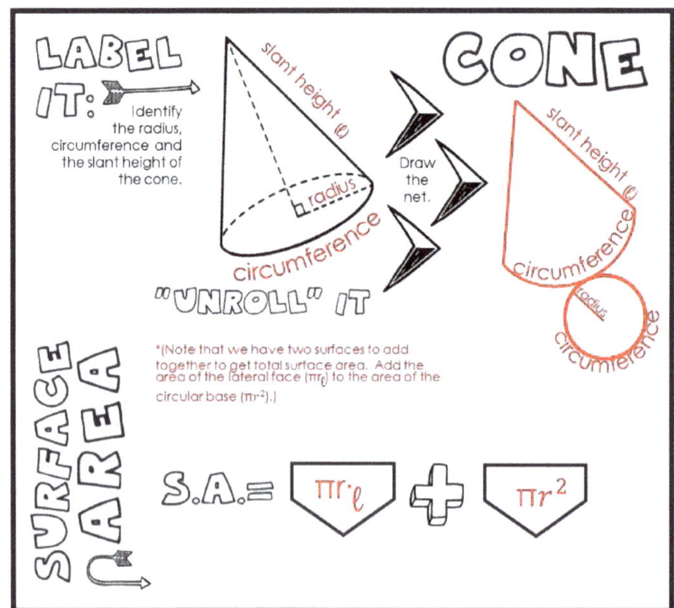
LABEL IT:
Identify the radius, circumference and the slant height of the cone.
CONE
slant height (ℓ)
radius
circumference
Draw the net.
"UNROLL" IT
SURFACE AREA
*(Note that we have two surfaces to add together to get total surface area. Add the area of the lateral face (πrℓ) to the area of the circular base (πr²).)
S.A.= πrℓ + πr²

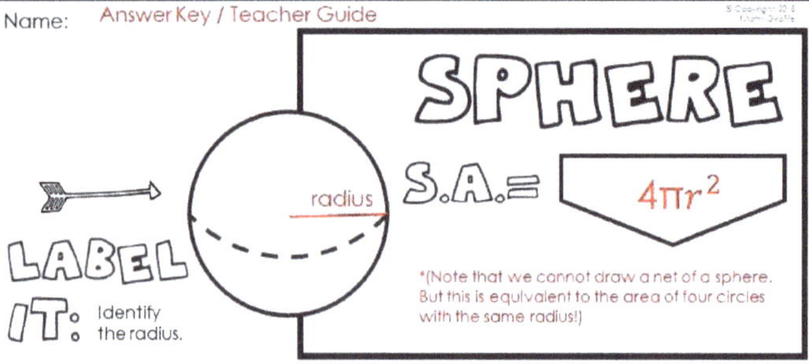
Name: Answer Key / Teacher Guide
SPHERE
radius
S.A.= 4πr²
LABEL IT: Identify the radius.
*(Note that we cannot draw a net of a sphere. But this is equivalent to the area of four circles with the same radius!)

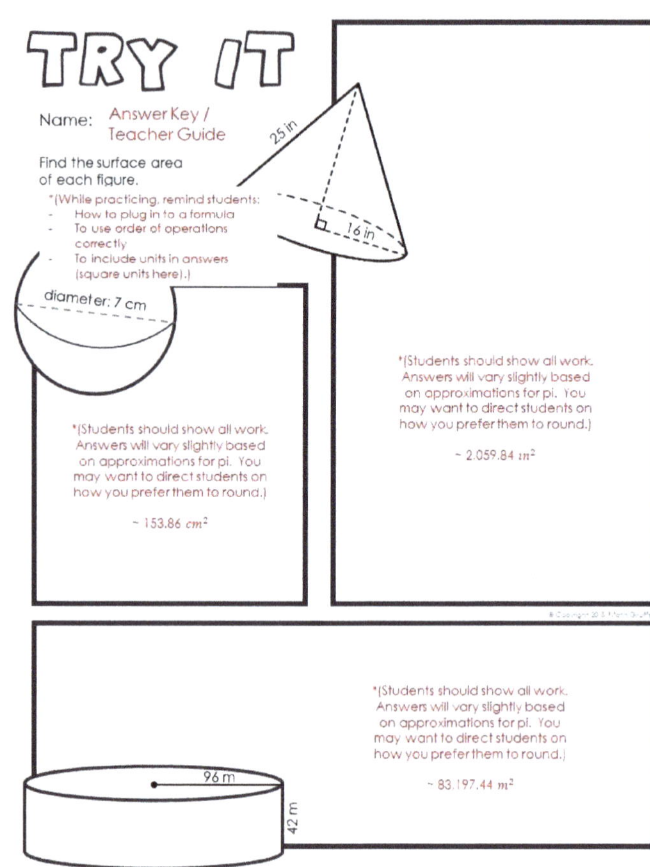
TRY IT
Name: Answer Key / Teacher Guide
Find the surface area of each figure.
*(While practicing, remind students:
- How to plug in to a formula
- To use order of operations correctly
- To include units in answers (square units here).)
25 in
16 in
diameter: 7 cm
*(Students should show all work. Answers will vary slightly based on approximations for pi. You may want to direct students on how you prefer them to round.)
~ 153.86 cm²
~ 2,059.84 in²
96 m
42 m
~ 83,197.44 m²

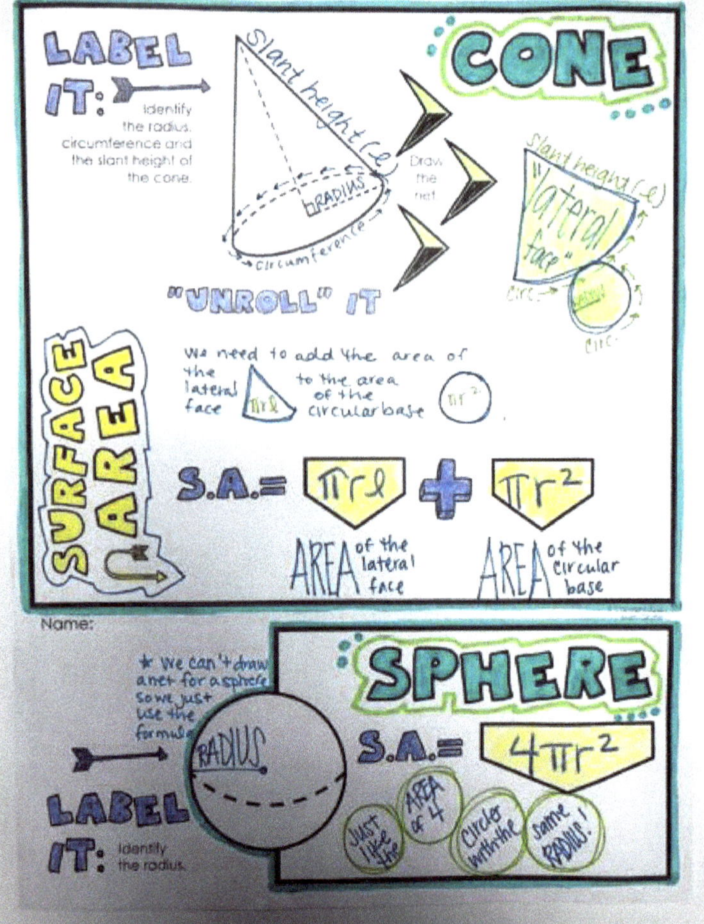
LABEL IT:
CONE
Slant height (ℓ)
RADIUS
circumference
"UNROLL" IT
"lateral face"
We need to add the area of the lateral face to the area of the circular base
SURFACE AREA
S.A.= πrℓ + πr²
AREA of the lateral face
AREA of the circular base
SPHERE
★ We can't draw a net for a sphere so we just use the formula
RADIUS
S.A.= 4πr²
LABEL IT:
Just like the AREA of 4 circles with the same RADIUS!

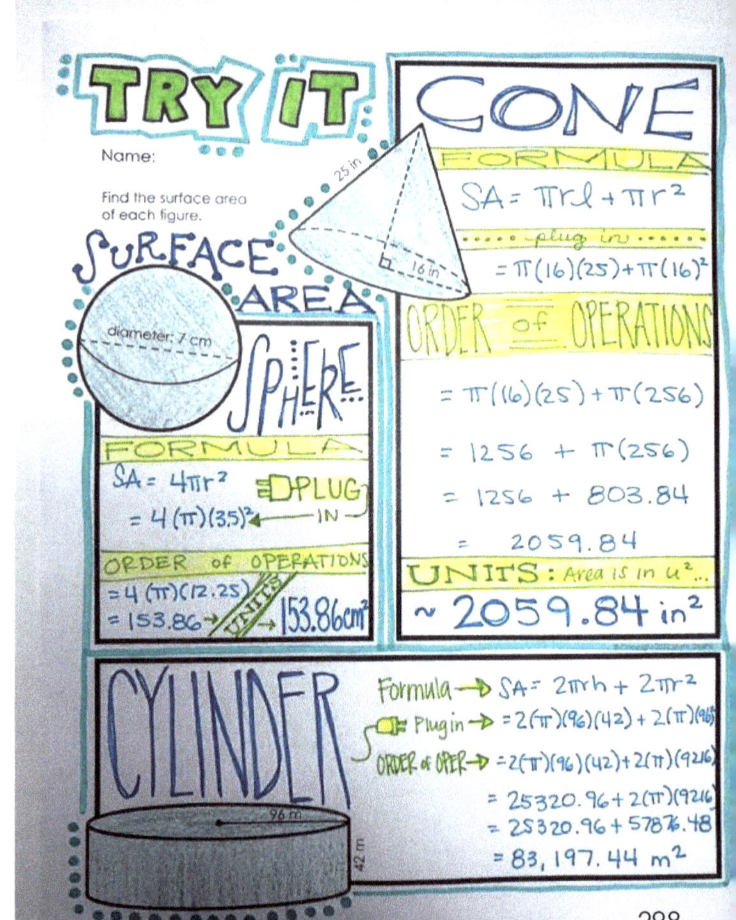
TRY IT
CONE
FORMULA
SA = πrℓ + πr²
plug in
= π(16)(25)+π(16)²
ORDER of OPERATIONS
= π(16)(25) + π(256)
= 1256 + π(256)
= 1256 + 803.84
= 2059.84
UNITS: Area is in u²
~ 2059.84 in²
SURFACE AREA
SPHERE
FORMULA
SA = 4πr²
PLUG IN
= 4(π)(3.5)²
ORDER of OPERATIONS
= 4(π)(12.25)
= 153.86
UNITS
153.86cm²
CYLINDER
Formula → SA = 2πrh + 2πr²
Plug in → = 2(π)(96)(42) + 2(π)(96)²
ORDER of OPER → = 2(π)(96)(42)+2(π)(9216)
= 25320.96 + 2(π)(9216)
= 25320.96 + 57876.48
= 83,197.44 m²

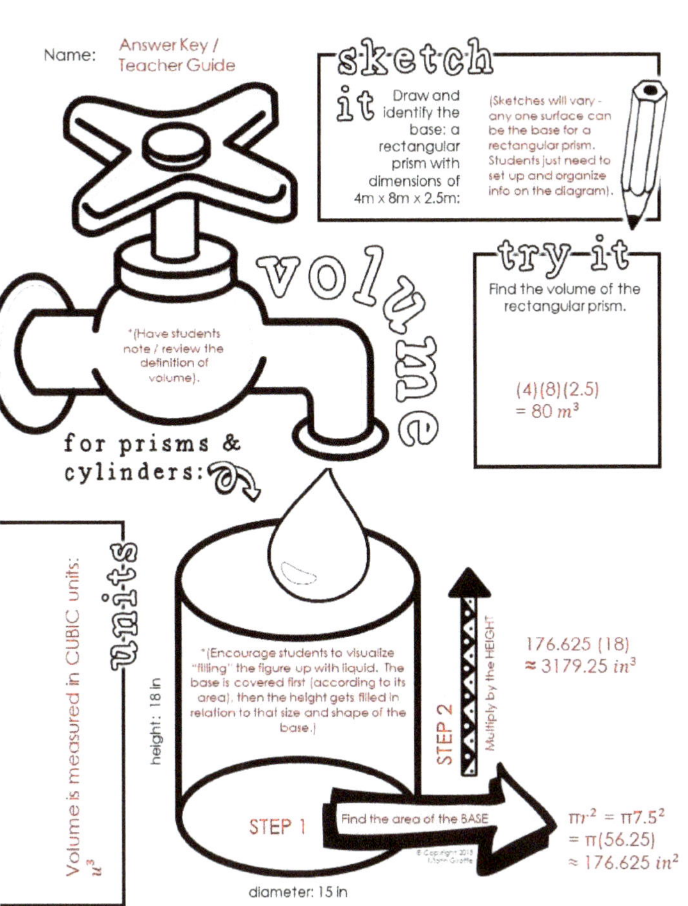

Name: Answer Key / Teacher Guide
sketch it
Draw and identify the base: a rectangular prism with dimensions of 4m x 8m x 2.5m:
(Sketches will vary - any one surface can be the base for a rectangular prism. Students just need to set up and organize info on the diagram).
volume
*(Have students note / review the definition of volume).
for prisms & cylinders:
try it
Find the volume of the rectangular prism.
(4)(8)(2.5)
= 80 m^3
units
Volume is measured in CUBIC units: u^3
*(Encourage students to visualize "filling" the figure up with liquid. The base is covered first (according to its area), then the height gets filled in relation to that size and shape of the base.)
height: 18 in
STEP 2
Multiply by the HEIGHT
176.625 (18)
≈ 3179.25 in^3
STEP 1
Find the area of the BASE
$\pi r^2 = \pi 7.5^2$
= π(56.25)
≈ 176.625 in^2
diameter: 15 in

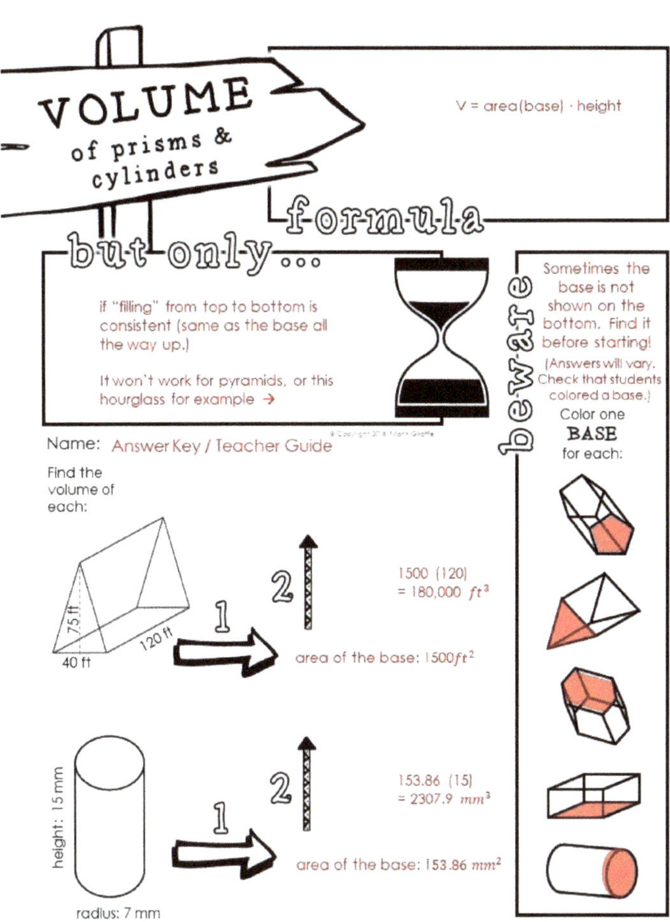

VOLUME
of prisms & cylinders
V = area(base) · height
formula
but only...
if "filling" from top to bottom is consistent (same as the base all the way up.)
It won't work for pyramids, or this hourglass for example →
beware
Sometimes the base is not shown on the bottom. Find it before starting!
(Answers will vary. Check that students colored a base.)
Color one BASE for each:
Name: Answer Key / Teacher Guide
Find the volume of each:
7.5 ft
40 ft
120 ft
1
2
1500 (120)
= 180,000 ft^3
area of the base: 1500ft^2
height: 15 mm
radius: 7 mm
153.86 (15)
= 2307.9 mm^3
area of the base: 153.86 mm^2

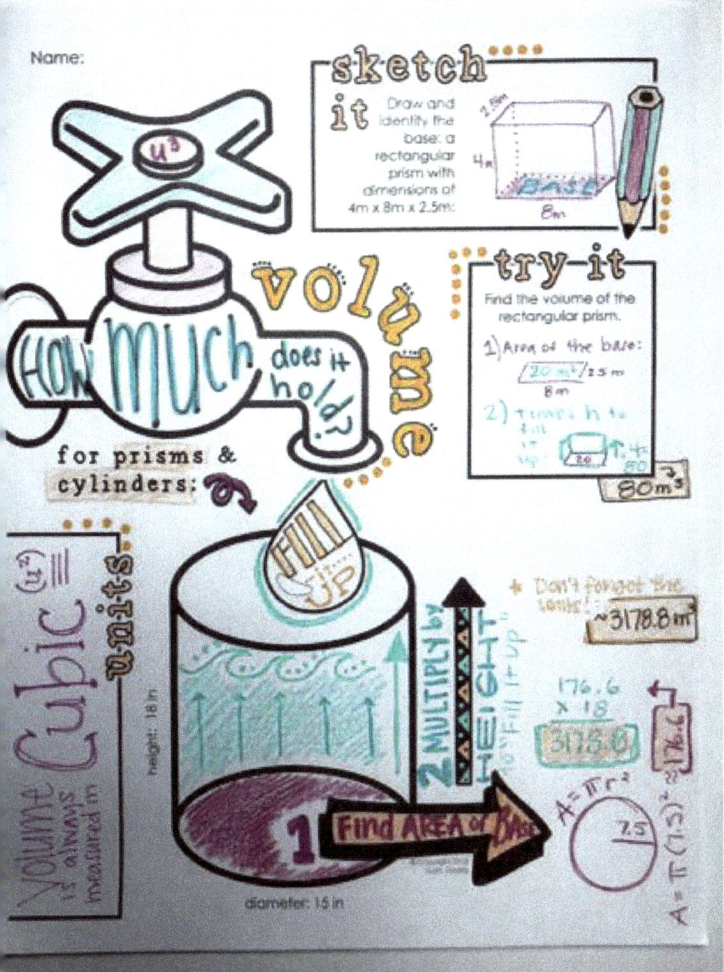

Name:
sketch it
Draw and identify the base: a rectangular prism with dimensions of 4m x 8m x 2.5m:
2.5m
4m
BASE
8m
volume
HOW MUCH does it hold?
for prisms & cylinders:
try it
Find the volume of the rectangular prism.
1) Area of the base:
20 m²
2) times h to fill it up
80 m^3
FILL it UP
Volume is always measured in Cubic (u^3) units
height: 18 in
2 MULTIPLY by HEIGHT
Don't forget the units!
~3178.8 in^3
176.6
x 18
3178.8
1 Find AREA of BASE
$A = \pi r^2$
7.5
$A = \pi (7.5)^2$
diameter: 15 in

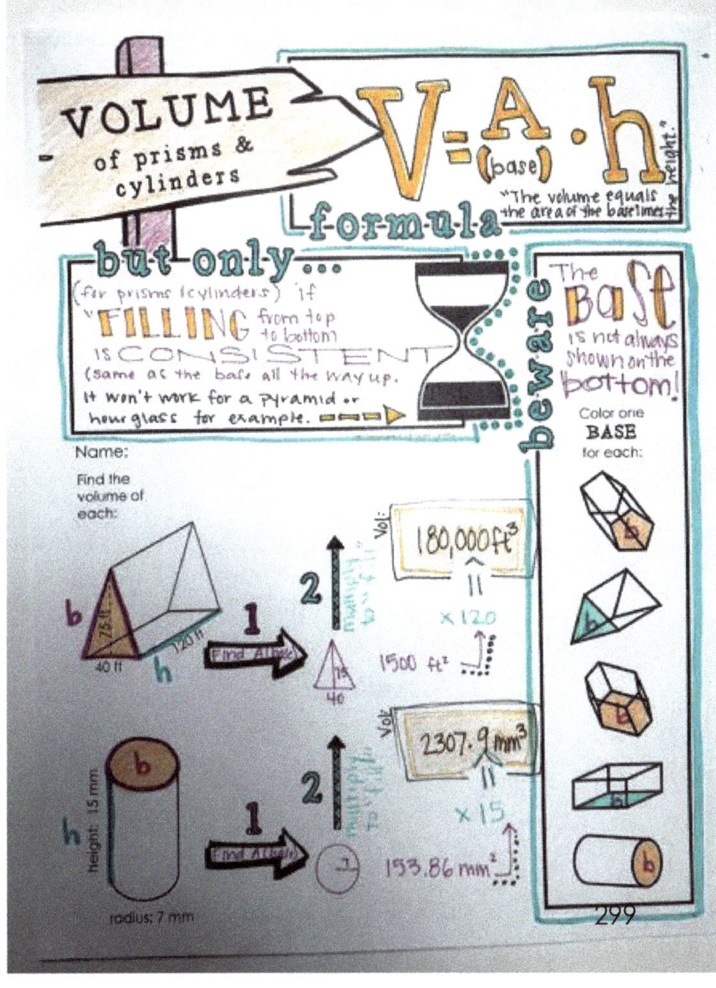

VOLUME
of prisms & cylinders
V = A(base) · h
"The volume equals the area of the base times the height."
formula
but only...
(for prisms & cylinders) if "FILLING" from top to bottom is CONSISTENT
(same as the base all the way up.
It won't work for a pyramid or hourglass for example.
beware
The Base is not always shown on the bottom!
Color one BASE for each:
Name:
Find the volume of each:
Vol: 180,000 ft^3
x 120
1500 ft^2
40 ft
120 ft
Find A(base)
Vol: 2307.9 mm^3
x 15
153.86 mm^2
height: 15 mm
radius: 7 mm
Find A(base)

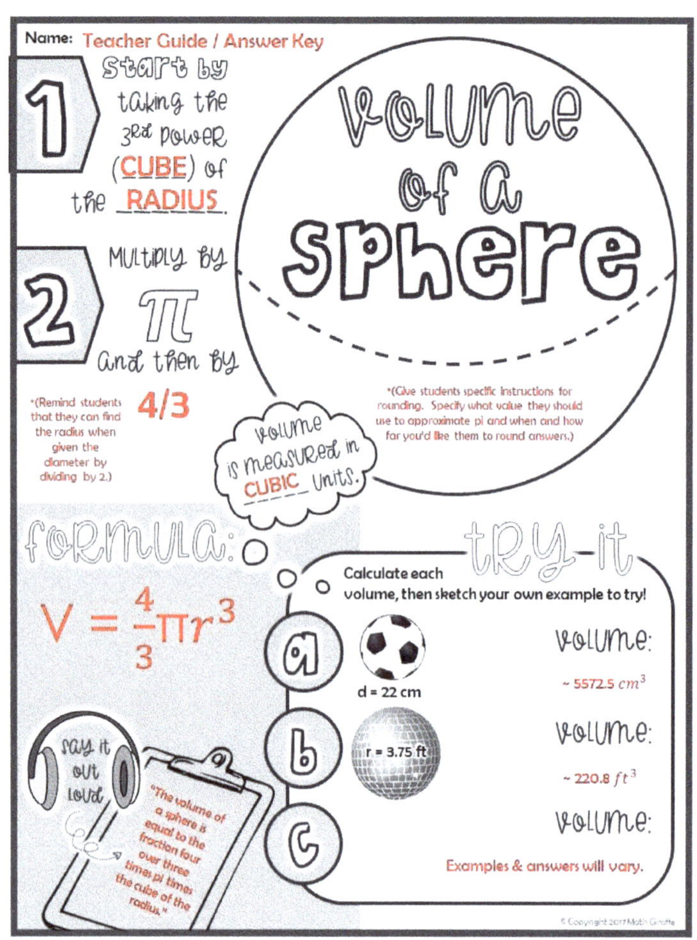
Name: Teacher Guide / Answer Key
Volume of a Sphere
1 Start by taking the 3rd power (CUBE) of the RADIUS.
2 Multiply by π and then by 4/3
*(Remind students that they can find the radius when given the diameter by dividing by 2.)
*(Give students specific instructions for rounding. Specify what value they should use to approximate pi and when and how far you'd like them to round answers.)
Volume is measured in CUBIC units.
formula:
$V = \frac{4}{3}\pi r^3$
try it
Calculate each volume, then sketch your own example to try!
a d = 22 cm Volume: ~ 5572.5 cm^3
b r = 3.75 ft Volume: ~ 220.8 ft^3
c Volume:
Examples & answers will vary.
say it out loud
"The volume of a sphere is equal to the fraction four over three times pi times the cube of the radius."
© Copyright 2017 Math Giraffe

Name: Teacher Guide / Answer Key
Volume of a Cone
1 Start by squaring the RADIUS.
2 Multiply by π
3 Multiply by 1/3 of the HEIGHT
HEIGHT
RADIUS
Label the height and the radius.
Volume is measured in CUBIC units.
full formula:
$V = \pi r^2 \frac{h}{3}$
try it
Calculate each volume, then sketch your own example to try!
a r = 1.3 ft h = 3 ft Volume: ~ 5.3 ft^3
b d = 4 cm h = 9 cm Volume: ~ 37.7 cm^3
c Volume:
Examples & answers will vary.
say it out loud
"The volume of a cone is equal to pi times the square of the radius times one third of the height."

Name:
Volume of a Sphere
Step 1 Start by taking the 3rd power (CUBE) of the radius.
Step 2 Multiply by π and then by 4/3
PI ≈ 3.14
Volume is measured in CUBED units.
ROUND final answers to the nearest TENTH
formula: 4/3πr³
★Remember, if given diameter, DIVIDE by 2 to find radius
try it
Calculate each volume, then sketch your own example to try!
a d = 22 cm 4/3 (3.14)(11³) 4/3 (3.14)(1331) Volume: ~5572.5 cm³
b r = 3.75 ft 4/3(3.14)(3.75³) 4/3(3.14)(52.73) Volume: ~220.8 ft³
c r=2m 4/3 (π)(2³) 4/3 (3.14)(8) Volume: ~33.5 m³
say it out loud
Four over three times pi (3.14) times the cube of the radius.
© Copyright 2017 Math Giraffe

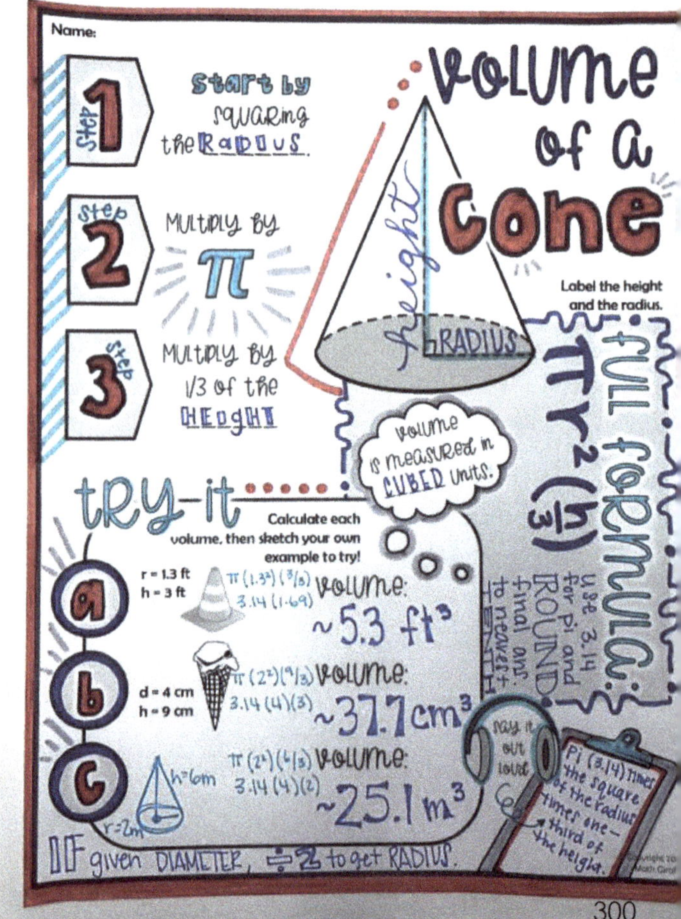
Name:
Volume of a Cone
Step 1 Start by squaring the Radius.
Step 2 Multiply by π
Step 3 Multiply by 1/3 of the HEIGHT
height
RADIUS
Label the height and the radius.
Volume is measured in CUBED units.
full formula: πr²(h/3)
Use 3.14 for pi and ROUND final ans. to nearest TENTH
try it
Calculate each volume, then sketch your own example to try!
a r = 1.3 ft h = 3 ft π (1.3²)(3/3) 3.14 (1.69) Volume: ~5.3 ft³
b d = 4 cm h = 9 cm π (2²)(9/3) 3.14 (4)(3) Volume: ~37.7 cm³
c h=6m r=2m π (2²)(6/3) 3.14 (4)(2) Volume: ~25.1 m³
If given DIAMETER, ÷2 to get RADIUS.
say it out loud
Pi (3.14) times the square of the radius times one-third of the height.

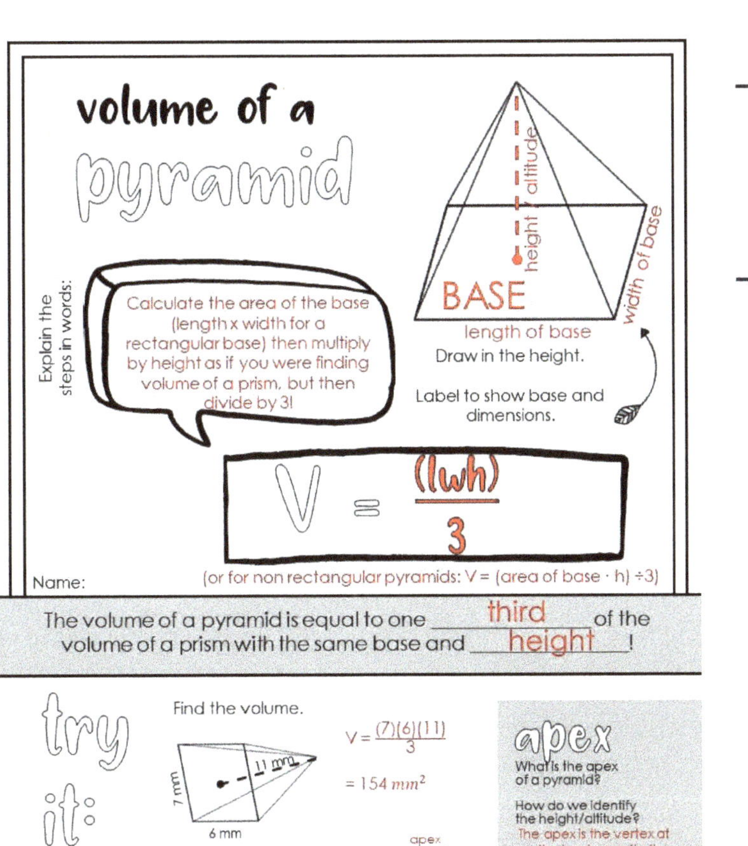

volume of a
pyramid
Explain the steps in words:
Calculate the area of the base (length x width for a rectangular base) then multiply by height as if you were finding volume of a prism, but then divide by 3!
height / altitude
BASE
length of base
width of base
Draw in the height.
Label to show base and dimensions.
V = (lwh) / 3
Name:
(or for non rectangular pyramids: V = (area of base · h) ÷3)

The volume of a pyramid is equal to one ___third___ of the the volume of a prism with the same base and ___height___!

try it:

Find the volume.

7 mm, 6 mm, 11 mm

$V = \frac{(7)(6)(11)}{3}$

$= 154\ mm^2$

apex

What is the apex of a pyramid?

How do we identify the height/altitude?

The apex is the vertex at the top (opposite the base). The height / altitude of a pyramid is the distance from the apex to the base.

apex

Reminder about units: Volume is represented in CUBIC units.

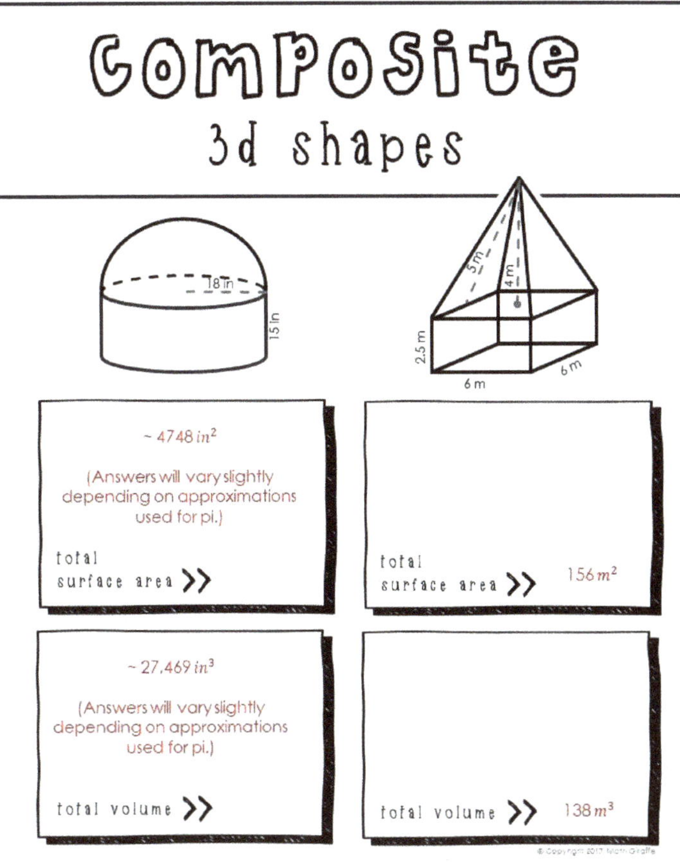

~ $4748\ in^2$ (Answers will vary slightly depending on approximations used for pi.) total surface area >>	total surface area >> $156\ m^2$
~ $27,469\ in^3$ (Answers will vary slightly depending on approximations used for pi.) total volume >>	total volume >> $138\ m^3$

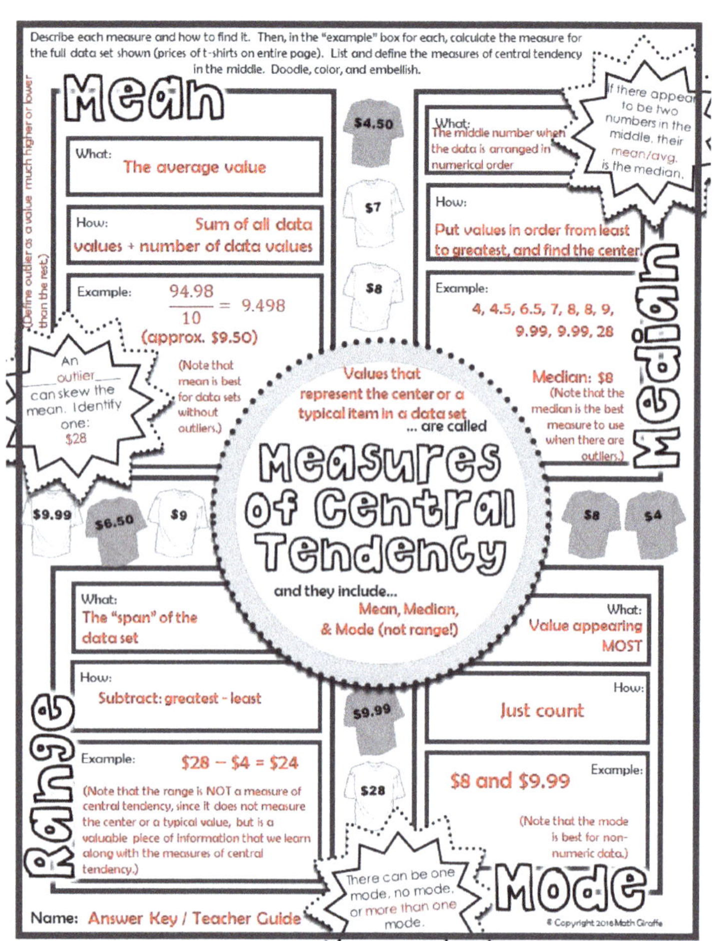

Describe each measure and how to find it. Then, in the "example" box for each, calculate the measure for the full data set shown (prices of t-shirts on entire page). List and define the measures of central tendency in the middle. Doodle, color, and embellish.
Mean
What: The average value
How: Sum of all data values ÷ number of data values
Example: 94.98 / 10 = 9.498 (approx. $9.50)
(Note that mean is best for data sets without outliers.)
An outlier can skew the mean. Identify one: $28
(Define outlier as a value much higher or lower than the rest.)
If there appear to be two numbers in the middle, their mean/avg. is the median.
What: The middle number when the data is arranged in numerical order
How: Put values in order from least to greatest, and find the center.
Example: 4, 4.5, 6.5, 7, 8, 8, 9, 9.99, 9.99, 28
Median: $8
(Note that the median is the best measure to use when there are outliers.)
Median
Values that represent the center or a typical item in a data set ... are called
Measures of Central Tendency
and they include... Mean, Median, & Mode (not range!)
$4.50
$7
$8
$9.99
$6.50
$9
$8
$4
$9.99
$28
Range
What: The "span" of the data set
How: Subtract: greatest - least
Example: $28 – $4 = $24
(Note that the range is NOT a measure of central tendency, since it does not measure the center or a typical value, but is a valuable piece of information that we learn along with the measures of central tendency.)
Mode
What: Value appearing MOST
How: Just count
Example: $8 and $9.99
(Note that the mode is best for non-numeric data.)
There can be one mode, no mode, or more than one mode.
Name: Answer Key / Teacher Guide
© Copyright 2016 Math Giraffe

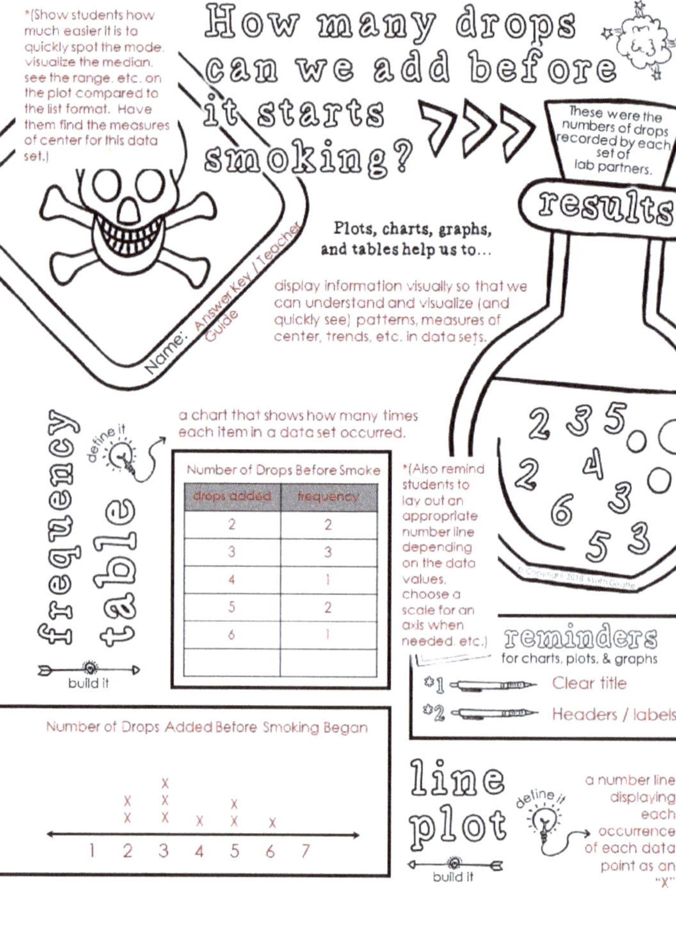

*(Show students how much easier it is to quickly spot the mode, visualize the median, see the range, etc. on the plot compared to the list format. Have them find the measures of center for this data set.)
How many drops can we add before it starts smoking?
These were the numbers of drops recorded by each set of lab partners.
results
Name: Answer Key / Teacher Guide
Plots, charts, graphs, and tables help us to...
display information visually so that we can understand and visualize (and quickly see) patterns, measures of center, trends, etc. in data sets.
2 3 5 2 4 6 3 5 3
frequency table
define it
a chart that shows how many times each item in a data set occurred.
build it
Number of Drops Before Smoke
drops added | frequency
2 | 2
3 | 3
4 | 1
5 | 2
6 | 1
*(Also remind students to lay out an appropriate number line depending on the data values, choose a scale for an axis when needed, etc.)
reminders for charts, plots, & graphs
*1 Clear title
*2 Headers / labels
Number of Drops Added Before Smoking Began
1 2 3 4 5 6 7
line plot
define it
a number line displaying each occurrence of each data point as an "X"
build it

Mean
AVERAGE
Best if NO outliers
Sum of ALL data values ÷ # of data values
94.98 / 10 = 9.498 or ~$9.50
MIDDLE #
If there appear to be two numbers in the middle, their MEAN is the median.
Put in order L→G, find center
Median
Measures of Central Tendency
mean, mode, + median *NOT RANGE
Range
"SPAN" of the DATA
SUBTRACT greatest MINUS least
28-4=24
*The range is NOT a meas. of Centr. Tend.
Just COUNT
Value appearing MOST
(best for non-numeric data)
$8 & 9.99 (both appear twice)
There can be one mode, no mode, or MORE THAN 1 mode.
Mode

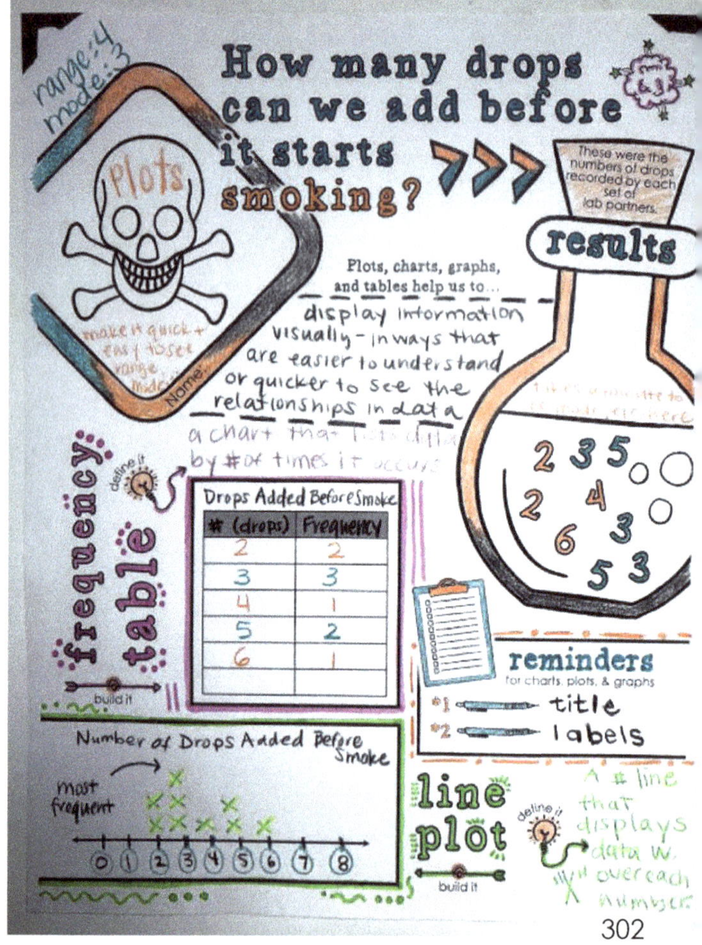

range: 4
mode: 3
How many drops can we add before it starts smoking?
These were the numbers of drops recorded by each set of lab partners.
results
Plots
Plots, charts, graphs, and tables help us to...
display information visually - in ways that are easier to understand or quicker to see the relationships in data.
2 3 5 2 4 6 3 5 3
frequency table
by # of times it occurs
Drops Added Before Smoke
(drops) | Frequency
2 | 2
3 | 3
4 | 1
5 | 2
6 | 1
reminders for charts, plots, & graphs
*1 title
*2 labels
Number of Drops Added Before Smoke
most frequent
0 1 2 3 4 5 6 7 8
line plot

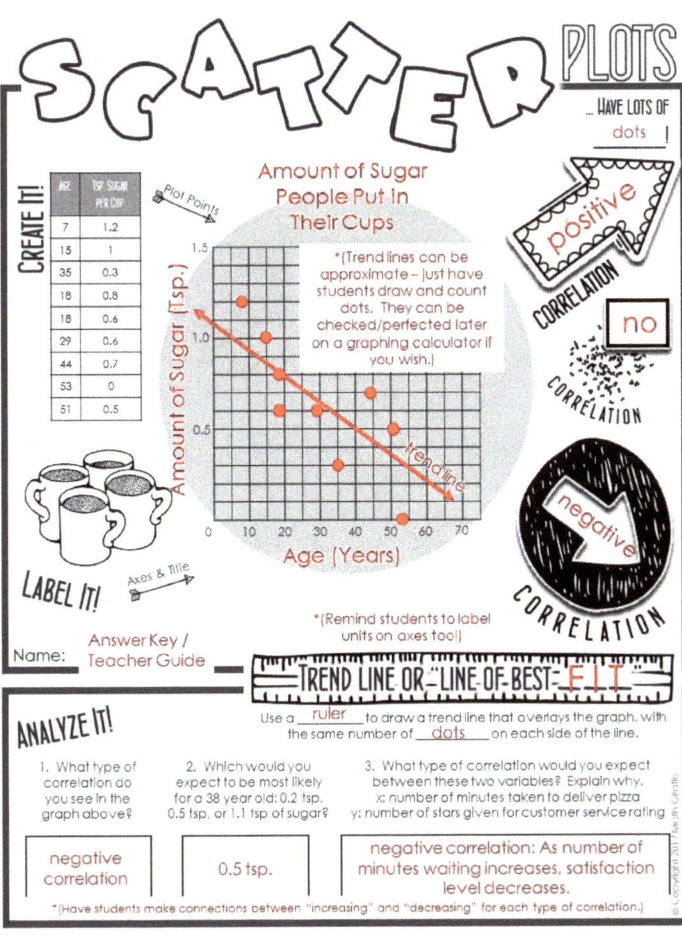
SCATTER PLOTS
...HAVE LOTS OF dots!
CREATE IT!
AGE
TSP. SUGAR PER CUP
7 1.2
15 1
35 0.3
18 0.8
18 0.6
29 0.6
44 0.7
53 0
51 0.5
Plot Points
Amount of Sugar People Put In Their Cups
*(Trend lines can be approximate – just have students draw and count dots. They can be checked/perfected later on a graphing calculator if you wish.)
Amount of Sugar (Tsp.)
Age (Years)
trend line
positive CORRELATION
no CORRELATION
negative CORRELATION
LABEL IT!
Axes & Title
*(Remind students to label units on axes too!)
Name: Answer Key / Teacher Guide
TREND LINE OR "LINE OF BEST FIT"
ANALYZE IT!
Use a ruler to draw a trend line that overlays the graph, with the same number of dots on each side of the line.
1. What type of correlation do you see in the graph above?
2. Which would you expect to be most likely for a 38 year old: 0.2 tsp, 0.5 tsp, or 1.1 tsp of sugar?
3. What type of correlation would you expect between these two variables? Explain why. x: number of minutes taken to deliver pizza y: number of stars given for customer service rating
negative correlation
0.5 tsp.
negative correlation: As number of minutes waiting increases, satisfaction level decreases.
*(Have students make connections between "increasing" and "decreasing" for each type of correlation.)
© Copyright 2017 Math Giraffe

BOX & WHISKER PLOTS
Name: Answer Key / Teacher Guide
display... The median and the spread of a data set
THE BOX represents the MIDDLE HALF of the data set. (goes from 1st quartile to the 3rd quartile)
THE WHISKERS show how wide the SPREAD of the data is (go from the box out in both directions to the highest and lowest data value)
5-NUMBER SUMMARY: minimum, 1st quartile (median of the lower half), median, 3rd quartile (median of the upper half), and maximum (see 5 arrows below for the 5 key numbers in the summary)
Q1
MEDIAN
Q3
Minimum (lowest data value)
1st (or LOWER) quartile: the median of the lower half
MEDIAN: Middle number in the data set (also called the 2nd quartile)
3rd (or UPPER) quartile: the median of the upper half
Maximum (highest data value)
QUARTILES
To divide the data into 4 quartiles, we first cut it in HALF, using the median as the center. Then, we find the median of each half.
*(Use this area to talk about the "CENTERS" – this helps students visualize finding the median (the central data point), then using that to cut the data in half and again find the median of each half. They can label each using the 5 number summary values, and see how the data is cut into 4 quartiles.)

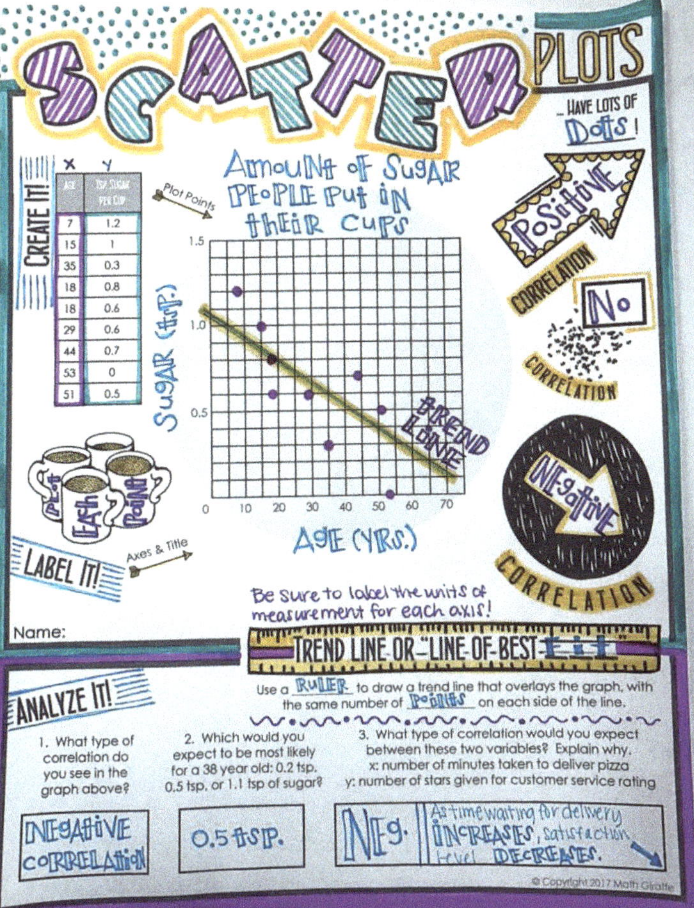
SCATTER PLOTS
...HAVE LOTS OF Dots!
CREATE IT!
X Y
Plot Points
Amount of Sugar People Put in Their Cups
SUGAR (tsp.)
AGE (YRS.)
TREND LINE
Positive CORRELATION
No CORRELATION
NEGATIVE CORRELATION
LABEL IT!
Axes & Title
Be sure to label the units of measurement for each axis!
Name:
TREND LINE OR "LINE OF BEST FIT"
ANALYZE IT!
Use a RULER to draw a trend line that overlays the graph, with the same number of POINTS on each side of the line.
1. What type of correlation do you see in the graph above?
2. Which would you expect to be most likely for a 38 year old: 0.2 tsp, 0.5 tsp, or 1.1 tsp of sugar?
3. What type of correlation would you expect between these two variables? Explain why. x: number of minutes taken to deliver pizza y: number of stars given for customer service rating
NEGATIVE CORRELATION
0.5 tsp.
Neg. As time waiting for delivery INCREASES, satisfaction level DECREASES.
© Copyright 2017 Math Giraffe

BOX & WHISKER PLOTS
THE WHISKERS
SPREAD
THE BOX
MIDDLE half
5-NUMBER SUMMARY
median
QUARTILES

BUILD A BOX & WHISKER PLOT
Create a box and whisker plot to display the data from the table.
HOUSE PRICES
House Prices in Stark County (in thousands)
$278.8 $154
$159 $230
$220 $245
$175.5 $389.9
$169 $89
$179 $122
$230 $456
$289 $269.9
$140 $189
$129 $215
$309 $165
*Remind students how to take the mean of the two middle numbers when finding the median of a data set with an even number of values.
1 Arrange the data in order from least to greatest
89 122 129 140 154 159 165 169 175.5 179 189
215 220 230 230 245 269.9 278.8 289 309 389.9 456
Find the median. Median: 202
2 Identify the lower and the upper quartiles by finding the median of each HALF of the data set (using the median as a divider to identify the two halves)
Q1: 159 Q3: 269.9
3 Create a number line and mark: A Median B Q1 & Q3 C Highest & lowest
Then draw in the box according to the quartiles
(Guide students to find an appropriate scale and markings for the number line if needed.)
HOUSE PRICES
$100,000 $200,000 $300,000 $400,000 $500,000
DON'T FORGET! A title and labels!
Name: Answer Key / Teacher Guide

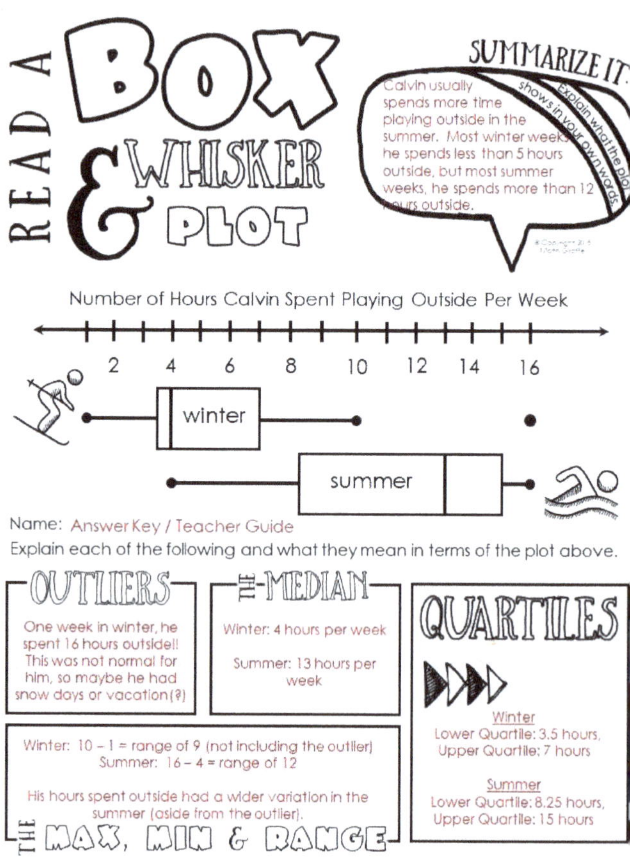
READ A BOX & WHISKER PLOT
SUMMARIZE IT: Explain what the plot shows in your own words.
Calvin usually spends more time playing outside in the summer. Most winter weeks he spends less than 5 hours outside, but most summer weeks, he spends more than 12 hours outside.
Number of Hours Calvin Spent Playing Outside Per Week
2 4 6 8 10 12 14 16
winter
summer
Name: Answer Key / Teacher Guide
Explain each of the following and what they mean in terms of the plot above.
OUTLIERS
One week in winter, he spent 16 hours outside!! This was not normal for him, so maybe he had snow days or vacation(?)
THE MEDIAN
Winter: 4 hours per week
Summer: 13 hours per week
QUARTILES
Winter
Lower Quartile: 3.5 hours, Upper Quartile: 7 hours
Summer
Lower Quartile: 8.25 hours, Upper Quartile: 15 hours
Winter: 10 – 1 = range of 9 (not including the outlier)
Summer: 16 – 4 = range of 12
His hours spent outside had a wider variation in the summer (aside from the outlier).
THE MAX, MIN & RANGE

BUILD A BOX & WHISKER PLOT
HOUSE PRICES
HOUSE PRICES
DON'T FORGET! TITLE & labels

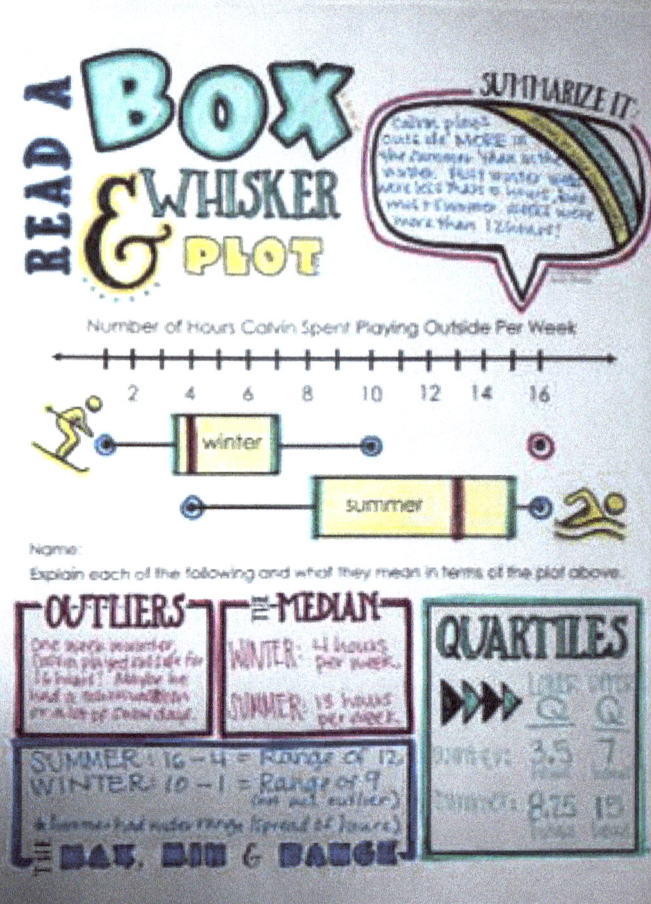
READ A BOX & WHISKER PLOT
SUMMARIZE IT:
Number of Hours Calvin Spent Playing Outside Per Week
2 4 6 8 10 12 14 16
winter
summer
Explain each of the following and what they mean in terms of the plot above.
OUTLIERS
THE MEDIAN
WINTER: 4 hours per week
SUMMER: 13 hours per week
QUARTILES
SUMMER: 16 – 4 = Range of 12
WINTER: 10 – 1 = Range of 9
THE MAX, MIN & RANGE

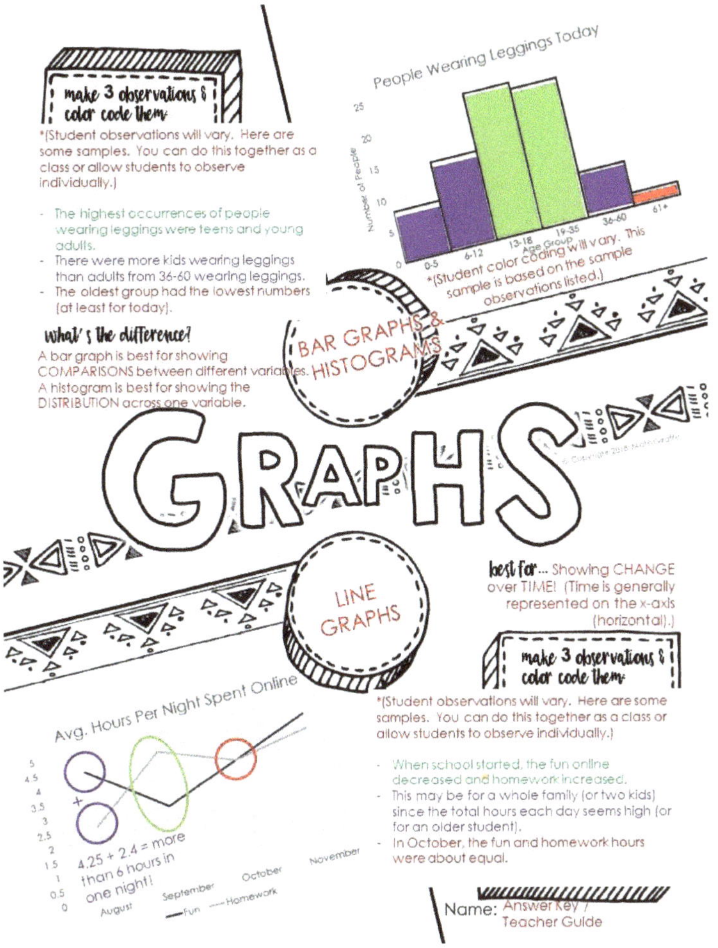

GRAPHS

make 3 observations & color code them

*(Student observations will vary. Here are some samples. You can do this together as a class or allow students to observe individually.)

- The highest occurrences of people wearing leggings were teens and young adults.
- There were more kids wearing leggings than adults from 36-60 wearing leggings.
- The oldest group had the lowest numbers (at least for today).

People Wearing Leggings Today

Number of People: 5, 10, 15, 20, 25

Age Group: 0-5, 6-12, 13-18, 19-35, 36-60, 61+

*(Student color coding will vary. This sample is based on the sample observations listed.)

BAR GRAPHS & HISTOGRAMS

what's the difference?

A bar graph is best for showing COMPARISONS between different variables. A histogram is best for showing the DISTRIBUTION across one variable.

LINE GRAPHS

best for... Showing CHANGE over TIME! (Time is generally represented on the x-axis (horizontal).)

make 3 observations & color code them

*(Student observations will vary. Here are some samples. You can do this together as a class or allow students to observe individually.)

- When school started, the fun online decreased and homework increased.
- This may be for a whole family (or two kids) since the total hours each day seems high (or for an older student).
- In October, the fun and homework hours were about equal.

Avg. Hours Per Night Spent Online

0, 0.5, 1, 1.5, 2, 2.5, 3, 3.5, 4, 4.5, 5

August, September, October, November

—Fun —Homework

4.25 + 2.4 = more than 6 hours in one night!

Name: Answer Key / Teacher Guide

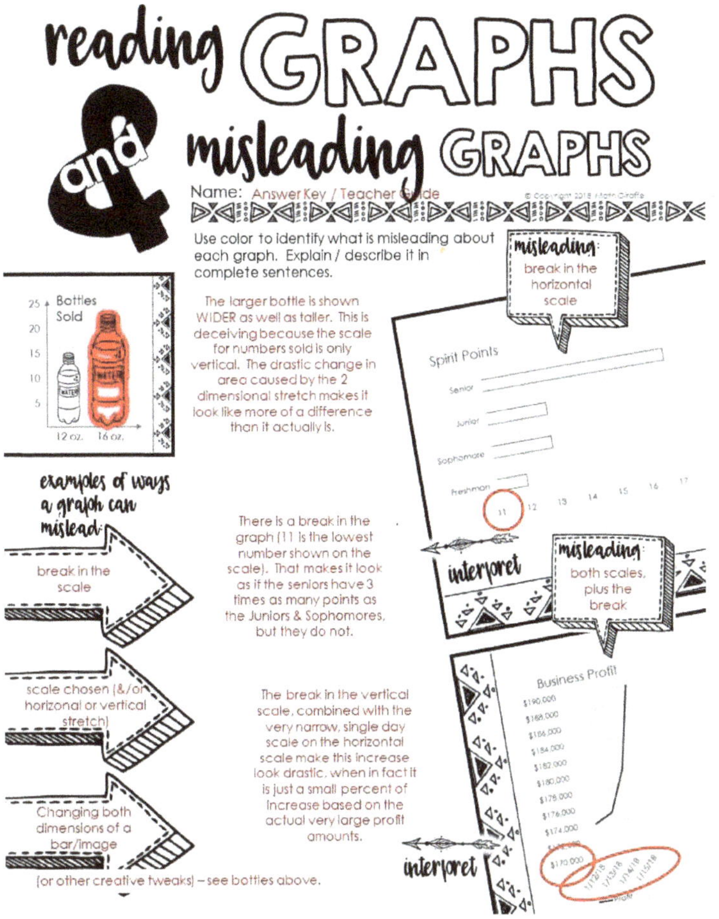

reading GRAPHS & misleading GRAPHS

Name: Answer Key / Teacher Guide

Use color to identify what is misleading about each graph. Explain / describe it in complete sentences.

Bottles Sold

5, 10, 15, 20, 25

12 oz. 16 oz.

The larger bottle is shown WIDER as well as taller. This is deceiving because the scale for numbers sold is only vertical. The drastic change in area caused by the 2 dimensional stretch makes it look like more of a difference than it actually is.

misleading: break in the horizontal scale

Spirit Points

Senior, Junior, Sophomore, Freshman

11, 12, 13, 14, 15, 16, 17

examples of ways a graph can mislead:

- break in the scale
- scale chosen (&/or horizontal or vertical stretch)
- Changing both dimensions of a bar/image

(or other creative tweaks) – see bottles above.

interpret

There is a break in the graph (11 is the lowest number shown on the scale). That makes it look as if the seniors have 3 times as many points as the Juniors & Sophomores, but they do not.

misleading: both scales, plus the break

Business Profit

$170,000, $172,000, $174,000, $176,000, $178,000, $180,000, $182,000, $184,000, $186,000, $188,000, $190,000

interpret

The break in the vertical scale, combined with the very narrow, single day scale on the horizontal scale make this increase look drastic, when in fact it is just a small percent of increase based on the actual very large profit amounts.

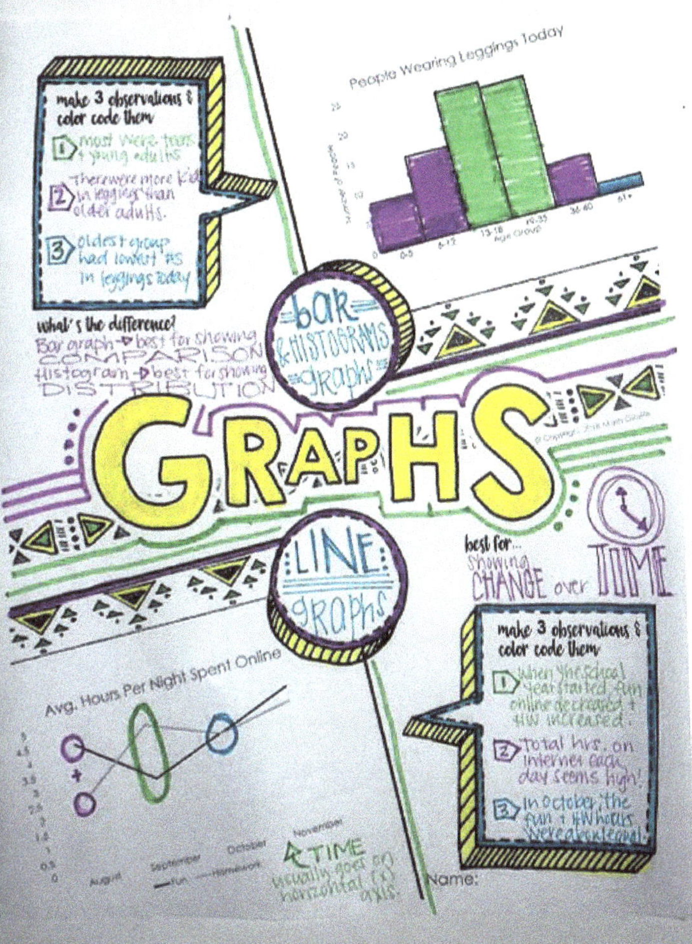

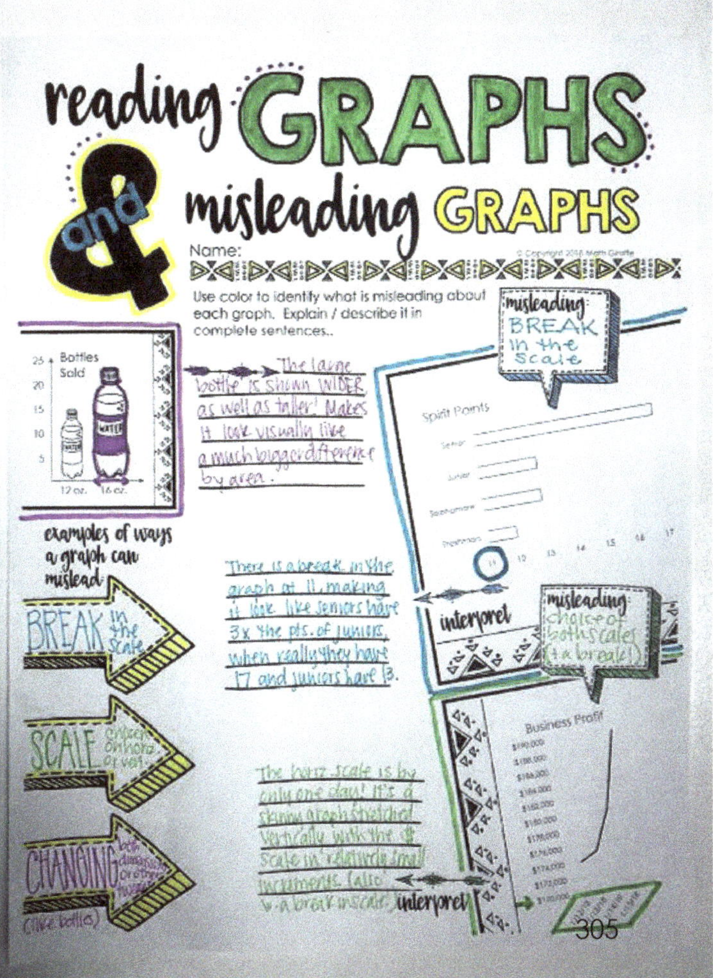

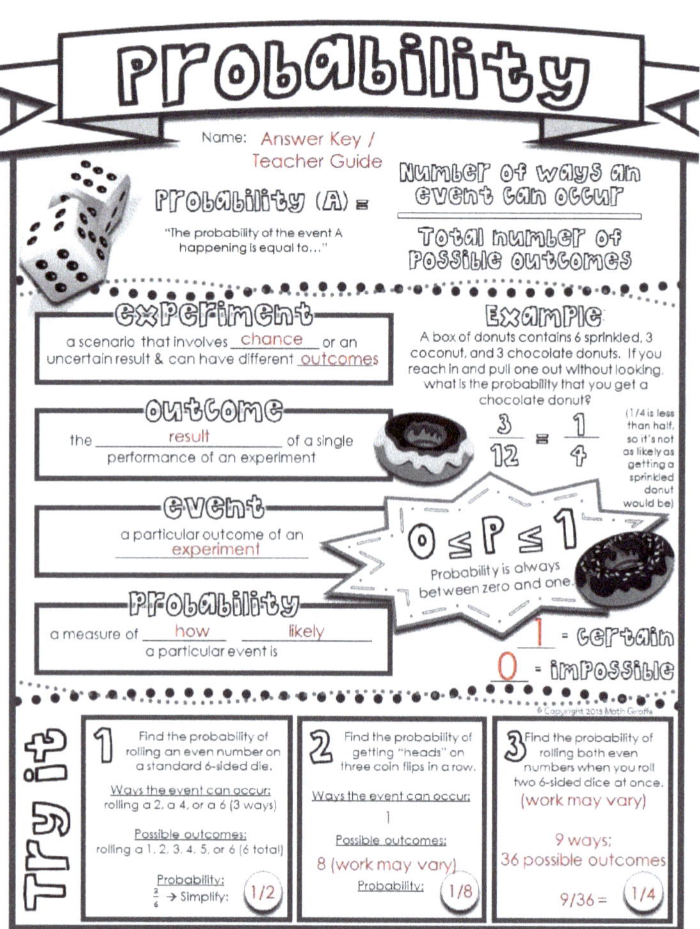
Probability
Name: Answer Key / Teacher Guide
Probability (A) =
"The probability of the event A happening is equal to…"
Number of ways an event can occur
Total number of possible outcomes
experiment
a scenario that involves chance or an uncertain result & can have different outcomes
outcome
the result of a single performance of an experiment
event
a particular outcome of an experiment
probability
a measure of how likely a particular event is
Example
A box of donuts contains 6 sprinkled, 3 coconut, and 3 chocolate donuts. If you reach in and pull one out without looking, what is the probability that you get a chocolate donut?
3/12 = 1/4
(1/4 is less than half, so it's not as likely as getting a sprinkled donut would be)
0 ≤ P ≤ 1
Probability is always between zero and one.
1 = certain
0 = impossible
Try it
1 Find the probability of rolling an even number on a standard 6-sided die.
Ways the event can occur: rolling a 2, a 4, or a 6 (3 ways)
Possible outcomes: rolling a 1, 2, 3, 4, 5, or 6 (6 total)
Probability: 3/6 → Simplify: 1/2
2 Find the probability of getting "heads" on three coin flips in a row.
Ways the event can occur: 1
Possible outcomes: 8 (work may vary)
Probability: 1/8
3 Find the probability of rolling both even numbers when you roll two 6-sided dice at once.
(work may vary)
9 ways; 36 possible outcomes
9/36 = 1/4

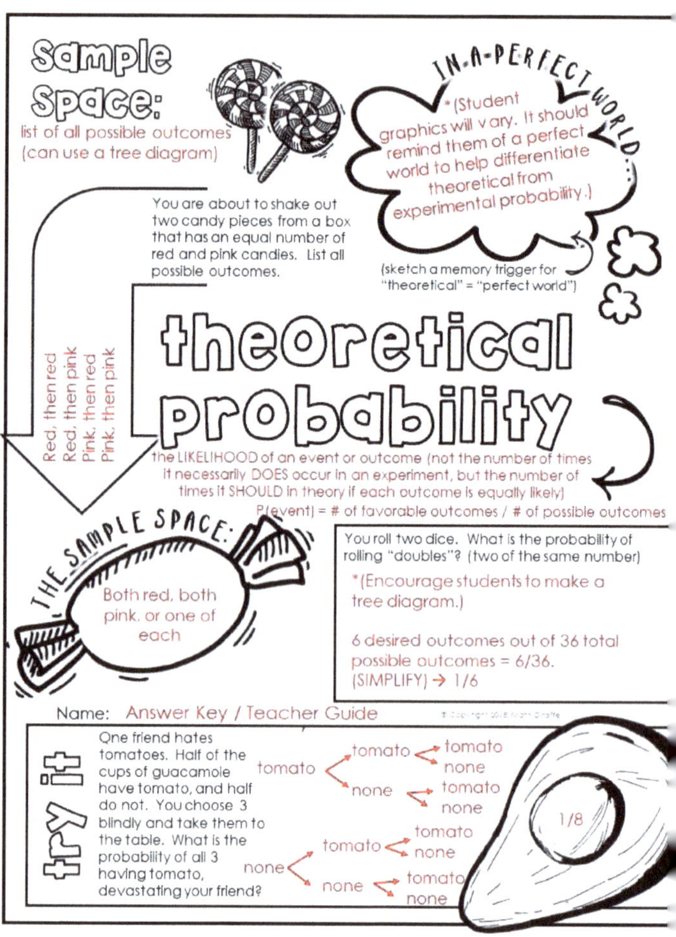
Sample Space:
list of all possible outcomes (can use a tree diagram)
IN-A-PERFECT WORLD
*(Student graphics will vary. It should remind them of a perfect world to help differentiate theoretical from experimental probability.)
(sketch a memory trigger for "theoretical" = "perfect world")
You are about to shake out two candy pieces from a box that has an equal number of red and pink candies. List all possible outcomes.
Red, then red
Red, then pink
Pink, then red
Pink, then pink
theoretical probability
the LIKELIHOOD of an event or outcome (not the number of times it necessarily DOES occur in an experiment, but the number of times it SHOULD in theory if each outcome is equally likely)
P(event) = # of favorable outcomes / # of possible outcomes
THE SAMPLE SPACE:
Both red, both pink, or one of each
You roll two dice. What is the probability of rolling "doubles"? (two of the same number)
*(Encourage students to make a tree diagram.)
6 desired outcomes out of 36 total possible outcomes = 6/36. (SIMPLIFY) → 1/6
Name: Answer Key / Teacher Guide
Try it
One friend hates tomatoes. Half of the cups of guacamole have tomato, and half do not. You choose 3 blindly and take them to the table. What is the probability of all 3 having tomato, devastating your friend?
tomato, tomato, tomato, none, none, tomato, none
none, tomato, tomato, none, none, tomato, none
1/8

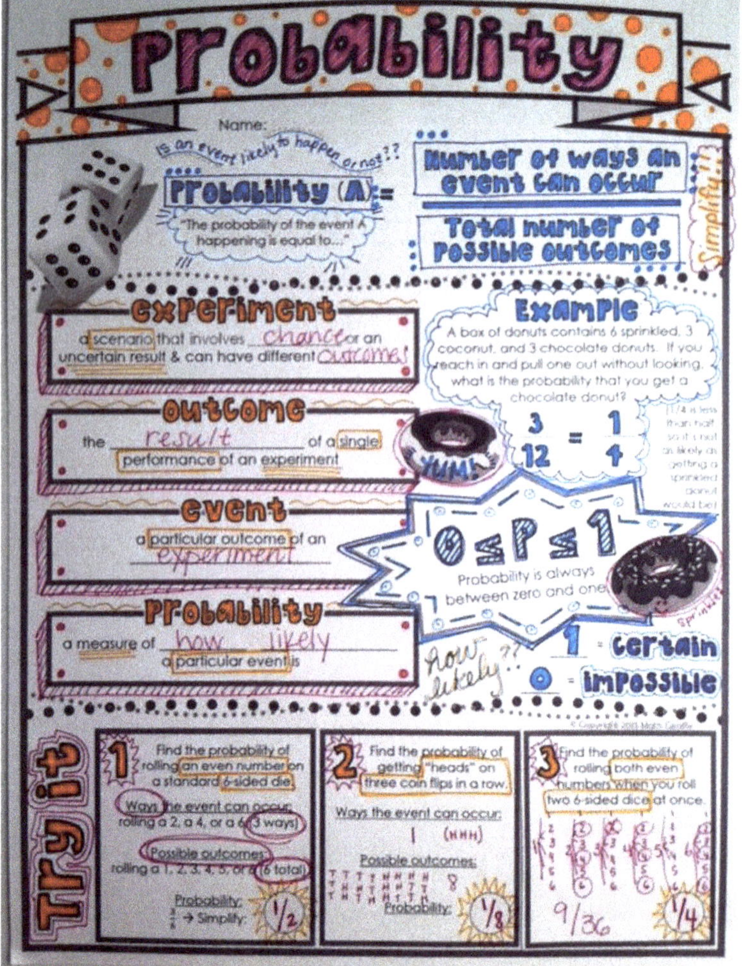
Probability
Is an event likely to happen or not??
Simplify!!
YUM!
How likely??

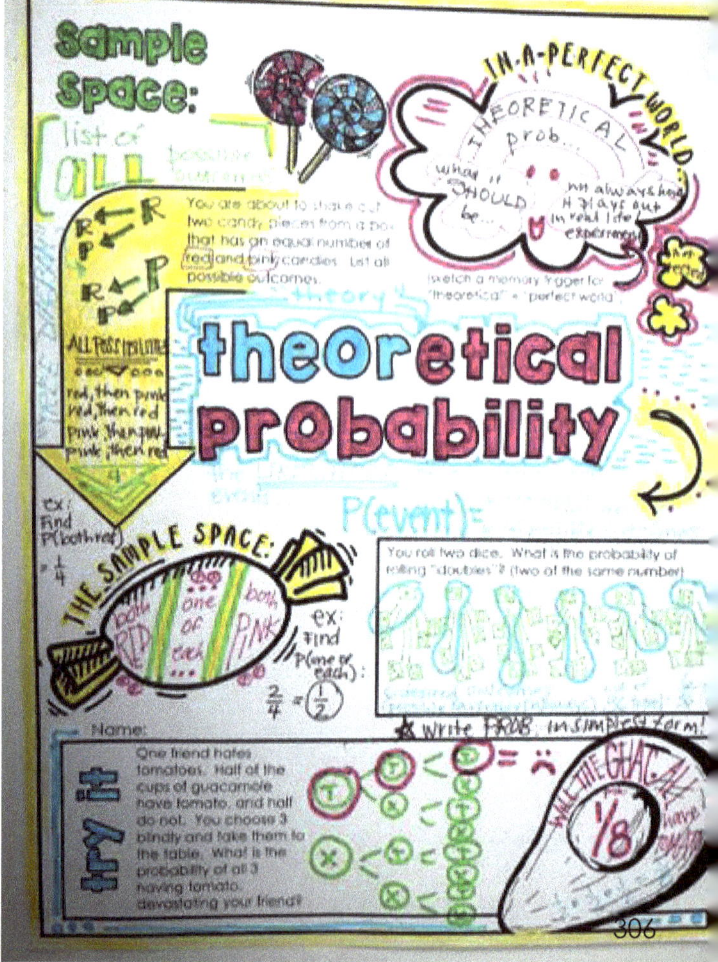
Sample Space:
IN-A-PERFECT WORLD
THEORETICAL prob.
ALL POSSIBILITIES
ex: Find P(both red) 1/4
P(event) =
ex: Find P(one of each): 2/4 = 1/2
Write PROB. in simplest form!
SAVE THE GUAC!
1/8

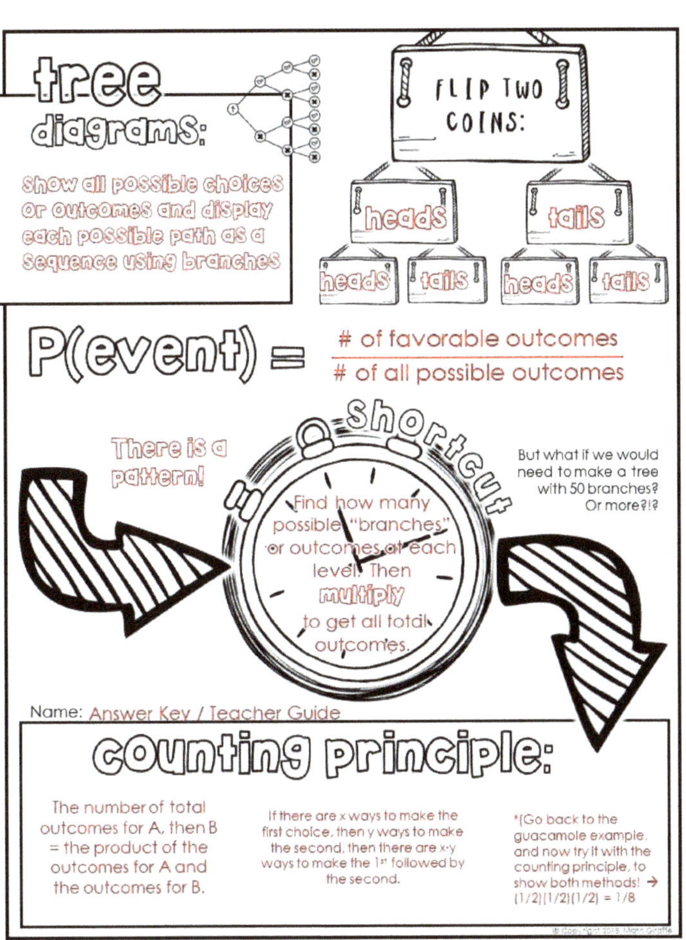

tree diagrams:
show all possible choices or outcomes and display each possible path as a sequence using branches
FLIP TWO COINS:
heads
tails
heads
tails
heads
tails
P(event) =
# of favorable outcomes
# of all possible outcomes
shortcut
There is a pattern!
Find how many possible "branches" or outcomes at each level. Then multiply to get all total outcomes.
But what if we would need to make a tree with 50 branches? Or more?!?
Name: Answer Key / Teacher Guide
counting principle:
The number of total outcomes for A, then B = the product of the outcomes for A and the outcomes for B.
If there are x ways to make the first choice, then y ways to make the second, then there are x·y ways to make the 1st followed by the second.
*(Go back to the guacamole example, and now try it with the counting principle, to show both methods! →
(1/2)(1/2)(1/2) = 1/8

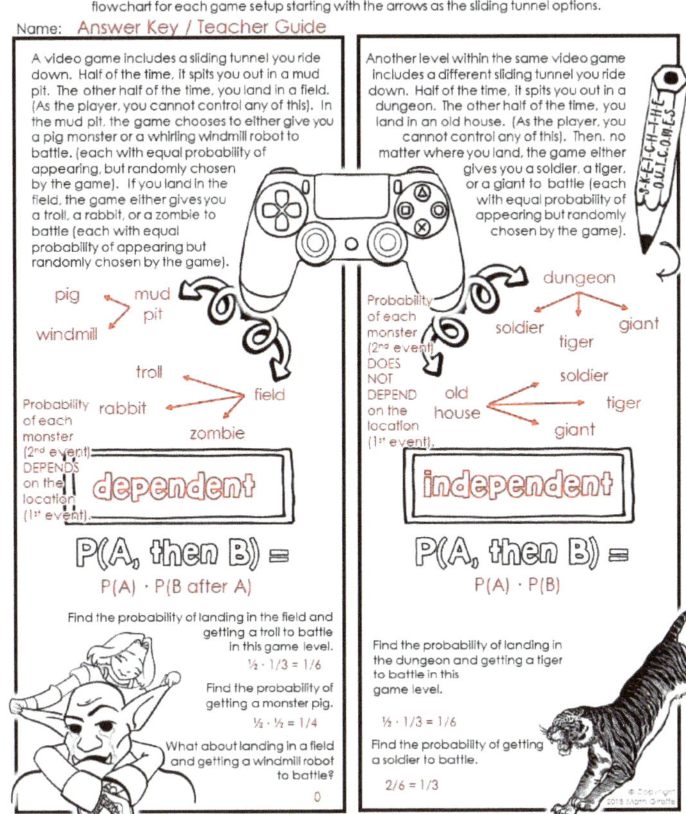

independent & dependent events
Read carefully, then determine whether the monster you battle DOES depend on where the slide spits you out, or whether it DOES NOT. Write the label "DEPENDENT" or "INDEPENDENT" in the box. Draw a flowchart for each game setup starting with the arrows as the sliding tunnel options.
Name: Answer Key / Teacher Guide
A video game includes a sliding tunnel you ride down. Half of the time, it spits you out in a mud pit. The other half of the time, you land in a field. (As the player, you cannot control any of this). In the mud pit, the game chooses to either give you a pig monster or a whirling windmill robot to battle. (each with equal probability of appearing, but randomly chosen by the game). If you land in the field, the game either gives you a troll, a rabbit, or a zombie to battle (each with equal probability of appearing but randomly chosen by the game).
Another level within the same video game includes a different sliding tunnel you ride down. Half of the time, it spits you out in a dungeon. The other half of the time, you land in an old house. (As the player, you cannot control any of this). Then, no matter where you land, the game either gives you a soldier, a tiger, or a giant to battle (each with equal probability of appearing but randomly chosen by the game).
SKETCH THE OUTCOMES
pig
windmill
mud pit
troll
rabbit
zombie
field
Probability of each monster (2nd event) DEPENDS on the location (1st event).
dependent
P(A, then B) =
P(A) · P(B after A)
Find the probability of landing in the field and getting a troll to battle in this game level.
½ · 1/3 = 1/6
Find the probability of getting a monster pig.
½ · ½ = 1/4
What about landing in a field and getting a windmill robot to battle?
0
dungeon
soldier
tiger
giant
old house
soldier
tiger
giant
Probability of each monster (2nd event) DOES NOT DEPEND on the location (1st event).
independent
P(A, then B) =
P(A) · P(B)
Find the probability of landing in the dungeon and getting a tiger to battle in this game level.
½ · 1/3 = 1/6
Find the probability of getting a soldier to battle.
2/6 = 1/3

tree diagrams:
Show possible choices or outcomes and display each possibility using branches for each possible outcome
FLIP TWO COINS:
H
T
H
T
H
T
P(event) =
# of favorable outcomes
# of all possible outcomes
Is there a pattern?
shortcut
MULTIPLY
But what if we would need to make a tree with 50 branches? Or more?!?
um... NO.
How can we more quickly determine # of poss. "branches" at each level?
Save TIME with the COUNTING PRINC.
Try this.
Name:
counting principle:
The # of outcomes for A then B equals the PRODUCT of outcomes for A & outcomes for B.
example: If there are 8 ways to make a 1st choice and then 5 ways to make a 2nd choice, there are 40 ways to make the 1st followed by the 2nd.
8(5)=40

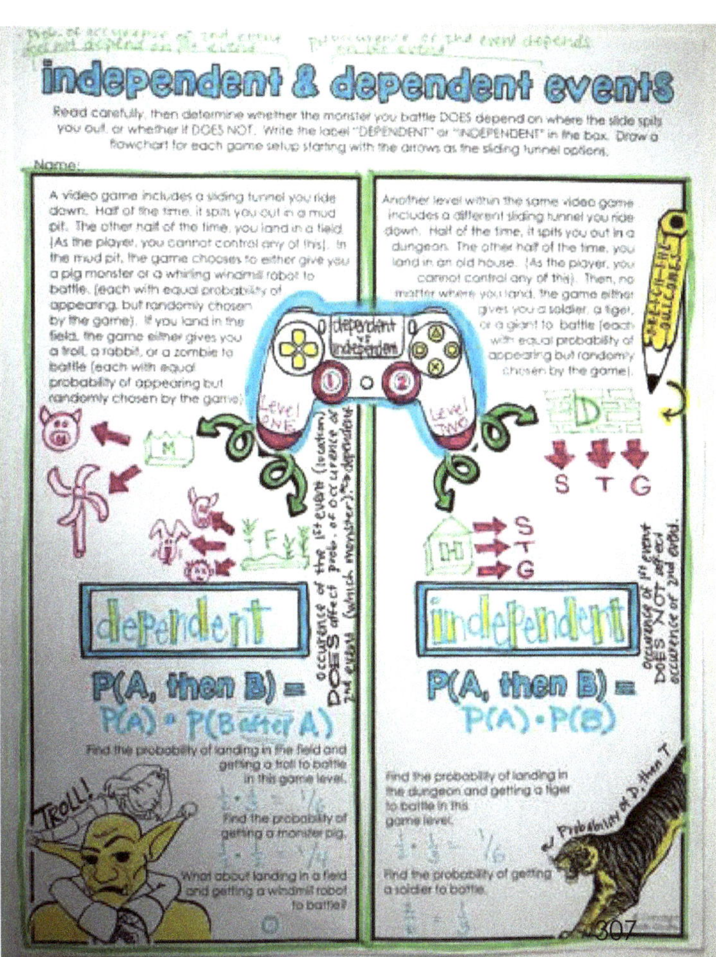

independent & dependent events
Read carefully, then determine whether the monster you battle DOES depend on where the slide spits you out, or whether it DOES NOT. Write the label "DEPENDENT" or "INDEPENDENT" in the box. Draw a flowchart for each game setup starting with the arrows as the sliding tunnel options.
Name:
dependent or independent
Level ONE
Level TWO
S
T
G
S
T
G
dependent
independent
P(A, then B) =
P(A) · P(B after A)
P(A, then B) =
P(A) · P(B)
TROLL!

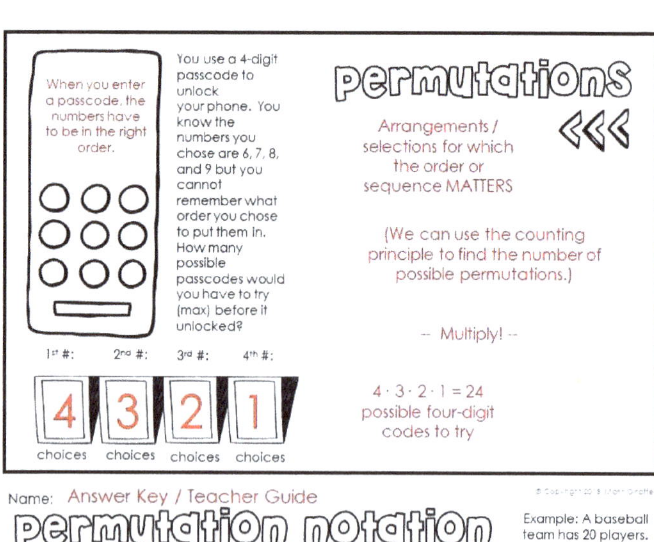
When you enter a passcode, the numbers have to be in the right order.
You use a 4-digit passcode to unlock your phone. You know the numbers you chose are 6, 7, 8, and 9 but you cannot remember what order you chose to put them in. How many possible passcodes would you have to try (max) before it unlocked?
permutations
Arrangements / selections for which the order or sequence MATTERS
(We can use the counting principle to find the number of possible permutations.)
-- Multiply! --
1st #: 2nd #: 3rd #: 4th #:
4 3 2 1
choices choices choices choices
4 · 3 · 2 · 1 = 24 possible four-digit codes to try

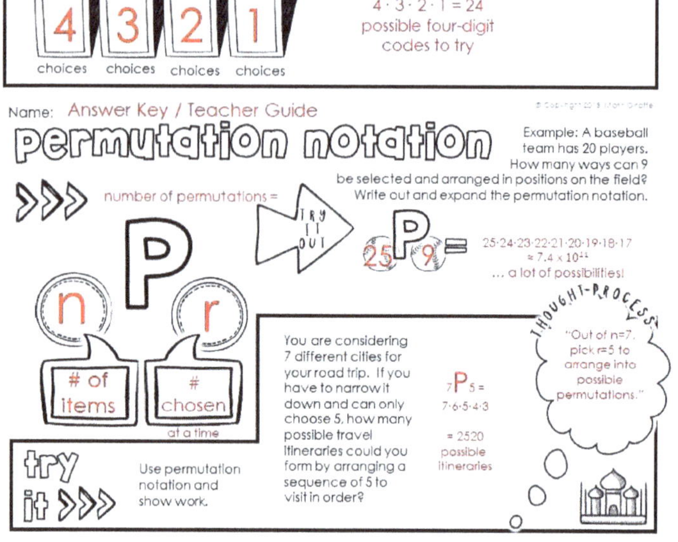
Name: Answer Key / Teacher Guide
permutation notation
number of permutations =
$_nP_r$
# of items
chosen at a time
Example: A baseball team has 20 players. How many ways can 9 be selected and arranged in positions on the field? Write out and expand the permutation notation.
TRY IT OUT
$_{25}P_9$ = 25·24·23·22·21·20·19·18·17 ≈ 7.4 × 10[11] ... a lot of possibilities!
You are considering 7 different cities for your road trip. If you have to narrow it down and can only choose 5, how many possible travel itineraries could you form by arranging a sequence of 5 to visit in order?
$_7P_5$ = 7·6·5·4·3 = 2520 possible itineraries
THOUGHT-PROCESS
"Out of n=7, pick r=5 to arrange into possible permutations."
try it
Use permutation notation and show work.

combinations
Arrangements / selections for which order / sequence does NOT matter
PIZZA TOPPINGS
- Pepperoni - Peppers
- Sausage - Tomato
- Mushrooms - Bacon
- Olives - Pineapple
How many possibilities are there for a pizza with two toppings?
To find the number of combinations, find the total # of permutations, then divide by the number of possible permutations for any group of r items.
notation
$_nC_r = \frac{_nP_r}{_rP_r}$
# of items
chosen
*(Have students test this manually by drawing lines between each possible pair of toppings on the menu systematically starting with the first two. Then show why this method saves time and how it is derived from that process.)
So for a pizza with two toppings, we represent it as:
$_8C_2 = \frac{_8P_2}{_2P_2} = \frac{8 \cdot 7}{2 \cdot 1} = \frac{56}{2} = 28$ possible combinations
Name: Answer Key / Teacher Guide
try it
Use combination notation and show work.
How many possibilities are there for a pizza with three toppings?
$_8C_3 = \frac{_8P_3}{_3P_3} = \frac{8 \cdot 7 \cdot 6}{3 \cdot 2 \cdot 1} = \frac{336}{6} = 56$ possible combinations

ENTER
permutations
arrangements / selections for which the ORDER / sequence MATTERS
We can use the COUNTING PRINCIPLE to find the total # of poss. permutations
4 3 2 1
24
permutation notation
# of permutations =
7P5
=7·6·5·4·3
=2520 possible itineraries!
"out of 7, pick 5"
try it

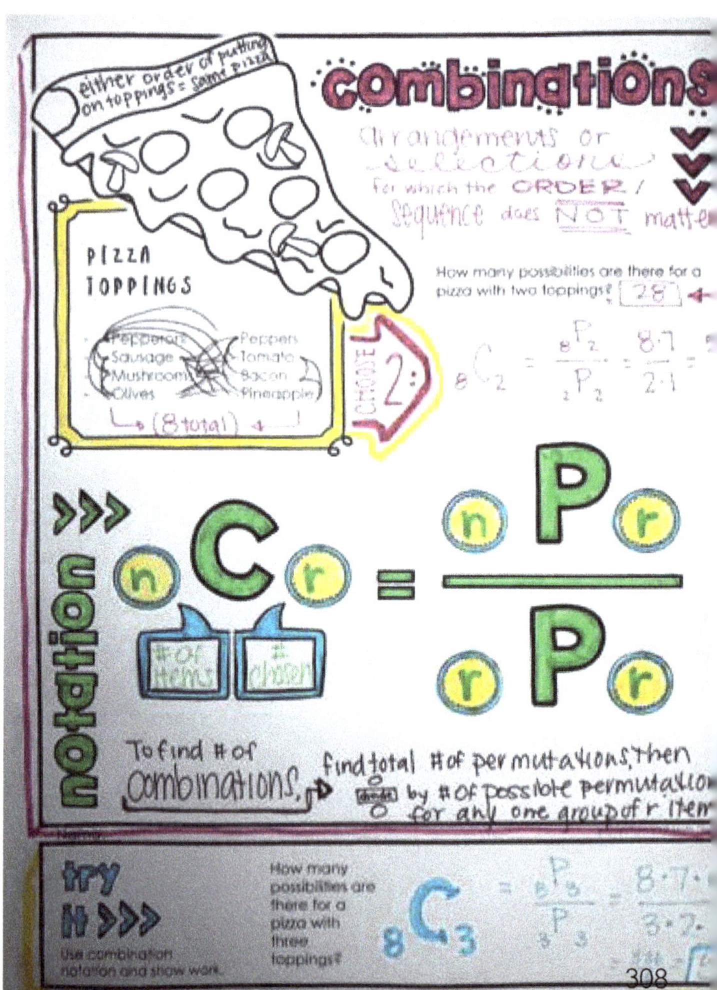
either order of putting on toppings = same pizza
combinations
arrangements or selections for which the ORDER / sequence does NOT matter
PIZZA TOPPINGS
(8 total)
CHOOSE 2
How many possibilities are there for a pizza with two toppings? 28
notation
To find # of combinations, find total # of permutations, then divide by # of possible permutations for any one group of r items
try it
Use combination notation and show work.
How many possibilities are there for a pizza with three toppings?

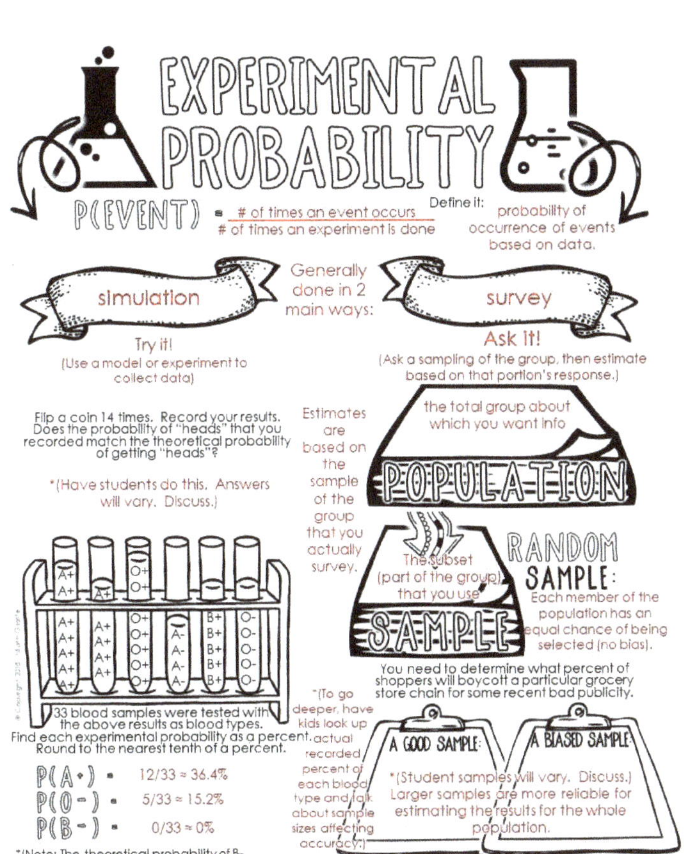
EXPERIMENTAL PROBABILITY
P(EVENT) = # of times an event occurs / # of times an experiment is done
Define it: probability of occurrence of events based on data.
simulation
Generally done in 2 main ways:
survey
Try it!
(Use a model or experiment to collect data)
Ask it!
(Ask a sampling of the group, then estimate based on that portion's response.)
Flip a coin 14 times. Record your results. Does the probability of "heads" that you recorded match the theoretical probability of getting "heads"?
*(Have students do this. Answers will vary. Discuss.)
Estimates are based on the sample of the group that you actually survey.
the total group about which you want info
POPULATION
The subset (part of the group) that you use
SAMPLE
RANDOM SAMPLE: Each member of the population has an equal chance of being selected (no bias).
You need to determine what percent of shoppers will boycott a particular grocery store chain for some recent bad publicity.
*(To go deeper, have kids look up actual recorded percent of each blood type and talk about sample sizes affecting accuracy.)
A GOOD SAMPLE
A BIASED SAMPLE
*(Student samples will vary. Discuss.) Larger samples are more reliable for estimating the results for the whole population.
33 blood samples were tested with the above results as blood types. Find each experimental probability as a percent. Round to the nearest tenth of a percent.
P(A+) = 12/33 ≈ 36.4%
P(O-) = 5/33 ≈ 15.2%
P(B-) = 0/33 ≈ 0%
*(Note: The theoretical probability of B- is actually 2%. What does that show about this experiment?) The sample may not include enough to be accurate. Also, experimental prob. Can only be as accurate as the data.
If 12 out of 150 surveyed would boycott, how many people in the total population of 160,000 shoppers might?
About 8%
= 12,800 people
Name: Answer Key / Teacher Guide

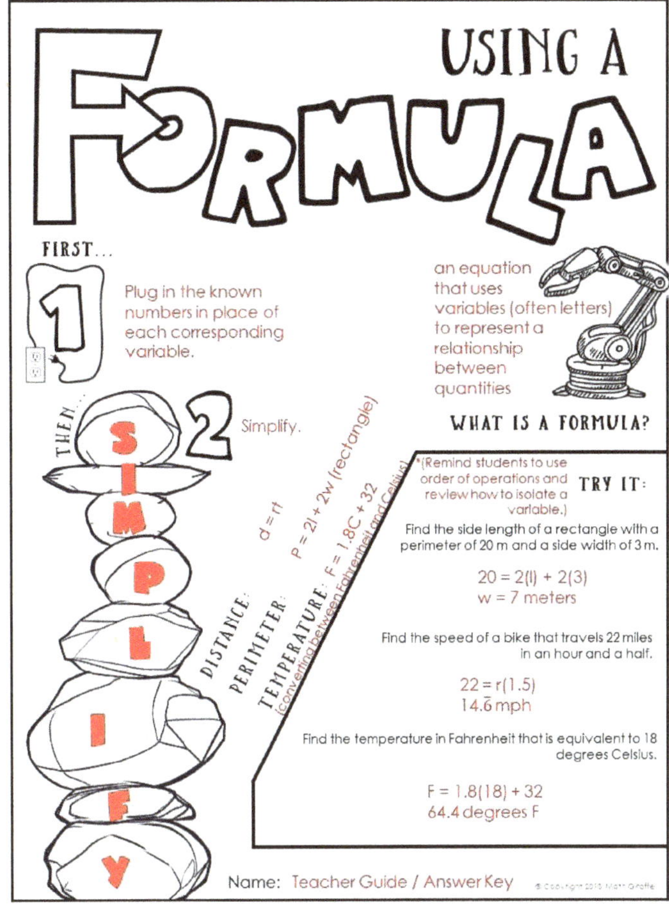
USING A FORMULA
FIRST...
1 Plug in the known numbers in place of each corresponding variable.
THEN...
2 Simplify.
SIMPLIFY
an equation that uses variables (often letters) to represent a relationship between quantities
WHAT IS A FORMULA?
DISTANCE: d = rt
PERIMETER: P = 2l + 2w (rectangle)
TEMPERATURE: F = 1.8C + 32 (converting between Fahrenheit and Celsius)
*(Remind students to use order of operations and review how to isolate a variable.)
TRY IT:
Find the side length of a rectangle with a perimeter of 20 m and a side width of 3 m.
20 = 2(l) + 2(3)
w = 7 meters
Find the speed of a bike that travels 22 miles in an hour and a half.
22 = r(1.5)
14.6 mph
Find the temperature in Fahrenheit that is equivalent to 18 degrees Celsius.
F = 1.8(18) + 32
64.4 degrees F
Name: Teacher Guide / Answer Key

EXPERIMENTAL PROBABILITY
based on data
P(EVENT) = # of times event occurs / # of times experiment is done
Define it: prob. of an event occurring based on actual data.
SIMULATION
SURVEY
ASK IT!
TRY IT!
Use a model or experiment to collect data.
Ask a sample of the group, then estimate prob. based on responses.
to collect data
Flip a coin 14 times. Record your results. Does the probability of "heads" that you recorded match the theoretical probability of getting "heads"?
P = 8/14 NO
TOTAL GROUP about which you want info (estimate about pop. are based on smaller group)
POPULATION
SUBSET
Part of the group that you use
SAMPLE
RANDOM SAMPLE:
Each member of popul. has equal chance of being selected (no bias)
FAIR
You need to determine what percent of shoppers will boycott a particular grocery store chain for some recent bad publicity.
33 blood samples were tested with the above results as blood types. Find each experimental probability as a percent. Round to the nearest tenth of a percent.
P(A+) = 12/33 ⇒ 36.4%
P(O-) = 5/33 ⇒ 15.2%
P(B-) = 0/33 ⇒ 0%
A GOOD SAMPLE
all people on the store's regional mailing list who've gotten shopper cards
A BIASED SAMPLE
5 people shopping now
*(Note: The theoretical probability of B- is actually 2%. What does that show about this experiment?) The sample is not large enough to be accurate. Experimental prob. ≠ theoretical sometimes!
Larger samples are more reliable!
If 12 out of 150 surveyed would boycott, how many people in the total population of 160,000 shoppers might? → 8% ≈ 12,800 people
Name:

USING A FORMULA
FIRST...
1 PLUG IN known #s for each variable
THEN...
2 Simplify to get the quantity you need.
SIMPLIFY
A FORMULA is an equation that uses VARIABLES to represent a RELATIONSHIP between quantities
XYZ
WHAT IS A FORMULA?
DISTANCE: d=rt
PERIMETER: P=2l+2w
TEMPERATURE: F=1.8C+32
TRY IT:
Find the side length of a rectangle with a perimeter of 20 m and a side width of 3 m.
P=2l+2w
replace
20 = 2(l) + 2(3)
simplify
20 = 2l + 6
14 = 2l
7m
Find the speed of a bike that travels 22 miles in an hour and a half.
d=rt
replace
22 = r(1.5)
14.6 = r
14.6 mph
Find temperature in Fahrenheit that is equivalent to 18 degrees Celsius.
F=1.8C+32
replace
F = 1.8(18) + 32
F = 32.4 + 32
64.4°F
Name:

LITERAL EQUATIONS
*(Remind students they already know these rules – no need to overthink it -- just don't break the laws and they will be fine!)
FOLLOW THE RULES
Properties of equality are LAW! Obey them.
Even with variables remaining unknown, we can manipulate an equation and isolate a variable.
TRANSFORMING A FORMULA
Color the variable you need to isolate, then solve.
COMMON MISTAKE: neglecting to use inverse operations, or breaking laws of equality (same operations to both sides – justice!)
JUST FOLLOW EQUATION LAWS!
Follow order of operations (backwards when UN- doing).
Do the same action to BOTH sides of the equation (showing work!)
Use inverse operations to move something to the other side of the =
HOW
Follow the laws while manipulating variables. (same as solving an equation with numbers!)
WHY
If we have to solve for a variable many times in a row, it is easier to transform the original BEFORE plugging in rather than doing that work each time.
SOLVING FOR A VARIABLE
"Solve for" means isolate!
TRY IT..
Solve for r: d = r · t
Divide by t to get r alone (isolated) on one side of the equals sign: d/t = r
Solve for b.
A=½b·h
A/h = ½ b
2A/h = b
Solve for h.
A=½b·h
A/b = ½ h
2A/b = h
Solve for k.
4k+j=m
4k = m - j
k = (m - j)/4
*(Remind students that dividing by ½ is the same as multiplying by 2. Dividing by 4 is the same as multiplying by ¼ (coefficient of 1 goes on top (numerator) and coefficient of 4 goes on the bottom.)
Name: Teacher Guide / Answer Key
© Copyright 2018 Math Giraffe

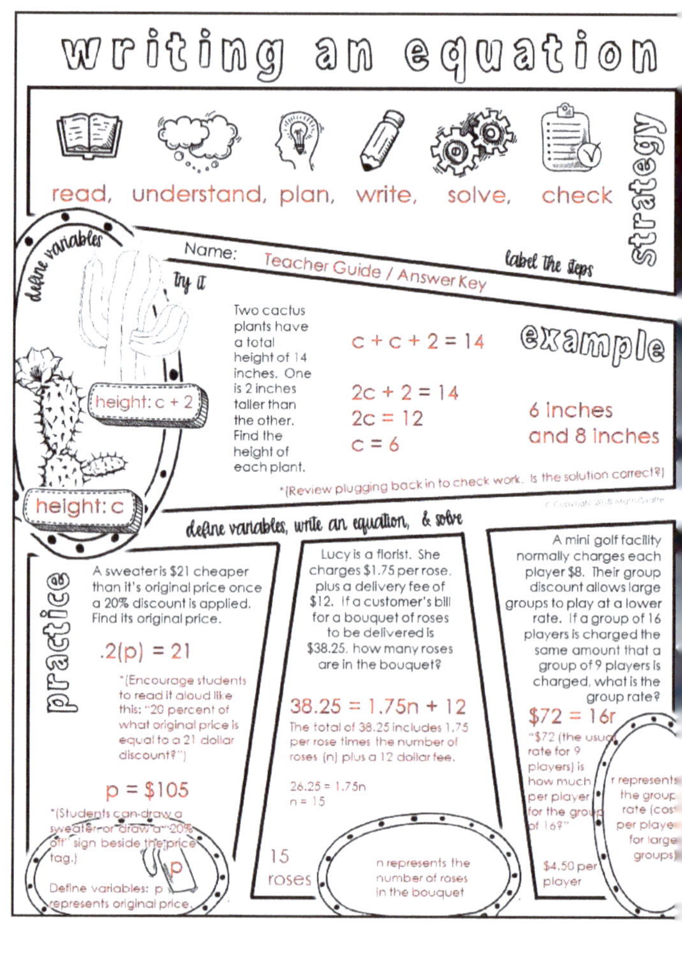
writing an equation
read, understand, plan, write, solve, check
strategy
Name: Teacher Guide / Answer Key
label the steps
define variables
Try it
Two cactus plants have a total height of 14 inches. One is 2 inches taller than the other. Find the height of each plant.
c + c + 2 = 14
2c + 2 = 14
2c = 12
c = 6
example
6 inches and 8 inches
height: c + 2
height: c
*(Review plugging back in to check work. Is the solution correct?)
define variables, write an equation, & solve
practice
A sweater is $21 cheaper than it's original price once a 20% discount is applied. Find its original price.
.2(p) = 21
*(Encourage students to read it aloud like this: "20 percent of what original price is equal to a 21 dollar discount?")
p = $105
*(Students can draw a sweater or draw a "20% off" sign beside the price tag.)
Define variables: p represents original price
Lucy is a florist. She charges $1.75 per rose, plus a delivery fee of $12. If a customer's bill for a bouquet of roses to be delivered is $38.25, how many roses are in the bouquet?
38.25 = 1.75n + 12
The total of 38.25 includes 1.75 per rose times the number of roses (n) plus a 12 dollar fee.
26.25 = 1.75n
n = 15
15 roses
n represents the number of roses in the bouquet
A mini golf facility normally charges each player $8. Their group discount allows large groups to play at a lower rate. If a group of 16 players is charged the same amount that a group of 9 players is charged, what is the group rate?
$72 = 16r
$4.50 per player

Properties of EQUALITY are the Laws
FOLLOW THE RULES
We already know how to isolate a variable.
LITERAL EQUATIONS
Even with variables (still unknown) in the equation, we can still manipulate!
TRANSFORMING A FORMULA
Color the variable you need to isolate, then solve.
COMMON MISTAKE
Breaking laws of EQUALITY (Do same thing to both sides!!) or neglecting to use inverse oper.
JUST FOLLOW EQUATION LAWS!
Follow order of operations FORWARD when DOING and BACKWARD when UN Doing
Do the same action to BOTH SIDES of the equation!
Use inverse oper. When moving to the other side of =
HOW manipulate variables to get the desired one alone on one side of =.
WHY
SOLVING FOR A VARIABLE
"Solve for" means ISOLATE
TRY IT..
Solve for r: d = r · t
DIVIDE BOTH SIDES by t to get r alone
d/t = r
Solve for b.
A=½b·h
2A/h = b
Solve for h.
A=½b·h
2A/b = h
Solve for k.
4k+j=m
Name:
© Copyright 2018 Math Giraffe

writing an equation
READ understand PLAN write SOLVE check
strategy
define variables
Name:
label the steps
Try it
Two cactus plants have a total height of 14 inches. One is 2 inches taller than the other. Find the height of each plant.
C + C + 2 = 14
2c + 2 = 14
2c = 12
c = 6
example
6 in. & 8 in.
height: C+2
height: C
State what your variable(s) represents
define variables, write an equation, & solve
practice
A sweater is $21 cheaper than it's original price once a 20% discount is applied. Find its original price.
.20(p) = 21
p = $105
plug back in to check!
Lucy is a florist. She charges $1.75 per rose, plus a delivery fee of $12. If a customer's bill for a bouquet of roses to be delivered is $38.25, how many roses are in the bouquet?
38.25 = 1.75r + 12
26.25 = 1.75(r)
r = 15
15 roses
A mini golf facility normally charges each player $8. Their group discount allows large groups to play at a lower rate. If a group of 14 players is charged the same amount that a group of 9 players is charged, what is the group rate?
$4.50 per player
represents group rate per player

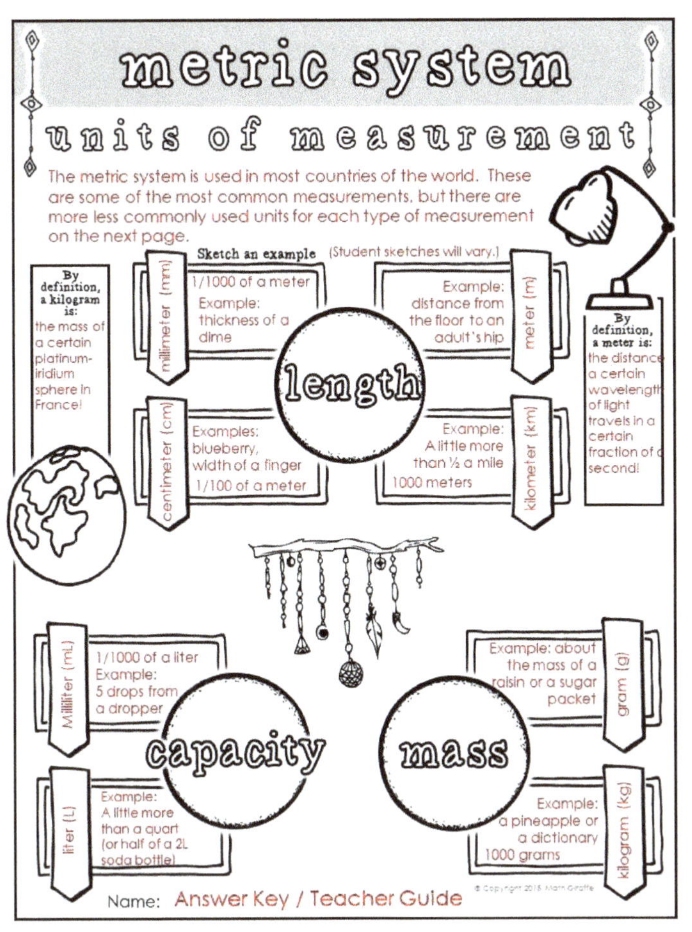
metric system
units of measurement
The metric system is used in most countries of the world. These are some of the most common measurements, but there are more less commonly used units for each type of measurement on the next page.
Sketch an example
(Student sketches will vary.)
By definition, a kilogram is: the mass of a certain platinum-iridium sphere in France!
millimeter (mm)
1/1000 of a meter
Example: thickness of a dime
meter (m)
Example: distance from the floor to an adult's hip
By definition, a meter is: the distance a certain wavelength of light travels in a certain fraction of a second!
length
centimeter (cm)
Examples: blueberry, width of a finger
1/100 of a meter
kilometer (km)
Example: A little more than ½ a mile
1000 meters
Milliliter (mL)
1/1000 of a liter
Example: 5 drops from a dropper
gram (g)
Example: about the mass of a raisin or a sugar packet
capacity
mass
liter (L)
Example: A little more than a quart (or half of a 2L soda bottle)
kilogram (kg)
Example: a pineapple or a dictionary
1000 grams
Name: Answer Key / Teacher Guide

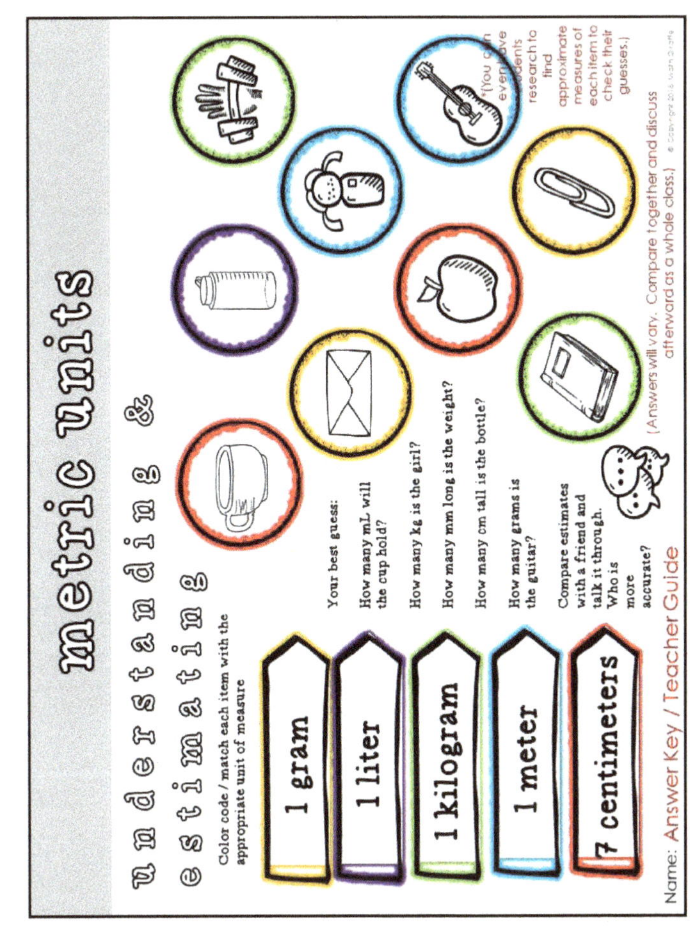
metric units
understanding & estimating
Color code / match each item with the appropriate unit of measure
1 gram
1 liter
1 kilogram
1 meter
7 centimeters
Your best guess:
How many mL will the cup hold?
How many kg is the girl?
How many mm long is the weight?
How many cm tall is the bottle?
How many grams is the guitar?
Compare estimates with a friend and talk it through. Who is more accurate?
(You can even have students research to find approximate measures of each item to check their guesses.)
(Answers will vary. Compare together and discuss afterward as a whole class.)
Name: Answer Key / Teacher Guide

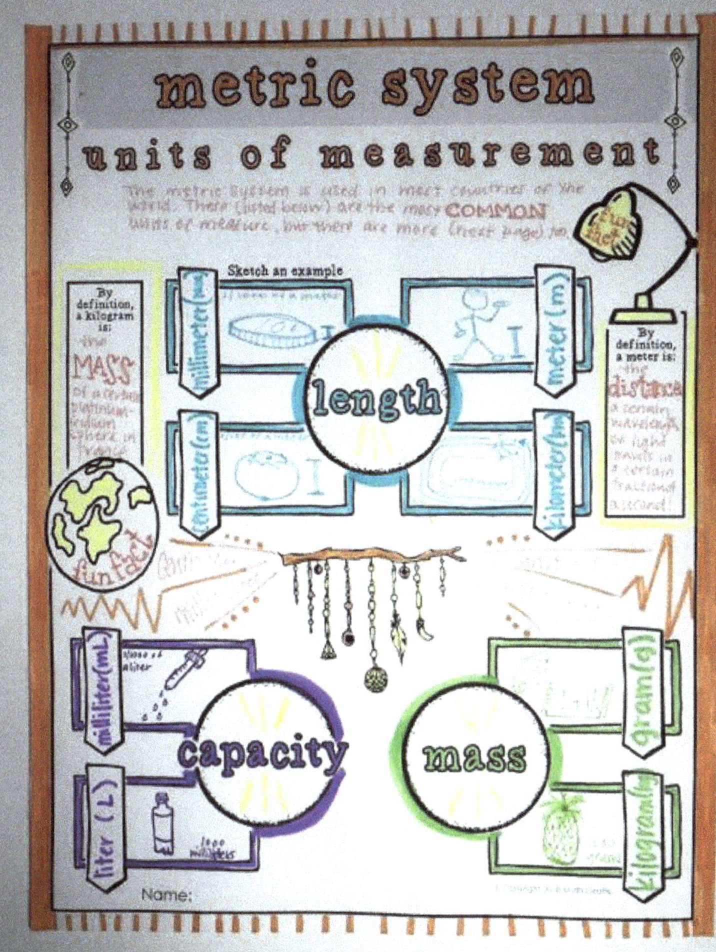
metric system
units of measurement
Sketch an example
length
capacity
mass
Name:

metric units
understanding & estimating
1 gram
1 liter
1 kilogram
1 meter
7 centimeters
Name:

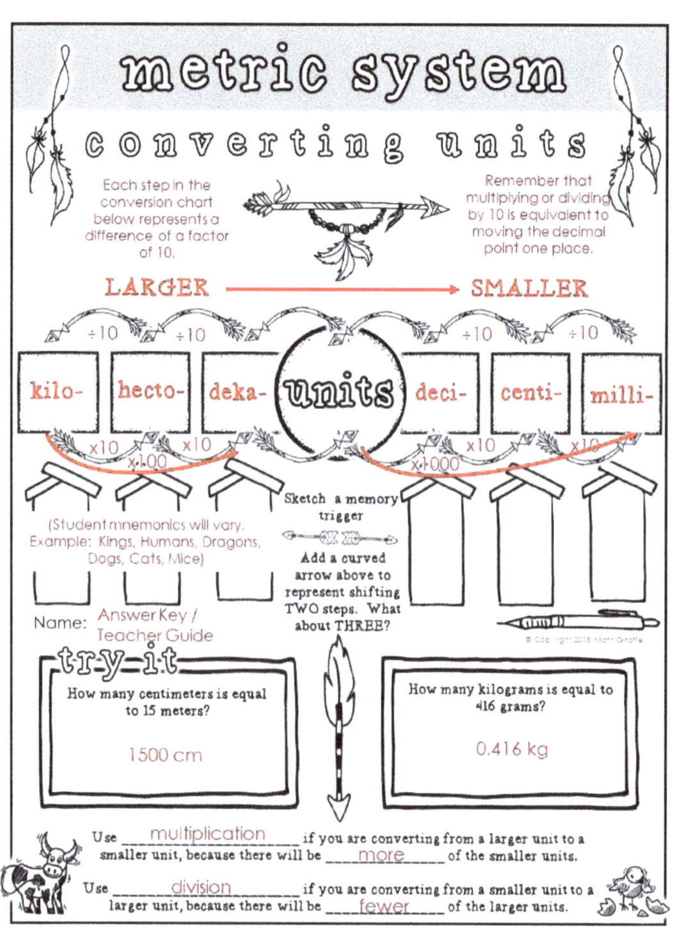

metric system
converting units
Each step in the conversion chart below represents a difference of a factor of 10.
Remember that multiplying or dividing by 10 is equivalent to moving the decimal point one place.
LARGER
SMALLER
÷10
÷10
÷10
÷10
kilo-
hecto-
deka-
units
deci-
centi-
milli-
x10
x10
x100
x10
x10
x1000
(Student mnemonics will vary. Example: Kings, Humans, Dragons, Dogs, Cats, Mice)
Sketch a memory trigger
Add a curved arrow above to represent shifting TWO steps. What about THREE?
Name: Answer Key / Teacher Guide
try-it
How many centimeters is equal to 15 meters?
1500 cm
How many kilograms is equal to 416 grams?
0.416 kg
Use multiplication if you are converting from a larger unit to a smaller unit, because there will be more of the smaller units.
Use division if you are converting from a smaller unit to a larger unit, because there will be fewer of the larger units.

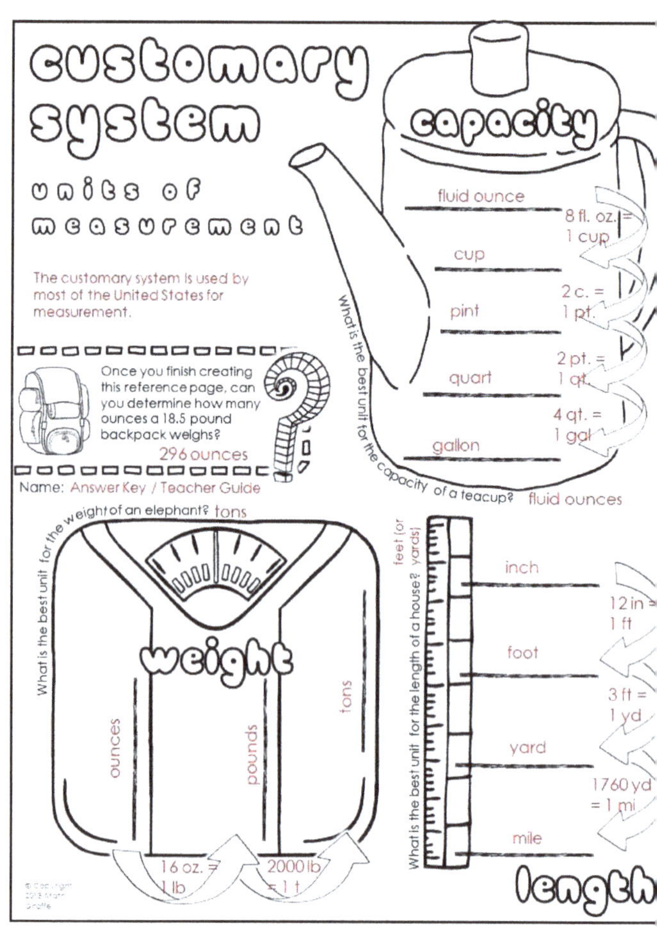

customary system
units of measurement
The customary system is used by most of the United States for measurement.
capacity
fluid ounce
cup
pint
quart
gallon
8 fl. oz. = 1 cup
2 c. = 1 pt.
2 pt. = 1 qt.
4 qt. = 1 gal
What is the best unit for the capacity of a teacup? fluid ounces
Once you finish creating this reference page, can you determine how many ounces a 18.5 pound backpack weighs?
296 ounces
Name: Answer Key / Teacher Guide
What is the best unit for the weight of an elephant? tons
weight
ounces
pounds
tons
16 oz. = 1 lb
2000 lb = 1 t
What is the best unit for the length of a house? feet (or yards)
inch
foot
yard
mile
12 in = 1 ft
3 ft = 1 yd
1760 yd = 1 mi
length

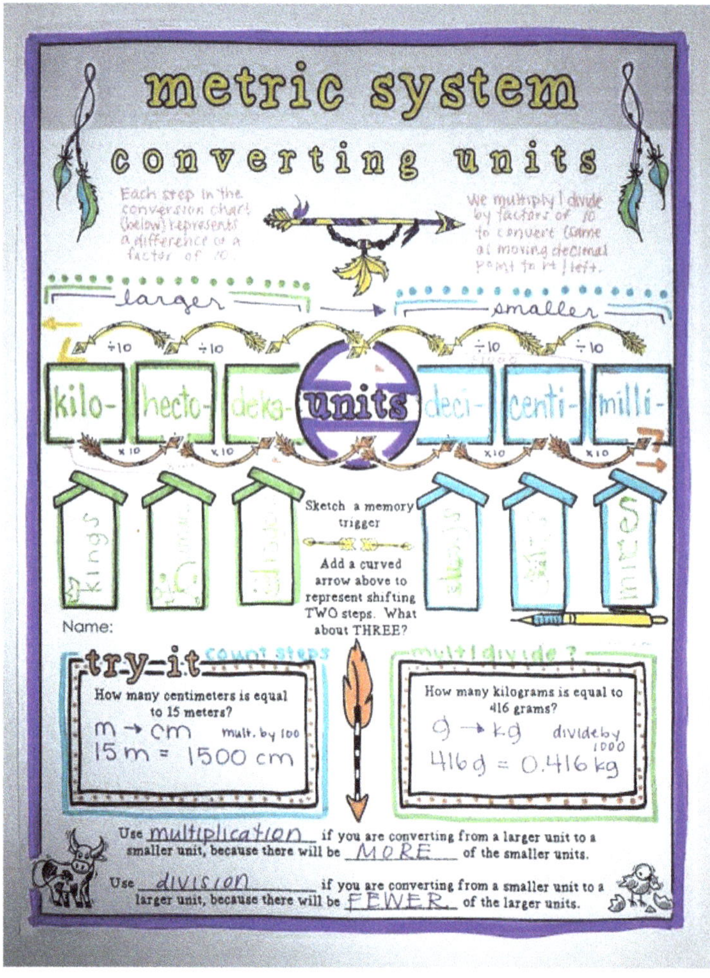

metric system
converting units
larger
smaller
kilo-
hecto-
deka-
units
deci-
centi-
milli-
Sketch a memory trigger
Add a curved arrow above to represent shifting TWO steps. What about THREE?
Name:
try-it
How many centimeters is equal to 15 meters?
m → cm mult. by 100
15 m = 1500 cm
How many kilograms is equal to 416 grams?
g → kg divide by 1000
416 g = 0.416 kg
Use multiplication if you are converting from a larger unit to a smaller unit, because there will be MORE of the smaller units.
Use division if you are converting from a smaller unit to a larger unit, because there will be FEWER of the larger units.

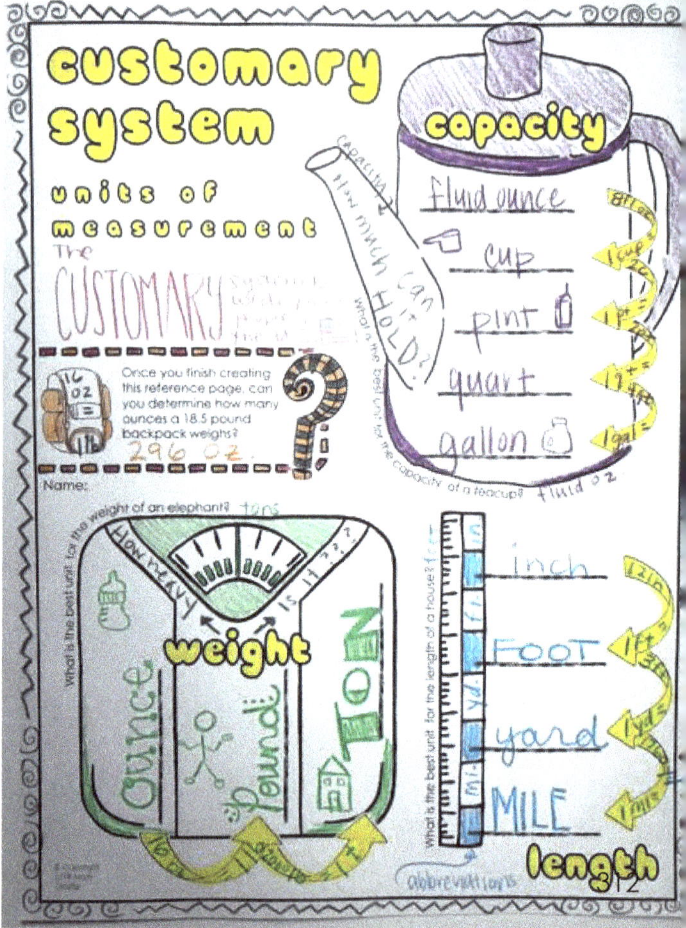

customary system
units of measurement
capacity
fluid ounce
cup
pint
quart
gallon
Once you finish creating this reference page, can you determine how many ounces a 18.5 pound backpack weighs?
296 oz.
Name:
What is the best unit for the weight of an elephant? tons
weight
ounce
pound
TON
inch
FOOT
yard
MILE
length

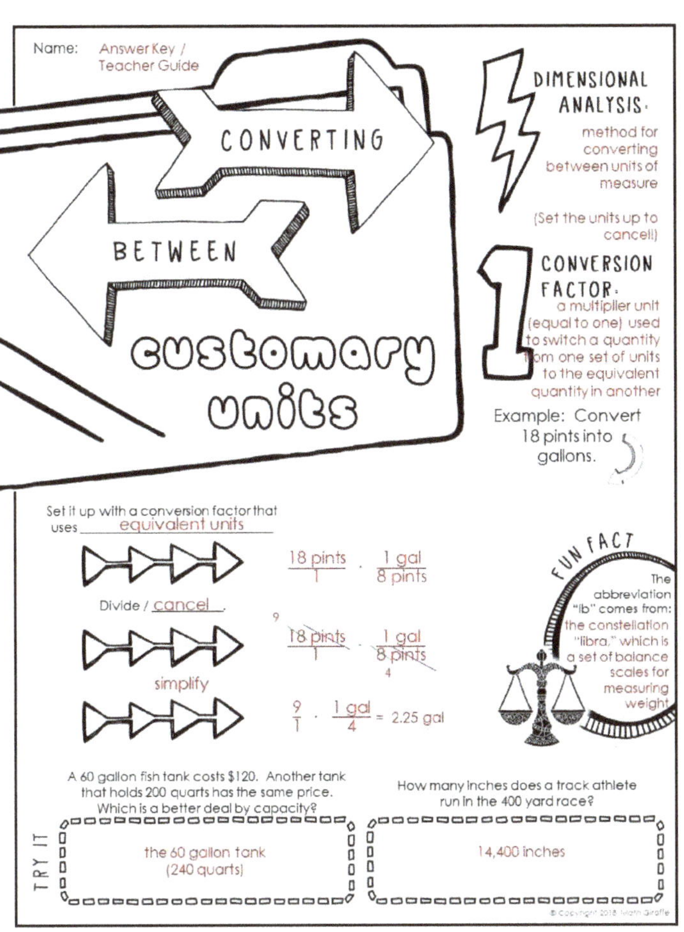

Name: Answer Key / Teacher Guide
CONVERTING
BETWEEN
customary units
DIMENSIONAL ANALYSIS: method for converting between units of measure
(Set the units up to cancel!)
1 CONVERSION FACTOR: a multiplier unit (equal to one) used to switch a quantity from one set of units to the equivalent quantity in another
Example: Convert 18 pints into gallons.
Set it up with a conversion factor that uses equivalent units
18 pints / 1 · 1 gal / 8 pints
Divide / cancel.
18 pints / 1 · 1 gal / 8 pints
simplify
9/1 · 1 gal / 4 = 2.25 gal
FUN FACT
The abbreviation "lb" comes from: the constellation "libra," which is a set of balance scales for measuring weight
A 60 gallon fish tank costs $120. Another tank that holds 200 quarts has the same price. Which is a better deal by capacity?
How many inches does a track athlete run in the 400 yard race?
TRY IT
the 60 gallon tank (240 quarts)
14,400 inches
© Copyright 2018 Math Giraffe

Units of Capacity
1 cup
8 fluid ounces
1 pint
2 cups
16 fluid ounces
1 ounce
(Encourage students to draw pictures within each shape to represent and remember.)
(You may want to also include fractional parts. Examples: 1 ounce is 1/8 of a cup. 1 pint is ½ of a quart, etc. – see sample photo)
(This page makes a great reference for students to keep. They can check work or count it out to build the concept if they have drawn the conversions like in the sample.)
1 gallon
4 quarts
8 pints
16 cups
128 fluid ounces
1 quart
2 pints
4 cups
32 fluid ounces
Name:
© Copyright 2017 Math Giraffe

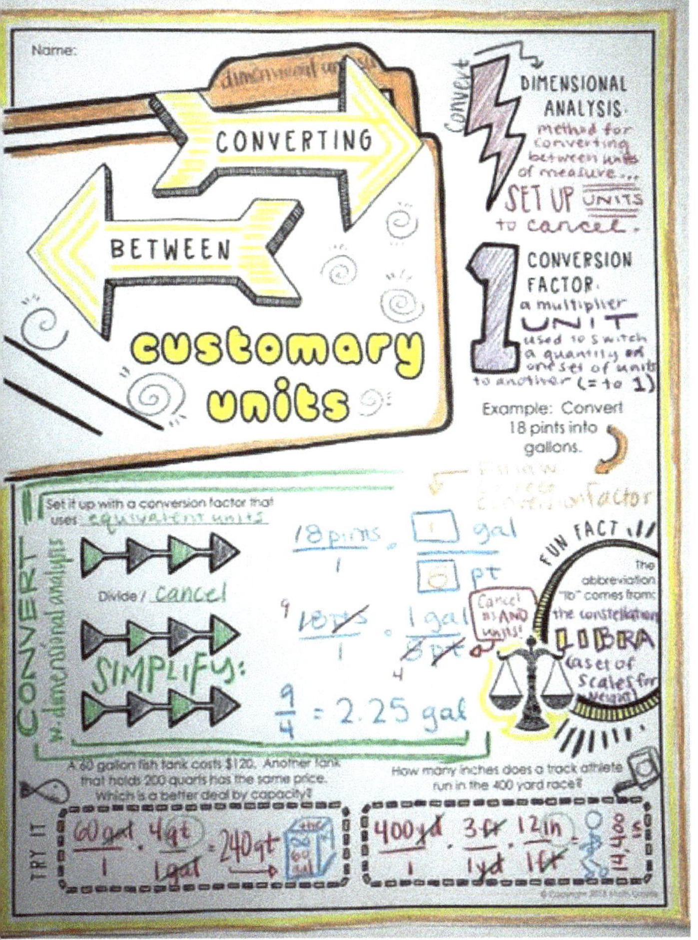

Name:
CONVERTING
BETWEEN
customary units
DIMENSIONAL ANALYSIS: method for converting between units of measure...
SET UP UNITS to cancel.
1 CONVERSION FACTOR: a multiplier UNIT used to switch a quantity from one set of units to another (= to 1)
Example: Convert 18 pints into gallons.
Set it up with a conversion factor that uses equivalent units
Divide / cancel
SIMPLIFY:
9/4 = 2.25 gal
CONVERT w/ dimensional analysis
FUN FACT
The abbreviation "lb" comes from: the constellation LIBRA (a set of scales for weight)
A 60 gallon fish tank costs $120. Another tank that holds 200 quarts has the same price. Which is a better deal by capacity?
How many inches does a track athlete run in the 400 yard race?
TRY IT
60 gal / 1 · 4 qt / 1 gal = 240 qt
400 yd / 1 · 3 ft / 1 yd · 12 in / 1 ft

Units of Capacity
1 cup
= 8 OUNCES
1 pint
= 16 OZ.
= 2 CUPS
= 1/8 of CUP
= 1/16 of PINT
= 1/32 of QUART
1 ounce
1 gallon
= 4 QUARTS = 8 PINTS = 16 CUPS
= 128 OUNCES
1 quart
= 2 PINTS
= 4 CUPS
= 32 FL. OZ.
= 1/4 of a GAL.
Name:
© Copyright 2017 Math Giraffe

DEFINE EACH:
principal
the initial amount invested or deposited
interest rate
the portion (%) of your initial principal deposit that you will get as interest
(This amount is added to your account by the bank.)
interest
amount of money you earn back in addition to your principal (in return for keeping your money there)
MAP OUT HOW IT WORKS:
You deposit an initial principal amount into the bank.
The bank takes your principal and invests it.
They earn profits on your money.
The bank keeps the rest of the profit and reinvests.
Your percent interest is calculated
The interest is added to your principal in your account.
NAME: Answer Key / Teacher Guide
Simple interest
EXPLAIN IT:
When you save money in the bank, you get some interest back (a small percent of the amount you put into the bank) as a reward for keeping your money there.
FILL IT IN:
I = Prt
INTEREST
PRINCIPAL
RATE
TIME
as a decimal
in years

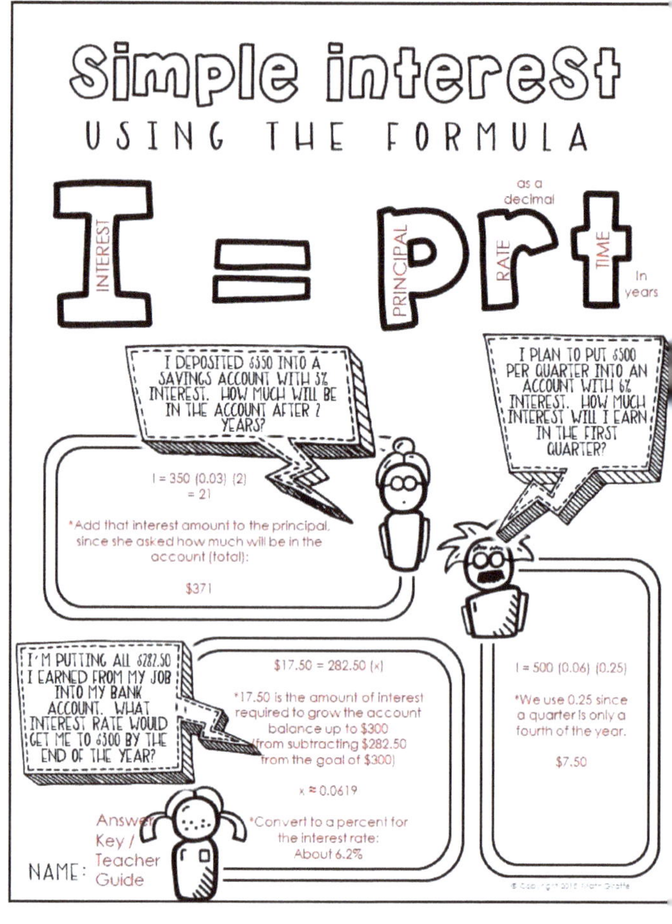
Simple interest
USING THE FORMULA
I = Prt
INTEREST
PRINCIPAL
RATE
TIME
as a decimal
in years
I DEPOSITED $350 INTO A SAVINGS ACCOUNT WITH 3% INTEREST. HOW MUCH WILL BE IN THE ACCOUNT AFTER 2 YEARS?
I = 350 (0.03) (2)
= 21
*Add that interest amount to the principal, since she asked how much will be in the account (total):
$371
I PLAN TO PUT $500 PER QUARTER INTO AN ACCOUNT WITH 6% INTEREST. HOW MUCH INTEREST WILL I EARN IN THE FIRST QUARTER?
I = 500 (0.06) (0.25)
*We use 0.25 since a quarter is only a fourth of the year.
$7.50
I'M PUTTING ALL $282.50 I EARNED FROM MY JOB INTO MY BANK ACCOUNT. WHAT INTEREST RATE WOULD GET ME TO $300 BY THE END OF THE YEAR?
$17.50 = 282.50 (x)
*17.50 is the amount of interest required to grow the account balance up to $300 (from subtracting $282.50 from the goal of $300)
x ≈ 0.0619
*Convert to a percent for the interest rate: About 6.2%
NAME: Answer Key / Teacher Guide

DEFINE EACH:
principal
the initial $ DEPOSITED
START
interest rate
%
of your $ that you will get as interest
(added to your account by the bank)
interest
$ a bank pays you in return for keeping your $ there
MAP OUT HOW IT WORKS:
You DEPOSIT your PRINCIPAL
the BANK invests the $
The bank PROFITS from the $
The bank keeps the rest of the $
Your portion of the INTEREST is calculated
You get INTEREST added to your principal.
NAME:
Simple interest
EXPLAIN IT: when you save money in the bank, they reward you with interest, since they profit off of your money.
FILL IT IN:
amount of INTEREST
PRINCIPAL
RATE
decimal
TIME
in years

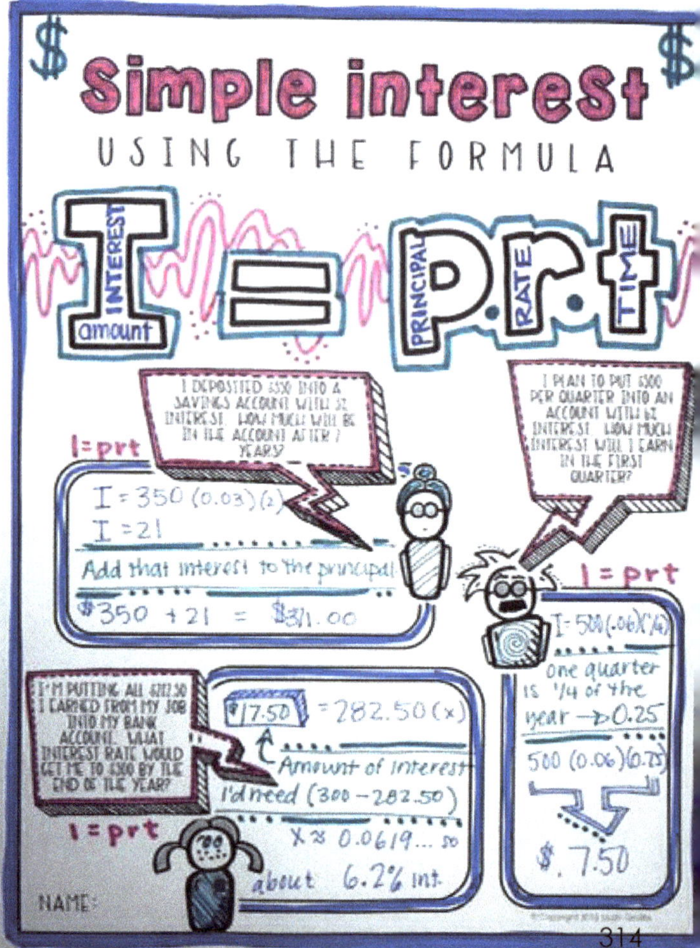
Simple interest
USING THE FORMULA
I = P.r.t
INTEREST
amount
PRINCIPAL
RATE
TIME
I DEPOSITED $350 INTO A SAVINGS ACCOUNT WITH 3% INTEREST. HOW MUCH WILL BE IN THE ACCOUNT AFTER 2 YEARS?
I=prt
I = 350 (0.03)(2)
I = 21
Add that interest to the principal
$350 + 21 = $371.00
I PLAN TO PUT $500 PER QUARTER INTO AN ACCOUNT WITH 6% INTEREST. HOW MUCH INTEREST WILL I EARN IN THE FIRST QUARTER?
I=prt
One quarter is 1/4 of the year → 0.25
500 (0.06)(0.25)
$.7.50
I'M PUTTING ALL $282.50 I EARNED FROM MY JOB INTO MY BANK ACCOUNT. WHAT INTEREST RATE WOULD GET ME TO $300 BY THE END OF THE YEAR?
17.50 = 282.50 (x)
Amount of interest I'd need (300 − 282.50)
x ≈ 0.0619... so about 6.2% int.
I=prt
NAME:

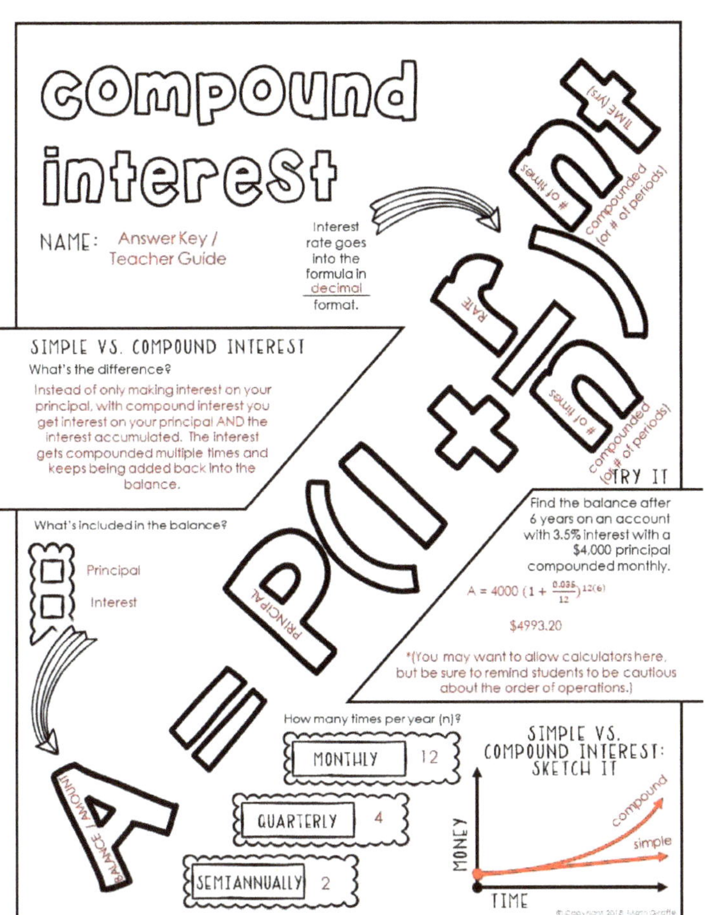
compound interest
NAME: Answer Key / Teacher Guide
Interest rate goes into the formula in decimal format.
A = P(1 + r/n)^nt
BALANCE / AMOUNT
PRINCIPAL
RATE
# of times compounded (or # of periods)
# of times compounded (or # of periods)
TIME (yrs)
SIMPLE VS. COMPOUND INTEREST
What's the difference?
Instead of only making interest on your principal, with compound interest you get interest on your principal AND the interest accumulated. The interest gets compounded multiple times and keeps being added back into the balance.
What's included in the balance?
Principal
Interest
TRY IT
Find the balance after 6 years on an account with 3.5% interest with a $4,000 principal compounded monthly.
A = 4000 (1 + 0.035/12)^12(6)
$4993.20
*(You may want to allow calculators here, but be sure to remind students to be cautious about the order of operations.)
How many times per year (n)?
MONTHLY 12
QUARTERLY 4
SEMIANNUALLY 2
SIMPLE VS. COMPOUND INTEREST: SKETCH IT
MONEY
TIME
compound
simple
© Copyright 2018 Math Giraffe

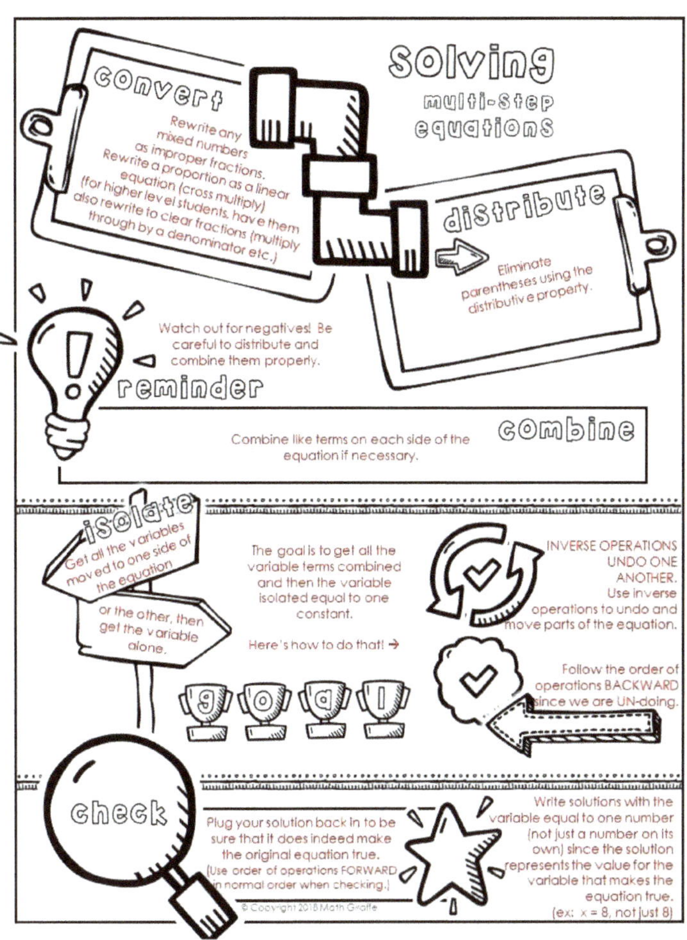
Solving multi-step equations
convert
Rewrite any mixed numbers as improper fractions. Rewrite a proportion as a linear equation (cross multiply) (for higher level students, have them also rewrite to clear fractions (multiply through by a denominator etc.)
distribute
Eliminate parentheses using the distributive property.
Watch out for negatives! Be careful to distribute and combine them properly.
reminder
combine
Combine like terms on each side of the equation if necessary.
isolate
Get all the variables moved to one side of the equation
or the other, then get the variable alone.
The goal is to get all the variable terms combined and then the variable isolated equal to one constant.
Here's how to do that! →
goal
INVERSE OPERATIONS UNDO ONE ANOTHER. Use inverse operations to undo and move parts of the equation.
Follow the order of operations BACKWARD since we are UN-doing.
check
Plug your solution back in to be sure that it does indeed make the original equation true. (Use order of operations FORWARD in normal order when checking.)
Write solutions with the variable equal to one number (not just a number on its own) since the solution represents the value for the variable that makes the equation true. (ex: x = 8, not just 8)
© Copyright 2018 Math Giraffe

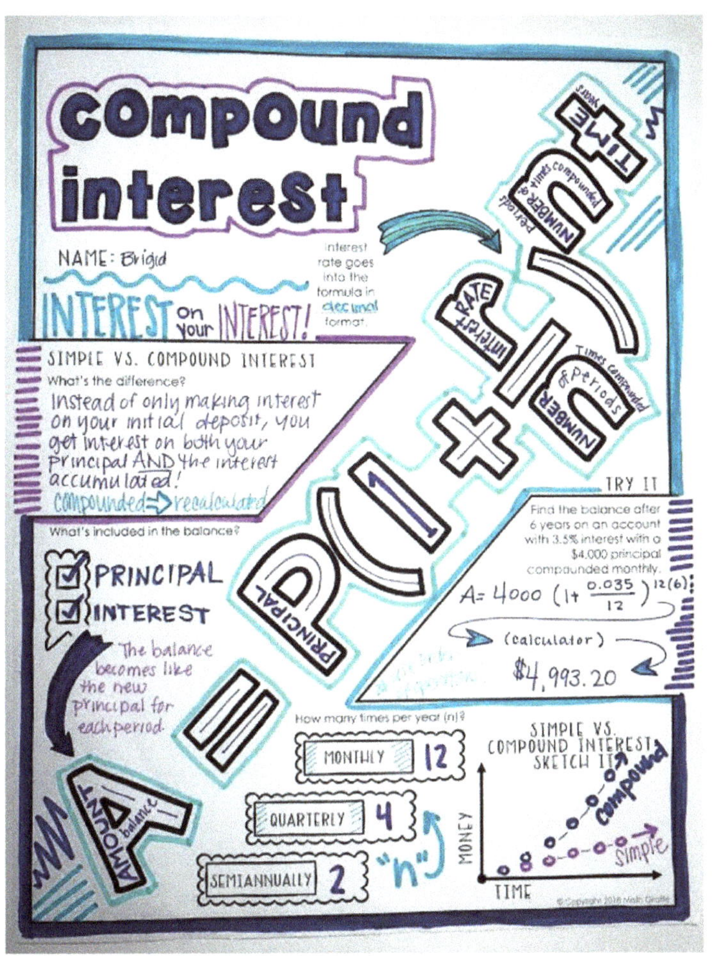
compound interest
NAME: Brigid
INTEREST on your INTEREST!
Interest rate goes into the formula in decimal format.
SIMPLE VS. COMPOUND INTEREST
What's the difference?
Instead of only making interest on your initial deposit, you get interest on both your principal AND the interest accumulated!
compounded ⇒ recalculated
What's included in the balance?
PRINCIPAL
INTEREST
The balance becomes like the new principal for each period
AMOUNT balance
PRINCIPAL
interest RATE
NUMBER of times compounded
NUMBER of periods
TIME years
TRY IT
Find the balance after 6 years on an account with 3.5% interest with a $4,000 principal compounded monthly.
A = 4000 (1+ 0.035/12)^12(6)
(calculator)
$4,993.20
How many times per year (n)?
MONTHLY 12
QUARTERLY 4
SEMIANNUALLY 2
"n"
SIMPLE VS. COMPOUND INTEREST: SKETCH IT
MONEY
TIME
compound
simple
© Copyright 2018 Math Giraffe

Solving: multi-step equations
guide sheet
convert
Rewrite:
any mixed #s as improper fractions
any proportion as eq. (cross multiply)
distribute
ELIMINATE any parentheses by using the distributive Prop.
Watch for negatives
reminder
COMBINE like terms on both sides of the equation
combine
isolate
all variables moved to one side of the equation or the other... then isolate the variable (get it all alone.)
The goal is to get the variable on just one side, then alone = to a constant (#)
goal
Here's how to do that
INVERSE operations UNDO one another. Use them to move parts to other side of =.
Follow order of operations backward since we're UN-doing
BACKWARD PEMDAS
check
Plug your solution back in to be sure it does make the original equation true.
X=#
Write solutions in the right form (x=#) not just a number
© Copyright 2018 Math Giraffe

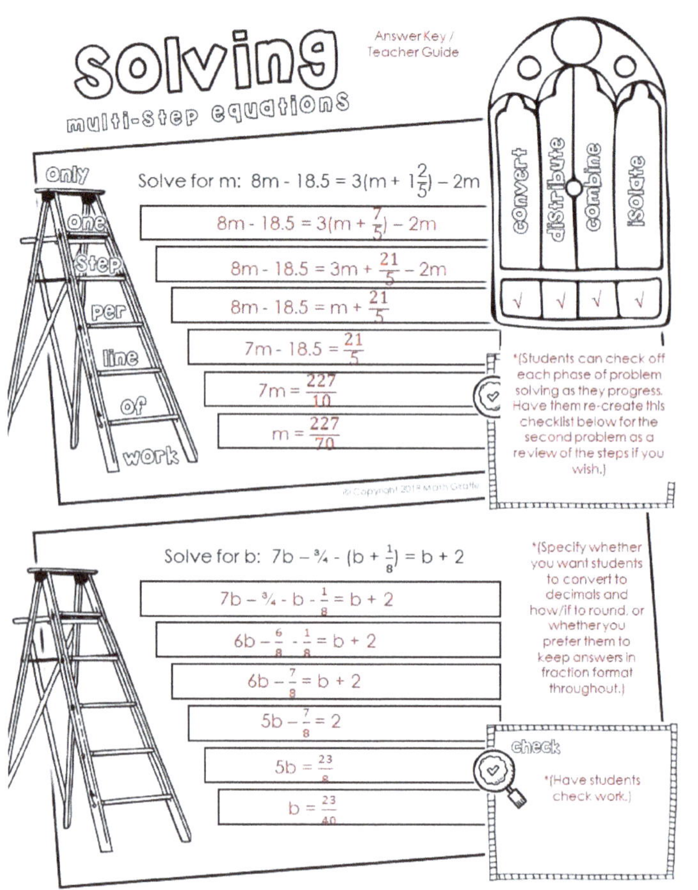
Solving
multi-step equations
Answer Key / Teacher Guide
Only One Step per line of work
Solve for m: $8m - 18.5 = 3(m + 1\frac{2}{5}) - 2m$
$8m - 18.5 = 3(m + \frac{7}{5}) - 2m$
$8m - 18.5 = 3m + \frac{21}{5} - 2m$
$8m - 18.5 = m + \frac{21}{5}$
$7m - 18.5 = \frac{21}{5}$
$7m = \frac{227}{10}$
$m = \frac{227}{70}$
convert
distribute
combine
isolate
*(Students can check off each phase of problem solving as they progress. Have them re-create this checklist below for the second problem as a review of the steps if you wish.)
Solve for b: $7b - ¾ - (b + \frac{1}{8}) = b + 2$
$7b - ¾ - b - \frac{1}{8} = b + 2$
$6b - \frac{6}{8} - \frac{1}{8} = b + 2$
$6b - \frac{7}{8} = b + 2$
$5b - \frac{7}{8} = 2$
$5b = \frac{23}{8}$
$b = \frac{23}{40}$
*(Specify whether you want students to convert to decimals and how/if to round, or whether you prefer them to keep answers in fraction format throughout.)
check
*(Have students check work.)
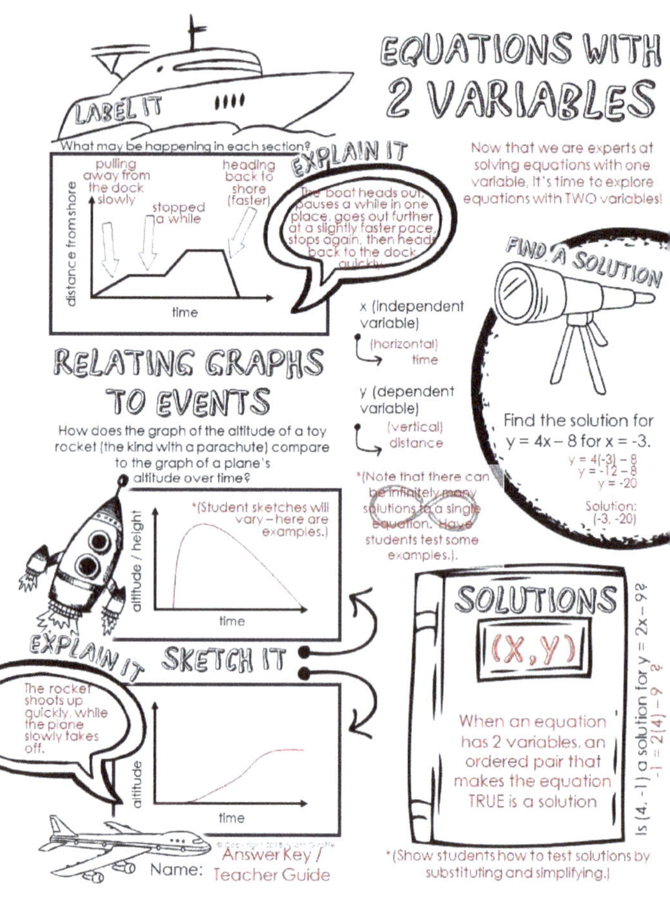
EQUATIONS WITH 2 VARIABLES
LABEL IT
What may be happening in each section?
pulling away from the dock slowly
stopped a while
heading back to shore (faster)
distance from shore
time
EXPLAIN IT
The boat heads out, pauses a while in one place, goes out further at a slightly faster pace, stops again, then heads back to the dock quickly.
Now that we are experts at solving equations with one variable, it's time to explore equations with TWO variables!
FIND A SOLUTION
x (independent variable)
(horizontal) time
y (dependent variable)
(vertical) distance
Find the solution for $y = 4x - 8$ for $x = -3$.
$y = 4(-3) - 8$
$y = -12 - 8$
$y = -20$
Solution: (-3, -20)
*(Note that there can be infinitely many solutions to a single equation. Have students test some examples.)
RELATING GRAPHS TO EVENTS
How does the graph of the altitude of a toy rocket (the kind with a parachute) compare to the graph of a plane's altitude over time?
*(Student sketches will vary – here are examples.)
altitude / height
time
EXPLAIN IT
SKETCH IT
The rocket shoots up quickly, while the plane slowly takes off.
altitude
time
SOLUTIONS
(X, Y)
When an equation has 2 variables, an ordered pair that makes the equation TRUE is a solution
Is (4, -1) a solution for $y = 2x - 9$?
$-1 \overset{?}{=} 2(4) - 9$
Answer Key / Teacher Guide
Name:
*(Show students how to test solutions by substituting and simplifying.)

Solving
multi-step equations
Check off each solving step...
Solve for m: $8m - 18.5 = 3(m + 1\frac{2}{5}) - 2m$
convert
distribute
combine
isolate
check
Solve for b: $7b - ¾ - (b + \frac{1}{8}) = b + 2$
check
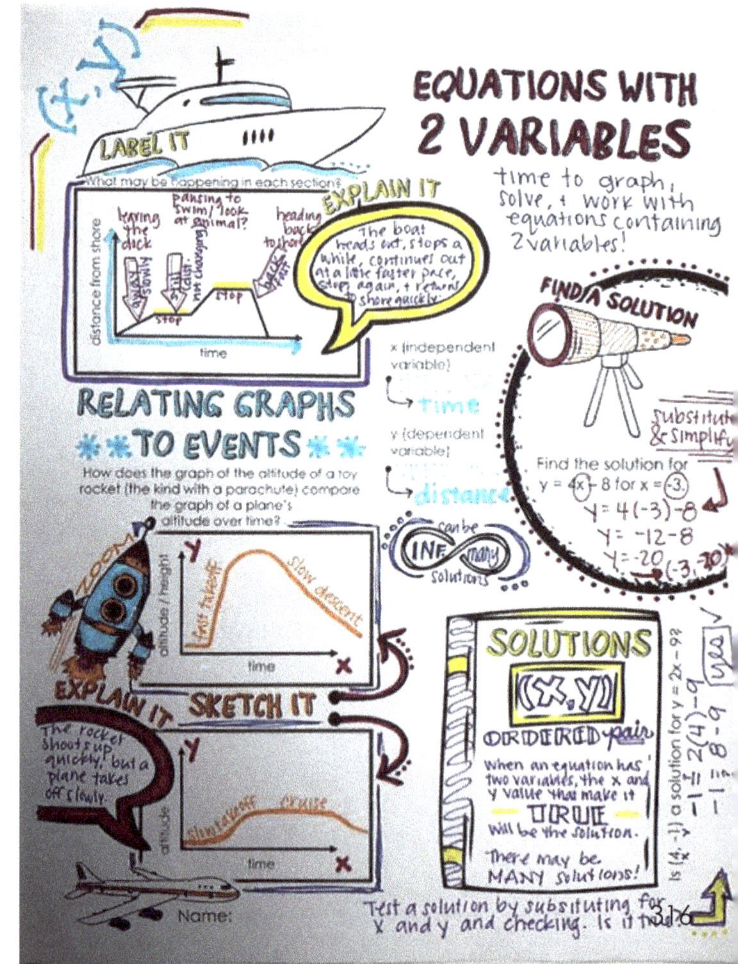
EQUATIONS WITH 2 VARIABLES
LABEL IT
What may be happening in each section?
EXPLAIN IT
time to graph, solve, + work with equations containing 2 variables!
FIND A SOLUTION
x (independent variable)
y (dependent variable)
RELATING GRAPHS TO EVENTS
How does the graph of the altitude of a toy rocket (the kind with a parachute) compare the graph of a plane's altitude over time?
Find the solution for $y = 4x - 8$ for $x = -3$.
EXPLAIN IT
SKETCH IT
SOLUTIONS
ORDERED pair
Is (4, -1) a solution for $y = 2x - 9$?
Name:
316

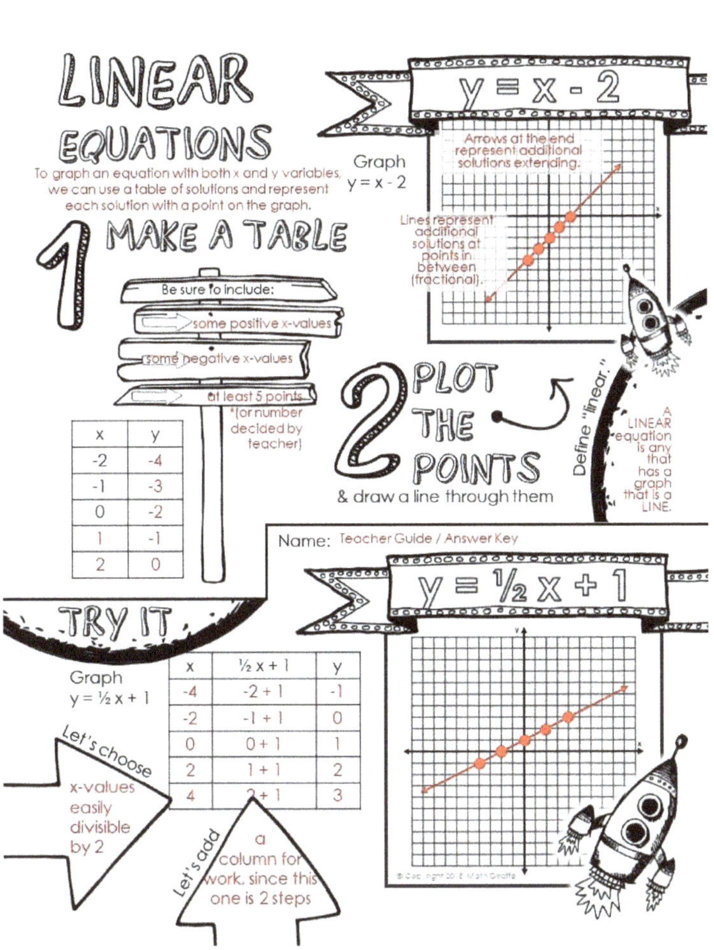

LINEAR EQUATIONS

To graph an equation with both x and y variables, we can use a table of solutions and represent each solution with a point on the graph.

1 MAKE A TABLE

Be sure to include:

- some positive x-values
- some negative x-values
- at least 5 points *(or number decided by teacher)

x	y
-2	-4
-1	-3
0	-2
1	-1
2	0

$y = x - 2$

Graph $y = x - 2$

Arrows at the end represent additional solutions extending.

Lines represent additional solutions at points in between (fractional).

2 PLOT THE POINTS

& draw a line through them

Define "linear."

A LINEAR equation is any that has a graph that is a LINE.

Name: Teacher Guide / Answer Key

TRY IT

$y = \frac{1}{2}x + 1$

Graph $y = \frac{1}{2}x + 1$

x	½ x + 1	y
-4	-2 + 1	-1
-2	-1 + 1	0
0	0 + 1	1
2	1 + 1	2
4	2 + 1	3

Let's choose x-values easily divisible by 2

Let's add a column for work, since this one is 2 steps

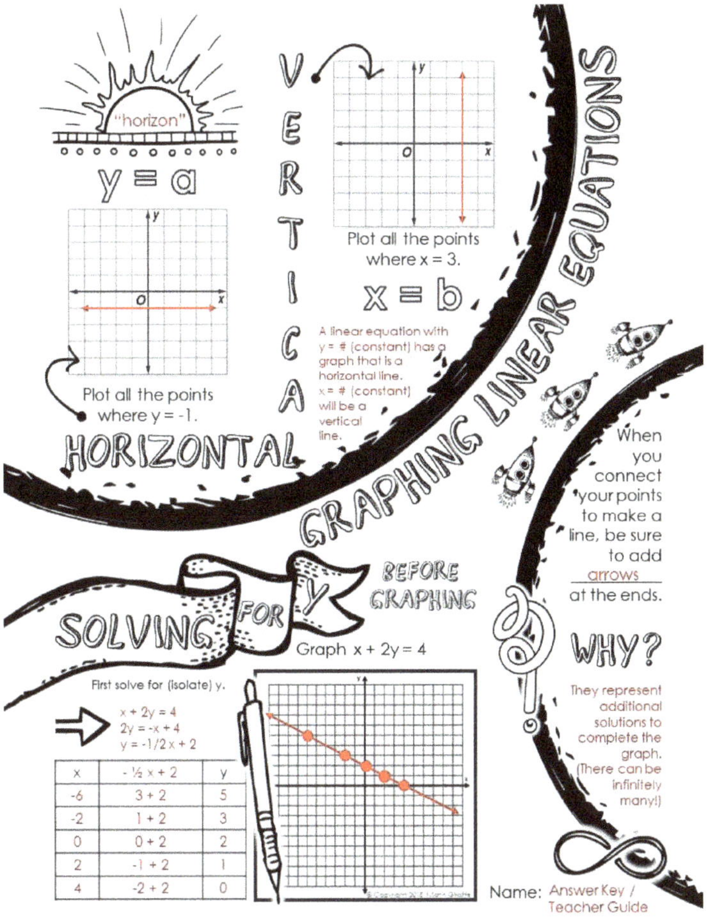

GRAPHING LINEAR EQUATIONS

HORIZONTAL

"horizon"

$y = a$

Plot all the points where $y = -1$.

VERTICAL

Plot all the points where $x = 3$.

$x = b$

A linear equation with y = # (constant) has a graph that is a horizontal line. x = # (constant) will be a vertical line.

When you connect your points to make a line, be sure to add arrows at the ends.

WHY?

They represent additional solutions to complete the graph. (There can be infinitely many!)

SOLVING FOR Y BEFORE GRAPHING

Graph $x + 2y = 4$

First solve for (isolate) y.

$x + 2y = 4$
$2y = -x + 4$
$y = -1/2x + 2$

x	- ½ x + 2	y
-6	3 + 2	5
-2	1 + 2	3
0	0 + 2	2
2	-1 + 2	1
4	-2 + 2	0

Name: Answer Key / Teacher Guide

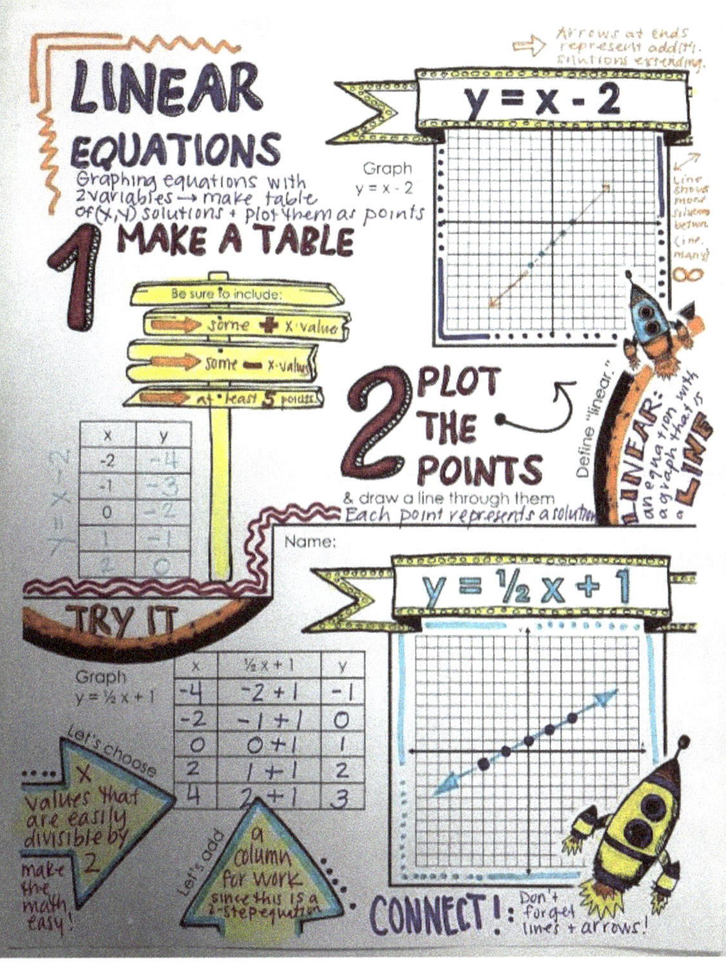

LINEAR EQUATIONS
Graphing equations with 2 variables → make table of (x,y) solutions + plot them as points
1 MAKE A TABLE
Be sure to include:
some + x-value
some – x-value
at least 5 points
Arrows at ends represent add'l solutions extending.
y = x - 2
Graph y = x - 2
2 PLOT THE POINTS
& draw a line through them
Each point represents a solution
Define "linear."
LINEAR: an equation with a graph that is a LINE
TRY IT
y = ½ x + 1
Graph y = ½ x + 1
Let's choose x values that are easily divisible by 2
make the math easy!
Let's add a column for work since this is a 2-step equation
CONNECT!: Don't forget lines + arrows!

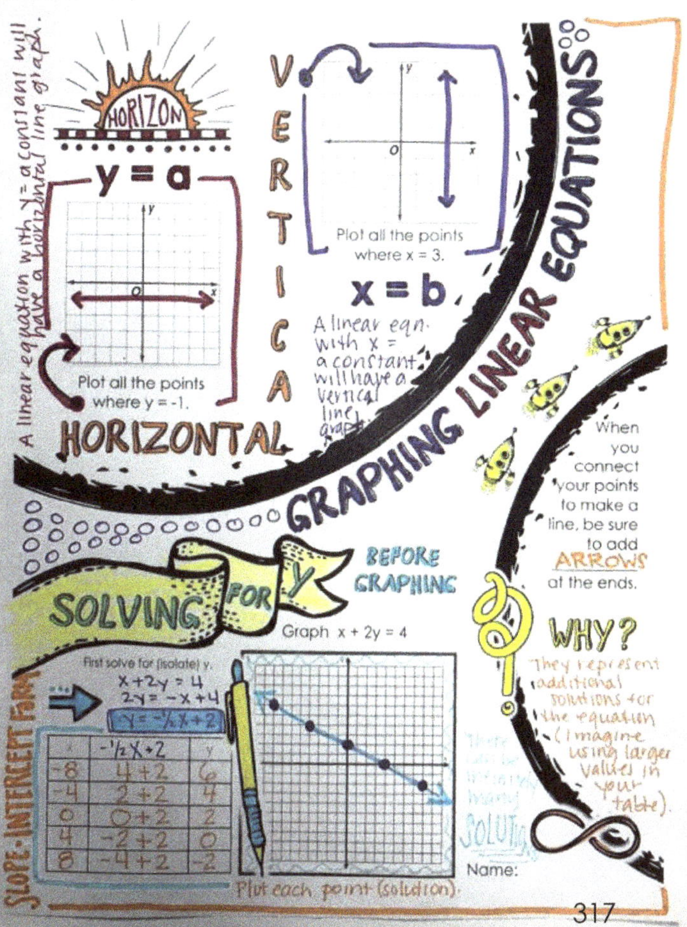

A linear equation with y = a constant will have a horizontal line graph.
HORIZON
y = a
Plot all the points where y = -1.
HORIZONTAL
VERTICAL
Plot all the points where x = 3.
x = b
A linear eqn. with x = a constant will have a vertical line graph.
GRAPHING LINEAR EQUATIONS
When you connect your points to make a line, be sure to add ARROWS at the ends.
WHY?
They represent additional solutions for the equation (Imagine using larger values in your table).
SOLVING FOR Y BEFORE GRAPHING
Graph x + 2y = 4
First solve for (isolate) y.
x+2y = 4
2y = -x + 4
y = -½x + 2
SLOPE-INTERCEPT FORM
Plot each point (solution).
Name:

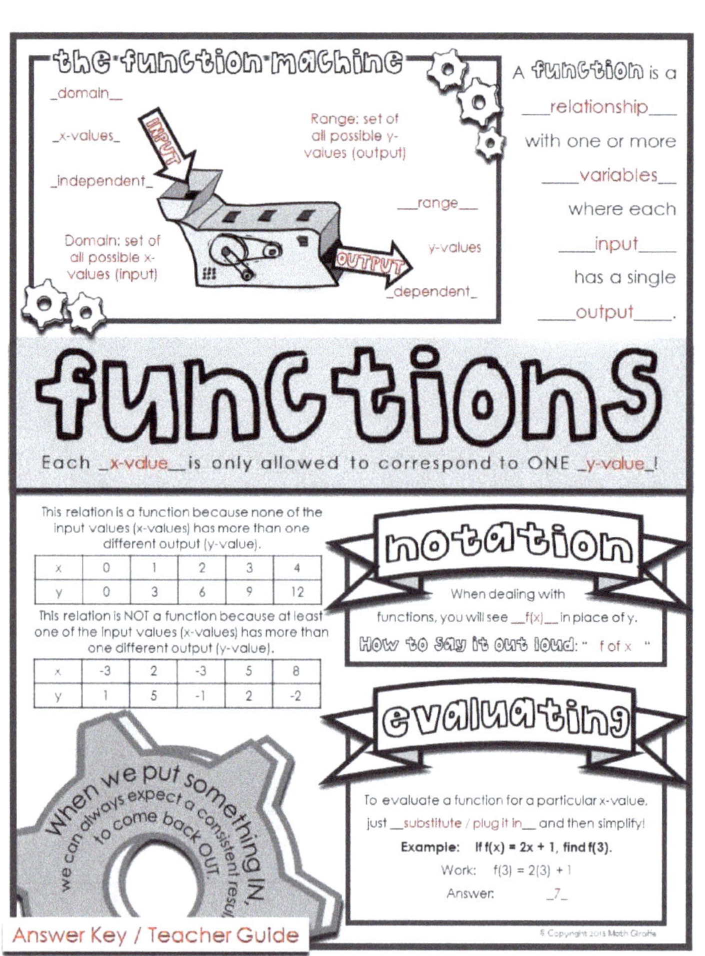

the function machine
domain
x-values
independent
Range: set of all possible y-values (output)
INPUT
__range__
y-values
OUTPUT
Domain: set of all possible x-values (input)
dependent
A function is a ___relationship___ with one or more ___variables___ where each ___input___ has a single ___output___.
functions
Each _x-value_ is only allowed to correspond to ONE _y-value_!
This relation is a function because none of the input values (x-values) has more than one different output (y-value).
x | 0 | 1 | 2 | 3 | 4
y | 0 | 3 | 6 | 9 | 12
This relation is NOT a function because at least one of the input values (x-values) has more than one different output (y-value).
x | -3 | 2 | -3 | 5 | 8
y | 1 | 5 | -1 | 2 | -2
notation
When dealing with functions, you will see __f(x)__ in place of y.
How to say it out loud: " f of x "
evaluating
To evaluate a function for a particular x-value, just __substitute / plug it in__ and then simplify!
Example: If f(x) = 2x + 1, find f(3).
Work: f(3) = 2(3) + 1
Answer: _7_
When we put something IN, we can always expect a consistent result to come back OUT.
Answer Key / Teacher Guide

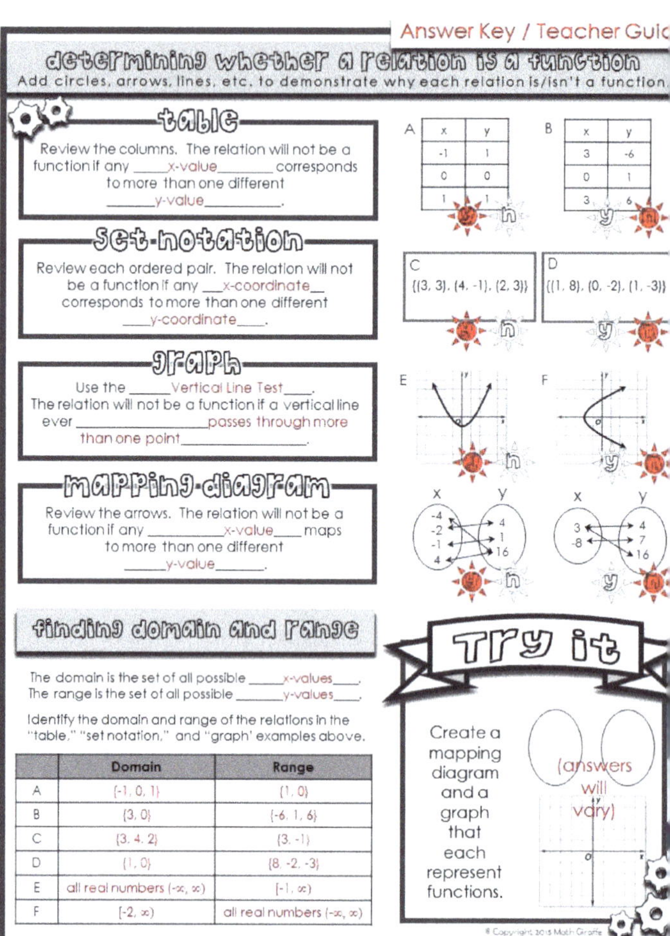

Answer Key / Teacher Guide
determining whether a relation is a function
Add circles, arrows, lines, etc. to demonstrate why each relation is/isn't a function.
table
Review the columns. The relation will not be a function if any ____x-value______ corresponds to more than one different ______y-value__________.
set-notation
Review each ordered pair. The relation will not be a function if any ___x-coordinate__ corresponds to more than one different ____y-coordinate____.
graph
Use the _____Vertical Line Test____.
The relation will not be a function if a vertical line ever _____________passes through more than one point_______________.
mapping-diagram
Review the arrows. The relation will not be a function if any _________x-value___ maps to more than one different ______y-value______.
A
x | y
-1 | 1
0 | 0
1 | 1
B
x | y
3 | -6
0 | 1
3 | 6
C {(3, 3), (4, -1), (2, 3)}
D {(1, 8), (0, -2), (1, -3)}
E
F
finding domain and range
The domain is the set of all possible _____x-values____.
The range is the set of all possible ______y-values____.
Identify the domain and range of the relations in the "table," "set notation," and "graph" examples above.
| Domain | Range
A | {-1, 0, 1} | {1, 0}
B | {3, 0} | {-6, 1, 6}
C | {3, 4, 2} | {3, -1}
D | {1, 0} | {8, -2, -3}
E | all real numbers (-∞, ∞) | [-1, ∞)
F | [-2, ∞) | all real numbers (-∞, ∞)
Try it
Create a mapping diagram and a graph that each represent functions.
(answers will vary)

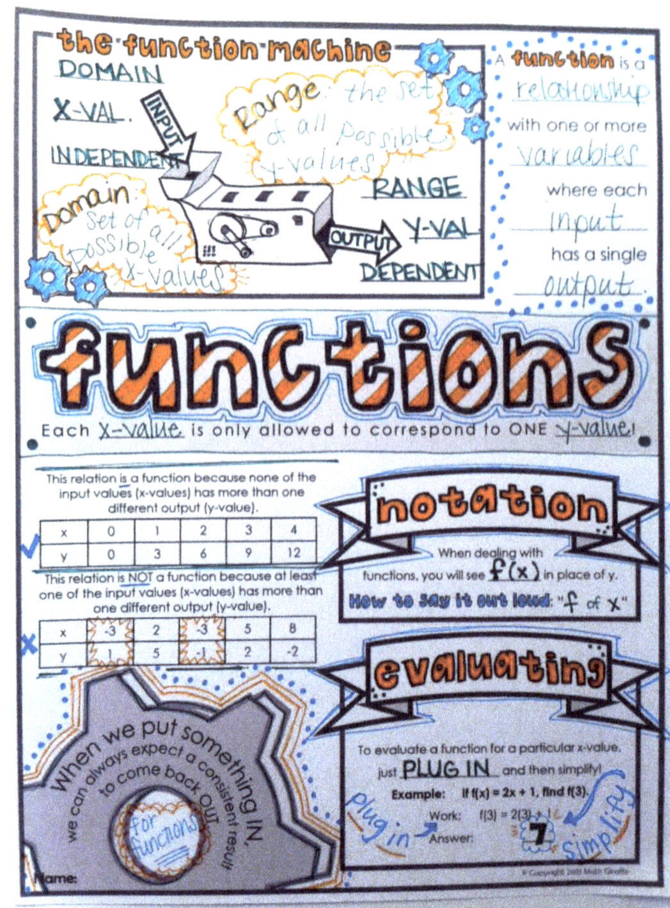

the function machine
DOMAIN
X-VAL
INDEPENDENT
Range: the set of all possible y-values
RANGE
Y-VAL
DEPENDENT
Domain Set of all possible x-values
A function is a relationship with one or more variables where each input has a single output.
functions
Each X-value is only allowed to correspond to ONE Y-value!
notation
When dealing with functions, you will see f(x) in place of y.
How to say it out loud: "f of x"
evaluating
To evaluate a function for a particular x-value, just PLUG IN and then simplify!
Example: If f(x) = 2x + 1, find f(3).
Work: f(3) = 2(3) + 1
Answer: 7
plug in
simplify
for functions
Name:

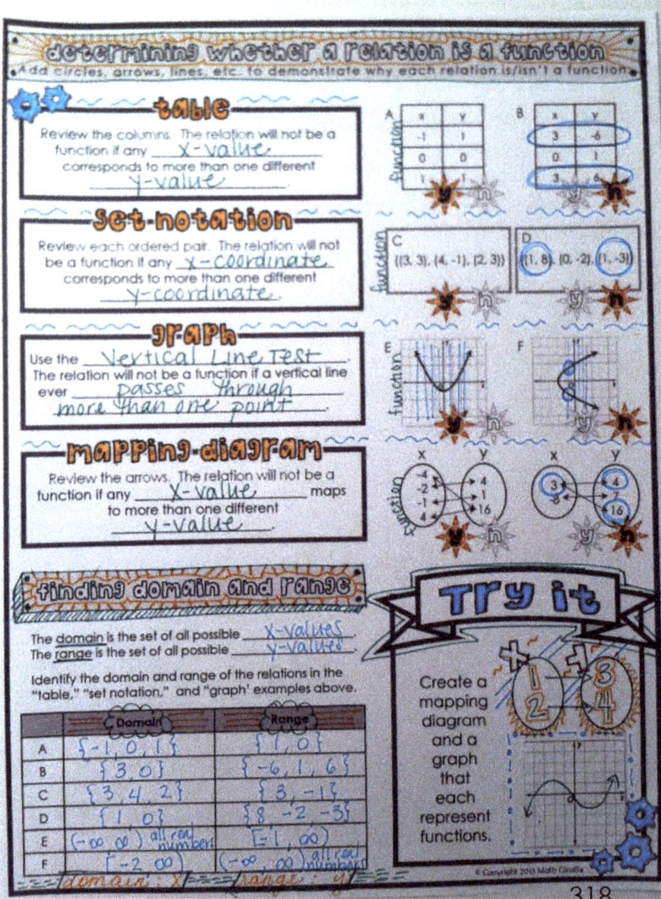

determining whether a relation is a function
table
Review the columns. The relation will not be a function if any X-value corresponds to more than one different Y-value
set-notation
Review each ordered pair. The relation will not be a function if any X-coordinate corresponds to more than one different Y-coordinate.
graph
Use the Vertical Line Test
The relation will not be a function if a vertical line ever passes through more than one point
mapping-diagram
Review the arrows. The relation will not be a function if any X-value maps to more than one different Y-value
finding domain and range
The domain is the set of all possible X-values
The range is the set of all possible Y-values
| Domain | Range
A | {-1, 0, 1} | {1, 0}
B | {3, 0} | {-6, 1, 6}
C | {3, 4, 2} | {3, -1}
D | {1, 0} | {8, -2, -3}
E | (-∞, ∞) all real numbers | [-1, ∞)
F | [-2, ∞) | (-∞, ∞) all real numbers
domain: x range: y
Try it
Create a mapping diagram and a graph that each represent functions.

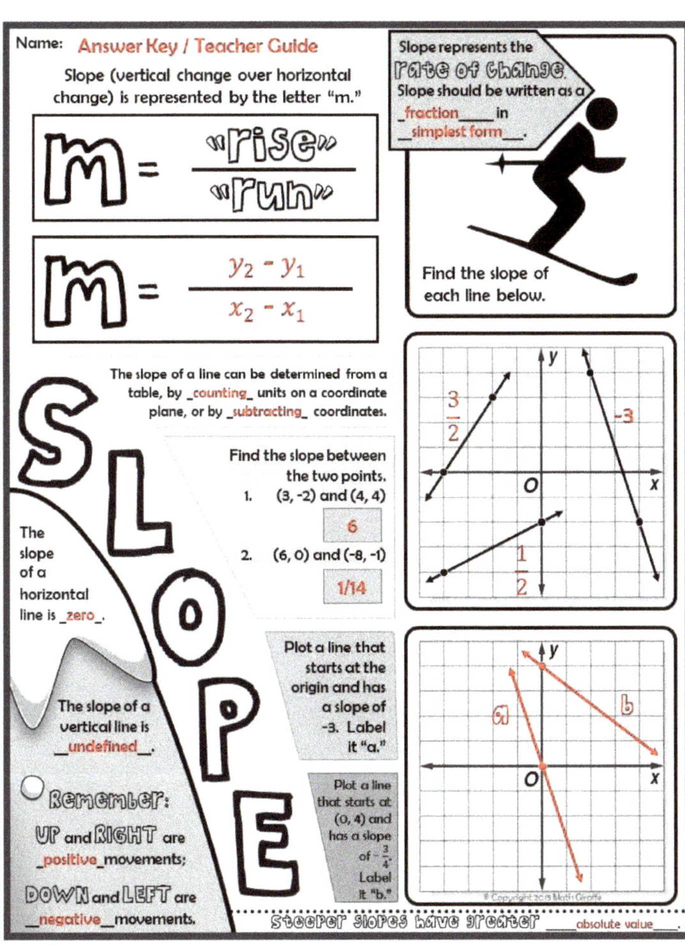

Name: Answer Key / Teacher Guide

Slope (vertical change over horizontal change) is represented by the letter "m."

$m = \frac{\text{"rise"}}{\text{"run"}}$

$m = \frac{y_2 - y_1}{x_2 - x_1}$

Slope represents the rate of change. Slope should be written as a _fraction_ in _simplest form_.

Find the slope of each line below.

The slope of a line can be determined from a table, by _counting_ units on a coordinate plane, or by _subtracting_ coordinates.

SLOPE

Find the slope between the two points.

1. (3, -2) and (4, 4) — 6
2. (6, 0) and (-8, -1) — 1/14

The slope of a horizontal line is _zero_.

The slope of a vertical line is _undefined_.

Plot a line that starts at the origin and has a slope of -3. Label it "a."

Plot a line that starts at (0, 4) and has a slope of $-\frac{3}{4}$. Label it "b."

Remember:

UP and RIGHT are _positive_ movements;

DOWN and LEFT are _negative_ movements.

Steeper slopes have greater _absolute value_.

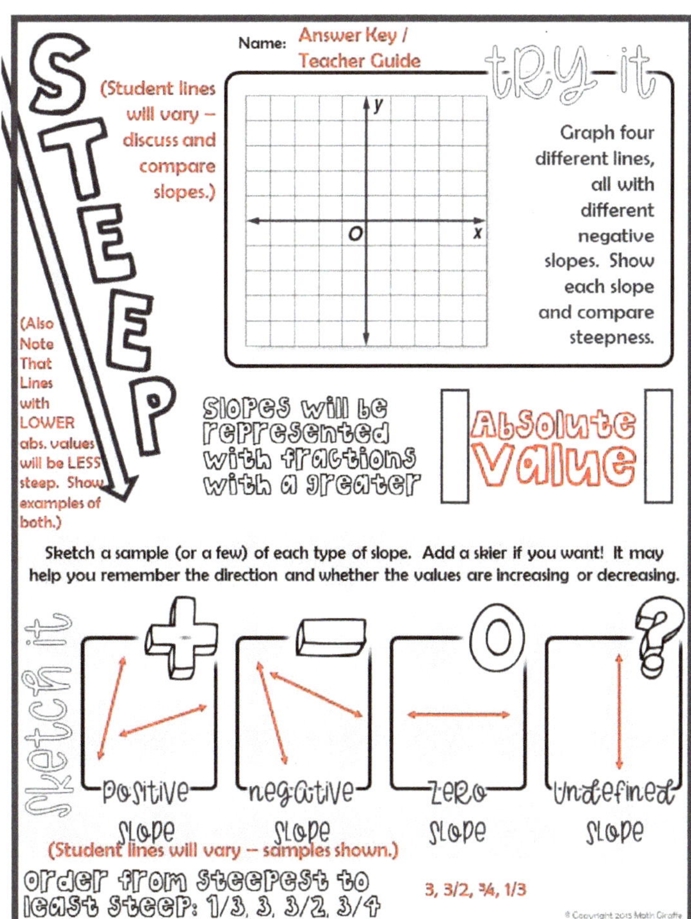

Name: Answer Key / Teacher Guide

STEEP

try-it

(Student lines will vary – discuss and compare slopes.)

Graph four different lines, all with different negative slopes. Show each slope and compare steepness.

(Also Note That Lines with LOWER abs. values will be LESS steep. Show examples of both.)

Slopes will be represented with fractions with a greater Absolute Value

Sketch a sample (or a few) of each type of slope. Add a skier if you want! It may help you remember the direction and whether the values are increasing or decreasing.

Sketch it

+	−	0	?
positive slope	negative slope	zero slope	undefined slope

(Student lines will vary -- samples shown.)

Order from steepest to least steep: 1/3, 3, 3/2, 3/4 — 3, 3/2, 3/4, 1/3

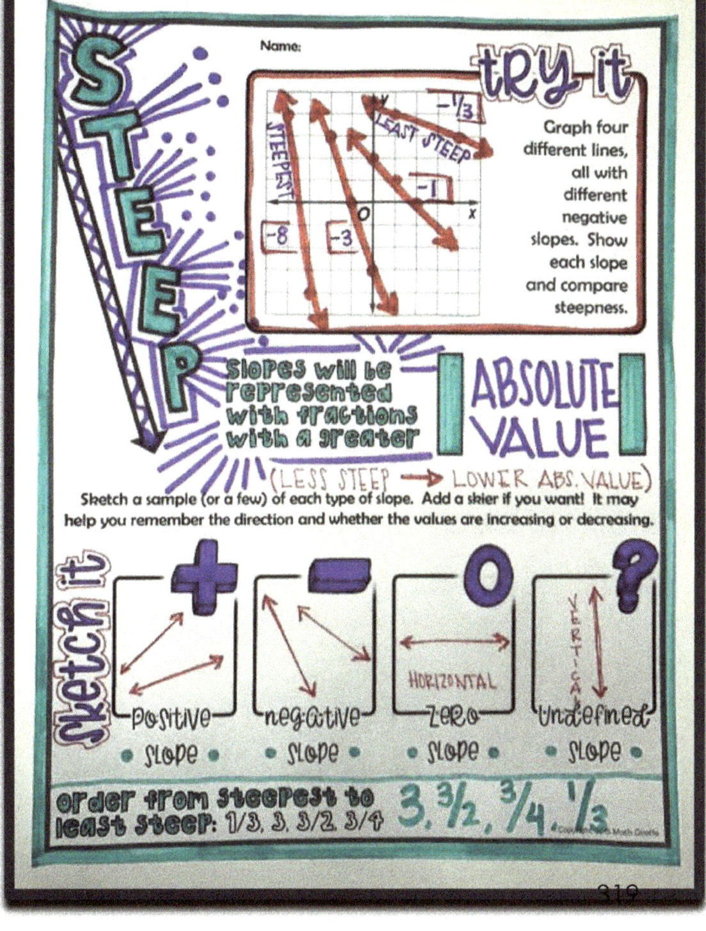

The SLOPE is...
the rate of change (rise over run).
Slope-Intercept Form
A Y-INTERCEPT is...
the point where the line crosses the y-axis.
a formula for the equation of a line using its slope and its y-intercept
slope
*(Remind students of the slope formula)
y-intercept
y = mx + b
variable
variable
steps for graphing
Try graphing the line y = -3x + 2.
1 Start by placing a point at the y-intercept
2 From that point, count out the slope to find the next point.
3 connect the two points to form a line.
Be careful about positive & negative directions
*(Remind students that a slope of -3 is the same as -3/1 so they should count down three and over to the right 1 from the y-int.)
Name: Answer Key / Teacher Guide
(Remind students to add arrows at the ends to show that the line continues.)

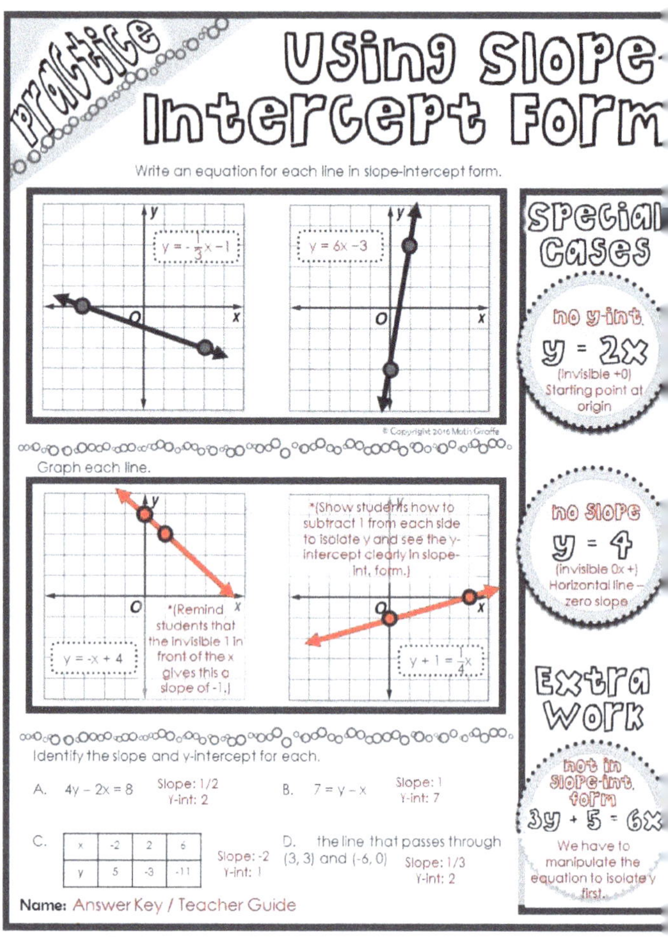
practice
Using Slope-Intercept Form
Write an equation for each line in slope-intercept form.
y = -1/3x - 1
y = 6x - 3
Special Cases
no y-int.
y = 2x
(invisible +0)
Starting point at origin
Graph each line.
*(Show students how to subtract 1 from each side to isolate y and see the y-intercept clearly in slope-int. form.)
*(Remind students that the invisible 1 in front of the x gives this a slope of -1.)
y = -x + 4
y + 1 = 1/4x
no slope
y = 4
(invisible 0x +)
Horizontal line - zero slope
Extra Work
Identify the slope and y-intercept for each.
A. 4y - 2x = 8 Slope: 1/2 Y-int: 2
B. 7 = y - x Slope: 1 Y-int: 7
C. Slope: -2 Y-int: 1
D. the line that passes through (3, 3) and (-6, 0) Slope: 1/3 Y-int: 2
not in slope-int. form
3y + 5 = 6x
We have to manipulate the equation to isolate y first.
Name: Answer Key / Teacher Guide

x	-2	2	6
y	5	-3	-11

The SLOPE is...
the RATE of CHANGE
Slope-Intercept Form
A Y-INTERCEPT is... the POINT where the line crosses the Y-AXIS
a formula for the equation of a LINE using its SLOPE and its y-intercept
SLOPE
$\frac{y_2 - y_1}{x_2 - x_1}$
y-INTERCEPT
y = mx + b
VARIABLE
VARIABLE
START
COUNT
DOWN 3 RIGHT 1
steps for graphing
Try graphing the line y = -3x + 2.
1 Start by placing a point at the y-int
2 From that point, count out the slope to find the next point.
3 connect the two points to form a line.
Be careful about +/- directions
Name:

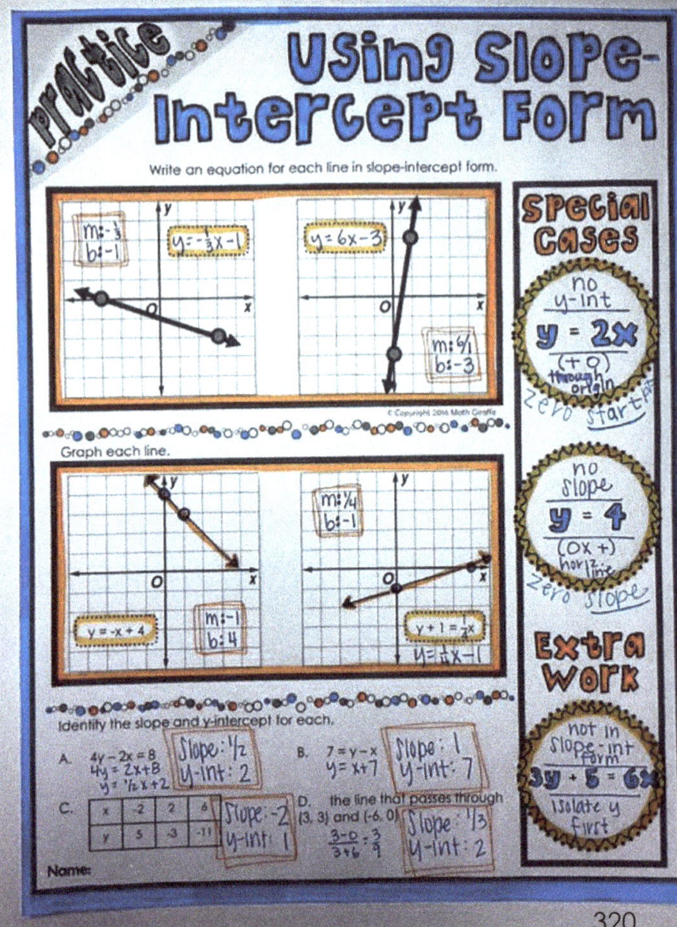
practice
Using Slope-Intercept Form
Write an equation for each line in slope-intercept form.
m: -1/3 b: -1
y = -1/3x - 1
y = 6x - 3
m: 6/1 b: -3
Special Cases
no y-int
y = 2x
(+0) through origin
zero start pt
Graph each line.
m: 1/4 b: -1
y = -x + 4
m: -1 b: 4
y + 1 = 1/4x
y = 1/4x - 1
no slope
y = 4
(0x +) horiz. line
zero slope
Extra Work
Identify the slope and y-intercept for each.
A. 4y - 2x = 8
4y = 2x + 8
y = 1/2x + 2
Slope: 1/2 y-int: 2
B. 7 = y - x
y = x + 7
Slope: 1 y-int: 7
C. Slope: -2 y-int: 1
D. the line that passes through (3, 3) and (-6, 0)
(3 - 0)/(3 + 6) = 3/9
Slope: 1/3 y-int: 2
not in slope-int form
3y + 5 = 6x
Isolate y first
Name:

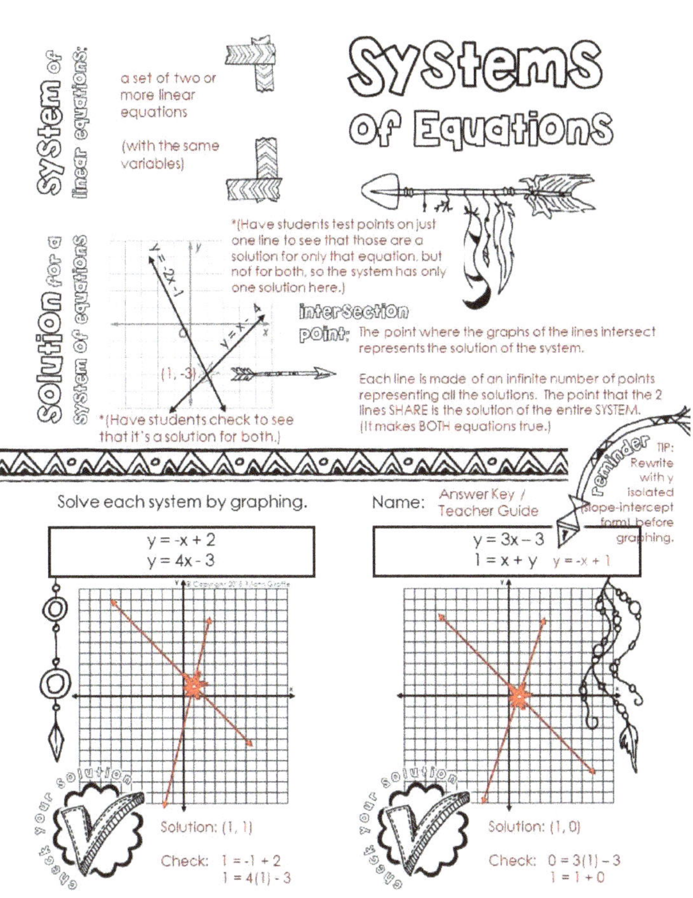

Systems of Equations
System of linear equations:
a set of two or more linear equations
(with the same variables)
Solution for a system of equations
*(Have students test points on just one line to see that those are a solution for only that equation, but not for both, so the system has only one solution here.)
intersection point:
The point where the graphs of the lines intersect represents the solution of the system.
Each line is made of an infinite number of points representing all the solutions. The point that the 2 lines SHARE is the solution of the entire SYSTEM. (It makes BOTH equations true.)
*(Have students check to see that it's a solution for both.)
(1, -3)
y = -2x - 1
y = x - 4
reminder
TIP: Rewrite with y isolated (slope-intercept form) before graphing.
Solve each system by graphing.
Name: Answer Key / Teacher Guide
y = -x + 2
y = 4x - 3
y = 3x – 3
1 = x + y
y = -x + 1
check your solution
Solution: (1, 1)
Check: 1 = -1 + 2
1 = 4(1) - 3
Solution: (1, 0)
Check: 0 = 3(1) – 3
1 = 1 + 0

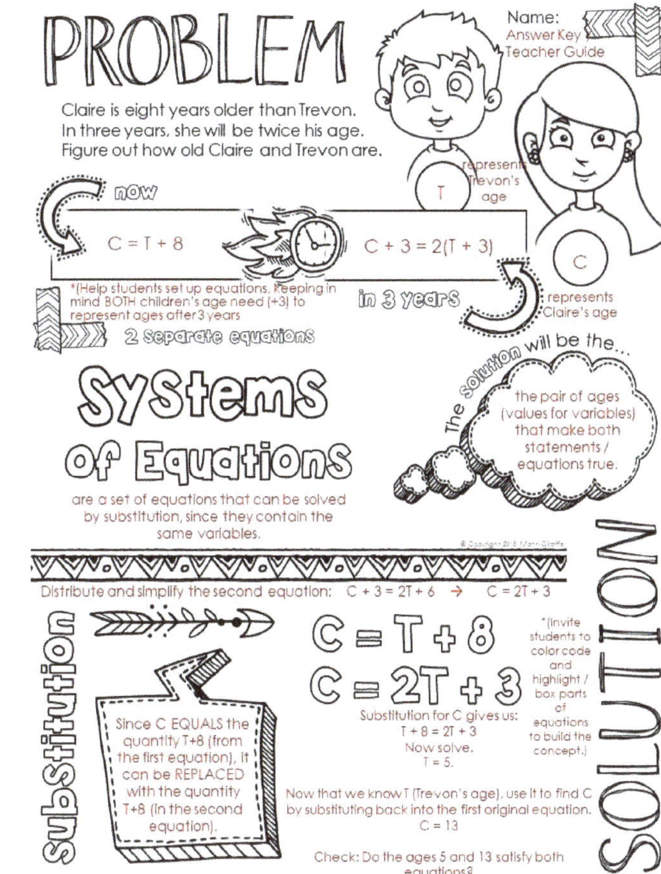

PROBLEM
Name: Answer Key Teacher Guide
Claire is eight years older than Trevon. In three years, she will be twice his age. Figure out how old Claire and Trevon are.
represents Trevon's age
T
C
represents Claire's age
now
C = T + 8
C + 3 = 2(T + 3)
*(Help students set up equations, keeping in mind BOTH children's age need (+3) to represent ages after 3 years
in 3 years
2 separate equations
The solution will be the...
the pair of ages (values for variables) that make both statements / equations true.
Systems of Equations
are a set of equations that can be solved by substitution, since they contain the same variables.
Distribute and simplify the second equation: C + 3 = 2T + 6 → C = 2T + 3
Substitution
C = T + 8
C = 2T + 3
Since C EQUALS the quantity T+8 (from the first equation), it can be REPLACED with the quantity T+8 (in the second equation).
Substitution for C gives us: T + 8 = 2T + 3
Now solve. T = 5.
*(Invite students to color code and highlight / box parts of equations to build the concept.)
Now that we know T (Trevon's age), use it to find C by substituting back into the first original equation. C = 13
Check: Do the ages 5 and 13 satisfy both equations?
SOLUTION

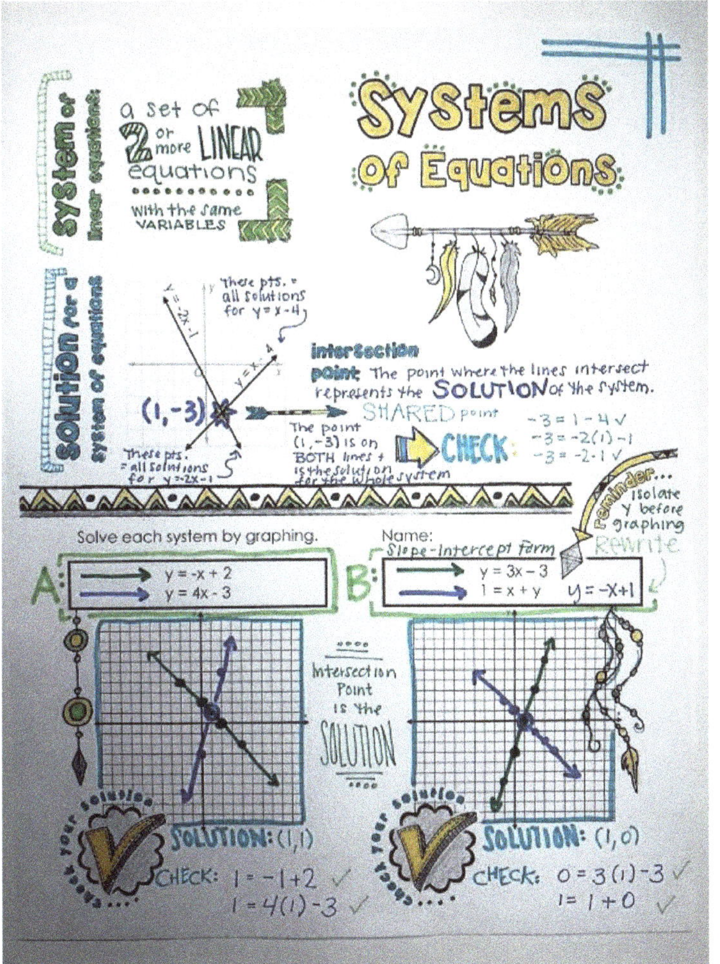

Systems of Equations
System of linear equations:
a set of 2 or more LINEAR equations
with the same VARIABLES
Solution for a system of equations
These pts. = all solutions for y = x - 4
intersection point: The point where the lines intersect represents the SOLUTION of the system.
(1, -3)
SHARED point
The point (1, -3) is on BOTH lines + is the solution for the whole system
CHECK
-3 = 1 - 4 ✓
-3 = -2(1) - 1
-3 = -2 - 1 ✓
These pts. = all solutions for y = -2x - 1
reminder... isolate y before graphing
Solve each system by graphing.
Name:
Slope-intercept form
Rewrite
A
y = -x + 2
y = 4x - 3
B
y = 3x – 3
1 = x + y
y = -x + 1
Intersection Point is the SOLUTION
check your solution
SOLUTION: (1, 1)
CHECK: 1 = -1 + 2 ✓
1 = 4(1) - 3 ✓
SOLUTION: (1, 0)
CHECK: 0 = 3(1) - 3 ✓
1 = 1 + 0 ✓

PROBLEM
Name:
define variables
Claire is eight years older than Trevon. In three years, she will be twice his age. Figure out how old Claire and Trevon are.
Use a system. Write 2 equations w. the same variables.
Trevon's Age
T
Claire's Age
C
now
C = T + 8
C + 3 = 2(T + 3)
in 3 years
2 separate equations
The solution will be the...
values for the ages (variables) that make BOTH equations / statements TRUE
Systems of Equations
Then we can solve our system using SUBSTITUTION.
Since they both have the same variables...
Distribute & simplify 2nd equation: C + 3 = 2T + 6
Substitution
C = T + 8
C = 2T + 3
Since C is EQUAL TO T+8 we can then REPLACE with T+8
T + 8 = 2T + 3
Now, we can solve - just one variable!
T + 8 = 2T + 3
8 = T + 3
5 = T
So now we know Trevon's age is 5.
Use it to find Claire's age.
C = 5 + 8
C = 13
Check It! ✓
SOLUTION

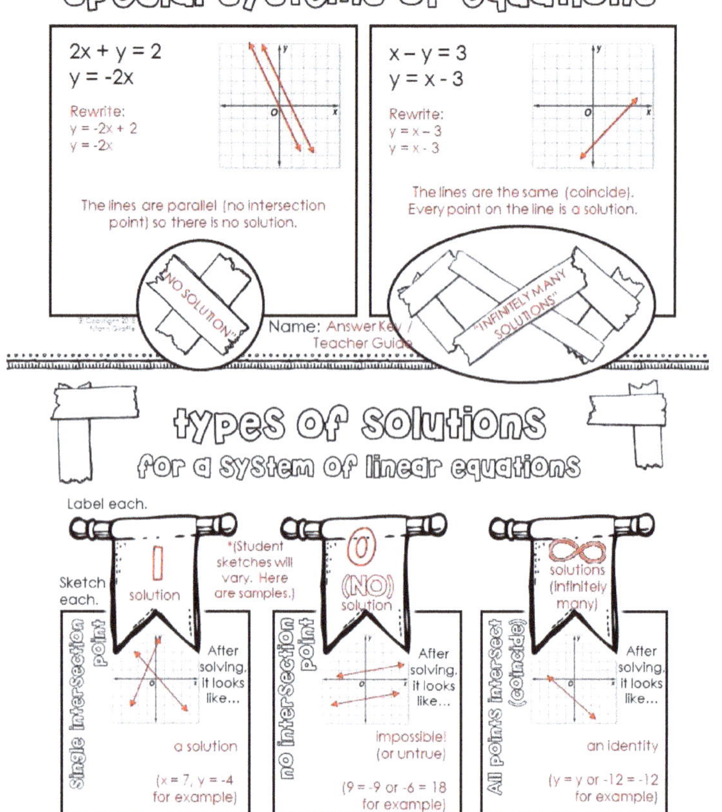
Special Systems of equations
2x + y = 2
y = -2x
Rewrite:
y = -2x + 2
y = -2x
The lines are parallel (no intersection point) so there is no solution.
"NO SOLUTION"
x – y = 3
y = x - 3
Rewrite:
y = x – 3
y = x - 3
The lines are the same (coincide). Every point on the line is a solution.
"INFINITELY MANY SOLUTIONS"
Name: Answer Key / Teacher Guide
types of solutions
for a system of linear equations
Label each.
Sketch each.
1 solution
*(Student sketches will vary. Here are samples.)
0 (NO) solution
∞ solutions (infinitely many)
single intersection point
After solving, it looks like...
a solution
(x = 7, y = -4 for example)
no intersection point
After solving, it looks like...
impossible! (or untrue)
(9 = -9 or -6 = 18 for example)
All points intersect (coincide)
After solving, it looks like...
an identity
(y = y or -12 = -12 for example)

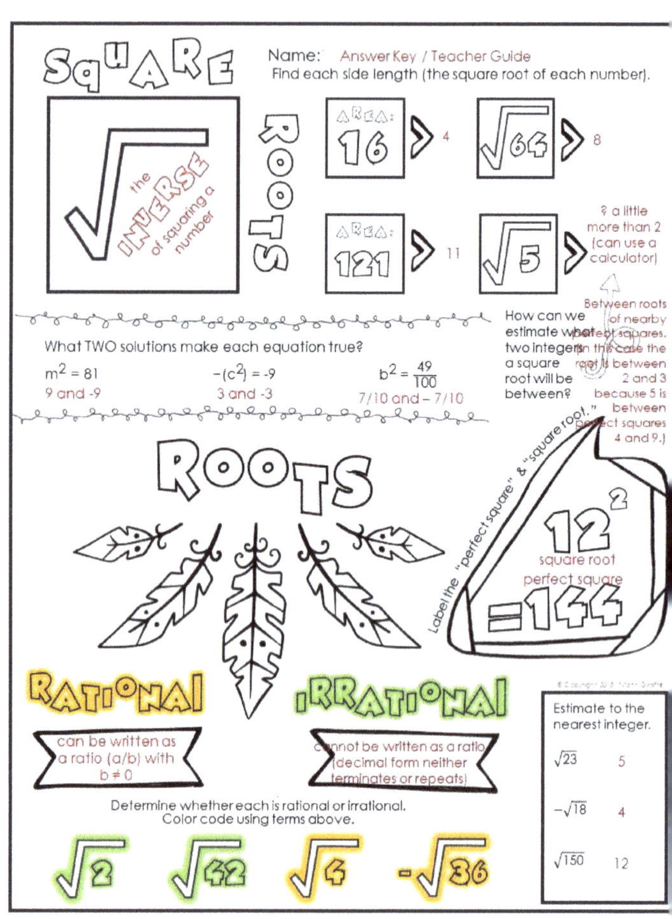
SQUARE ROOTS
Name: Answer Key / Teacher Guide
Find each side length (the square root of each number).
the INVERSE of squaring a number
AREA: 16 > 4
√64 > 8
AREA: 121 > 11
√5 > a little more than 2 (can use a calculator)
What TWO solutions make each equation true?
m² = 81
9 and -9
-(c²) = -9
3 and -3
b² = 49/100
7/10 and – 7/10
How can we estimate what two integers a square root will be between?
Between roots of nearby perfect squares. (In this case the root is between 2 and 3 because 5 is between perfect squares 4 and 9.)
ROOTS
Label the "perfect square" & "square root."
12²
square root
perfect square
=144
RATIONAL
can be written as a ratio (a/b) with b ≠ 0
IRRATIONAL
cannot be written as a ratio (decimal form neither terminates or repeats)
Determine whether each is rational or irrational. Color code using terms above.
√2 √42 √4 -√36
Estimate to the nearest integer.
√23 5
-√18 4
√150 12

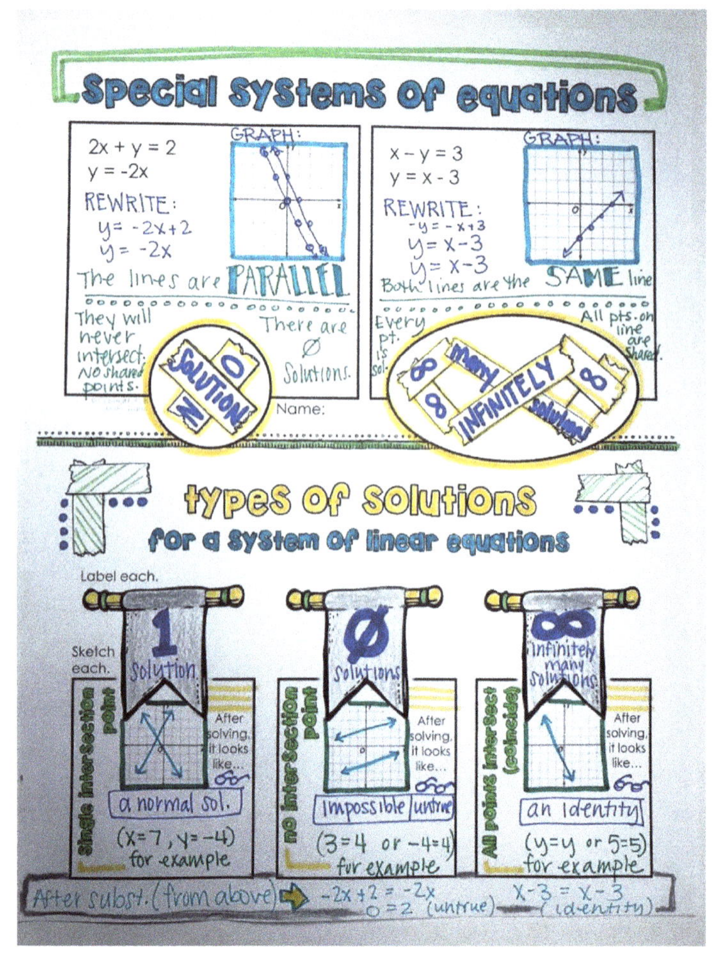
Special Systems of equations
2x + y = 2
y = -2x
GRAPH:
REWRITE:
y = -2x+2
y = -2x
The lines are PARALLEL
They will never intersect. No shared points.
There are Ø Solutions.
0 SOLUTION
Name:
x – y = 3
y = x - 3
GRAPH:
REWRITE:
-y = -x+3
y = x-3
y = x-3
Both lines are the SAME line
Every pt. is sol.
All pts. on line are shared.
∞ many INFINITELY solutions ∞
types of solutions
for a system of linear equations
Label each.
Sketch each.
1 solution
Ø solutions
∞ infinitely many solutions
single intersection point
After solving, it looks like...
a normal sol.
(x=7, y=-4) for example
no intersection point
After solving, it looks like...
impossible/untrue
(3=4 or -4=4) for example
All points intersect (coincide)
After solving, it looks like...
an identity
(y=y or 5=5) for example
After subst. (from above) ➡ -2x+2 = -2x 0=2 (untrue) X-3 = X-3 (identity)

SQUARE ROOTS
Name:
Find each side length (the square root of each number).
the INVERSE of squaring a number
side:
AREA: 16 > 4
√64 > 8
side:
AREA: 121 > 11
√5 > more than 2
√4 = 2 √9 = 3
What TWO solutions make each equation true? ➡ POS & NEG!
m² = 81
9 and -9
-(c²) = -9
3 and -3
b² = 49/100
7/10 and -7/10
How can we estimate what two integers a square root will be between?
use roots of the NEARBY perfect squares
ROOTS
Label the "perfect square" & "square root."
sq rt
12²
=144
RATIONAL
IRRATIONAL
Determine whether each is rational or irrational. Color code using terms above.
Irr. √2 √42 √4 -√36 rational
Estimate to the nearest integer.
√23 ~5
-√18 ~4
√150 ~12

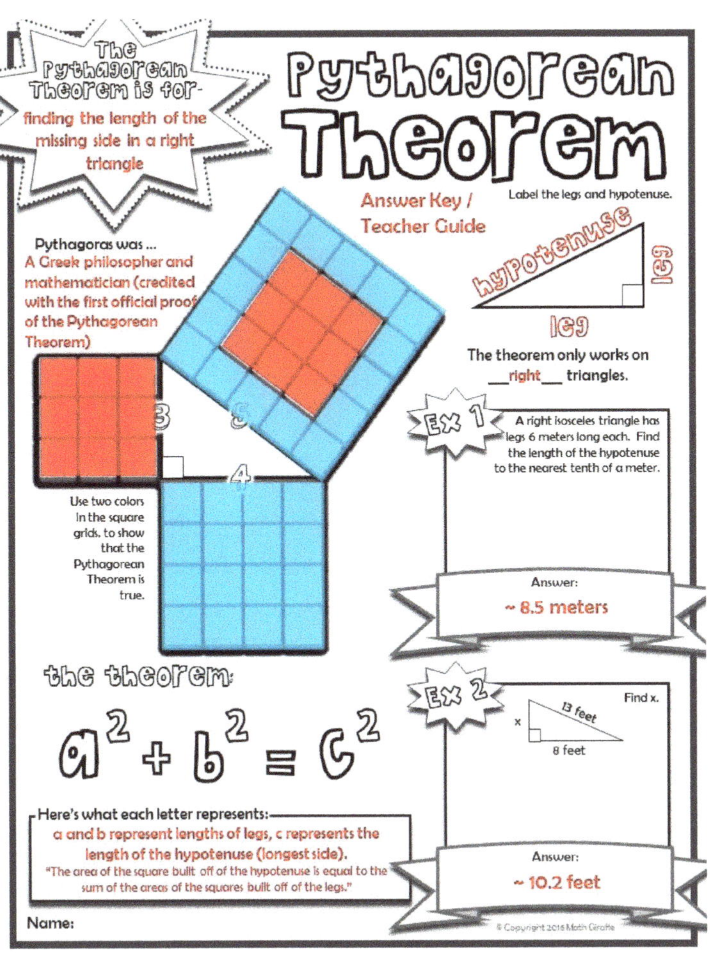
The Pythagorean Theorem is for- finding the length of the missing side in a right triangle
Pythagorean Theorem
Answer Key / Teacher Guide
Label the legs and hypotenuse.
hypotenuse
leg
leg
Pythagoras was ...
A Greek philosopher and mathematician (credited with the first official proof of the Pythagorean Theorem)
3
5
4
Use two colors in the square grids. to show that the Pythagorean Theorem is true.
The theorem only works on __right__ triangles.
Ex 1
A right isosceles triangle has legs 6 meters long each. Find the length of the hypotenuse to the nearest tenth of a meter.
Answer:
~ 8.5 meters
Ex 2
Find x.
13 feet
x
8 feet
Answer:
~ 10.2 feet
the theorem:
$a^2 + b^2 = c^2$
Here's what each letter represents:
a and b represent lengths of legs, c represents the length of the hypotenuse (longest side).
"The area of the square built off of the hypotenuse is equal to the sum of the areas of the squares built off of the legs."
Name:
© Copyright 2016 Math Giraffe

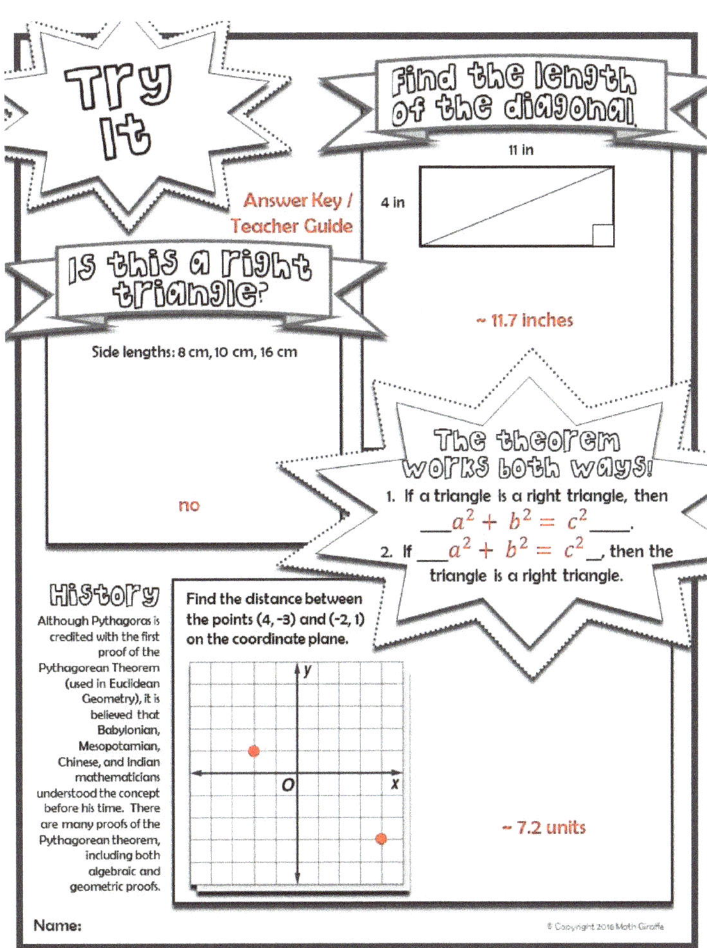
Try It
Answer Key / Teacher Guide
Find the length of the diagonal
11 in
4 in
~ 11.7 inches
Is this a right triangle?
Side lengths: 8 cm, 10 cm, 16 cm
no
The theorem works both ways!
1. If a triangle is a right triangle, then $a^2 + b^2 = c^2$.
2. If $a^2 + b^2 = c^2$, then the triangle is a right triangle.
History
Although Pythagoras is credited with the first proof of the Pythagorean Theorem (used in Euclidean Geometry), it is believed that Babylonian, Mesopotamian, Chinese, and Indian mathematicians understood the concept before his time. There are many proofs of the Pythagorean theorem, including both algebraic and geometric proofs.
Find the distance between the points (4, -3) and (-2, 1) on the coordinate plane.
y
O
x
~ 7.2 units
Name:
© Copyright 2016 Math Giraffe

The Pythagorean Theorem is for- finding SIDE lengths in a Right Δ.
Pythagorean Theorem
Label the legs and hypotenuse.
Pythagoras was ... a GREEK philosopher & MATHEMATICIAN
HYPOTENUSE
leg
leg
The theorem only works on right triangles.
3
5
4
Use two colors in the square grids. to show that the Pythagorean Theorem is true.
Ex 1
A right isosceles triangle has legs 6 meters long each. Find the length of the hypotenuse to the nearest tenth of a meter.
$6^2 + 6^2 = c^2$
$36 + 36 = c^2$
$72 = c^2$
$c = \sqrt{72}$
Answer:
Ex 2
the theorem:
$a^2 + b^2 = c^2$
Here's what each letter represents:
a and b → LEGS
c → HYPOT.
Name:

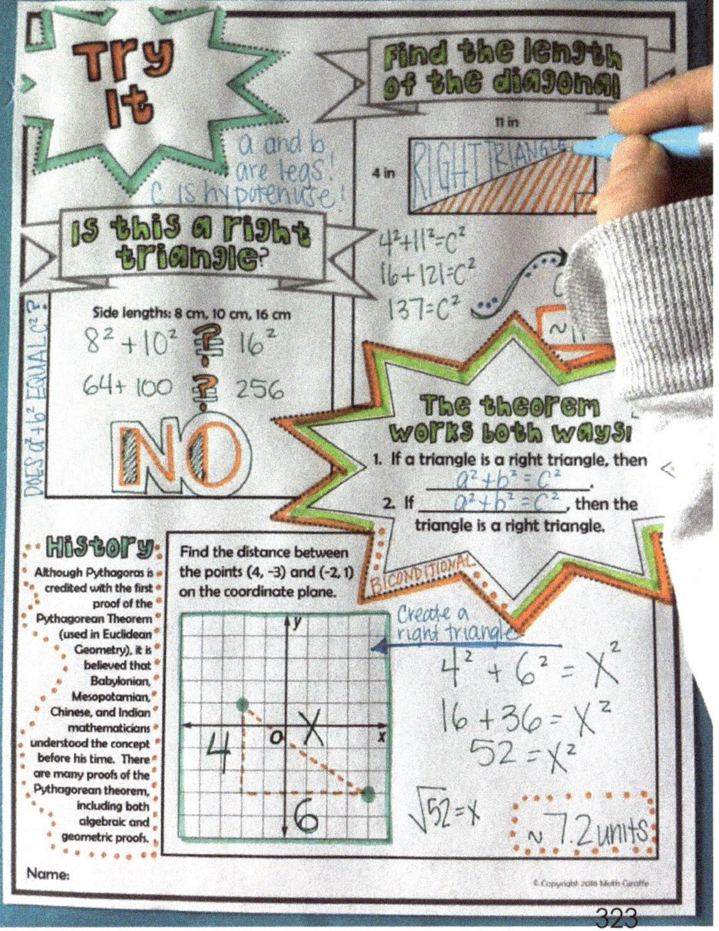
Try It
a and b are legs!
c is hypotenuse!
Find the length of the diagonal
11 in
4 in
RIGHT TRIANGLE
$4^2 + 11^2 = c^2$
$16 + 121 = c^2$
$137 = c^2$
Is this a right triangle?
Side lengths: 8 cm, 10 cm, 16 cm
$8^2 + 10^2 \overset{?}{=} 16^2$
$64 + 100 \overset{?}{=} 256$
NO
DOES $a^2 + b^2$ EQUAL c^2?
The theorem works both ways!
1. If a triangle is a right triangle, then $a^2 + b^2 = c^2$.
2. If $a^2 + b^2 = c^2$, then the triangle is a right triangle.
BICONDITIONAL
History
Although Pythagoras is credited with the first proof of the Pythagorean Theorem (used in Euclidean Geometry), it is believed that Babylonian, Mesopotamian, Chinese, and Indian mathematicians understood the concept before his time. There are many proofs of the Pythagorean theorem, including both algebraic and geometric proofs.
Find the distance between the points (4, -3) and (-2, 1) on the coordinate plane.
Create a right triangle
$4^2 + 6^2 = x^2$
$16 + 36 = x^2$
$52 = x^2$
$\sqrt{52} = x$
~ 7.2 units
4
6
X
Name:
© Copyright 2016 Math Giraffe

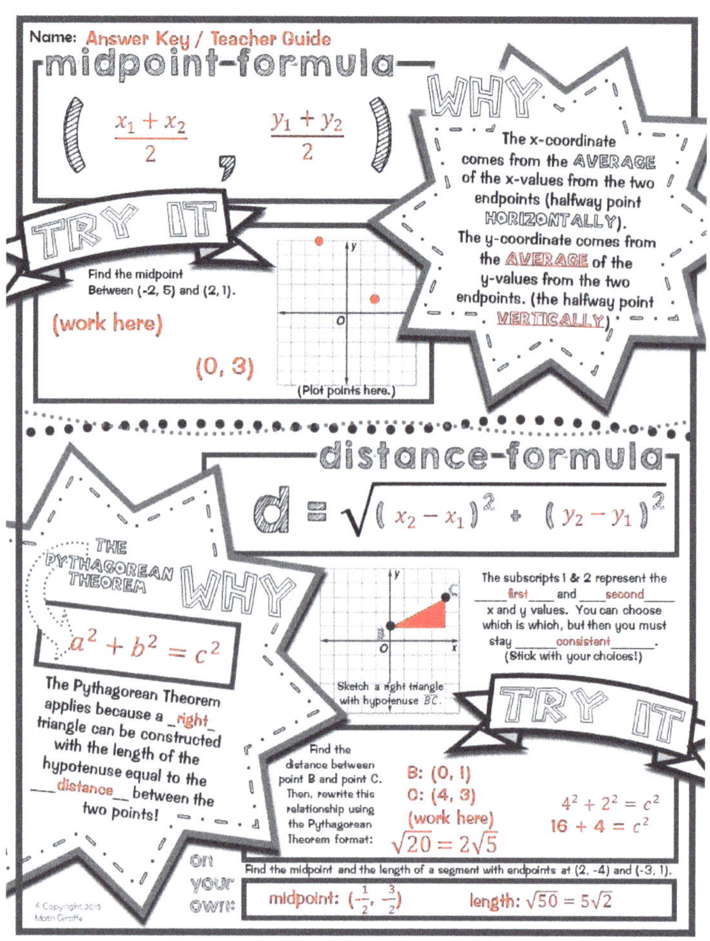
Name: Answer Key / Teacher Guide
midpoint-formula
$\left(\frac{x_1 + x_2}{2}, \frac{y_1 + y_2}{2} \right)$
WHY
The x-coordinate comes from the AVERAGE of the x-values from the two endpoints (halfway point HORIZONTALLY).
The y-coordinate comes from the AVERAGE of the y-values from the two endpoints. (the halfway point VERTICALLY)
TRY IT
Find the midpoint Between (-2, 5) and (2, 1).
(work here)
(0, 3)
(Plot points here.)
distance-formula
$d = \sqrt{(x_2 - x_1)^2 + (y_2 - y_1)^2}$
THE PYTHAGOREAN THEOREM
WHY
$a^2 + b^2 = c^2$
The Pythagorean Theorem applies because a _right_ triangle can be constructed with the length of the hypotenuse equal to the __distance__ between the two points!
The subscripts 1 & 2 represent the ___first___ and ___second___ x and y values. You can choose which is which, but then you must stay ___consistent___. (Stick with your choices!)
Sketch a right triangle with hypotenuse $\overline{BC}$.
TRY IT
Find the distance between point B and point C. Then, rewrite this relationship using the Pythagorean Theorem format:
B: (0, 1)
C: (4, 3)
(work here)
$\sqrt{20} = 2\sqrt{5}$
$4^2 + 2^2 = c^2$
$16 + 4 = c^2$
on your own:
Find the midpoint and the length of a segment with endpoints at (2, -4) and (-3, 1).
midpoint: $(-\frac{1}{2}, -\frac{3}{2})$
length: $\sqrt{50} = 5\sqrt{2}$
© Copyright 2015 Math Giraffe

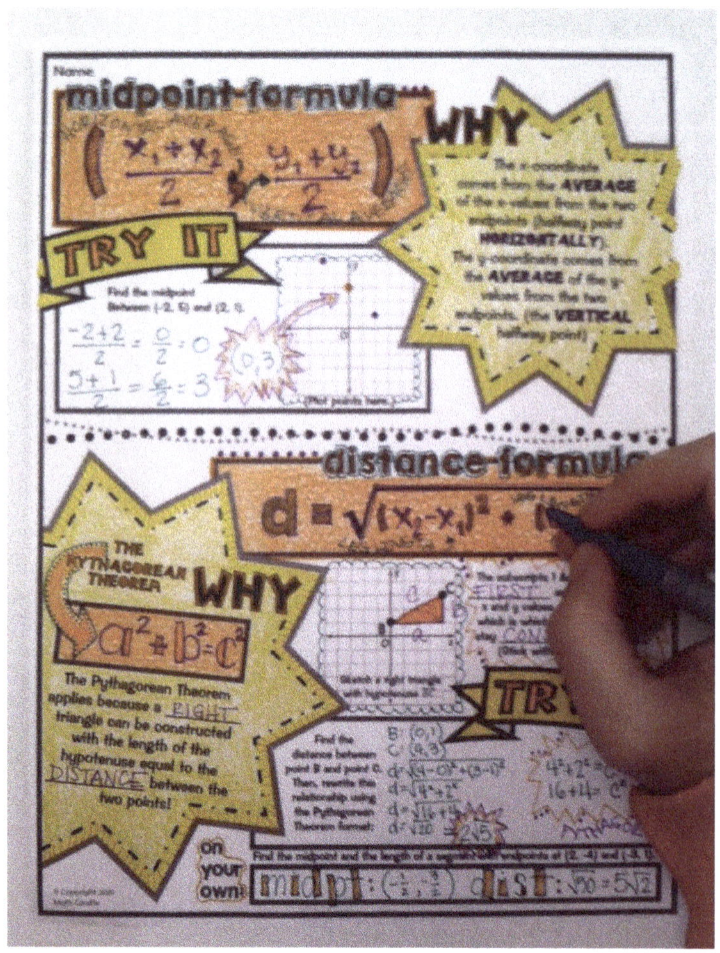

“My 8th grade summer school who are mostly boys loved coloring the notes. I'm pleasantly surprised!

“I'm hooked on this concept of using doodling to increase the brain’s ability to soak up details. It's been very successful in class and most of my students seemed to like how colorful they could be.”

“Keeps them extremely focused!”

“My students don't complain about taking notes when they are in a Doodle Notes format. Many of them will line all of their colors up on their desk, and they are ready to learn. Thank you for making these.”

“What a great resource to use with my students. It seemed like no matter what I had them write with their pencils it wasn't sticking...give them a doodle and some color-BAM! more kids are relating back to it. Thank you so much for this resource.”

“Mildly (okay majorly) addicted to the doodle notes that Math Giraffe has produced”

“Thank you, my kids love to do the Doodle Notes and they refer back to them in their journal when they need it. To me that is the biggest compliment possible.”

“One of my students interrupted me to yell, "I finally get it!!!" He then went on to say that he wished I had started using them earlier in the year”

“I just did this with my kids the first time and they LOVED it! I am also discovering that they are remembering the material much better than me just lecturing!”

“I teach 6th grade and have bought all these products that are covered in this grade. I want to use these for various strategies. As intro to lesson, additional support, focus tool, review. Just so great.”

“I love teaching with Doodle Notes! My students are always actively engaged the entire time!”

“My special education students love these. Color coding and decorating the notes gives them more time to interact with the notes without just reading them 10 times or extracting information. This makes it much more interesting for them, and my spatially-inclined students LOVE the chance to draw and doodle.”

“My students loved this! The coloring helped relax the students while they took notes.”

“The color coding helps them to make connections.”

“The best purchase I've made so far! You helped my students comprehend a very difficult concept. Wow! Thank you”

“Doodling helps my daughter's stress level during math lessons!”

CPSIA information can be obtained
at www.ICGtesting.com
Printed in the USA
LVHW070415011220
673106LV00045B/1359

9 781733 335416